Ionic Liquid Devices

Smart Materials

Series editors:
Hans-Jörg Schneider, *Saarland University, Germany*
Mohsen Shahinpoor, *University of Maine, USA*

Titles in this series:

1: Janus Particle Synthesis, Self-Assembly and Applications
2: Smart Materials for Drug Delivery: Volume 1
3: Smart Materials for Drug Delivery: Volume 2
4: Materials Design Inspired by Nature
5: Responsive Photonic Nanostructures: Smart Nanoscale Optical Materials
6: Magnetorheology: Advances and Applications
7: Functional Nanometer-Sized Clusters of Transition Metals: Synthesis, Properties and Applications
8: Mechanochromic Fluorescent Materials: Phenomena, Materials and Applications
9: Cell Surface Engineering: Fabrication of Functional Nanoshells
10: Biointerfaces: Where Material Meets Biology
11: Semiconductor Nanowires: From Next-Generation Electronics to Sustainable Energy
12: Supramolecular Materials for Opto-Electronics
13: Photocured Materials
14: Chemoresponsive Materials: Stimulation by Chemical and Biological Signals
15: Functional Metallosupramolecular Materials
16: Bio-Synthetic Hybrid Materials and Bionanoparticles: A Biological Chemical Approach Towards Material Science
17: Ionic Polymer Metal Composites (IPMCs): Smart Multi-Functional Materials and Artificial Muscles Volume 1
18: Ionic Polymer Metal Composites (IPMCs): Smart Multi-Functional Materials and Artificial Muscles Volume 2
19: Conducting Polymers: Bioinspired Intelligent Materials and Devices
20: Smart Materials for Advanced Environmental Applications
21: Self-cleaning Coatings: Structure, Fabrication and Application
22: Functional Polymer Composites with Nanoclays
23: Bioactive Glasses: Fundamentals, Technology and Applications
24: Smart Materials for Tissue Engineering: Fundamental Principles

25: Smart Materials for Tissue Engineering: Applications
26: Biobased Smart Polyurethane Nanocomposites: From Synthesis to Applications
27: Inorganic Two-dimensional Nanomaterials: Fundamental Understanding, Characterizations and Energy Applications
28: Ionic Liquid Devices

How to obtain future titles on publication:
A standing order plan is available for this series. A standing order will bring delivery of each new volume immediately on publication.

For further information please contact:
Book Sales Department, Royal Society of Chemistry, Thomas Graham House, Science Park, Milton Road, Cambridge, CB4 0WF, UK
Telephone: +44 (0)1223 420066, Fax: +44 (0)1223 420247
Email: booksales@rsc.org
Visit our website at www.rsc.org/books

Ionic Liquid Devices

Edited by

Ali Eftekhari
Ulster University, Belfast, UK
Email: eftekhari@elchem.org

ROYAL SOCIETY OF CHEMISTRY

THE QUEEN'S AWARDS FOR ENTERPRISE: INTERNATIONAL TRADE 2013

Smart Materials No. 28

Print ISBN: 978-1-78801-181-5
PDF eISBN: 978-1-78801-183-9
EPUB eISBN: 978-1-78801-208-9
ISSN: 2046-0066

A catalogue record for this book is available from the British Library

The Royal Society of Chemistry is a charity, registered in England and Wales, Number 207890, and a company incorporated in England by Royal Charter (Registered No. RC000524), registered office: Burlington House, Piccadilly, London W1J 0BA, UK, Telephone: +44 (0) 207 4378 6556.

For further information see our web site at www.rsc.org

Printed in the United Kingdom by CPI Group (UK) Ltd, Croydon, CR0 4YY, UK

Foreword

Ionic liquids may change the world! Since ionic liquids have the potential to show unique physico-chemical properties that cannot be realised by ordinary molecular liquids, they will be able to widen the science of liquids. This book, *Ionic Liquids Devices*, has been published as a collection of the latest results on the science of ionic liquids, especially those with the keywords of "interaction" and "design". Since many ionic liquids have negligible vapour pressure, you will find that the electron microscopy of "wet" samples has been carried out first with ionic liquids. In addition, an efficient carbon dioxide absorption device was found to be conceivable due to the use of ionic liquids. This pushed us to design other gas sensors. Surprising developments in these techniques are introduced in this book.

There are some interesting devices awaiting collaboration with ionic liquids. Readers of this book will be surprised by the advancement of the science of ionic liquids. A few known technologies are summarized in this book, but readers will also find the expanded application areas of these devices with the aid of ionic liquids. For example, known sensors have been improved and the area of application has been widened considerably by the use of ionic liquids. Health science has also been developed by the use of ionic liquids. Wearable sensors and drug delivery systems are now in our hands with the study of functional ionic liquids and their biocompatibility. These developments are based on the study of the interaction between ionic liquids and biomaterials. Many scientists have stated that ionic liquids are a kind of salt and they should be hazardous for biological systems. Actually, many ionic liquids are toxic for biosystems, but there are some exceptions. Some ionic liquids can be prepared from biomolecules such as amino acids. Since the bioactivity of these ionic liquids is extremely important, in the discussion of biocompatibility and biological application of ionic liquids, it is also commented on in this book.

Smart Materials No. 28
Ionic Liquid Devices
Edited by Ali Eftekhari

Published by the Royal Society of Chemistry, www.rsc.org

The scientific future is yours and it should be upgraded by the contribution of ionic liquids. Enjoy this book and apply ionic liquids to your study and/or research!

Hiroyuki Ohno

Preface

Ionic liquids (ILs) have been the subject of many research projects in various disciplines during the last few years. This emerging attention is because of an incredible versatility in design. Unfortunately, ILs have been simply considered as alternative solvents, but their potentials are way beyond simple solvents. In practice, they have been unable to replace the conventional solvents for two essential reasons: high viscosity and cost. Owing to the interactive nature of mobile ions, a high viscosity is somehow a characteristic feature of ILs. Although mass production reduces the cost of ILs for practical applications, the purification process is still time-consuming and expensive since ILs are prone to get contaminated easily.

The present book aims to introduce other types of IL applications in which these issues are no longer hindering problems. In fact, high viscosity can be an advantage in a series of applications in which the liquidity of IL is not the key feature but the ion mobility. In this case, the liquidity of ILs should be controlled in designing smart devices. In many of these applications of ILs as smart materials, the purity is not a major requirement, rather the material response is the target, which most contaminations cannot affect.

This book is the first of its kind and, thus, is far from being a comprehensive resource covering all potential applications. In fact, this field has not been properly classified yet, and many of relevant works are somehow lost in the literature. We simply wish to attract the attention of researchers from quite different disciplines to new and innovative potentials of ILs.

I have had the opportunity to collaborate with leading authors to prepare a meaningful collection of different applications. The first two chapters summarize the possibilities of utilizing some characterization techniques for inspecting ILs in novel applications. Then, the ionic characteristics of ILs are reviewed to have a better understanding of the IL flexibility in interactive systems. Common applications of ILs will be the bridge to the chapters

Smart Materials No. 28
Ionic Liquid Devices
Edited by Ali Eftekhari

Published by the Royal Society of Chemistry, www.rsc.org

devoted to various types of sensors and actuators. The final chapters introduce groundbreaking biomedical applications of ILs.

In addition to editors and authors who contribute scientifically, there are numerous hardworking backstage people who make the publication of a book possible. These sincere efforts should be especially acknowledged, but it will be a long list. However, I would like to specifically thank Ms Catriona Clarke who has devoted much time to this project and worked closely with me to enhance the book as a whole. In addition, I wish to thank Professor Mohsen Shahinpoor, the editor of the *RSC Smart Materials* book series, whose helpful discussions initiated the possibility of considering this emerging topic.

The present book is the result of the teamwork of various people who wished to emphasize the potentials and merits of ILs in smart applications. Definitely, this volume is not perfect and free from errors, but it is the starting point. As the editor, I appreciate any comment or feedback from the readers to improve possible forthcoming volumes.

Ali Eftekhari

Contents

Smart Materials No. 28
Ionic Liquid Devices
Edited by Ali Eftekhari

Published by the Royal Society of Chemistry, www.rsc.org

CHAPTER 1

Novel Analytical Techniques for Smart Ionic Liquid Materials

TETSUYA TSUDA*[a], CHIH-YAO CHEN[a] AND CHARLES L. HUSSEY*[b]

[a]Department of Applied Chemistry, Graduate School of Engineering, Osaka University, 2-1 Yamada-oka, Suita, Osaka 565-0871, Japan; [b]Department of Chemistry and Biochemistry, The University of Mississippi, University, MS 38677-1848, USA
*E-mail: ttsuda@chem.eng.osaka-u.ac.jp, chclh@olemiss.edu

1.1 Introduction

As described elsewhere, an ionic liquid (IL), which is sometimes called a room-temperature ionic liquid (RTIL), a room-temperature molten salt (RTMS), or an ambient-temperature molten salt (ATMS), has many unique properties. But, the most important point is that one IL combines nearly all of these features. For this reason, many scientists and engineers are keeping an eye on ILs as liquid materials and functional reaction media for supporting the development of future technologies,[1–11] *e.g.*, electrolytes for next generation secondary batteries and PEM fuel cells, functional solvents for organic synthesis and nanoparticle preparation, extraction solvents for rare metal ions and CO_2, and lubricants for precision instruments. Some scientists have attempted to establish novel analytical techniques that combine ILs with analysis equipment operating under vacuum conditions, *e.g.*, scanning electron microscopes (SEM), transmission electron microscopes (TEM),

Smart Materials No. 28
Ionic Liquid Devices
Edited by Ali Eftekhari

Published by the Royal Society of Chemistry, www.rsc.org

energy dispersive X-ray analysis (EDX), electron diffraction (ED), and X-ray photoemission spectroscopy (XPS). The negligible vapor pressure and anti-static properties of ILs enables analytical techniques that were previously considered to be impossible. Now these techniques have become widely recognized and accepted as powerful tools to reveal various chemical reaction processes in IL and ionic conformation change at the interface between IL and other phases. In this chapter, we divide these cutting-edge techniques into roughly three categories, SEM, TEM, and XPS, as described below.

1.2 SEM Observations with ILs

SEM is firmly established as a powerful tool for obtaining a three-dimensional surface image of specimens, and it is suitable for observing relatively large objects on a micrometer scale. The entire microscope has to be held at high vacuum for several reasons: First, gas in the gun assembly would interfere with electron emission and degrade the electron source. Second, electron beams will be scattered by any gas in the chamber, degrading column performance. Finally, the ionization of gas could cause electrical discharge and destroy the detector. These conditions imply that specimens must be vacuum tolerant, so they are traditionally dry solids. Wet specimens are typically frozen or desiccated before observation. This means that it is not easy for us to directly observe variations in the sample during dynamic experiments. For non-conductive/insulating specimens, creating a thin conductive coating (tens of nanometres) of a metal or carbon, although not mandatory, can prevent charging and improve the secondary electron signal. The most common vacuum coating methods are sputtering and thermal evaporation.

The extremely low vapor pressure (and thermal stability) of ILs match up well to the requirements of electron microscopy. The introduction of ILs into SEM chambers was first suggested in 2006 by Kuwabata *et al.* (Figure 1.1).[12] It has been reported that ILs act as electrically conductive materials with high fluidity and can be directly observed by SEM without the accumulation of electron charges. Since this milestone work, many publications have followed. It is important to note that other solvents that possess negligible volatility such as silicone oil are also compatible with the vacuum conditions. However, the images for samples coated with these oils are distorted and fluctuate greatly owing to the build-up of charge on the sample. The distinct behavior observed for ILs could be attributed to their ability to solvate electrons and allow them to move in the liquid.[13] It has been reported that even if the accelerating voltage is reduced to 1 kV, clear images are obtained for ILs, suggesting that such low energies are sufficient for injecting electrons into ILs.[12] This unique property of ILs is useful as an alternative way to afford electrical conductivity to non-conductive materials. Over the past decade, the application of ILs for SEM observation can be divided into the following two categories: (i) as pre-treatment reagents to make the sample vacuum tolerant and electrically conductive and (ii) as media for (electro)chemical reactions proceeding under vacuum conditions.

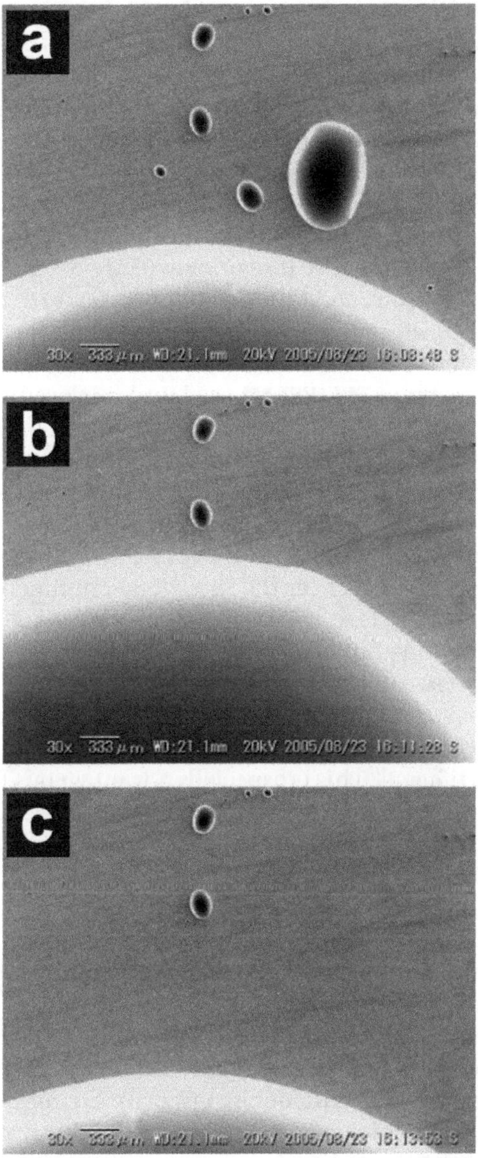

Figure 1.1 (a) SEM image of $[C_4mim][PF_6]$ IL droplets, (b) after tilting the sample stage at *ca.* 10°, and (c) after returning from the tilting condition to the original position. Reproduced from ref. 12 with permission from The Chemical Society of Japan.

1.2.1 ILs as Pre-treatment Reagents

Specimen preparation is decisive with any microscopic technique, the basic concern being that the specimen prepared is truly representative of the sample of interest. Although the preparation of SEM specimens is easy in comparison to TEM because there is no strict requirement to make the specimen exceedingly thin, it could still be challenging for specific samples. For instance, most biological specimens are made up largely of water and non-dense tissue materials. Thus, the sample must first be chemically fixed with aldehyde, dehydrated through an acetone or alcohol series, and then dried at the critical point (in order to mitigate specimen deformation due to tension during drying). Finally it must be coated with a conductive film prior to SEM observation.[14] This conventional protocol is laborious and time-consuming, not to mention that it introduces unwanted artifacts and/or contaminants throughout the whole course of sample preparation that are mistakenly identified as features of the specimen. Although environmental SEM (ESEM) used under relaxed vacuum conditions can examine wet or hydrated samples, the images are of limited resolution.[15]

ILs offer a simple yet efficient approach to address these issues. Tsuda and co-workers have showed that dropping a predetermined amount of diluted ILs onto such samples allows clear visualization of the specimen close to its native state (Figure 1.2).[16–19] Thus, in this case, ILs fulfil the important tasks of forming a conductive layer on the biological specimen and, at the same time, preserving its structure.[16] For some types of samples with hard outer layers (*e.g.*, exoskeletons or cell walls) that negate the need for fixation, the pre-treatment process can be completed within a couple of minutes. An additional benefit of using ILs as pre-treatment agents over common protocols is the fact that ILs keep the specimens wet and they minimize drying induced or vacuum-related artifacts.[16] This is especially advantageous for biological specimen that are sensitive to environmental conditions, *e.g.*, red blood cells[20] and drug or vaccine carriers.[21] Hyono *et al.*[20] have reported that red blood cells treated by ILs maintain their size and shape well even under high vacuum conditions, whereas those pre-treated by conventional procedures show shrinkage and deformation. In addition, IL pretreatment has recently proven to be a safe alternative to the characterization of infectious pathogens by avoiding the risk of generating aerosols during the sputter coating process.[22]

Notably, the selection and optimized usage of IL (concentration, amount, processing period and temperature, *etc.*) differ for each type of sample. Generally, hydrophilic ILs of moderate viscosity are found to be more suitable for pre-treating biological specimens than the hydrophobic ILs.[19,23] This is why 1-butyl-3-methylimidazolium tetrafluoroborate ($[C_4mim][BF_4]$) and 1-ethyl-3-methylimidazolium acetate ($[C_2mim][CH_3CO_2]$) are widely adopted.[16,17,19,22] In addition, choline- and lactate-based ILs usually show a favorable biocompatibility.[16,19,20] Image contrast and resolution are found to be enhanced by the combination of IL pre-treatment and Pt-blue staining, or with osmium tetroxide-coated coverslips.[24] Thus far, IL pre-treatment methods have proven to work with several kinds of insulating materials (rock,[25] ceramic,[26] mineral,[27] *etc.*), and are not limited to biological specimens.

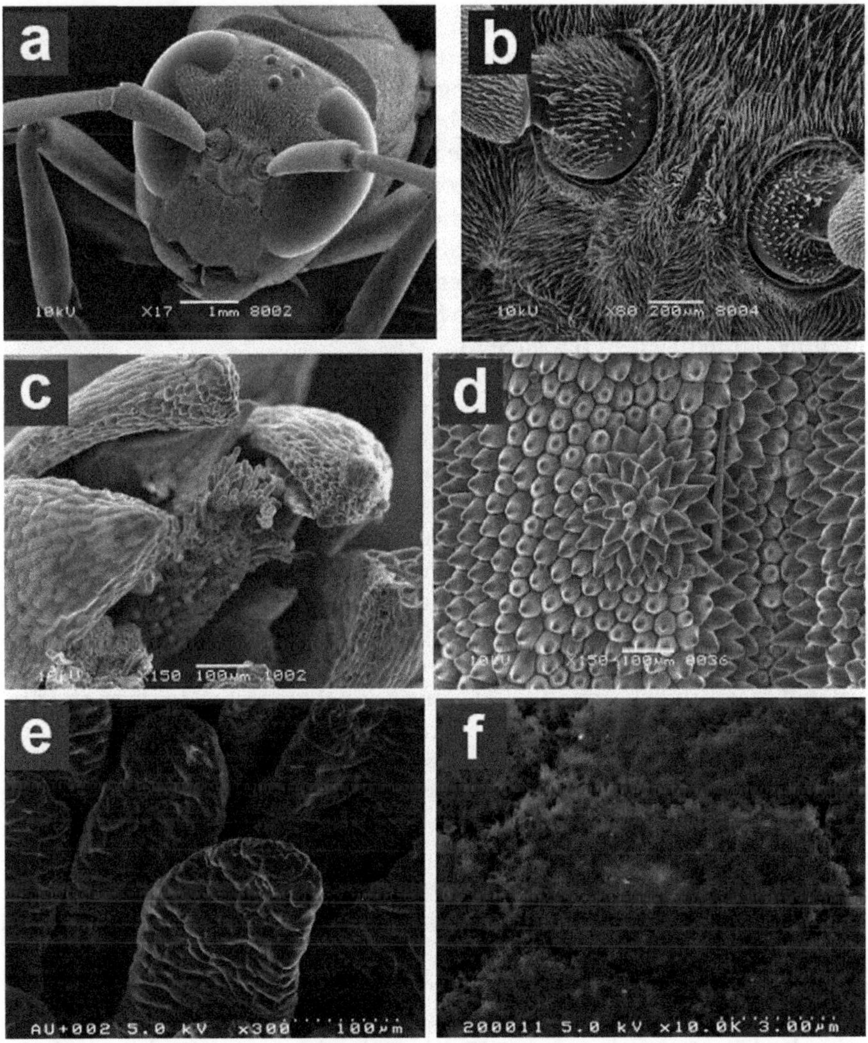

Figure 1.2 SEM images acquired after pretreatment in ILs: (a) head and (b) antennal fossae of a yellow jacket wasp, (c) petal and (d) stamen of flower (Asteraceae), (e) villi of mouse small intestine, and (f) microvillus on absorptive epithelial cell of mouse (magnified image of (e)). The ILs used are 5 vol% [C_4mim][TFSA] in ethanol for (a)–(d) and 5 vol% [C_2mim][BF_4] in water for (e) and (f). The immersion times are (a) and (b) 60 s, (c) and (d), 30 s, and (e) and (f) 600 s. Reproduced from ref. 16 with permission from John Wiley and Sons. Copyright 2011 Wiley-VCH Verlag GmbH & Co. KGaA, Weinheim.

1.2.2 ILs as Reaction Media and Electrolytes for Microscale Reactions

Dynamic processes associated with liquids, such as flow, diffusion, self-assembly, crystallization, and material synthesis, play crucial roles over a wide range of scientific and technological applications. However, their direct visualization by electron microscopy has been challenging due to the evaporation of the liquids. Again, ILs provide the platform to extend this knowledge. In this respect, Kashin *et al.*[28] monitored acid-catalyzed biomass conversion reactions by SEM and investigated how self-organized structures of binary IL/water and IL/alcohol systems influence the conversion efficiency and selectivity. Likewise, Kim *et al.*[29] investigated the Brownian motion of nanomaterials dispersed within an IL thin film, and Horigome *et al.*[30] reported the swelling of microgels in ILs *via* SEM.

It is well known that ILs can be utilized as an advantageous electrolyte for a broad set of electrochemical reactions.[31] This has inspired researchers to extract detailed and time-resolved information in realistic environments (*in situ* or *operando*). As an early demonstration of *in situ* SEM, Arimoto and co-workers[32] studied Ag electrodeposition from ILs and revealed the growth of granular and dendritic forms of the deposits. Being aware of the high potential of various families of ILs in energy-related applications,[33] this same group recently focused on the anodic electrode reaction in lithium-ion batteries (LIBs).[34,35] Silicon is one of the most promising active anode materials for any future LIBs as it offers pre-eminent theoretical capacity, nearly an order of magnitude beyond that of conventional graphite. However, the dramatic volume change (~300%) experienced by Si alloys/dealloys with lithium can induce severe pulverization and subsequent loss of electrical connectivity.[36] Nanostructuring of Si to facilitate strain relaxation is generally regarded as an effective strategy. This is attested *via in situ* SEM techniques coupled with specially designed cells (Figure 1.3). The fracture resistance and volume variation characteristics of Si show strong size and shape dependencies (Figure 1.4).[35] Among the samples studied, the thin flake form of Si outperforms microparticles and nanoparticles (not shown here) in terms of reversibility and retention of capacity. Serious damage of the Si thin flake is not recognized during the charge–discharge process. In addition, the feasibility of using backscattered electrons (BSEs) has also been examined to track the elemental distributions during operation.[35] Similar SEM results have been obtained by other groups. Chen *et al.*[37] found that SnO_2 particles with dimensions greater than a few hundred nanometers behave differently from their smaller counterparts (below 100 nm). During lithiation, the larger SnO_2 underwent irreversible volume changes together with the growth of cracks and extrusions, while no cracks were observed for the smaller particles. The formation of the extrusion, albeit the mechanism remains unclear, suggests that the lithiation is inhomogeneous within the larger SnO_2 particles.[37] For these experiments optimizing the beam energy and using appropriate probe

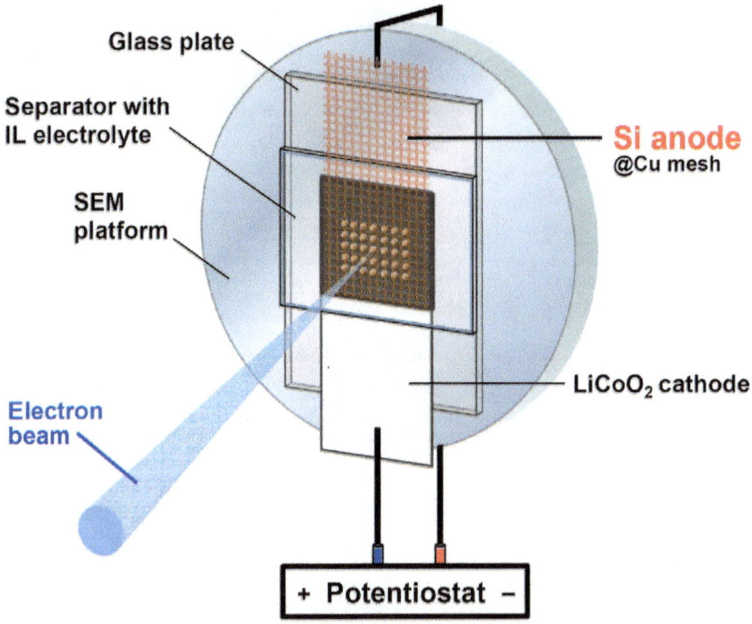

Figure 1.3 Schematic diagram of experimental setup inside a SEM for *in situ* SEM observation of electrode materials for Li-ion batteries.

currents, sometimes unavoidably accompanied by sacrifices in resolution and contrast, are prerequisite. Moreover, the amount and wetting condition of IL electrolytes on the working electrode largely affect the electrode performance and image quality. Abundant electrolyte enables the electrode to function well. However, it may disturb observations of electrode reactions because the electron beam cannot penetrate the thick IL. Maintaining a thin and sufficient IL electrolyte during reactions is of significance. As a rule of thumb, the penetration depth of the primary electron beam into ILs is estimated to be at least 1 μm.[17]

With the aim to enhance the mechanical stability of Si, Xia *et al.*[38] prepared Cu–Si core–shells by lithography-based methods. *In situ* SEM observations in an IL electrolyte disclosed that the sample can accommodate ~250% volumetric expansion without detectable crack formation, highlighting the importance of both rational material design and precise characterization. Likewise, the *in situ* SEM method can be applied to active cathode materials.[39,40] Miller *et al.*[39] studied the microstructural evolution of a single $Li(Ni_{0.8}Co_{0.15}Al_{0.05})O_2$ particle during charge/discharge by using a focused ion beam SEM (FIB-SEM). The cross section of the sample exhibits progressive intergranular separation caused by the anisotropic changes in lattice parameters associated with the lithiation/delithiation process. These results suggest that the physical separation and isolation of grains may contribute to

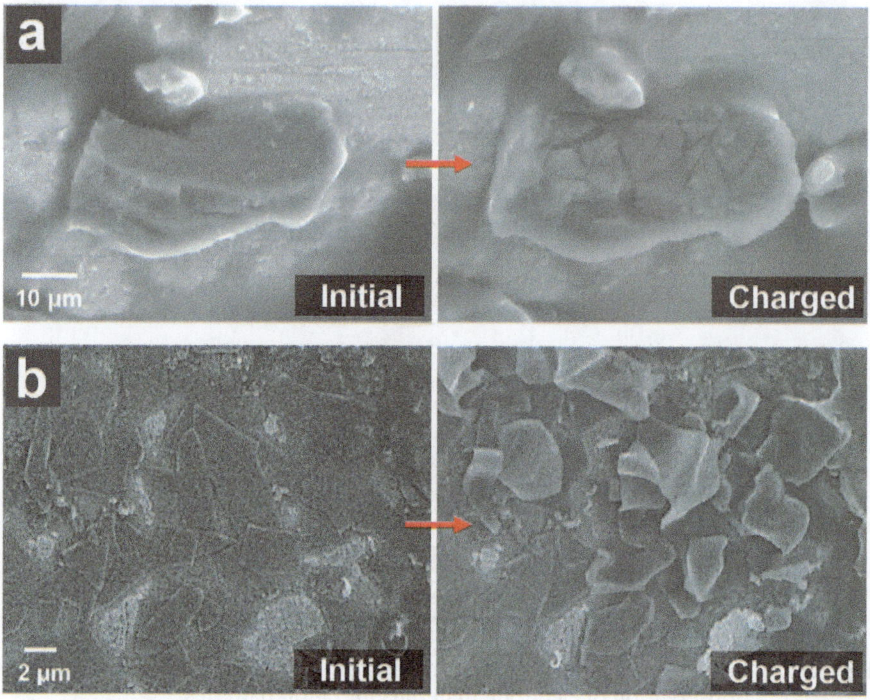

Figure 1.4 *In situ* SEM images of the charge process of (a) a Si microparticle and (b) a Si thin flake. The IL electrolyte used is 1.0 mol L^{-1} Li[TFSA] in [C_2mim][FSA]. Reproduced from ref. 35 with permission from Macmillan Publishers Limited: Scientific Reports. Copyright 2016.

performance degradation of practical LIBs. Another example of new insight that *in situ* observations can provide is electrolyte penetration behavior, which undoubtedly influences the utilization and performance efficiency of the electrode. This information is hardly achievable by other methods.

Most previous studies about SEM observations of LIB components examined the tested electrode materials by post-mortem methods. Hence, extreme care must be taken to avoid contamination, deformation and air/moisture exposure of the specimens during the cell disassembly, electrode washing, and drying.[41] Meanwhile, the physical and chemical states of electrodes during reaction are continuously renewed (non-equilibrium states). *In situ* studies are therefore preferred to deepen the understanding of time- and potential-resolved changes. Indeed, Bridel and co-workers[42] have pointed out that the electrode morphology evolves in a very different manner from that expected based on *ex situ* analysis. On the other hand, caution has to be exercised in interpreting the *in situ* results. As electrochemical cells are multifaceted systems in which critical components are relevantly influenced, researchers should be aware of the extent to which the cell setup for *in situ* experiments relates to practical applications.[41]

1.3 TEM Observations with ILs

TEM is a distinguished characterization tool that can provide morphological, structural, and compositional information at nanoscale or even atomic resolution. Besides the EDX that is commonly used for elemental analysis, many other analytical methods have been employed in conjunction with TEM, including electron diffraction that allows nanocrystal structure determinations, electron energy loss spectroscopy (EELS) that probes electronic excited states of materials, and Lorentz imaging that identifies micromagnetic structures. As a result, TEM has become the technique for addressing great challenges in materials science. Despite its versatility, TEM generally requires sophisticated skills for sample preparation. Because the sample has to be stable in vacuum and sufficiently thin (usually less than 100 nm) for electron beam penetration to form an image of reasonable resolution, preparing such specimens with minimal artifacts is a big challenge. This is even truer for liquid samples. Cryo-TEM[43] is one of the approaches used to clear this hurdle, but additional freezing devices and temperature-sensitive preparation processes are required. Moreover, the images obtained at such extreme temperatures may not reflect the native state at room temperature.

With the recent advances in instrumentation, two approaches have been developed to circumvent the difficulties associated with viewing liquids by TEM. The first is to entrap a liquid film between two electron transparent membranes (*e.g.*, silicon nitride[44] or graphene[45]) that can withstand the pressure difference between a liquid stage and the high vacuum in the chamber. The "closed" liquid cell makes imaging liquids by TEM a routine task and has attracted tremendous interest in diverse fields. Exciting progress has been made over the past few years, much of which is reviewed in ref. 46 and 47. The second approach is to use the "open" cell configuration, which utilizes ILs without the hermetic component. As already stated, ILs themselves can be directly imaged by SEM;[12] they can also be viewed by TEM as long as their thicknesses are carefully managed. Another occasion is when the object of interest lies in a solid material around the ILs. For example, consider the case of an electrochemical reaction. In this example there is no requirement for the thickness and amount of the ILs that need to be present. The advantages of using the open cell over the closed cell are manifold: (i) sample preparation is simple; (ii) higher resolution is achievable due to the absence of a membrane; (iii) imaging can performed in a standard TEM; and (iv) the field of view is much larger. Below we discuss IL-based TEM observations and related experiments.

1.3.1 ILs Observed by TEM

Not surprisingly, great efforts have been devoted to investigating the microstructure of ILs because the structure of these materials strongly relates to their physicochemical properties.[48] The existence of clusters in ILs has been reported both experimentally and theoretically.[49,50] Most of them are deduced from indirect evidence based on the results of spectroscopic analyses. With the aid of a super-aligned carbon nanotube network as the substrate, Chen *et al.*[51] fabricated an ultrathin freestanding IL film and directly observed well-ordered clusters within the IL for the first time (Figure 1.5). Generally, homogeneous liquid films show no obvious contrast for TEM imaging. The key to identifying IL clusters is the thickness of the as-obtained IL film (tens of nanometers) ensuring high resolution. Freeze–fracture TEM (FF-TEM) experiments have also been used to reveal the existence of spherical aggregates in IL mixtures. The size of these heterogeneous domains is found to be several tens of nanometers.[52,53] The unusually large local domains that were observed, however, are likely related to the sample preparation/pre-treatment for FF-TEM observation, rather than the real bulk nanostructure.[48]

1.3.2 ILs as Pre-treatment Reagents

Nanotechnology is an emerging field of science dealing with the creation and utilization of materials with ranges from approximately 1 to 100 nm. Wet chemistry synthetic methods provide an easier way to synthesize nanomaterials of precisely controlled size and shape as well as homogeneity. Nevertheless, nanomaterials are only kinetically stable. Characterization of such samples, especially in the colloidal state, is often plagued with issues of drying-induced artifacts, such as agglomeration or deformation. ILs have proven valuable in this context. Maddikeri *et al.*[54] have shown that polymer nanostructures formed by spontaneous self-assembly can be visualized by conventional TEM. Renz *et al.*[55] compared the TEM images of hydrogel nanoparticles prepared by different methods. They showed that a hydrogel specimen prepared by traditional drop cast methods is totally destroyed. In contrast, for the specimen dispersed in an ultrathin IL film, the morphology is preserved. It is postulated that the IL replaces the water of the hydrogel and thus preserves the overall morphology. This new method of using ILs as an embedding matrix is very easy and can be completed in less than 10 min. It involves merely spreading the sample solution and IL onto a holey TEM grid, wiping out excess IL, and then waiting for solvent evaporation. This method is also applicable to inorganic nanoparticles. It has been shown that ILs prevent the aggregation of spinel ferrites synthesized by mini-emulsion upon solvent evaporation and lead to TEM images of improved quality in comparison to those prepared by a known method.[56] For biological samples, Kuwabata and co-workers[57] have shown that ILs-induced lipid vesicle fusion can be directly observed by TEM without aggregation. They recently found that IL treatment enlarges liposome size in certain cases and that vesicle morphology is kept under vacuum conditions.

It is well known that ILs themselves serve as promising media for synthesizing nanoparticles. In addition to their wide electrochemical window and good thermal/chemical stability, one unique property of ILs is that they form extended hydrogen-bonding systems in the liquid state that facilitate an electrostatic and steric colloid-type stabilization of nanomaterials.[49,58] To gain

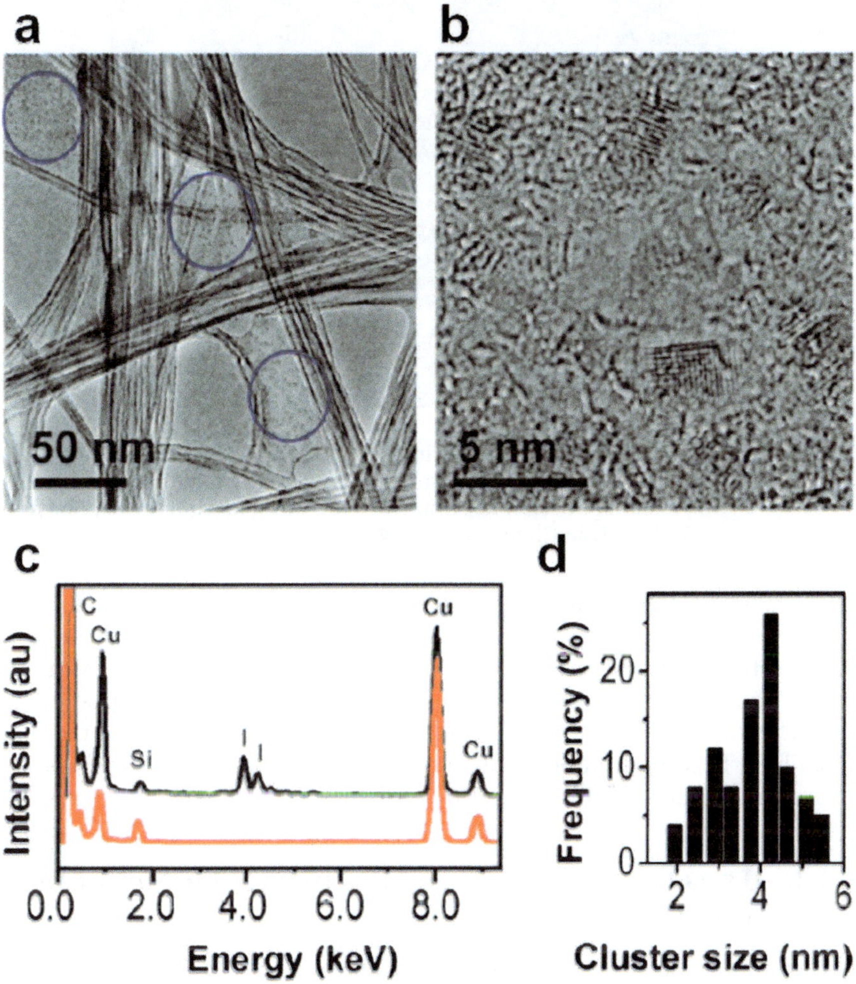

Figure 1.5 High resolution TEM images of freestanding [C_4mim]I IL thin films: (a) TEM image of a thin film supported by nanosized holes on a carbon nanotube network. The blue circles mark the location of the freestanding [C_4mim]I on the grid. (b) High-resolution TEM image showing the existence of well-ordered clusters in the [C_4mim]I thin film. (c) Energy dispersive X-ray spectra from the [C_4mim]I thin film (black) and carbon nanotube based TEM grid (red). The observed Si and Cu signals are from the carbon nanotube network and sample holder. (d) Histogram of the cluster size distribution at 25 °C, obtained from measurements of 80 clusters. Reprinted with permission from ref. 51. Copyright 2011 American Chemical Society.

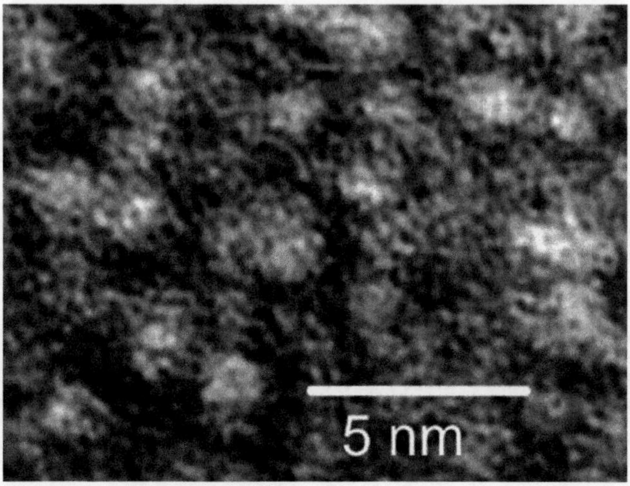

Figure 1.6 TEM micrograph (negative image, underfocus) of Pt(0) in the $[C_4mim]$ $[PF_6]$ ionic liquid showing the contrast density fluctuation around the metal nanoparticles. Reprinted with permission from ref. 59. Copyright 2003 American Chemical Society.

further insight, Dupont and co-workers[59] synthesized Pt nanoparticles in an IL and re-dispersed them into the same IL for TEM examination. Strong contrast fluctuation is observed in the IL with Pt nanoparticles, whereas no feature or contrast density variations can be found at neat liquid phase (Figure 1.6). This result reflects the fact that liquid molecules around the nanoparticles have become immobilized, and it shows the interactions of ILs with as-prepared nanoparticles.

1.3.3 ILs as Reaction Media and Electrolytes for Nanoscale Reactions

The combination of ILs and TEM has led to a better understanding of dynamic processes in liquids from a microscopic viewpoint. Some recent attempts, namely *in situ* TEM, conducted with a conventional TEM and standard holder are described here. For example, the nanostructures and motion of polymer assemblies (micelles and vesicles) dissolved in ILs have been observed directly.[60] Similarly, the cellulose dissolution process in the ILs has been explored.[61] It is also reported that the temperature-dependent local structure of CoFe Prussian blue analogues stabilized in an IL can be resolved at the atomic level. The realization of such high resolution is attributed to the use of the appropriate IL.[62] In addition, ILs can alleviate electron beam damage during TEM observations, too.[60,62] These experiments are conducted by simply loading the sample/ILs solutions onto the TEM grid, a stable and freestanding film is therefore formed due to the characteristic surface tension of ILs. In this respect, Mizoguchi and co-workers[63,64] have

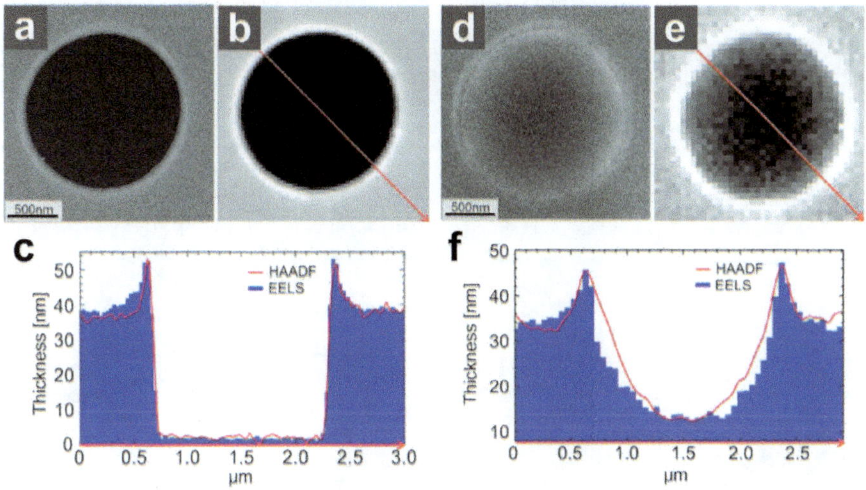

Figure 1.7 Thickness analysis of (a–c) a no-liquid hole and (d–f) thin liquid film of [C_2mim][TFSA]. (a, d) High-angle annular dark-field imaging scanning TEM (HAADF-STEM) images. (b, e) STEM-EELS thickness mapping images. (c, f) Thickness profile on the red line in (b) and (e). Reprinted from *Ultramicroscopy*, T. Miyata, T. Mizoguchi, Fabrication of thin TEM sample of ionic liquid for high-resolution ELNES measurements, Copyright (2016) with permission from Elsevier.[63]

carefully characterized the IL film formed inside the holes of a TEM grid by EELS. Based on the comparison of the mean free path of the inelastic scattered electrons (IMFP), zero-loss peak and spectrum intensity between the IL and amorphous carbon, they found the film thickness, in this case, [C_2mim][TFSA], to be in the range 10.2–13.8 nm (Figure 1.7). Because this value is much less than the IMFP, TEM can achieve the measurements without any loss of spatial/energy resolution.[63] In addition to the selection and concentration adjustment of ILs, the great variety of grid specifications, such as pore size and shape, material, and support membrane (if necessary), are believed to help optimize the IL film thickness and the image resolution.[63]

Tsuda and co-workers[65,66] have shown that irradiation of certain types of ILs containing metal salts leads to metal nanoparticle generation inside the ILs by partial radiolysis followed by reduction. They made an extended effort to investigate the essential steps in the nucleation of metal nanoparticles as well as the growth kinetics.[67] Furthermore, Ostwald ripening and coalescence processes are also disclosed. The high spatial and temporal resolution allows the clear observation of lattice fringes that provide intriguing information correlating coalescence within crystal states (Figure 1.8). One of the key considerations for the successful generation of nanoparticles is the use of a suitable IL mixture that enables nanoparticle formation at a moderate rate. An interesting study reported by Kimura *et al.*[68] reveals that the early stages of nucleation of $NaClO_3$ in an IL involves simultaneous formation and

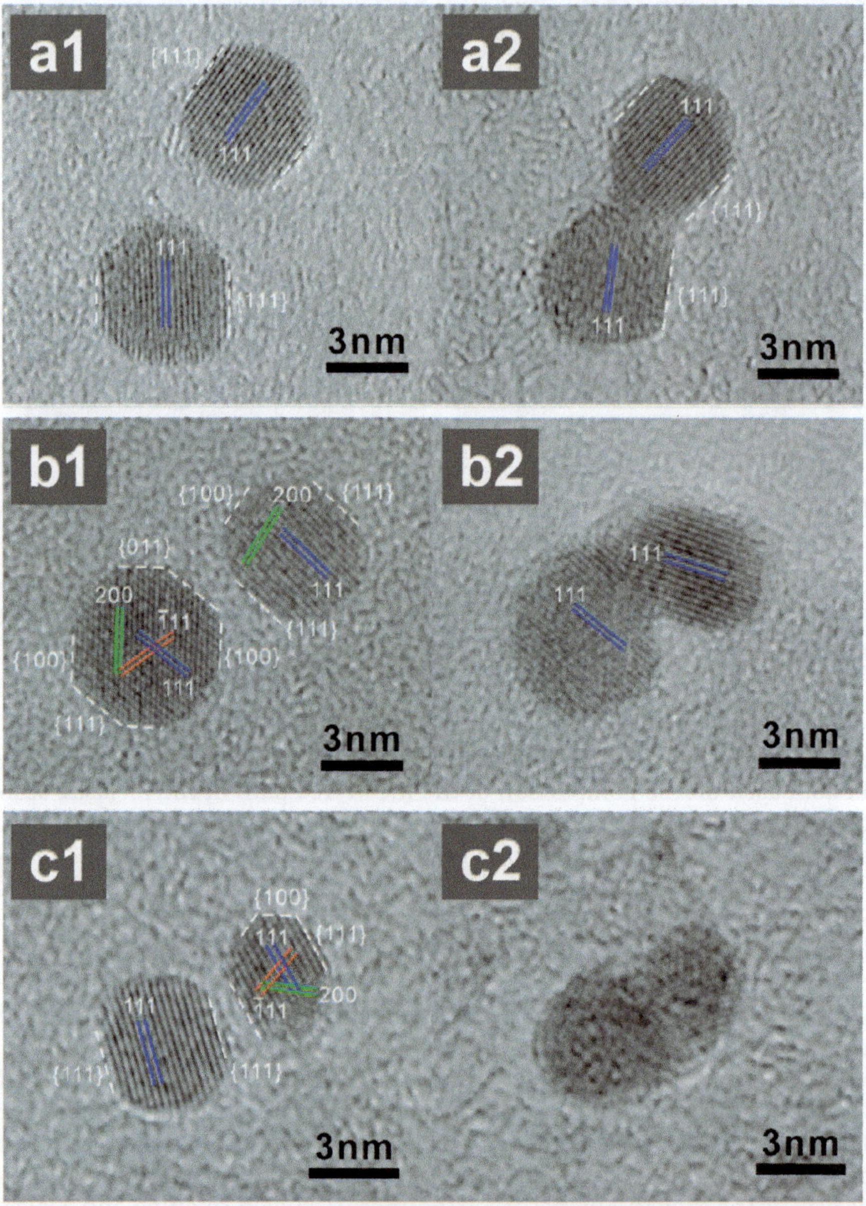

Figure 1.8 TEM images of Au nanoparticles taken prior (a1, b1, and c1) and during coalescence (a2, b2, and c2). Facets are displayed in the figures with broken lines, and indices are given in braces whenever possible. An IL mixture consisted of a 50 vol% mixture of [C_4mim][TFSA] and [$N_{3,1,1,1}$][TFSA] containing 0.1 mol L^{-1} $NaAuCl_4$, diluted with ethanol; a concentration of 2 vol% is used. ($N_{3,1,1,1}$: trimethylpropylammonium). Reprinted with permission from ref. 67. Copyright 2014 American Chemical Society.

dissolution of pre-nucleation clusters. The local density fluctuations provoke nano-volumes with higher concentration where pre-critical nuclei can form. They were also able to identify the metastable phases by using electron diffraction techniques. Recently, single Au atoms formed in the IL, and their movement, have been reported by Mizoguchi and Miyata.[69]

Huang *et al.*[70] have used *in situ* TEM techniques with an open cell to study electrode reactions in battery materials. The nanobattery built in the TEM consists of an ionic liquid electrolyte, a single SnO_2 nanowire anode, and bulk $LiCoO_2$ as the cathode. The nanowire anode electrode is loaded on a piezo-positioner and driven to approach the cathode. This strategy provides atomic level spatial resolution and analytical capability for studying the lithiation/delithiation mechanisms of LIB electrode materials.[70] After this pioneering work, considerable follow up studies dealing with other electrode materials (*e.g.*, Si,[71] Ge,[72] $LiMn_2O_4$,[73] graphene,[74] and carbon nanotubes[75]) were reported. The typical IL electrolyte is Li[TFSA]–[C_4mpyr][TFSA] (C_4mpyr: 1-butyl-1-methylpyrrolidinium; TFSA: bis(trifluoromethylsulfonyl)amide). For interested readers there are several reviews to consult for additional details about this topic.[76,77]

1.4 XPS with ILs

In 2005, the first application of XPS analysis to IL-polymer composites and ILs themselves was carried out by Crespo[78] and Licence.[79] After these milestone studies, various aspects of chemistry in ILs and surface investigations at the interface between different phases, *e.g.*, the vacuum phase and IL layer, have been studied extensively using XPS. Information at the molecular level can also be obtained by grazing emission of X-ray to the specimen. This approach is called angle-resolved XPS (ARXPS). Figure 1.9 shows a schematic illustration of the experimental arrangement associated with ARXPS measurements.[80] In this technique, surface sensitivity can be controlled with the X-ray incidence angle. For example, at 0°, information can be obtained in the near-surface region (information depth (ID): 7–9 nm). If needed, information can be collected at the topmost surface layers (at 70°, ID: 2–3 nm; at 80°, ID: 1–1.5 nm). Note that at 80° 65% of the XPS intensity arises from the first 0.3–0.5 nm. Because of this, the surface composition and the arrangement of the molecules in the topmost layer are explored on a molecular scale. Of course, the ID depends on the kinetic energy used during the measurements.

Recently, a different approach was proposed that uses a high energy Ag L_α' X-ray source called energy-resolved XPS (ERXPS), which is useful for obtaining the information on the IL–vacuum interface structure without changing the emission angle.[81] Like conventional X-ray sources, the Ag L_α' source also provides a robust chemical state analysis for ionic liquids and enables characterization by higher energy core photoelectron emissions, *e.g.*, S 1s and P 1s. However, the signal intensity is significantly reduced in comparison to the longstanding conventional X-ray source, Al K_α. Further information about the measurement principles is described in ref. 81. In this article, the authors find that both depth profiling techniques, ARXPS and ERXPS, are

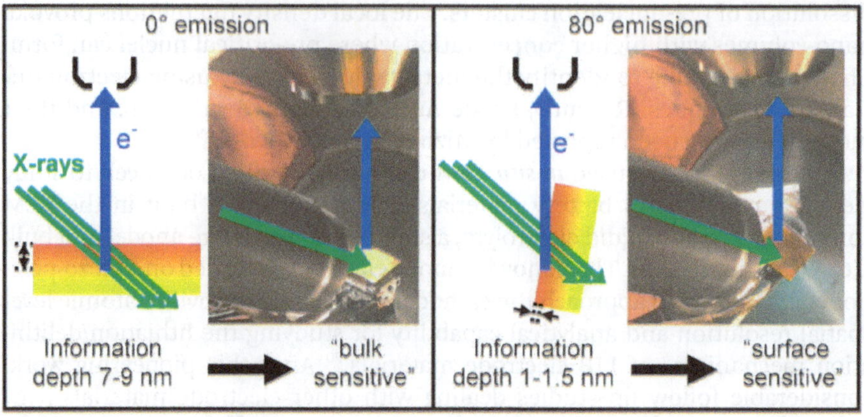

Figure 1.9 Schematic illustration of the experimental geometries used for angle-resolved XPS measurements: (left) Normal emission (0°) corresponding to "bulk-sensitive" geometry and (right) grazing emission (80°) corresponding to "surface-sensitive" geometry. The incident X-rays are indicated in green and the emitted photoelectrons in blue. Reproduced from ref. 80 with permission from the PCCP Owner Societies.

supportive and additive to each other. Taken together, these developments show that XPS has become recognized as a useful analytical method for IL science. Several specific examples will be introduced below. But in this chapter these illustrations are confined to recent work that has appeared after publication of excellent review articles written by Professor Hans-Peter Steinrück and Professor Peter Licence, who are experts in this field.[80,82]

1.4.1 Bulk Composition, Interionic Interaction, and Surface Composition in ILs

Currently, it is well known that information about bulk composition is readily and precisely obtained by regular XPS measurements without grazing emission of X-rays. Information on a wide variety of ILs has been systematically collected to date. Recent target ILs are the tetraalkylammonium,[83] tetraalkylphosphonium,[83] dialkylpyrrolidinium,[84] alkylpyridinium,[85] and guanidinium cations,[86] coupled with commonly-used anions. New insight concerning the electronic environments of all of the components of such ILs has been obtained. As an example, high resolution XP spectra of the C 1s region with component fittings for bis(methyloctylamino)dimethylguanidinium cation ($[(C_8C_1)_2dmg]^+$)-based ILs are shown in Figure 1.10.[86] XP spectra can be obtained without difficulty, but it is not easy to find the appropriate peak fitting model. In this case, a four-component model is applied to understand the XP spectra. We can expect to find that the carbon C_{core}, which is directly bound to three electron-withdrawing nitrogen atoms, has the highest binding energy. The second highest binding energy is the first carbon of the alkyl-chain that is bound to a single nitrogen atom, C_{hetero}. An intermediate carbon

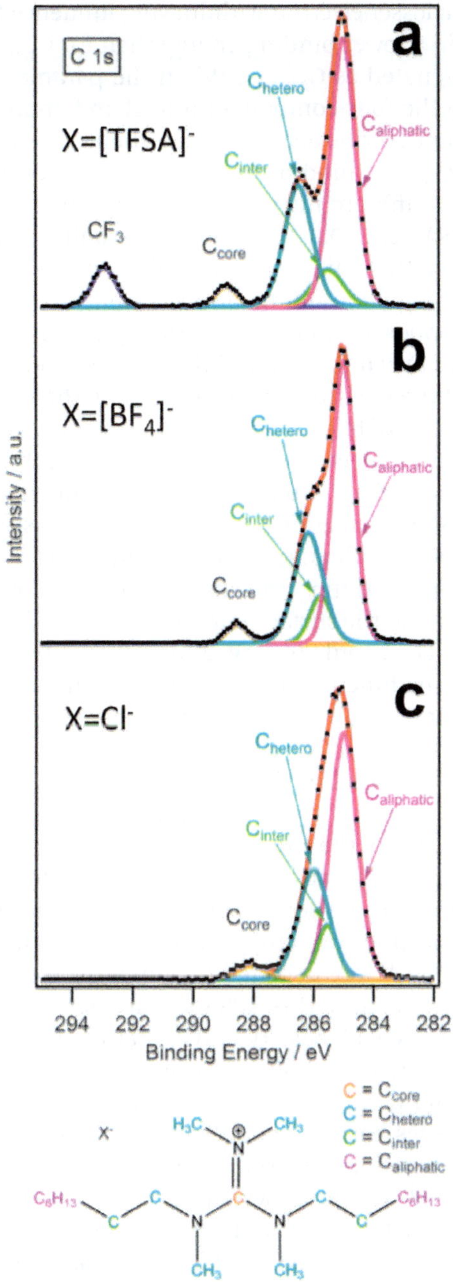

Figure 1.10 High resolution XP spectra of the C 1s region with component fittings for (a) $[(C_8C_1)_2dmg][TFSA]$, (b) $[(C_8C_1)_2dmg][BF_4]$, and (c) $[(C_8C_1)_2dmg]Cl$. Intensities are normalized to the N_{cation} 1s photoemission peak for $[(C_8C_1)_2dmg][BF_4]$. All XP spectra are charge corrected to $C_{aliphatic}$ = 285.0 eV. Reproduced from ref. 86 with permission from the PCCP Owner Societies.

environment under a lesser electron-withdrawing influence from the nitrogen atom, C_{inter}, results in a lower binding energy than for C_{core} and C_{hetero}. Other carbons can be designated as $C_{aliphatic}$. When the parameter constraints are considered by using the four-component model and appropriate fitting, the calculated model can describe accurately the chemical environments of the cation. In Figure 1.11,[86] the binding energies for guanidinium-based ILs are compared with some different nitrogen-based cationic structures. Note that the energies for $[TFSA]^-$ at 399.3–399.5 eV are virtually unchanged. It is concluded by the authors that this could be explained by considering that the positive charge of the guanidinium cation is well distributed across three nitrogen atoms, as opposed to just one in the pyrrolidinium cation, reflecting the weak bond between the guanidinium cation and the anion through the delocalisation of the positive charge around the three nitrogen centers.

ARXPS is a powerful tool to examine the liquid structure at the IL–vacuum interface structure on a molecular scale. By using this technique, we can even understand the impact of the cation core and the functional moiety of the cation on cation–anion interactions and surface composition.[87,88] Steinrück *et al.*[88] have discussed the influence of substituents and functional groups on the organic cations in relation to the surface composition in ILs. From the XPS results, they find that the surface enrichment of the functional groups at the IL–vacuum interface is only observed for systems that have a very weak interaction between the functional group and the ionic head group.

The target is not only organic cations – the impact of metal cations such as Ag^+, Al^{3+}, Ni^{2+}, Cu^{2+}, and Zn^{2+} on the surface structure of the ILs with their metal bis(trifluoromethanesulfonyl)amide ($[TFSA]^-$) salts is also discussed.[89] The surface concentrations of the solvated metal ions can be estimated by normalization of the area of the characteristic peaks for each metal (Al 2p; Ag 3d; Cu 3p; Ni 3p; Zn 3p) at normal and grazing emission to the F 1s peak intensity measured under the same conditions. A typical example is exhibited in Figure 1.12. Nickel and zinc are slightly enriched at the surface, but the other three metals, especially Al, are depleted in comparison to the bulk. Another interesting approach related to this XPS measurement is to characterize the anionic species in Lewis acid–base type ILs, that is, the so-called 1st-generation ILs composed of metal halides and organic halide salts. In $SnCl_2$–1-alkyl-3-methylimidazolium chloride ILs, the anionic species are found to be Cl^- and $[SnCl_3]^-$ in Lewis basic ILs and $[SnCl_3]^-$ and $[Sn_2Cl_5]^-$ in Lewis acidic ILs.[90] At neutral compositions, the only anion is $[SnCl_3]^-$. There is no apparent evidence of $[SnCl_4]^-$ in the IL system. Afterwards, the characterization of different types of metal chloride–1-methyl-3-octylimidazolium chloride ($[C_8mim]Cl$) ILs was performed by similar XPS measurements.[91] The metal chlorides are $FeCl_2$, $FeCl_3$, $CoCl_2$, $NiCl_2$, and $ZnCl_2$. In addition to the anion speciation described above, hydrogen bond basicity and Gutmann acceptor numbers (Figure 1.13) of the ILs are also estimated from the XPS binding energies of the N 1s, C_{hetero} 1s, and Zn $2p_{3/2}$ photoelectron peaks. These are very useful new approaches to characterize the liquid properties of ILs.

To know the chemical structures at the IL–solid interface, several approaches have been proposed based on the combination of different

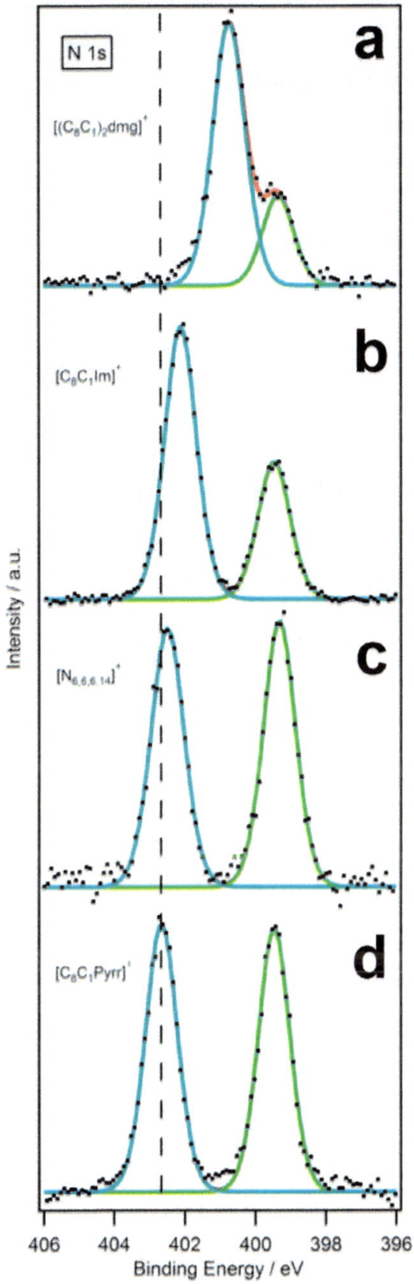

Figure 1.11 N 1s XP spectra of (a) $[(C_8C_1)_2dmg][TFSA]$, (b) $[C_8mim][TFSA]$, (c) $[N_{6,6,6,14}][TFSA]$, and (d) $[C_8mpyr][TFSA]$. Intensities are normalized to the F 1s peak for $[C_8C_1Pyrr][TFSA]$. XP spectra were charge corrected by referencing the aliphatic C 1s photoemission peak to 285 eV. Reproduced from ref. 86 with permission from the PCCP Owner Societies.

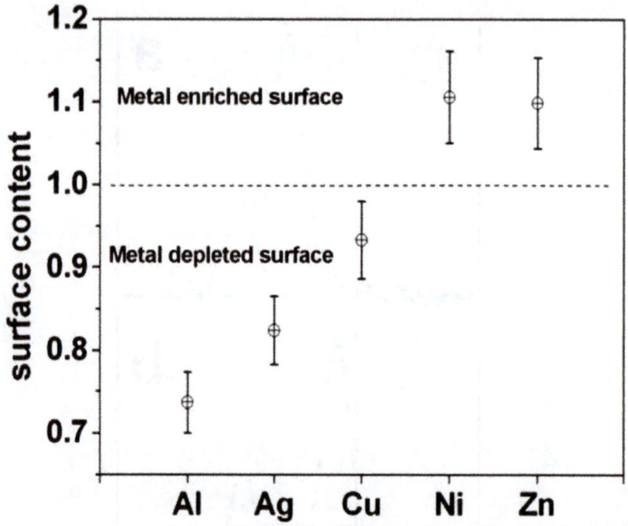

Figure 1.12 Metal surface content as determined by ARXPS peaks. Values lower than 1 mean that the surface is depleted of metal with respect to the bulk, while values larger than 1 means enrichment. Reprinted from *Surface Science*, **648**, S. Caporali, M. Pedio, C. Chiappe, C. S. Pomelli, R. G. Acres, U. Bardi, Surface study of metal-containing ionic liquids by means of photoemission and absorption spectroscopies, 360–365, Copyright (2016) with permission from Elsevier.

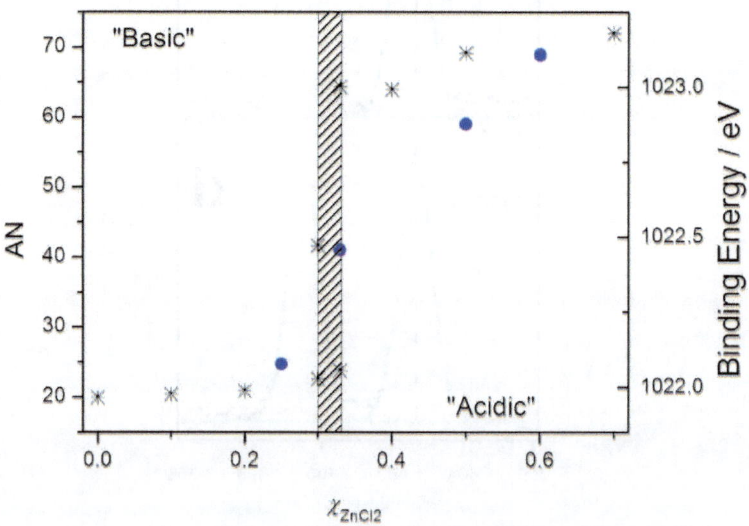

Figure 1.13 Plots of Gutmann acceptor numbers (AN) calculated (black stars) and Zn $2p_{3/2}$ binding energies (blue circles) against the mole fraction of $ZnCl_2$, x_{ZnCl_2}, in $[C_8mim]Cl–ZnCl_2$ ILs. The neutral zone is shown as a shaded region between x_{ZnCl2} = 0.3 and 0.33. Reproduced from ref. 91 with permission from The Royal Society of Chemistry.

analytical techniques, such as scanning tunneling microscopy, FT-IR, zeta potential measurements, thermogravimetric analysis, atomic force microscopy, XPS, and DFT calculations.[92–94] Because the combination generates a high synergistic effect, it is a useful analytical approach to identify chemical/physical reactions on catalysts and electrodes within the IL layer. Just to mention one example (Figure 1.14), Behm *et al.*[92] have succeeded in the

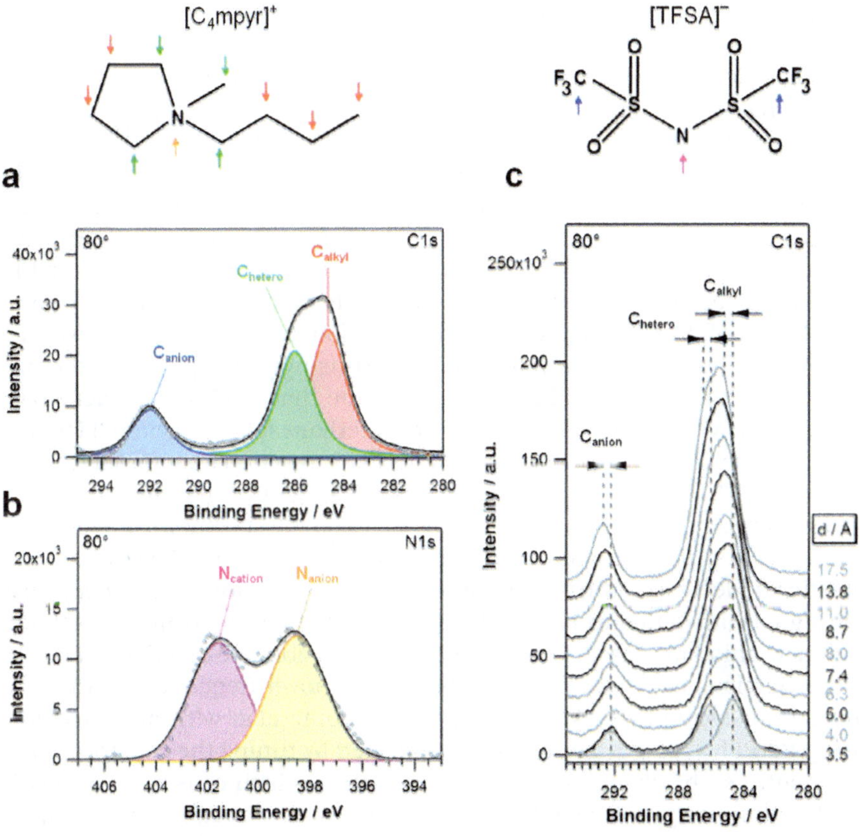

Figure 1.14 Core level XP spectra for C 1s (a) and N 1s (b) in sub-monolayer [C_4mpyr][TFSA] on Ag(111) at room temperature (nominal thickness ~3.5 Å). Stick presentations of the molecules are placed above the spectra. The color-coded arrows refer to the corresponding color-coded XP peaks. The experimental peak areas are close to the nominal stoichiometry of the molecules, indicating that anions and cations are adsorbed on the surface with a ratio of 1 : 1. (c) Sequence of C 1s spectra for an increasing coverage of [C_4mpyr][TFSA] on Ag(111). Dashed lines indicate the binding energies of the anion and cation related peaks in the limits of small coverages (adsorbed species with metal–adsorbate bonds) and of high coverages (17.5 Å film), reflecting condensed ion species (no metal–ion bonds). Reprinted with permission from ref. 92. Copyright 2013 American Chemical Society.

identification of the cations and anions in the adlayer on Ag(111) by comparison of the experimental images and simulated STM images based on DFT calculations. It is in perfect agreement with the data derived from the ARXPS measurements that anions and cations adsorb on the Ag(111) one-on-one.

1.4.2 *In situ* XPS Monitoring of Chemical Reactions in ILs

Monitoring of chemical reactions by *in situ* XPS measurement has great potential, because we can directly gain information about chemical and physical reactions. Of course, the target reaction has to proceed under vacuum conditions, and the final product should be stable *in vacuo* as well. Several research groups have reported various chemical reactions, *e.g.*, different organic chemical reactions,[95–99] fluorination reactions,[100,101] and electrochemical reactions.[102–104] Steinrück *et al.*[95] have demonstrated *in situ* XPS monitoring for the organic synthesis process in the IL. In early work, ILs were used as both reaction media and reactive substrates. In more depth, ILs containing organic cations with an amine functional groups were alkylated by another IL based on the 4-chlorobutylsulfonate anion to yield a zwitterionic salt. Usage of ARXPS revealed that the reaction occurs in the near surface region, and the conversion rate is 56%. Surprisingly, it is possible to monitor even a simple gas/liquid acid–base reaction that is a Brønsted acid–base reaction between [C$_8$mim]Cl and trifluoromethanesulfonic acid (H[TfO]) by XPS.[96] The reaction is shown in Scheme 1.1. This reaction was conducted with gaseous H[TfO], which is favorable for use in a vacuum chamber because of its low vapor pressure (≈10^1 hPa at 298 K). Clear XP spectra for F 1s, O 1s, S 2p, N 1s, C 1s, and Cl 2p during the reaction were obtained. The Cl 2p signal drastically decreased during the progress of the reaction, and, by contrast, the signals attributed to F 1s, O 1s, S 2p, and C 1s of the trifluoromethanesulfonate anion increased. Further detailed discussions are described in ref. 96. Villar-Garcia *et al.*[99] have reported that the turnover frequency (TOF) of the Suzuki cross-coupling reaction of phenylboronic acid with aryl bromides using Pd(PPh$_3$)$_4$ as a catalyst can be controlled by tuning the electronic environment, *i.e.*, by changing the IL composition, which causes variations in the binding energy for the Pd 3d$_{5/2}$. As we have seen throughout this chapter, the binding energy is determined by XPS. Figure 1.15 indicates the TOF of the Suzuki cross-coupling reaction described above as a function of the Pd 3d$_{5/2}$ binding energies in different IL mixtures under the same conditions.

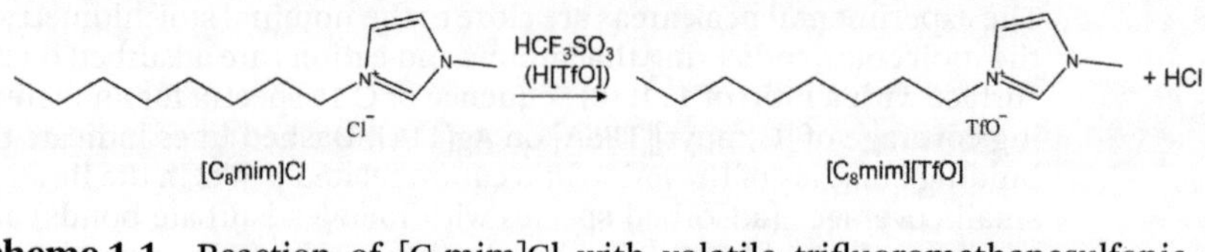

Scheme 1.1 Reaction of [C$_8$mim]Cl with volatile trifluoromethanesulfonic acid (H[TfO]) to form an IL, [C$_8$mim][TfO], and HCl that evaporates under UHV conditions.

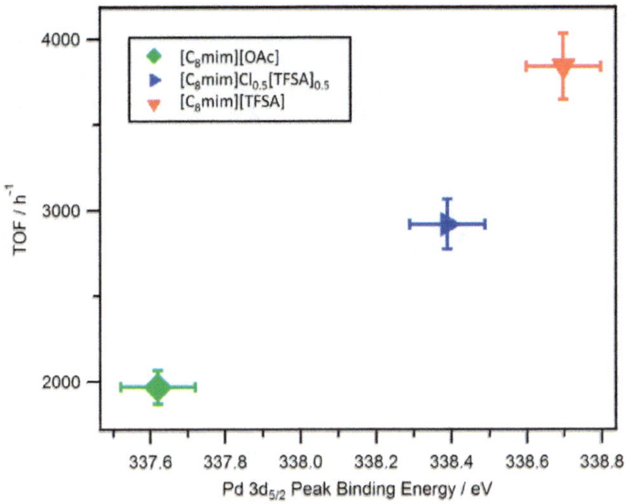

Figure 1.15 Pd $3d_{5/2}$ binding energies of the solutions produced after mixing aqueous Na_2CO_3/aryl bromide/$Pd(PPh_3)_4$/NaCl/[C_8mim][TFSA], aqueous Na_2CO_3/aryl bromide/$Pd(PPh_3)_4$/NaCl/[C_8mim][OAc] and aqueous Na_2CO_3/aryl bromide/$Pd(PPh_3)_4$/NaCl/[C_8mim]$Cl_{0.5}$[TFSA]$_{0.5}$ *vs* turnover frequency (TOF) of the Suzuki cross-coupling reaction of phenylboronic acid with the aryl bromide in these mixtures at the same conditions. Reproduced from ref. 99 with permission from The Royal Society of Chemistry.

Undoubtedly the TOF is dependent on the binding energy. This thought-provoking result will make a big impact on organic chemistry in ILs because such fine tuning of the catalyst is not easy in conventional solvents.

One of the features of ILs is their wide electrochemical window. At the reduction limiting potential, organic cations are reduced and/or decomposed. Meanwhile, at the oxidation limiting potential, anions are oxidized. The wide electrochemical window means that there is a large gap between these two limiting potentials. It is important to know the reactions at the potentials for designing IL electrolytes for electrochemical technologies, but most of them still remain unclear, especially anion oxidation reactions with fluorine release. Recently, two research groups have reported the chemical reactions of fluoroanions, $[BF_4]^-$ and $[SbF_6]^-$, in ILs.[100,101] The reactions reported in these articles are not electrochemical, but we think it is possible to learn more if *in situ* electrochemical XPS techniques are employed.

As mentioned above, several *in situ* electrochemical XPS techniques with ILs have already been reported.[102–104] One of the electrochemical cells used for XPS is depicted in Figure 1.16. Detailed information about the cell is available in ref. 102. We think that electrochemical cells for *in situ* SEM and TEM observations may also have applicability similar to that for *in situ* XPS because the cell design concept is very similar. Conversely, the beam irradiation techniques used for ARXPS would be an effective approach for *in situ*

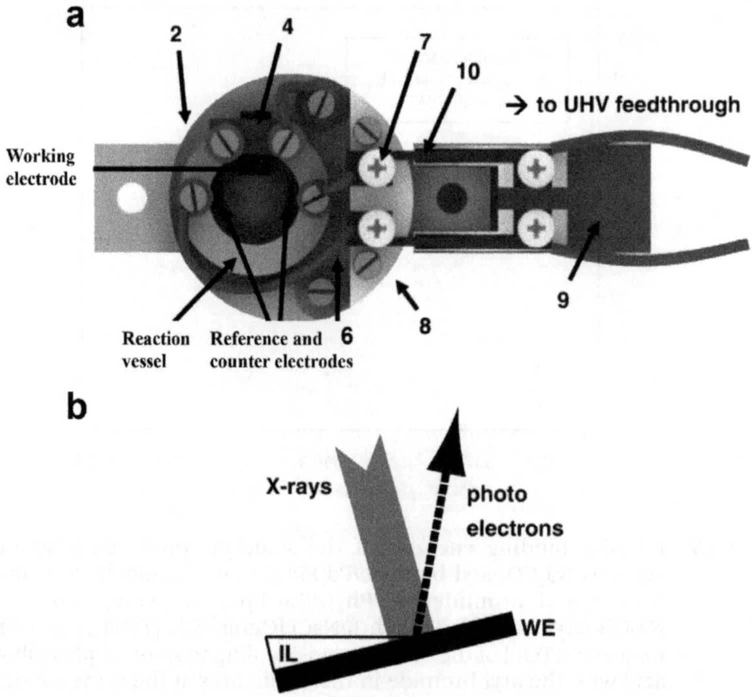

Figure 1.16 (a) Sketch of an *in situ* XPS cell, and (b) sketch of the sample area (interphase) probed by XPS. Reprinted from *Electrochemistry Communications*, **13** (6), D. Weingarth, A. Foelske-Schmitz, A. Wokaun, R. Kötz, *In situ* electrochemical XPS study of the Pt/[EMIM][BF_4] system, 619–622, Copyright (2011) with permission from Elsevier.[102]

EDX measurements for electrochemical devices that we have reported previously.[18,32,105,106] During the *in situ* electrochemical XPS measurements, a binding energy shift of ionic species in the ILs is reported.[102] In fact, the binding energy shift also occurs under open circuit potential. It is considered to be an X-ray induced shift, but this unfavorable behavior is greatly improved by using a working electrode with a large capacitance, such as an activated carbon electrode.[103]

1.5 Conclusion

As all readers of this book know, a great number of scientific investigations and technological applications with ILs have been reported to date. This is undoubtedly due to their desirable physicochemical properties. However, most researchers who use ILs often neglect the opportunities provided by the negligible vapor pressure and antistatic properties of these solvents. In fact, it is difficult for us to recall devices and chemical processes that take advantage of such features. This is not the case with the analytical technologies that require vacuum conditions. In this chapter, we have introduced some vacuum analytical technologies useful for interrogating ILs. ILs are very useful liquid materials and reaction media for SEM, TEM, EDX, and XPS exploration. By employing novel analytical technologies, we can readily know the morphology of a variety of hydrous specimens and directly observe/monitor various chemical reactions. In addition, if the analysis equipment is equipped with electrical feed-through terminals, we can carry out *in situ* analyses of the electrochemical reactions in ILs directly in the vacuum chamber where these techniques are carried out. We expect that the usefulness of IL-based vacuum analytical techniques will become widely recognized.

References

1. *Ionic Liquids in Synthesis*, ed. P. Wasserscheid and T. Welton, Wiley-VCH Verlag, Weinheim, 2nd edn, 2007.
2. *Electrodeposition from Ionic Liquids*, ed. F. Endres, A. P. Abbott and D. R. MacFarlane, Wiley-VCH Verlag, Weinheim, 2008.
3. T. Tsuda and C. L. Hussey in *Modern Aspects of Electrochemistry Vol. 45*, ed. R. E. White, Springer Science+Business Media, New York, 2009, pp. 63–174.
4. T. Torimoto, T. Tsuda, K. Okazaki and S. Kuwabata, *Adv. Mater.*, 2010, **22**, 1196.
5. J. Dupont and J. D. Scholten, *Chem. Soc. Rev.*, 2010, **39**, 1780.
6. *Electrochemical Aspects of Ionic Liquids*, ed. H. Ohno, Wiley-Interscience, New Jersey, 2nd edn, 2011.
7. U. A. Rana, M. Forsyth, D. R. MacFarlane and J. M. Pringle, *Electrochim. Acta*, 2012, **84**, 213.
8. H. Wang, G. Gurau and R. D. Rogers, *Chem. Soc. Rev.*, 2012, **41**, 1519.
9. T.-P. Fellinger, A. Thomas, J. Yuan and M. Antonietti, *Adv. Mater.*, 2013, **25**, 5838.
10. Q. Zhang and J. M. Shreeve, *Chem. Rev.*, 2014, **114**, 10527.
11. S. Ravula, S. N. Baker, G. Kamath and G. A. Baker, *Nanoscale*, 2015, **7**, 4338.
12. S. Kuwabata, A. Kongkanand, D. Oyamatsu and T. Torimoto, *Chem. Lett.*, 2006, **35**, 600.
13. E. W. Castner and J. F. Wishart, *J. Chem. Phys.*, 2010, **132**, 120901.
14. M. J. Dykstra and L. E. Reuss, *Biological Electron Microscopy: Theory, Techniques, and Troubleshooting*, Springer, US, New York, 2nd edn, 2003.
15. A. M. Donald, *Nat. Mater.*, 2003, **2**, 511.
16. T. Tsuda, N. Nemoto, K. Kawakami, E. Mochizuki, S. Kishida, T. Tajiri, T. Kushibiki and S. Kuwabata, *ChemBioChem*, 2011, **12**, 2547.
17. T. Tsuda, E. Mochizuki, S. Kishida, H. Sakagami, S. Tachibana, M. Ebisawa, N. Nemoto, Y. Nishimura and S. Kuwabata, *Electrochemistry*, 2012, **80**, 308.

18. T. Tsuda, E. Mochizuki, S. Kishida, N. Nemoto, Y. Ishigaki and S. Kuwabata, Room-Temperature Ionic Liquid-Based SEM/EDX Techniques for Biological Specimens and *in situ* Electrode Reaction Observation, in *Molten Salts Chemistry and Technology*, ed. M. Gaune-Escard and G. M. Haarberg, John Wiley & Sons Ltd, Chichester, UK, 2014.
19. Y. Asahi, J. Miura, T. Tsuda, S. Kuwabata, K. Tsunashima, Y. Noiri, T. Sakata, E. Ebisu and M. Hayashi, *AMB Express*, 2015, **5**, 6.
20. A. Hyono, T. Yonezawa, K. Kawai, S. Abe, M. Fujihara, H. Azuma and S. Wakamoto, *Surf. Interface Anal.*, 2014, **46**, 425.
21. F. Shima, T. Akagi, M. Shudo, E. Mochizuki, T. Tsuda, S. Kuwabata and M. Akashi, *Polym. J.*, 2015, **47**, 631.
22. C. G. Golding, L. L. Lamboo, D. R. Beniac and T. F. Booth, *Sci. Rep.*, 2016, **6**, 26516.
23. Y. Ishigaki, Y. Nakamura, T. Takehara, T. Kurihara, H. Koga, T. Takegami, H. Nakagawa, N. Nemoto, N. Tomosugi, S. Kuwabata and S. Miyazawa, *Microsc. Res. Tech.*, 2011, **74**, 1104.
24. A. Dwiranti, L. Y. Lin, E. Mochizuki, S. Kuwabata, A. Takaoka, S. Uchiyama and K. Fukui, *Microsc. Res. Tech.*, 2012, **75**, 1113.
25. S. Imashuku, T. Kawakami, L. Ze and J. Kawai, *Microsc. Microanal.*, 2012, **18**, 365.
26. C. Takahashi, D. K. Pattanayak, T. Shirai and M. Fuji, *Ceram. Int.*, 2013, **39**, 1065.
27. C. Takahashi, T. Shirai and M. Fuji, *Mater. Chem. Phys.*, 2013, **141**, 657.
28. A. S. Kashin, K. I. Galkin, E. A. Khokhlova and V. P. Ananikov, *Angew. Chem., Int. Ed.*, 2016, **55**, 2161.
29. P. Y. Kim, A. E. Ribbe, T. P. Russell and D. A. Hoagland, *ACS Nano*, 2016, **10**, 6257.
30. K. Horigome, T. Ueki and D. Suzuki, *Polym. J.*, 2016, **48**, 273.
31. M. Armand, F. Endres, D. R. MacFarlane and H. Ohno, *Nat. Mater.*, 2009, **8**, 621.
32. S. Arimoto, H. Kageyama, T. Torimoto and S. Kuwabata, *Electrochem. Commun.*, 2008, **10**, 1901.
33. D. R. MacFarlane, N. Tachikawa, M. Forsyth, J. M. Pringle, P. C. Howlett, G. D. Elliott, J. H. Davis, M. Watanabe, P. Simon and C. A. Angell, *Energy Environ. Sci.*, 2014, 7, 232.
34. T. Tsuda, T. Kanetsuku, T. Sano, Y. Oshima, K. Ui, M. Yamagata, M. Ishikawa and S. Kuwabata, *Microscopy*, 2015, **64**, 159.
35. C. Y. Chen, T. Sano, T. Tsuda, K. Ui, Y. Oshima, M. Yamagata, M. Ishikawa, M. Haruta, T. Doi, M. Inaba and S. Kuwabata, *Sci. Rep.*, 2016, **6**, 36153.
36. X. Su, Q. L. Wu, J. C. Li, X. C. Xiao, A. Lott, W. Q. Lu, B. W. Sheldon and J. Wu, *Adv. Energy Mater.*, 2014, **4**, 1300882.
37. D. Chen, S. Indris, Z. Schulz, B. Gamer and R. Monig, *J. Power Sources*, 2011, **196**, 6382.
38. X. X. Xia, C. V. Di Leo, X. W. Gu and J. R. Greer, *ACS Energy Lett.*, 2016, **1**, 492.

39. D. J. Miller, C. Proff, J. G. Wen, D. P. Abraham and J. Bareno, *Adv. Energy Mater.*, 2013, **3**, 1098.
40. E. Strelcov, J. Cothren, D. Leonard, A. Y. Borisevich and A. Kolmakov, *Nanoscale*, 2015, **7**, 3022.
41. P. P. R. M. L. Harks, F. M. Mulder and P. H. L. Notten, *J. Power Sources*, 2015, **288**, 92.
42. J. S. Bridel, T. Azais, M. Morcrette, J. M. Tarascon and D. Larcher, *J. Electrochem. Soc.*, 2011, **158**, A750.
43. M. Adrian, J. Dubochet, J. Lepault and A. W. McDowall, *Nature*, 1984, **308**, 32.
44. M. J. Williamson, R. M. Tromp, P. M. Vereecken, R. Hull and F. M. Ross, *Nat. Mater.*, 2003, **2**, 532.
45. J. M. Yuk, J. Park, P. Ercius, K. Kim, D. J. Hellebusch, M. F. Crommie, J. Y. Lee, A. Zettl and A. P. Alivisatos, *Science*, 2012, **336**, 61.
46. N. de Jonge and F. M. Ross, *Nat. Nanotech.*, 2011, **6**, 695.
47. F. M. Ross, *Science*, 2015, **350**, aaa9886.
48. R. Hayes, G. G. Warr and R. Atkin, *Chem. Rev.*, 2015, **115**, 6357.
49. J. Dupont, *J. Braz. Chem. Soc.*, 2004, **15**, 341.
50. S. M. Chen, S. J. Zhang, X. M. Liu, J. Q. Wang, J. J. Wang, K. Dong, J. Sun and B. H. Xu, *Phys. Chem. Chem. Phys.*, 2014, **16**, 5893.
51. S. M. Chen, K. Kobayashi, R. Kitaura, Y. Miyata and H. Shinohara, *ACS Nano*, 2011, **5**, 4902.
52. N. Li, S. H. Zhang, L. Q. Zheng, B. Dong, X. W. Li and L. Yu, *Phys. Chem. Chem. Phys.*, 2008, **10**, 4375.
53. J. G. Li, Y. F. Hu, S. F. Sun, S. Ling and J. Z. Zhang, *J. Phys. Chem. B*, 2012, **116**, 6461.
54. R. R. Maddikeri, S. Colak, S. P. Gido and G. N. Tew, *Biomacromolecules*, 2011, **12**, 3412.
55. P. Renz, M. Kokkinopoulou, K. Landfester and I. Lieberwirth, *Macromol. Chem. Phys.*, 2016, **217**, 1879.
56. A. Antonello, G. Jakob, P. Dolcet, R. Momper, M. Kokkinopoulou, K. Landfester, R. Munoz-Espi and S. Gross, *Chem. Mater.*, 2016, **29**(3), 985.
57. E. H. Hayakawa, E. Mochizuki, T. Tsuda, K. Akiyoshi, H. Matsuoka and S. Kuwabata, *PLoS One*, 2013, **8**, e85467.
58. M. Antonietti, D. B. Kuang, B. Smarsly and Z. Yong, *Angew. Chem., Int. Ed.*, 2004, **43**, 498.
59. C. W. Scheeren, G. Machado, J. Dupont, P. F. P. Fichtner and S. R. Texeira, *Inorg. Chem.*, 2003, **42**, 4738.
60. U. Mansfeld, S. Hoeppener and U. S. Schubert, *Adv. Mater.*, 2013, **25**, 761.
61. N. Luo, Y. X. Lv, D. X. Wang, J. M. Zhang, J. Wu, J. S. He and J. Zhang, *Chem. Commun.*, 2012, **48**, 6283.
62. M. Itoi, T. Jike, D. Nishio-Hamane, S. Udagawa, T. Tsuda, S. Kuwabata, K. Boukheddaden, M. J. Andrus and D. R. Talham, *J. Am. Chem. Soc.*, 2015, **137**, 14686.

63. T. Miyata and T. Mizoguchi, *Ultramicroscopy*, 2017, **178**, 81.
64. T. Miyata, M. Fukuyama, A. Hibara, E. Okunishi, M. Mukai and T. Mizoguchi, *Microscopy*, 2014, **63**, 377.
65. T. Tsuda, S. Seino and S. Kuwabata, *Chem. Commun.*, 2009, 6792.
66. T. Tsuda, T. Sakamoto, Y. Nishimura, S. Seino, A. Imanishi and S. Kuwabata, *Chem. Commun.*, 2012, **48**, 1925.
67. T. Uematsu, M. Baba, Y. Oshima, T. Tsuda, T. Torimoto and S. Kuwabata, *J. Am. Chem. Soc.*, 2014, **136**, 13789.
68. Y. Kimura, H. Niinomi and K. Tsukamoto, *J. Am. Chem. Soc.*, 2014, **136**, 1762.
69. T. Miyata and T. Mizoguchi, *Microsc. Microanal.*, 2016, **22**, 852.
70. J. Y. Huang, L. Zhong, C. M. Wang, J. P. Sullivan, W. Xu, L. Q. Zhang, S. X. Mao, N. S. Hudak, X. H. Liu, A. Subramanian, H. Y. Fan, L. A. Qi, A. Kushima and J. Li, *Science*, 2010, **330**, 1515.
71. M. T. McDowell, I. Ryu, S. W. Lee, C. M. Wang, W. D. Nix and Y. Cui, *Adv. Mater.*, 2012, **24**, 6034.
72. X. H. Liu, S. Huang, S. T. Picraux, J. Li, T. Zhu and J. Y. Huang, *Nano Lett.*, 2011, **11**, 3991.
73. S. Lee, Y. Oshima, E. Hosono, H. S. Zhou, K. Kim, H. M. Chang, R. Kanno and K. Takayanagi, *J. Phys. Chem. C*, 2013, **117**, 24236.
74. X. H. Liu, J. W. Wang, Y. Liu, H. Zheng, A. Kushima, S. Huang, T. Zhu, S. X. Mao, J. Li, S. L. Zhang, W. Lu, J. M. Tour and J. Y. Huang, *Carbon*, 2012, **50**, 3836.
75. Y. Liu, H. Zheng, X. H. Liu, S. Huang, T. Zhu, J. W. Wang, A. Kushima, N. S. Hudak, X. Huang, S. L. Zhang, S. X. Mao, X. F. Qian, J. Li and J. Y. Huang, *ACS Nano*, 2011, **5**, 7245.
76. C. M. Wang, W. Xu, J. Liu, D. W. Choi, B. Arey, L. V. Saraf, J. G. Zhang, Z. G. Yang, S. Thevuthasan, D. R. Baer and N. Salmon, *J. Mater. Res.*, 2010, **25**, 1541.
77. X. H. Liu and J. Y. Huang, *Energy Environ. Sci.*, 2011, **4**, 3844.
78. R. Fortunato, C. A. M. Afonso, J. Benavente, E. Rodrigez-Castellón and J. G. Crespo, *J. Membr. Sci.*, 2005, **256**, 216.
79. E. F. Smith, I. J. Villar-Garcia, D. Briggs and P. Licence, *Chem. Commun.*, 2005, 5633.
80. H.-P. Steinrück, *Phys. Chem. Chem. Phys.*, 2012, **14**, 5010.
81. R. K. Blundell, A. E. Delorme, E. F. Smith and P. Licence, *Phys. Chem. Chem. Phys.*, 2016, **18**, 6122.
82. K. R. J. Lovelock, I. J. Villar-Garcia, F. Maier, H.-P. Steinrück and P. Licence, *Chem. Rev.*, 2010, **110**, 5158.
83. R. K. Blundell and P. Licence, *Phys. Chem. Chem. Phys.*, 2014, **16**, 15278.
84. R. K. Blundell and P. Licence, *Chem. Commun.*, 2014, **50**, 12080.
85. S. Men, D. S. Mitchell, K. R. J. Lovelock and P. Licence, *ChemPhysChem*, 2015, **16**, 2211.
86. A. R. Santos, R. K. Blundell and P. Licence, *Phys. Chem. Chem. Phys.*, 2015, **17**, 11839.

87. S. Men, B. B. Hurisso, K. R. J. Lovelock and P. Licence, *Phys. Chem. Chem. Phys.*, 2014, **14**, 5229.
88. C. Kolbeck, I. Niedermaier, A. Deyko, K. R. J. Lovelock, N. Taccardi, W. Wei, P. Wasserscheid, F. Maier and H.-P. Steinrück, *Chem.–Eur. J.*, 2014, **20**, 3954.
89. S. Caporali, M. Pedio, C. Chiappe, C. S. Pomelli, R. G. Acres and U. Bardi, *Surf. Sci.*, 2016, **648**, 360.
90. M. Currie, J. Estager, P. Licence, S. Men, P. Nockemann, K. R. Seddon, M. Swadźba-Kwaśny and C. Terrade, *Inorg. Chem.*, 2013, **52**, 1710.
91. A. W. Taylor, S. Men, C. J. Clarke and P. Licence, *RSC Adv.*, 2013, **3**, 9436.
92. F. Buchner, K. Forster-Tonigold, B. Uhl, D. Alwast, N. Wagner, H. Farkhondeh, A. Groß and J. Behm, *ACS Nano*, 2013, **7**, 7773.
93. J. P. L. Perez, B. W. McMahon, S. Schneider, J. A. Boatz, T. W. Hawkins, P. D. McCrary, P. A. Beasley, S. P. Kelley, R. D. Rogers and S. L. Anderson, *J. Phys. Chem. C*, 2013, **117**, 5693.
94. Y. Yokota, Y. Mino, Y. Kanai, T. Utsunomiya, A. Imanishi and K. Fukui, *J. Phys. Chem. C*, 2015, **119**, 18467.
95. C. Kolbeck, I. Niedermaier, N. Taccardi, P. S. Schulz, F. Maier, P. Wasserscheid and H.-P. Steinrück, *Angew. Chem., Int. Ed.*, 2012, **51**, 2610.
96. I. Niedermaier, N. Taccardi, P. Wasserscheid, F. Maier and H.-P. Steinrück, *Angew. Chem., Int. Ed.*, 2013, **52**, 8904.
97. S. Men, K. R. J. Lovelock and P. Licence, *Chem. Phys. Lett.*, 2016, **645**, 53.
98. S. Men and J. Jiang, *Chem. Phys. Lett.*, 2016, **646**, 125.
99. I. J. Villar-Garcia, K. R. J. Lovelock, S. Men and P. Licence, *Chem. Sci.*, 2014, **5**, 2573.
100. K. L. Syres and R. G. Jones, *Langmuir*, 2015, **31**, 9799.
101. L. S. Longo Jr, E. F. Smith and P. Licence, *ACS Sustainable Chem. Eng.*, 2016, **4**, 5953.
102. D. Weingarth, A. Foelske-Schmitz, A. Wokaun and R. Kötz, *Electrochem. Commun.*, 2011, **13**, 619.
103. A. Foelske-Schmitz, D. Weingarth, A. Wokaun and R. Kötz, *ECS Electrochem. Lett.*, 2013, **2**, H13.
104. J. Kruusma, A. Tõnisoo, R. Pärna, E. Nõmmiste and E. Lust, *J. Electrochem. Soc.*, 2014, **161**, A1266.
105. T. Uematsu, J.-T. Han, T. Tsuda and S. Kuwabata, *J. Phys. Chem. C*, 2012, **116**, 20902.
106. T. Tsuda, M. Baba, Y. Sato, R. Sakao, K. Matsumoto, R. Hagiwara and S. Kuwabata, *Chem.–Eur. J.*, 2011, **17**, 11122.

CHAPTER 2

Electron Microscopy of Wet Materials Using Ionic Liquids

C. TAKAHASHI

Aichi Gakuin University, School of Pharmacy, Pharmaceutical Engineering, 1-100, Kusumoto-cho, Chikusa-ku, 464-8650, Nagoya, Japan
*E-mail: chisato@dpc.agu.ac.jp

2.1 Introduction

Ionic liquids have been used in different types of fields because of their unique properties such as high electrical conductivity, negligible vapor pressure, non-flammability, and high polarity.[1–3] This chapter mainly focuses on the use of electron microscopy (EM) with a room temperature (RT) ionic liquid (IL) (RTIL) for the pretreatment of hydrous materials. There are two main types of EM, namely, transmission electron microscopy (TEM) and scanning electron microscopy (SEM). TEM and SEM are well-established techniques developed by advancements in high-resolution performance imaging, high-resolution analysis equipment, and three-dimensional (3D) observational capabilities.[4–6] These characterization techniques are not only useful for obtaining information about the structural and morphological properties of a material, but also about their chemical composition. In recent years, *in situ* EM observation techniques have been developed for reaction fields because they allow for real-time phenomena to be visualized with a high resolution.[7–9] These techniques have allowed for the identification and characterization of metallic, semiconductor, and biological materials. However, it is

Smart Materials No. 28
Ionic Liquid Devices
Edited by Ali Eftekhari

Published by the Royal Society of Chemistry, www.rsc.org

difficult to observe the original morphology and structure of hydrous materials (such as biological materials and water-absorbed materials) because these materials can only be seen when wet and EM under high vacuum conditions dehydrates such materials. Furthermore, a conducting surface is required for non-conductive materials to reduce the charging and allow for a sufficient contrast to be obtained. Therefore, to establish a simple EM technique for hydrated materials, a sample treatment method using ILs was developed so as to take advantage of the properties of ILs.[10,11] This simple technique can shorten the sample preparation time and allow the hydrous materials to be observed without requiring any complicated processes or specific apparatus.

This chapter summarizes the EM observation techniques suited for revealing the fine morphology and structure and the detailed reaction behaviors of hydrated materials that cannot be obtained using conventional techniques.

2.2 EM Observation of Wet Materials Using Sample Preparation Techniques with ILs

In 2005, Scherson *et al.* reported that RTILs are stable under a high vacuum condition ($<5 \times 10^{-9}$ Torr),[12] resulting in considerable attention being paid to using them as solvents for vacuum technologies. This finding was a breakthrough in the scientific field as it negated the common theory that suggested that liquids are easily evaporated under vacuum conditions. ILs are able to provide a conductive and negligible vapor pressure to samples without any treatment. In addition, Wishart *et al.* reported that electrons injected into the ILs are stable and move into the liquid phase of the ILs even at high accelerating voltages.[13] On the basis of this discovery, Kuwabata *et al.* developed a SEM observation technique that insulates samples using ILs.[14] Star-like sands were observed using a hydrophobic IL; 1-ethyl-3-methylimidazolium bis(trifluoromethanesulfonyl) imide ([EtMeIm][TFSI]), whereas abrasive paper was observed using a hydrophobic IL; 1-butyl-3-methylimidazolium bis(trifluoromethylsulfonyl) imide ([BuMeIm][$(CF_3SO_2)_2N$]). The same research group subsequently claimed that EM observations of hydrous materials can be made using ILs.[15] Many researchers have further developed suitable sample preparation techniques for hydrous materials using various ILs, as summarized in Table 2.1.

2.2.1 Biological Materials

Evaluation of biological materials using EM is important for understanding their precise morphology and structure and also their bodily functions. However, biological materials in wet form are not suitable for EM observation because EM samples need to be maintained under vacuum condition. Therefore, the biological samples need to be prepared in a dried form to be observed *via* EM. Conventional sample pretreatment techniques for SEM and TEM consist of the following procedures: fixation, staining, dehydration,

Table 2.1 List of suitable ILs for sample pretreatment in EM analysis.

	IL
1	[EtMeIm][BF_4]
2	[BuMeIm][BF_4]
3	[EtMeIm][Lac]
4	[Ch][Lac]
5	[BuMeIm][TFSI]
6	Choline-like IL
7	[EtMeIm][AcO]
8	[$P_{4,4,4,1}$][DMP]
9	[BuMeIm][PF_6]

drying, and coating for SEM, and fixation, staining, dehydration, embedding, sectioning, and coating for TEM; thus, biological samples have to be dehydrated and dried during the sample preparation process, thereby making the observation of biological samples maintained in the liquid phase impossible. Specific apparatuses, such as cryomicrotomes, focused ion beam (FIB)-SEM, wet-SEM, environmental EM, and cryo-EM, have recently been developed to observe the fine structure and morphology of intact biological materials.[16–20] In addition, super-resolving microscopy and phase-contrast microscopy have also been developed for this purpose.[21,22] However, these sample preparation techniques are complicated and these apparatuses are expensive.

To overcome these issues, simple pretreatment techniques using ILs for the EM observation of biological materials have been intensively researched, and several studies where animal cells, microorganisms, plants, and seaweed have been observed using SEM or TEM after the materials were pretreated with an IL have been reported.[23–34] This section reviews the research done on several biological materials; the procedures for the individual biological samples are summarized in Table 2.2.

2.2.1.1 *Animal and Plant Cells and Chromosomes*

The EM observations of various samples derived from animal and plant (barley) cells, such as human lung epithelial cells, pancreatic carcinoma cells, mouse fibroblast cells, mouse villi, Indian muntjac spread chromosomes, human chromosomes, barley seed chromosomes, and HeLa S3 cell chromosomes have been reported to date.[23–28] On the basis of the results of these analyses, 1-ethyl-3-methylimidazolium tetrafluoroborate ([EtMeIm][BF_4]) was found to be suitable as an IL for human lung epithelial A549 cells; 1-ethyl-3-methylimidazolium lactate ([EtMeIm][Lac]) and choline lactate ([Ch][Lac]) were suitable for mouse fibroblast L929 cells; and 1-butyl-3-methylimidazolium tetrafluoroborate ([BuMeIm][BF_4]) was suitable for Indian muntjac spread chromosomes, polyamine (PA) isolated human chromosomes, barley seed chromosomes, and PA isolated HeLa S3 cell chromosomes (Table 2.2). The structural morphologies of

Table 2.2 Summary of representative IL-based EM techniques for biological materials.

Group	Sample	Microscopy	IL	Techniques	Comments
Ishigaki *et al.*[23–25]	Human lung epithelial A549 cells, pancreatic carcinoma cells	SEM	[EtMeIm][BF_4]	Accelerating voltage: SEM 0.9–1.2 kV, IL: 10.0–40.0 w/w%	Preheating of the IL and colloidal gold conjugated antibody treatment is useful for SEM imaging
Kuwabata and Tsuda *et al.*[26]	Mouse fibroblast L929 cells, plant (petal of *Asteraceae*), plant (pollen of *P. juliae, A. coronaria, L. paludosum, L. odoratus*)	SEM	[EtMeIm][Lac] [Ch][Lac] [BuMeIm][TFSI] [BuMeIm][BF_4]	Accelerating voltage: SEM 5.0–10.0 kV, IL: 0.5–10.0 v/v%	
Fukui *et al.*[27]	Chromosome (Indian muntjac spread, PA isolated human)	SEM	[BuMeIm][BF_4]	Accelerating voltage: SEM 1.5 kV, IL: 0.5–1.0 w/w%	Pt-blue staining is useful for SEM imaging
Fukui *et al.*[28]	Chromosome (barley seeds, HeLa S3 cells using the PA method)	FIB-SEM	[BuMeIm][BF_4]	Accelerating voltage: SEM 1.5–2.0 kV, IL: 0.5 w/w%	Pt-blue staining is useful for SEM imaging.
Asahi *et al.*[29]	Biofilm (*S. mutans*)	SEM	[EtMeIm][AcO] [Ch][Lac]	Accelerating voltage: SEM 5.0 kV, IL: 10.0 v/v%	
Takahashi *et al.*[30]	Biofilm and bacteria (*S. epidermidis*)	SEM TEM	[BuMeIm][BF_4]	Accelerating voltage: SEM 10.0 kV, TEM 200 kV, IL: 0.5 w/w%	
Kawai *et al.*[31,32]	Chicken-tissue, sponge cake, bacteria (*S. aureus*), plant (*C. sasanqua* containing pollen)	SEM	Choline-like IL	Accelerating voltage: SEM 15.0 kV, IL: 10.0 w/w%	
Beniac *et al.*[33]	Bacteria (*Leptospira biflexa, Salmonella Senftenberg*), virus (vaccinia, Ebola)	SEM	[BuMeIm][BF_4]	Accelerating voltage: SEM 4.0 kV, IL: 2.5 w/w%	Metal coated filter is useful for SEM imaging
Kawai *et al.*[31]	Seaweed	SEM	Choline-like IL	Accelerating voltage: SEM 15.0 kV, IL: IL : hydrous sample = 10 mL : 0.5 g	
Takahashi *et al.*[34]	Seaweed	SEM	[BuMeIm][BF_4]	Accelerating voltage: SEM 5.0 kV, IL: 30 mol% (toward water within sample)	Understanding the osmotic pressure of each sample is essential

the IL-treated human lung epithelial A549 cells and pancreatic carcinoma cells were observed using SEM, and the ultrastructure of microvilli, which can be used for the fabrication of a bridge between cells, have successfully been visualized.[23–25] Treatment of the sample with gold colloidal-conjugated antibodies was found to improve the image contrast as these antibodies can bind to the sample nonspecifically and act as a source of secondary electrons. The optimum accelerating voltage was found to be approximately 0.9–1.2 kV.

The mouse fibroblast L929 cells were selected for the SEM observation and the possibility of observations under high magnification was investigated.[26] By adding the glutaraldehyde (GA) fixation step before the IL sample preparation, SEM observation of the ultrafine structures could be pursued. The biocompatibility of ILs was found to be important for the SEM observations of sensitive animal cells. In addition, an appropriate immersion time in the IL was related to the osmotic pressure of the IL. The selection of suitable ILs and suitable immersion times for the samples in the ILs are essential for successfully observing biological materials using SEM.

FIB-SEM observations have been performed on a number of chromosomes in addition to the above-mentioned SEM observation; Indian muntjac spread chromosomes, PA isolated human chromosomes, barley seed chromosomes, and PA isolated HeLa S3 cell chromosomes were selected as the samples.[28] SEM observations for the Indian muntjac spread and PA isolated human chromosomes were made using both IL pretreatment and platinum blue staining, and a combination of these two procedures drastically enhanced the image contrast and resolution of the samples.[27] The optimum accelerating voltage for these chromosomes was approximately 1.5 kV. FIB-SEM observations for the barley seed and PA isolated HeLa S3 cell chromosomes were made using the IL sample preparation technique. As the IL penetrated the chromosomes, the cavities within the chromosomes were found to maintain their morphologies without drying, thereby confirming that ILs allowed for the chromosomes to be maintained in a liquid state in the FIB-SEM chamber.

2.2.1.2 Microorganisms: Bacteria and Viruses

EM observations using ILs have been studied for several types of bacteria, including *Streptococcus mutans*, *Staphylococcus aureus*, *Staphylococcus epidermidis*, *Leptospira biflexa*, and *Salmonella Senftenberg*, and several viruses, such as the vaccinia Ankara virus and the Ebola virus.[29–33] 1-Ethyl-3-methylimidazolium acetate ([EtMeIm][AcO]) and [Ch][Lac] were used as the ILs for the *S. mutans* cells, choline-like ILs for the *S. aureus* cells, and [BuMeIm][BF_4] for the *S. epidermidis* cells (Table 2.2).

Using the techniques shown in Table 2.2, the *S. mutans* biofilm, which was composed of cocci and extracellular polymeric substances (EPS), was observed using SEM.[29] The results showed that hydrophilic ILs are more suitable for SEM observations than hydrophobic ILs, and the optimum accelerating voltage was approximately 5 kV. In addition, to obtain a better SEM image of the *S. aureus* cells, various types of filter papers such as polycarbonate micropore membrane filters, hydrophilic cellulose membrane filters, normal filter papers, and newsprints were used as the sample mounting materials.[32] It was found that clearer images were obtained using the polycarbonate micropore membrane filter compared with those obtained using the other filter papers. In addition, a choline-like IL was synthesized for the sample preparation of hydrous materials without any morphological deformations. The choline-like IL exhibited high water-solubility and osmotic pressure as choline-like IL possesses a heterocyclic ring. These trials enabled the observation of shape-retaining *S. aureus* cells. The optimum accelerating voltage was approximately 15 kV.

SEM and TEM observations of *S. epidermidis* cells and the biofilm have been conducted by the author's research group, and the complex process of the biofilm formation was confirmed after different cultivation times (Figure 2.1).[30] Using TEM, the fibril, which plays an important role in cell adhesion, was characterized at a protein level (Figure 2.1b–h). A TEM analysis revealed that the fibril consists of three different types of protein assembly, and the curled protein assembly derived from the Aap protein seems to play an important role in extension during fibril formation. This result was the first TEM observation of protein monomers at high magnification. In this study, a field-emission-type SEM and TEM were used for observation at accelerating voltages of 5 and 200 kV, respectively. However, we did not observe electron beam damage under such harsh conditions because of the fibrils being prepared with an IL. In contrast, such high magnification observations have not been conducted using conventional methods. For the virus samples, [BuMeIm][BF_4] was used as an IL for SEM observations.[33] Following the procedure described in Table 2.2, drying processes such as shrinkage, wrinkling, or cracking can be drastically diminished for this method compared with those for the cases where conventional methods such as dehydrated sputter-coated SEM preparations and TEM negative staining method are used. Furthermore, it was found that metal-coated filters are effective for SEM observations using samples pretreated with ILs; the optimum accelerating voltage was approximately 4 kV.

2.2.1.3 Plant Cells, Petals, and Pollen

Reports have been published on EM observations using ILs for various plant cells, including the petals of Asteraceae, four kinds of pollen (*Primula juliae*, *Anemone coronaria*, *Leucoglossum paludosum*, and *Lathyrus odoratus*), and *Camellia-sasanqua*-containing pollen.[26,31] 1-Butyl-3-methylimidazolium bis(trifluoromethylsulfonyl)imide ([BuMeIm][TFSI]) and [BuMeIm][BF_4] were used as ILs for the Asteraceae petals and the pollens of *P. juliae*, *A. coronaria*, *L. paludosum*, and *L. odoratus*. Choline-like IL was used as an IL for *C. sasanqua*-containing pollen.

The detailed morphologies of the petals of Asteraceae and the pollen of *P. juliae*, *A. coronaria*, *L. paludosum*, and *L. odoratus* were observed using

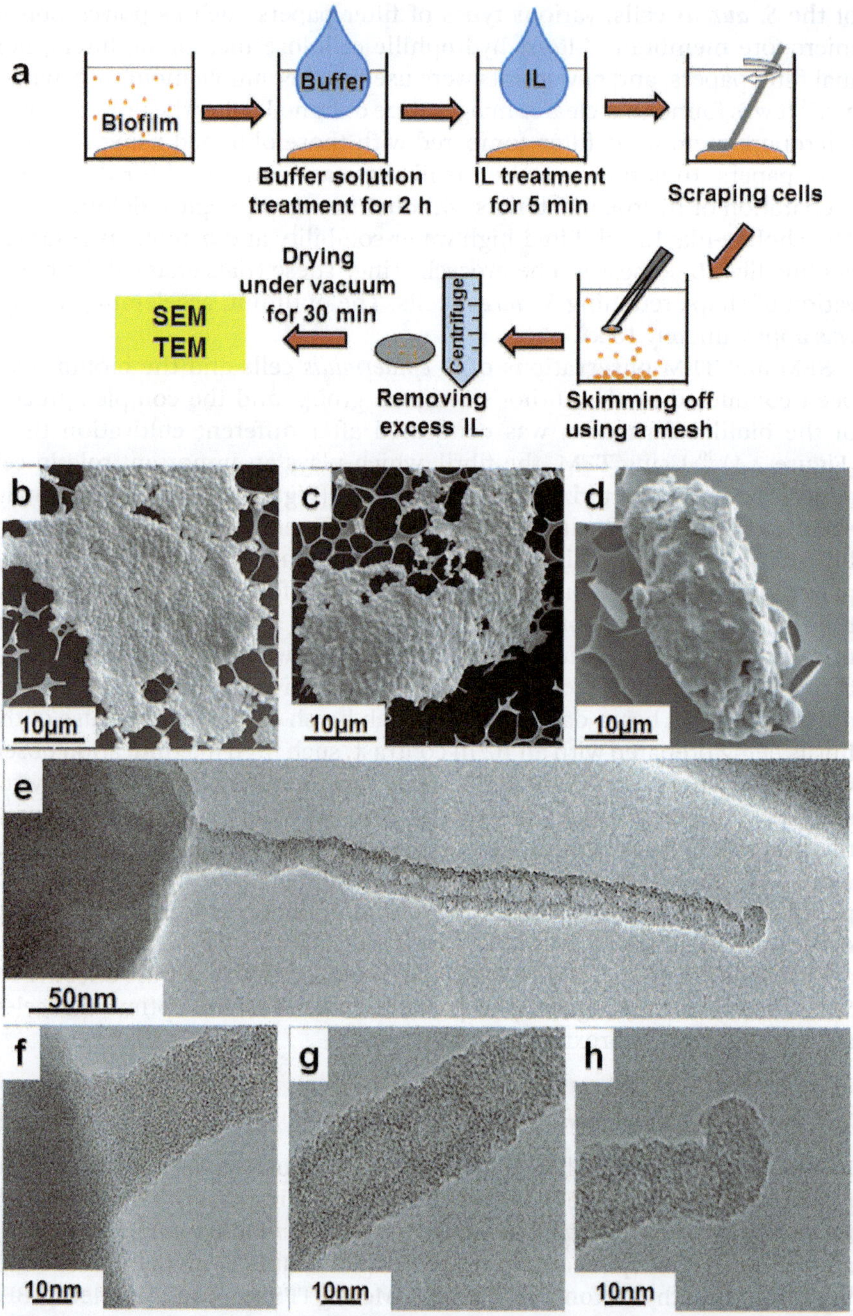

Figure 2.1 (a) Schematic of the pretreatment techniques using [BuMeIm][BF_4] as the IL for a *S. epidermidis* biofilm. SEM images of a biofilm of *S. epidermidis* prepared after different incubation times following pretreatment. The biofilms were incubated at 37 °C for (b) 12, (c) 16, and (d) 24 h. (e–h) Molecular architecture of fibril after 12 h of cultivation: (e) TEM

SEM with the above IL.[26] In these cases, the optimum accelerating voltage was approximately 5 kV. Using the choline-like IL, the morphologies of *C. sasanqua*-containing pollen were observed without any dissolution of the cell membranes; the optimum accelerating voltage was approximately 15 kV.[31] As described in Section 2.2.1.1, the immersion time depends on the types of ILs used. The choline-like IL showed a higher osmotic pressure than the imidazolium-based IL, indicating that a longer immersion time is required for imidazolium-based ILs than for choline-based ILs.

2.2.1.4 Seaweed

To optimize the IL-based SEM observation technique for hydrous biological materials, hydrous seaweed was selected as a sample. $[BuMeIm][BF_4]$ and choline-like IL were used as ILs for sample preparation. The possibility of SEM observations of hydrous seaweed using the IL ($[BuMeIm][BF_4]$) has been previously suggested as well.[15] Subsequently, choline-like ILs have been used as shape-retaining reagents because of their ability to strongly penetrate the cell membrane of the biological materials (Section 2.2.1.2). With the aid of the choline-like IL, the shape of the hydrous seaweed was observed at the micron scale; the optimum accelerating voltage was approximately 15 kV.[31] In this case, a hydrophilic imidazolium-type IL such as $[BuMeIm][BF_4]$ was not suitable for morphological observations. To solve this problem, a technique using $[BuMeIm][BF_4]$ has been developed by the author's research group.[34] To understand the effect of osmotic pressure on the deformation of the seaweed morphology, the pressure of the seaweed in sea water and that of the seaweed treated with the IL were calculated using the Van't Hoff equation. By maintaining the osmotic pressure of the seaweed treated with IL to be isotonic against the seaweed in sea water, morphological changes caused by the IL were successfully avoided (Figure 2.2). Similar to the optical microscope image, an image was obtained by the SEM observation using pretreatment with the IL (comparison between the images is shown in Figure 2.2b and c). The optimum accelerating voltage was approximately 5 kV. Furthermore, centrifugation at 10 000 rpm was found to be very useful for removing the excess IL from around the sample, which would otherwise have interfered with any detailed morphological observations. In addition, it should be noted that a treatment for 30 min is sufficient for a drying process under vacuum conditions, although sample preparations had been performed using the drying process for 24 h in a previously reported study. Thus, it was concluded that

---

image of a fibril, (f) high magnification image of protein subunits in a protein assembly anchored to the cell-wall surface, (g) protein subunits in the middle region of a fibril, and (h) protein subunits at the tip of a fibril. (Reproduced with permission from *Analytical and Bioanalytical Chemistry*, Electron microscopy of Staphylococcus epidermis fibril and biofilm formation using image-enhancing ionic liquid, **407** (6), 2014, C Takahashi, © Springer-Verlag Berlin Heidelberg 2014, ref. 30).

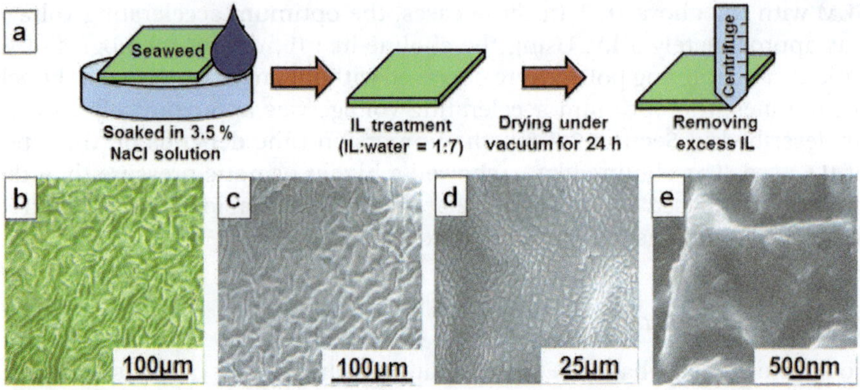

Figure 2.2 (a) Schematic of the pretreatment techniques using [BuMeIm][BF_4] as the IL for seaweed. (b) Optical microscope images of swelled seaweed after soaking in 3.5% NaCl solution for 10 min and (c–e) SEM images after soaking as in (b) and the subsequent IL treatment as in (a). (Reproduced with permission from John Wiley and Sons, Copyright © 2012 Wiley-VCH Verlag GmbH & Co. KGaA, Weinheim, ref. 34).

the exact morphology of the hydrous seaweed was visualized using the developed technique with the hydrophilic imidazolium-type IL.

2.2.2 Hydrated or Insulated Materials

In the field of materials research, characterization of the structures of hydrous organic and inorganic materials and their physical properties on a micron or nano-order scale is important for understanding the properties of materials used for various functional applications. However, characterization of hydrous materials using EM analysis is challenging, and dried inorganic materials are generally used for such characterization. Although some types of apparatus have been developed for EM analyses, as described in Section 2.2.1, they are still not commonly used. For conducting simple EM analyses for hydrous inorganic materials, IL-based EM techniques have been developed. Using IL pretreatment, polymers, minerals, ceramic green bodies, and nanotubes in wet states have been observed.[38–47] In this section, the research concerning a selection of organic and inorganic hydrous materials has been reviewed; the procedures for the respective hydrated or insulated materials are summarized in Table 2.3.

2.2.2.1 Polymer

For SEM observations using ILs for hydrated polymers, agar gel and a superabsorbent polymer have been studied.[35–38] These materials include a lot of water, unlike the biological materials described in the previous section. [BuMeIm][BF_4] was used as an IL for agar gel by the author's group,[35–37]

Table 2.3 Summary of representative IL-based EM techniques for hydrated or insulated materials.

Group	Sample	Microscopy	IL	Techniques	Comments
Takahashi *et al.*[35–37]	Agar gel	SEM	[BuMeIm][BF_4]	Accelerating voltage: SEM 5.0 kV, IL: 30 mol% IL (toward water within sample)	
Tsuda *et al.*[38]	Superabsorbent polymer	SEM	[EtMeIm][AcO] [$P_{4,4,4,1}$][DMP]	Accelerating voltage: SEM 5.0 kV, IL: raw	Understanding the osmotic pressure of each IL is essential for optimization of the IL pretreatment time
Takahashi *et al.*[39]	Montmorillonite swelled by water	SEMTEM (SAED)	[BuMeIm][BF_4]	Accelerating voltage: SEM 5.0 kV, TEM 120 kV, IL: 30 mol% IL (toward water within sample)	
Takahashi *et al.*[40–42]	IL intercalated montmorillonite	TEM (EDS, SAED)	[BuMeIm][BF_4]	Accelerating voltage: TEM 120 kV, IL: IL : dried sample = 10 : 1	
Takahashi *et al.*[43–45]	HAp	SEM TEM (SAED)	[BuMeIm][BF_4]	Accelerating voltage: SEM 5.0 kV, TEM 200 kV, IL: 30 mol% IL (toward water within sample)	
Schmuki *et al.*[46]	Nanotubes	SEM	[BuMeIm][BF_4] [BuMeIm][PF_5]	Accelerating voltage: SEM 10.0 kV, IL: IL diluted with water (no detailed information)	
Brodusch *et al.*[47]	Rare earth ore	SEM (EDS, EBSD)	[BuMeIm][BF_4]	Accelerating voltage: SEM 5.0 and 20.0 kV, IL: no detailed information	

whereas [EtMeIm][AcO] and tri-*n*-butylmethylphosphonium dimethylphosphate ([$P_{4,4,4,1}$][DMP]) were used for the superabsorbent polymer.[38]

Agar gel is widely used as a culturing media for microbes, in cataphoresis, and as a gelling agent in various ceramic forming processes. The characterization of complex network structures in wet conditions is necessary (especially in ceramic forming processes) as the main structure of a ceramic is decided during the drying process. Agar gel was therefore selected as a biological polymer for the SEM observation. The sample preparation technique using the IL was optimized according to the interaction between the water molecules and [BuMeIm][BF_4], because some reports have stated that the addition of a small amount of water to [BuMeIm][BF_4] results in changes in the IL properties.[48,49] By adjusting the water concentration in the agar gel to 15–30 mol%, the exact morphologies of the swelled agar gels under different solidification conditions were successfully visualized (Figure 2.3); the optimum accelerating voltage was approximately 5 kV.[35–37] A Raman measurement showed that [BuMeIm][BF_4] and water exist as BF_4^-–HOH–BF_4^- *via* weak hydrogen bonds even under vacuum conditions owing to its strong bonding ability. The weak bonding was stable even at 200 °C and under vacuum condition. These results suggest that the water molecules within the agar gel were displaced by the IL because of the formation of weak hydrogen bonds. Thus, the mechanism behind the SEM observation of agar gel in the presence of [BuMeIm][BF_4] has been revealed.

A superabsorbent polymer was also observed through SEM. This superabsorbent polymer is known to be covered with a uniform layer formed by a secondary crosslinking reaction after water absorption. When the superabsorbent polymer absorbs a large quantity of water, a number of cracks appear on the uniform layer. To confirm the availability of the ILs for the SEM observation, the morphologies of the cracks were observed. The treatments with neat [EtMeIm][AcO] and [$P_{4,4,4,1}$][DMP] provided a high contrast observation of the hydrous superabsorbent polymer; the optimum accelerating voltage was approximately 5 kV.[38] These ILs have low densities. Secondary electrons are much more easily emitted from the superabsorbent polymer particles owing to the primary electron bombardment than when an IL with high density is used to pretreat the polymer. This fact is reasonable for obtaining a clear SEM image. The results obtained from the SEM observation and a thermogravimetric analysis suggests that the water in the superabsorbent polymer can be replaced by an IL during the sample preparation using the IL. This information supports the above case of the agar gel. It was concluded that displacement of the IL and water within the sample allows for the exact morphology of the hydrous polymer to be observed.

2.2.2.2 Mineral

EM observations of rare earth ore and hydrous clay using ILs have been reported.[41,47] [BuMeIm][BF_4] was used as an IL for the EM analyses of rare earth ore and hydrous clay. The rare earth ore was chosen as an insulating material.

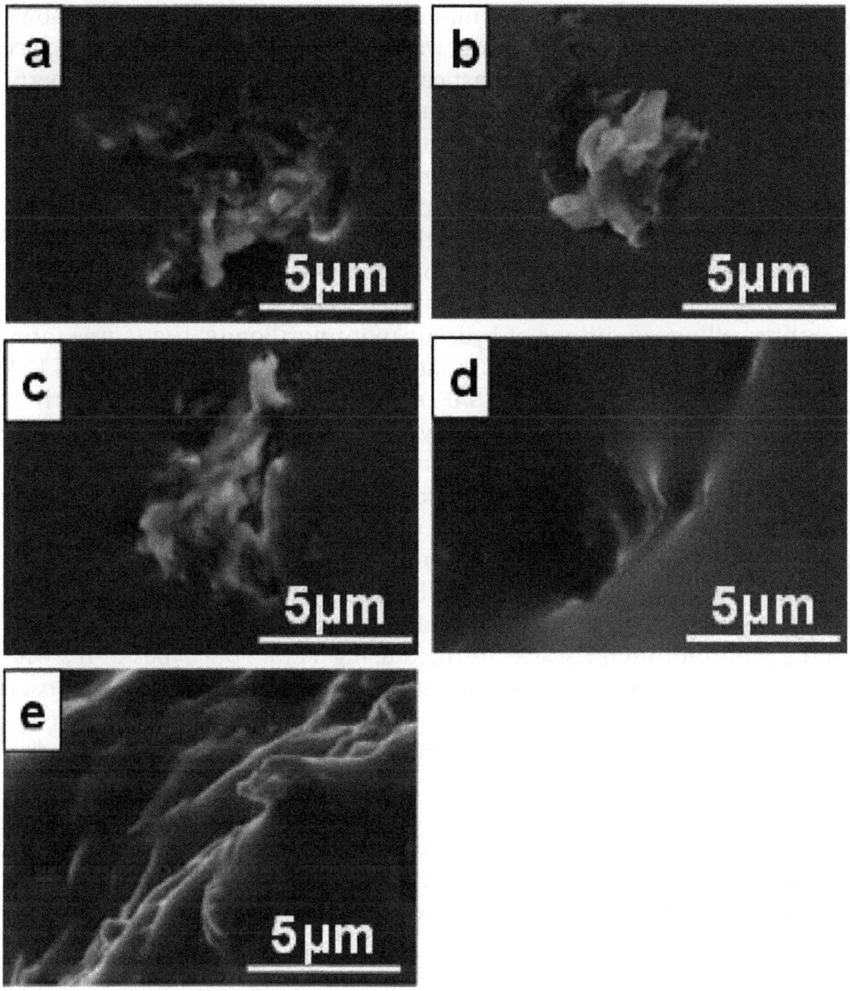

Figure 2.3 SEM images of agar gel pretreated with [BMIM][BF_4] solution at (a) 15, (b) 30, (c) 60, (d) 80, and (e) 95 mol% H_2O. (Reprinted from *Materials Chemistry and Physics*, **136** (2–3), C. Takahashi, T. Shirai, M. Fuji, FE-SEM observation, and mechanism of interaction of wet agar gel in various swelling conditions using hydrophilic ionic liquid, 816–822, Copyright (2012) with permission from Elsevier[36]).

Through the IL treatment of the insulating sample, stable secondary electron and backscattered electron imaging was performed using a wide range of accelerating voltages. Images with high contrast were confirmed even at low voltages, although the deposition of the conductive materials was not effective using conventional methods. Using the sample prepared with the IL, the structural characterization of the rare earth ore was successfully conducted *via* X-ray mapping and electron backscatter diffraction (EBSD) analyses. These results revealed a fine microstructure of zircon/fergusonite-Y aggregates encased in the matrix minerals. Montmorillonite is well known for its layered structures that assist water in entering the interlayer space and cause the mineral to swell drastically.[50] Owing to this specific property, various types of intercalation compounds have been studied for several decades. As such, the intercalation behavior in wet conditions needs to be examined. The author's group therefore conducted EM studies of swelled montmorillonite and IL-intercalated montmorillonite in a liquid state and developed a technique using ILs for the EM analyses of this mineral.[39–42] First, morphological and structural observations of the montmorillonite swelled by water were performed using pretreatment with the IL. Even under different swelling conditions of hydrous montmorillonite, the IL was intercalated within the montmorillonite interlayer in a fixed arrangement. EM, selected area electron diffraction (SAED), X-ray diffraction (XRD), inductively coupled plasma (ICP), and thermogravimetric-differential thermal analysis (TG-DTA) revealed that the ultrafine structure of hydrous montmorillonite has a different *d*-spacing value because the *d*-spacing of hydrous montmorillonite treated with IL changes with the size of the IL cation.[41] However, we found that the skeleton structure of the hydrous montmorillonite was retained. Thus, this technique would be useful for the characterization of hydrous minerals. Secondly, the intercalation behavior of various types of ILs into the montmorillonite was confirmed using TEM. Four types of ILs with different salts (imidazolium and ammonium) and cation sizes were chosen as the guest solvents. The crystal swelling behavior of each intercalated compound was observed with the aid of the ILs even under wet conditions.[40–42] The fabricated intercalated compounds showed a higher thermal stability and electrical conductivity than raw IL. In addition, through the measurement of the properties of the intercalated compounds, it was confirmed that [BuMeIm][BF_4] is highly selective toward montmorillonite.[42] This technique allows for the observation of not only hydrous minerals but also intercalated mineral compounds, including liquids.

2.2.2.3 Ceramic Green Body

Hydroxyapatite (HAp) [$Ca_{10}(PO_4)_6(OH)_2$], which is a ceramic green body, was selected for the EM analyses. [BuMeIm][BF_4] was used to prepare the HAp ceramic green bodies. HAp is a well-known biocompatible material and occurs naturally as one of the major components in bones and teeth. To utilize this useful type of ceramic for various applications, it is necessary to understand the complex structure of the HAp green bodies during the drying process. With this concept, the morphology and structure of the porous HAp green body was observed under wet conditions using a simple technique with an IL (Figure 2.4).[43–45] SEM images reveal that the as-prepared porous green body had a pore diameter of approximately 300–600 nm, which gradually decreased to approximately 200–400 nm during drying in a chamber where the relative humidity was decreased from 90% to 50%.[43,44] Then, a TEM analysis of the hydrous HAp green body was conducted to understand its crystal nanostructure at the

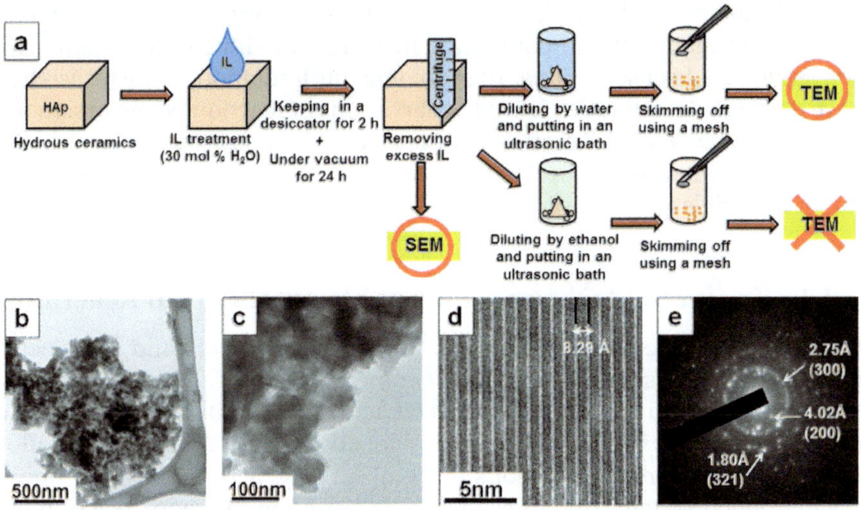

Figure 2.4 (a) Schematic of the pretreatment techniques using [BuMeIm][BF_4] as the IL for hydrous HAp green bodies. (b–e) TEM images (b), bright-field image (c), high-resolution image (d), and SAED pattern (e) of the as-prepared porous HAp green body following the IL treatment. (Reproduced from ref. 45 with permission from the Royal Society of Chemistry.)

nanoscale level.[45] During the drying process, the as-prepared sample took on a particle-morphology with particles approximately 20–150 nm in diameter. This diameter gradually contracted to approximately 10–100 nm upon drying. Some parts of the crystal structure remained swollen in its hydrous condition, which could be successfully observed with the aid of the IL. We did not find any structural changes being induced by the IL treatment (for comparison, we checked the X-ray diffractometer (XRD) results as well as the *d*-spacing derived from the SAED pattern). To optimize the TEM analysis technique with the IL, the effect of the solvent on the imaging of the nanostructured HAp green body was also determined. It was found that water was a suitable solvent for the hydrous HAp green body, whereas ethanol was not an appropriate one. When the HAp sample was treated with the IL, the weak hydrogen bonding among the IL–water–Hap polymer complex network structure acted as a conducting media. In addition, the formation of weak hydrogen bonds between the IL and water prevents the porous HAp body from dehydrating in the EM chamber. It was found that choosing the suitable IL, IL content, and solvent for the pretreatment is an important factor for IL-based EM imaging.

2.3 Direct EM Observations of the Reaction Behavior

EM observations of hydrous materials using the sample pretreatment with IL have been conducted as described above. As a further development, we focused on conducting direct EM observations of the reaction behaviors of the hydrous materials. In general, real time observations and characterizations of hydrated or liquid-containing materials are difficult to obtain using conventional EM observations. To overcome this limitation, some researchers have developed techniques for EM observations of reaction behaviors. In this section, *in situ* SEM observations of the electrochemical reactions in the ILs and EM analyses of the interactions between different drug products and target materials are reviewed.

2.3.1 *In situ* SEM Observations of Electrochemical Reactions

In the field of electrochemistry, it is important to understand the electrochemical reactions and chemical phenomena to design and develop functional materials such as electrochemical actuators and batteries. Therefore, electrochemical systems have been developed with the aid of ILs that allow for the electrochemical reactions to be observed.[51] On the basis of one of the developed systems an *in situ* SEM observation of the redox reaction of a polypyrrole film, which is a conducting polymer, has been successfully performed.[52] It was confirmed that the thickness of the polypyrrole film could be changed *via* the polarization of the film-deposited Pt electrode in the IL. In addition, an EDS analysis was conducted for the polarized electrode immersed in the IL. The combination of an *in situ* SEM observation and an EDS analysis is thus useful for understanding the electrochemical reactions of various materials. Subsequently, real-time SEM observations of the metal deposition from the ILs has been performed using a similar type of electrochemical system.[53,54] The growth of metal dendrites was clearly observed through *in situ* SEM observations, whereas it was difficult to obtain similar results using conventional SEM. An EDS analysis further showed that the dendrites were composed of metal. It has recently been reported that *in situ* SEM techniques for real-time visualization of Si active materials for lithium ion battery cells have been developed.[55] The morphological and elemental distribution variations in the Si anodes during the charge/discharge processes were determined using those techniques. These observations have revealed that the different morphologies of Si-active materials during the charge/discharge processes are related to the size and shape of the Si active materials.

2.3.2 Direct EM Observation of the Interaction Between Drugs and Target Materials

In pharmaceutical research, quantitative analyses and macroscopic observations are often performed during the design of the drugs. However, the morphological structure of the drugs has become increasingly complex; therefore, it is essential to characterize them at the nanoscale level. In addition, the development of a precise process that takes into account the potential interactions of the drug products with the pathogenic agents has become increasingly necessary. With the aim of realizing this, *S. epidermidis* biofilms were treated with polymeric nanoparticles (NPs) in this study, which can be used as nanocarriers for drug delivery systems (DDS); we then observed these biofilms using EM with ILs.[56–60] [BuMeIm][BF_4] was used as an IL for the EM observations. Following the procedure described in Table 2.4 and Figure 2.5, two types of sample preparation techniques were performed for the SEM and TEM observations.

First, the sample for the SEM observation was prepared as shown in Figure 2.1. The optimized observation technique was useful for revealing the antibacterial activity of the polymeric NPs against the biofilm. The EDS analysis was also effective for confirming the chemical composition. These results showed that various types of polymeric NPs, such as surface-modified NPs, drug-incorporated NPs, and metal-decorated NPs, possess different antibacterial activities against the biofilm.[56–58,60]

Second, the technique for TEM observation was developed because there were some issues related to the EM observation of deformable composite materials (Figure 2.5). For example, high magnification SEM imaging is difficult to achieve using the method in which the sample is pretreated with the IL. Furthermore, polymeric NPs are not suitable for conventional TEM observations because the electron beam irradiation causes damage to the NPs. The author's group found that the combination of the sample preparation technique using the IL and TEM observations conducted using a cooling holder allows nano-structural observations to be made and eliminates the need for specialized equipment (such as cryo-TEM) as well as the need for cryo-transfer procedures and biological sample preparation techniques.[59] With the help of annular dark-field (ADF) scanning TEM (STEM) imaging, the antibacterial activity of polymeric NPs against individual bacterial cells beneath the biofilm has been visualized without slicing the samples into thin sections (Figure 2.5).[58,60] The information obtained by ADF-STEM imaging also revealed the bodily function and detoxification mechanisms of the polymeric NPs against the bacterial cells (Figure 2.5g). In addition, the limitation of the developed observation technique was confirmed by *in situ* STEM observations, and the exact structure of the sample was found to be maintained

Table 2.4 Summary of representative IL-based EM techniques for composite biological materials.

Group	Sample	Microscopy	IL	Techniques	Comments
Takahashi *et al.*[56–60]	Biofilm (*S. epidermidis*) treated with polymeric nanoparticles	SEM (EDS), ADF-STEM (SAED)	[BuMeIm] [BF_4]	Accelerating voltage: SEM 5.0–10.0 kV, TEM 200 kV, IL: 0.5 w/w%	Use of a cooling holder is effective

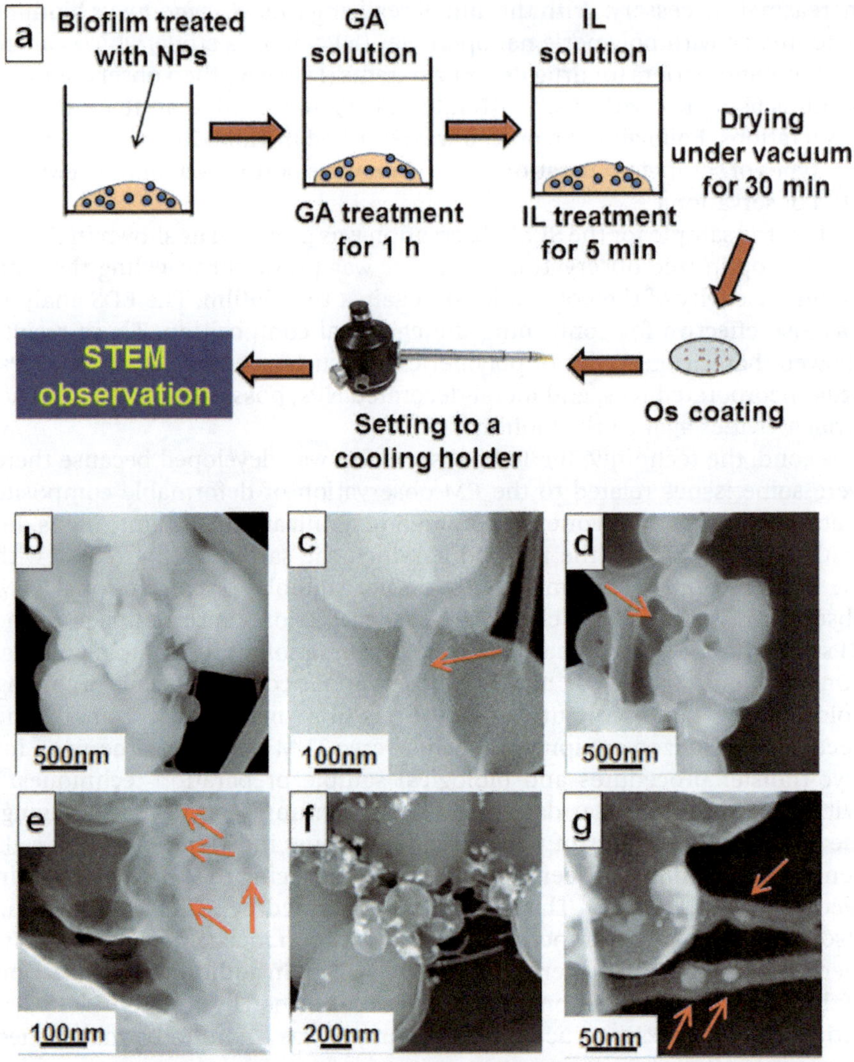

Figure 2.5 (a) Techniques used to prepare a biofilm sample treated with polymeric NPs for STEM observation. [BuMeIm][BF_4] was used as the IL for pretreatment of the sample. Reproduced with permission from C. Takahashi, A microscopy method for scanning transmission electron microscopy imaging of the antibacterial activity of polymeric nanoparticles on a biofilm with an ionic liquid, Journal of Biomedical Materials Research, John Wiley and Sons[59] Copyright © 2016 Wiley-VCH Verlag GmbH & Co. KGaA, Weinheim. ADF-STEM images of *S. epidermidis* biofilms treated with surface-modified PLGA NPs, (b, c), drug-incorporated PLGA NPs (d, e), and Ag-incorporated PLGA NPs (f, g) for 2 h using the techniques shown in (a). The red arrow in (c) indicates the EPS removal from the biofilm surface by the surface-modified PLGA NPs; those in (d) and (e) indicate the bacterial cells that were attacked by drug-incorporated PLGA NPs; and those in (g) indicate silver nanoparticles intruding

for 25 min under electron beam irradiation at a total current flow of 40 pA.[59] TEM observations of the potential interaction between the polymeric NPs and the biofilm on a nanoscale level would be useful in preparing suitable NPs for drug delivery.

2.4 Optimization of IL-based EM Observation Techniques

To establish a simple pretreatment technique with the aid of ILs for EM observations, the pretreatment techniques were optimized; suitable EM conditions have been summarized for three types of hydrous samples, namely, (A) biological materials (Figure 2.6), (B) hydrated or insulated materials (Figure 2.7), and (C) composite biological materials (Figure 2.8).

(A) Biological materials (Figure 2.6):
   (i) Animal cells, microorganisms, and plants (materials were maintained by osmotic pressure): (Step 1) Samples were treated with PBS or GA solution. (Step 2) Samples were treated with the suitable IL. (Step 3) Samples were dried under vacuum. (Step 4) To remove the excess IL from the sample, a centrifuge or Kimwipe paper was used. The ILs suitable for the EM observations of animal and plant cells and chromosomes are [EtMeIm][BF_4], [EtMeIm][Lac], [Ch][Lac], [BuMeIm][BF_4], and choline-like IL; [EtMeIm][AcO], [Ch][Lac], [BuMeIm][BF_4], and choline-like IL were found suitable for the EM observations of microorganisms including bacteria and viruses; and [BuMeIm][TFSI], [BuMeIm][BF_4], and choline-like IL were suitable for the EM observations of plant petals and pollen. [BuMeIm][TFSI], [BuMeIm][BF_4], choline-like IL, and [EtMeIm][AcO] are commercially available. Notably, GA fixation is effective for EM observation of the ultrafine structure of biological materials. Metal coating and staining may also be effective for obtaining high contrast images.
   (ii) Seaweed (materials were swelled by liquid and maintained this way *via* osmotic pressure): (Step 1) Samples were treated with a suitable IL (commercially available choline-like IL and [BuMeIm][BF_4]) while maintaining the osmotic pressure. (Step 2) Samples were dried under vacuum. (Step 3) To remove the excess IL from the sample, a centrifuge or Kimwipe paper was used.

---

into the tube-like structure of the bacterial cells. (Reprinted from Materials Science and Engineering C, 72, C. Takahashi, N. Matsubara, Y. Akachi, N. Ogawa, G. Kalita, T. Asaka, M. Tanemura, Y. Kawashima, H. Yamamoto, Visualization of silver-decorated poly (DL-lactide-*co*-glycolide) nanoparticles and their efficiency against Staphylococcus epidermidis, 143–149, Copyright (2017) with permission from Elsevier[60]).

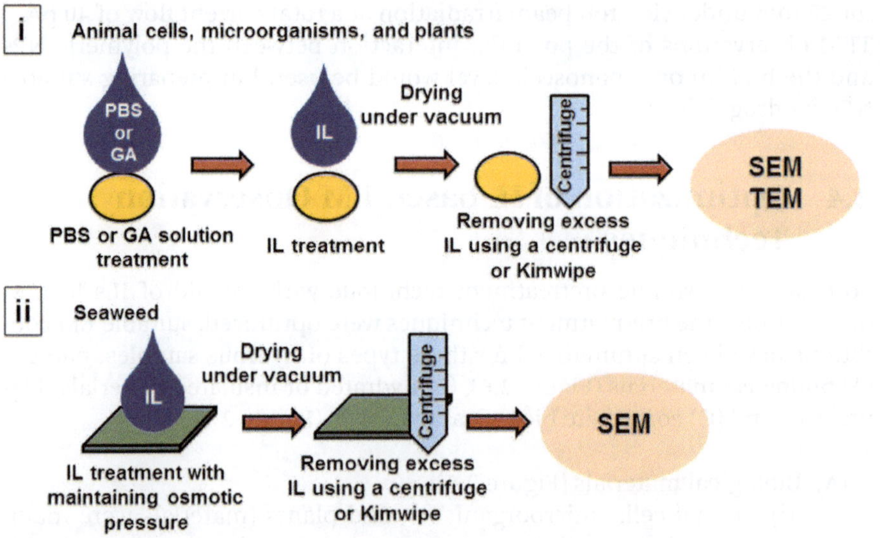

Figure 2.6 Observation techniques for biological materials.

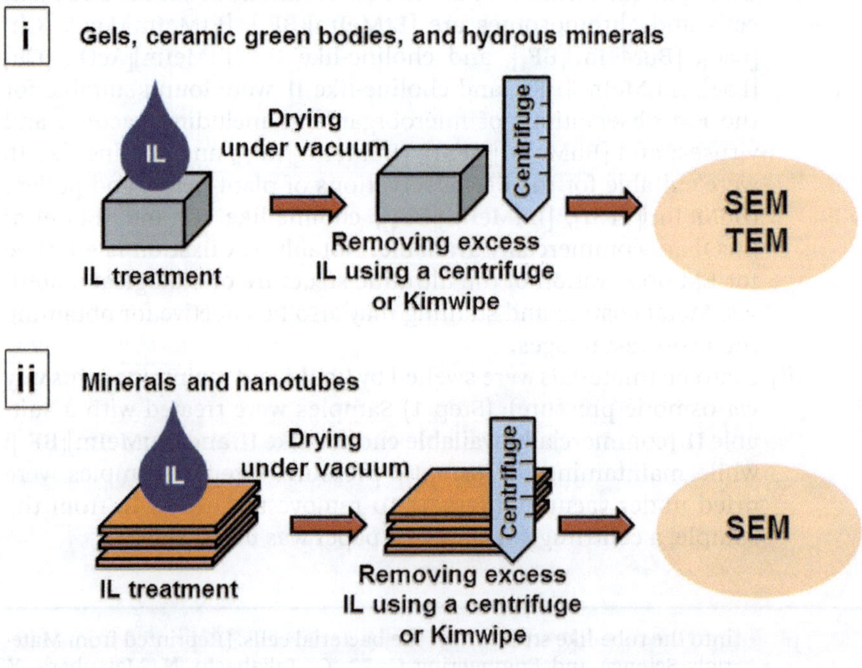

Figure 2.7 Observation techniques for hydrated or insulated materials.

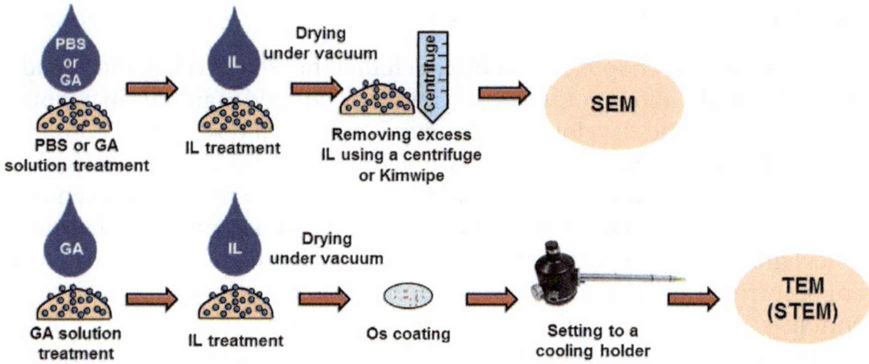

Figure 2.8 Observation techniques for composite biological materials.

(B) Hydrated or insulated materials (Figure 2.7):
   (i) Gels, ceramic green bodies, and hydrous minerals (materials contain significant amounts of water): (Step 1) Samples were treated with a suitable IL. (Step 2) Samples were dried under vacuum. (Step 3) To remove the excess IL from the sample, a centrifuge or Kimwipe paper was used. Suitable ILs were [BuMeIm][BF_4], [EtMeIm][AcO], and [$P_{4,4,4,1}$][DMP]. To observe the detailed morphology and structure of the intact samples, it is important to understand the osmotic pressure of each IL and the water content within the sample. The IL [BuMeIm][BF_4] is commercially available.
   (ii) Minerals and nanotubes (materials that have no conductive properties): (Step 1) Samples were treated with a suitable IL. (Step 2) Samples were dried in vacuum. (Step 3) To remove the excess IL from the sample, a centrifuge or Kimwipe paper was used. The suitable ILs were commercially available [BuMeIm][BF_4] and 1-butyl-3-methylimidazolium hexafluorophosphate ([BuMeIm][PF_6]).

(C) Composite biological materials (Figure 2.8):
   (i) For SEM observation: (Step 1) Samples were treated with PBS or GA solution. (Step 2) Samples were treated with a suitable IL. (Step 3) Samples were dried under vacuum. (Step 4) To remove the excess IL from the sample, a centrifuge or Kimwipe paper was used. For TEM observation: (Step 1) The samples were treated with the GA solution. (Step 2) The samples were treated with a suitable IL. (Step 3) The samples were dried under vacuum and were coated using Os. (Step 4) The samples were placed in a cooling holder. The suitable IL for both the SEM and TEM observations was commercially available [BuMeIm][BF_4]. Notably, it was not essential to use the cooling holder for non-deformable materials.

2.5 Conclusions

On the basis of this chapter, it can be concluded that the developed methodology utilizing ILs allows for the observation of hydrated materials in various fields using EM. The developed technique is simple and will enable intact, fine morphological and structural observations and direct observations of hydrous materials in a liquid state to be made without any complex sample pretreatments and without the use of specialized equipment; this is because the negligible vapor pressure derived from the IL can prevent drying artifacts from occurring during the EM observations. Suitable sample pretreatment techniques and EM conditions were identified to be different for different types of hydrous samples. Thus, we need to select a suitable technique depending on the type of the sample being observed. Although applications using IL for EM have only started to gain attention in the last two decades, this technique will contribute to the further development of science and technology in many fields. The techniques using ILs for EM observations and analyses presented in this chapter will certainly provide better information for scientists who engage in IL research using EM equipment.

References

1. J. S. Wilkes and M. J. Zaworotko, *J. Chem. Soc., Chem. Commun.*, 1992, **13**, 965.
2. S. Werner, M. Haumann and P. Wasserscheid, *Annu. Rev. Chem. Biomol. Eng.*, 2010, **1**, 203.
3. H. Niedermeyer, J. P. Hallett, I. J. Villar-Garcia, P. A. Hunt and T. Welton, *Chem. Soc. Rev.*, 2012, **41**, 7780.
4. S. Muto, J. Rusz, K. Tatsumi, R. Adam, S. Arai, V. Kocevski, P. M. Oppeneer, D. E. Bürgler and C. M. Schneider, *Nat. Commun.*, 2014, **5**, 3138.
5. A. Rigort and J. M. Plitzko, *Arch. Biochem. Biophys.*, 2015, **81**, 122.
6. M. Beck and W. Baumeister, *Trends Cell Biol.*, 2016, **26**, 825.
7. H. Saka, T. Mima, Y. Takeuchi, T. Marukawa, S. Arai, K. Kuroda, K. Kishita and T. Kamino, *Microsc. Microanal.*, 2006, **12**, 764.
8. T. V. Vineesh, K. M. Praveen, C. Takahashi, G. Kalita, S. Alwarappan, D. K. Pattanayak and T. N. Narayanan, *Adv. Energy Mater.*, 2015, **5**, 1500658.
9. M. L. Taheri, E. A. Stach, I. Arslan, P. A. Crozier, B. C. Kabius, T. LaGrange, A. M. Minor, S. Takeda, M. Tanase, J. B. Wagner and R. Sharma, *Ultramicroscopy*, 2016, **170**, 86.
10. T. Torimoto, T. Tsuda, K. Okazaki and S. Kuwabata, *Adv. Mater.*, 2010, **19**, 1196.
11. S. Kuwabata, T. Tsuda and T. Torimoto, *J. Phys. Chem. Lett.*, 2010, **1**, 3177.
12. M. Johnston, J. J. Lee, G. S. Chottiner, B. Miller, T. Tsuda, C. L. Hussey and D. A. Scherson, *J. Phys. Chem. B*, 2005, **109**, 11296.
13. J. F. Wishart and P. Neta, *J. Phys. Chem. B*, 2003, **107**, 7261.

14. S. Kuwabata, A. Kongkanand, D. Oyamatsu and T. Torimoto, *Chem. Lett.*, 2006, **35**, 600.
15. S. Arimoto, M. Sugimura, H. Kageyama, T. Torimoto and S. Kuwabata, *Electrochim. Acta*, 2008, **53**, 6228.
16. T. Kodaka, T. Toko, K. Debari, H. Hisamitsu, A. Omori and S. Kawata, *J. Electron Microsc.*, 1992, **41**, 381.
17. T. Tanabe, *J. Electron Microsc.*, 2010, **60**, 35.
18. K. Ohta, S. Sadayama, A. Togo, R. Higashi, R. Tanoue and K. Nakamura, *Micron*, 2012, **43**, 612.
19. S. Takeda and H. Yoshida, *Microscopy*, 2013, **62**, 193.
20. A. Leforestier, P. Levitz, T. Preat, P. Guttmann, L. J. Michot and P. Tchénio, *J. Struct. Biol.*, 2014, **188**, 177.
21. R. Danev, B. Buijsse, M. Khoshouei, J. M. Plitzko and W. Baumeister, *Proc. Natl. Acad. Sci. U. S. A.*, 2014, **111**, 15635.
22. M. Booth, D. Andrade, D. Burke, B. Patton and M. Zurauskas, *Microscopy*, 2015, **64**, 251.
23. Y. Ishigaki, Y. Nakamura, T. Takehara, N. Nemoto, T. Kurihara, H. Koga, H. Nakagawa, T. Takegami, N. Tomosugi, S. Miyazawa and S. Kuwabata, *Microsc. Res. Tech.*, 2011, **74**, 415.
24. Y. Ishigaki, Y. Nakamura, T. Takehara, T. Shimasaki, T. Tatsuno, F. Takano, Y. Ueda, Y. Motoo, T. Takegami, H. Nakagawa, S. Kuwabata, N. Nemoto, N. Tomosugi and S. Miyazawa, *Microsc. Res. Tech.*, 2011, **74**, 1024.
25. Y. Ishigaki, Y. Nakamura, T. Takehara, T. Kurihara, H. Koga, T. Takegami, H. Nakagawa, N. Nemoto, N. Tomosugi, S. Kuwabata and S. Miyazawa, *Microsc. Res. Tech.*, 2011, **74**, 1104.
26. T. Tsuda, N. Nemoto, K. Kawakami, E. Mochizuki, S. Kishida, T. Tajiri, T. Kushibiki and S. Kuwabata, *Chem. Bio. Chem.*, 2011, **12**, 2547.
27. A. Dwiranti, L. Lin, E. Mochizuki, S. Kuwabata, A. Takaoka, S. Uchiyama and K. Fukui, *Microsc. Res. Tech.*, 2012, **75**, 1113.
28. T. Hamano, A. Dwiranti, K. Kaneyoshi, S. Fukuda, R. Kometani, M. Nakao, H. Takata, S. Uchiyama, N. Ohmido and K. Fukui, *Microsc. Microanal.*, 2014, **20**, 1340.
29. Y. Asahi, J. Miura, T. Tsuda, S. Kuwabata, K. Tsunashima, Y. Noiri, T. Sakata, S. Ebisu and M. Hayashi, *AMB Express*, 2015, **5**, 6.
30. C. Takahashi, G. Kalita, N. Ogawa, K. Moriguchi, M. Tanemura, Y. Kawashima and H. Yamamoto, *Anal. Bioanal. Chem.*, 2015, **407**, 1607.
31. K. Kawai, K. Kaneko, H. Kawakami and T. Yonezawa, *Langmuir*, 2011, **27**, 7353.
32. K. Kawai, K. Kaneko, H. Kawakami, T. Narushima and T. Yonezawa, *Colloids Surf., B*, 2013, **102**, 9.
33. C. G. Golding, L. L. Lamboo, D. R. Beniac and T. F. Booth, *Sci. Rep.*, 2016, **6**, 26516.
34. C. Takahashi, T. Shirai and M. Fuji, *Microsc. Res. Tech.*, 2013, **76**, 66.
35. C. Takahashi, T. Shirai and M. Fuji, *Adv. Technol. Mater. Mater. Process. J.*, 2011, **13**, 39.

36. C. Takahashi, T. Shirai and M. Fuji, *Mater. Chem. Phys.*, 2012, **136**, 816.
37. C. Takahashi, T. Shirai and M. Fuji, *Mater. Chem. Phys.*, 2012, **133**, 565.
38. T. Tsuda, E. Mochizuki, S. Kishida, K. Iwasaki, K. Tsunashima and S. Kuwabata, *PLoS One*, 2014, **9**, e91193.
39. C. Takahashi, T. Shirai and M. Fuji, *Mater. Chem. Phys.*, 2013, **141**, 657.
40. C. Takahashi, T. Shirai and M. Fuji, *Mater. Chem. Phys.*, 2012, **135**, 681.
41. C. Takahashi, T. Shirai, Y. Hayashi and M. Fuji, *Solid State Ionics*, 2013, **241**, 53.
42. C. Takahashi, T. Shirai and M. Fuji, *Solid State Ionics*, 2014, **267**, 16.
43. C. Takahashi, D. K. Pattanayak, T. Shirai and M. Fuji, *J. Eur. Ceram. Soc.*, 2012, **33**, 629.
44. C. Takahashi, D. K. Pattanayak, T. Shirai and M. Fuji, *Ceram. Int.*, 2013, **39**, 1065.
45. C. Takahashi, D. K. Pattanayak, T. Shirai and M. Fuji, *RSC Adv.*, 2014, **4**, 27322.
46. P. Roy, T. Dey and P. Schmuki, *Electrochem. Solid-State Lett.*, 2010, **13**, E11.
47. N. Brodusch, K. Waters, H. Demers and R. Gauvin, *Microsc. Res. Tech.*, 2014, **77**, 225.
48. H. Abe, Y. Yoshimura, Y. Imai, T. Goto and H. Matsumoto, *J. Mol. Liq.*, 2009, **150**, 16.
49. B. Sun and P. Wu, *J. Phys. Chem. B*, 2010, **114**, 9209.
50. F. Bergaya, B. K. G. Theng and G. Lagaly, *Handbook of Clay Science*, Elsevier, Amsterdam, 2006.
51. S. Kuwabata, T. Tsuda and T. Torimoto, *J. Phys. Chem. Lett.*, 2010, **1**, 3177.
52. S. Arimoto, D. Oyamatsu, T. Torimoto and S. Kuwabata, *ChemPhysChem*, 2008, **9**, 763.
53. S. Arimoto, H. Kageyama, T. Torimoto and S. Kuwabata, *Electron. Commun.*, 2008, **10**, 1901.
54. T. Tsuda, E. Mochizuki, S. Kishida, H. Sakagami, Y. Tachibana, M. Ebisawa, N. Nemoto, Y. Nishimura and S. Kuwabata, *Electrochemistry*, 2012, **80**, 308.
55. C. Y. Chen, T. Sano, T. Tsuda, K. Ui, Y. Oshima, M. Yamagata, M. Ishikawa, M. Haruta, T. Doi, M. Inaba and S. Kuwabata, *Sci. Rep.*, 2016, **6**, 36153.
56. C. Takahashi, N. Ogawa, Y. Kawashima and H. Yamamoto, *Microscopy*, 2015, **64**, 169.
57. C. Takahashi, S. Saito, A. Suda, N. Ogawa, Y. Kawashima and H. Yamamoto, *RSC Adv.*, 2015, **5**, 71709.
58. C. Takahashi, Y. Akachi, N. Ogawa, K. Moriguchi, T. Asaka, M. Tanemura, Y. Kawashima and H. Yamamoto, *Med. Mol. Morphol.*, 2017, **50**, 9–16.
59. C. Takahashi, S. Muto and H. Yamamoto, *J. Biomed. Mater. Res., Part B*, 2017, DOI: 10.1002/jbm.b.33680.
60. C. Takahashi, N. Matsubara, Y. Akachi, N. Ogawa, G. Kalita, T. Asaka, M. Tanemura, Y. Kawashima and H. Yamamoto, *Mater. Sci. Eng., C*, 2017, **72**, 143.

CHAPTER 3

Ionic Liquid-based Surfactants: A Step Forward

IDAIRA PACHECO-FERNÁNDEZ, PROVIDENCIA GONZÁLEZ-HERNÁNDEZ, VERÓNICA PINO*, JUAN H. AYALA AND ANA M. AFONSO*

Universidad de La Laguna, Departamento de Química, Unidad Departamental de Química Analítica, Avenida Astrofísico Francisco Sánchez s/n, La Laguna, Tenerife, 38206, Spain
*E-mail: aafonso@ull.edu.es, veropino@ull.edu.es

3.1 Introduction

Ionic liquids (ILs) are a class of ionic compounds with melting points below 100 °C, thus remaining liquids at common working temperatures. Their structures usually include bulky organic cations paired with organic or inorganic anions. These non-molecular solvents present interesting and unique physicochemical properties, such as high thermal stability, negligible vapour pressure at room temperature, non-flammability, high ionic conductivity, and a wide electrochemical window of stability. Interestingly, simple modifications in their structural composition are accompanied by changes in their properties, such as melting point, viscosity, density, hydrophobicity, and miscibility with water or with a variety of organic solvents. Therefore, ILs can be tailored by simple changes in their structures and, consequently, they can be tuned for task-specific applications.[1,2] For this reason, ILs have been tagged

Smart Materials No. 28
Ionic Liquid Devices
Edited by Ali Eftekhari

Published by the Royal Society of Chemistry, www.rsc.org

as "designer solvents", and have been considered interesting alternatives to conventional organic solvents, given their impressive solvation ability for a number of compounds with an organic or inorganic nature.[3] Therefore, ILs have been used in a wide variety of applications, particularly in synthesis, catalysis, electrochemistry, separation science, chemical engineering, and materials science.[4,5]

It has been reported that ILs possessing a long alkyl chain in their structures present surface active properties when dissolved in water at concentrations higher than a critical concentration value: the critical micelle concentration (CMC).[6] The surfactant character can be associated with the IL anion, the cation, or both. Therefore, IL-based surfactants constitute a new class of surfactant systems, which exhibit high surface activity. It is important to highlight that most authors consider more appropriate terms to be IL-based surfactants or surface active ILs (SAILs), rather than IL-surfactants, because many IL-based surfactants present melting points above 100 °C.[6] In other words, neat ILs are used in the design of novel and task-specific IL-based surfactants.

The same inherent versatility of ILs has been observed for IL-based surfactants. Consequently, simple modifications in the structure of the anion/cation provide significant changes in the micellar properties of the resulting IL-based surfactants. For example, their miscibility is tuned by modifying the IL anion moiety. ILs containing a certain cation paired with BF_4^- are miscible in water, while ILs composed with the same cation and the Br^- anion are partially miscible with water, and those ILs maintaining the cation but paired with PF_6^- anion are immiscible with water.[7] Furthermore, an increase in the length of the alkyl chain of an IL-based surfactant, independent of the position in which the alkyl chain is attached to the cationic ring, is generally accompanied by a decrease of the CMC values, and by an increase in the aggregation number (N_{agg}) of the micelle. It is also demonstrated that dicationic IL-based surfactants present lower CMC values than their monocationic analogues.[8]

The first IL was discovered in 1914,[9] but the knowledge and usage of IL-based surfactants are much more recent. The first characterization studies of IL-based surfactants were carried out almost simultaneously by two independent teams: Sirieix-Plénet *et al.*[7] and Bowers *et al.*,[10] both in 2004. Sirieix-Plénet *et al.* evaluated the micellar behaviour of the IL-based surfactant 1-decyl-3-methylimidazolium bromide (C_{10}MIm-Br) in water by conductivity studies. Bowers *et al.* studied the aggregation behaviour of three IL-based surfactants (C_4MIm-BF_4, C_8MIm-Cl, and C_8MIm-I) in water by surface tension, conductivity, and small-angle neutron scattering (SANS) measurements.[10] Since then, a large number of studies have been devoted to the synthesis and characterization of these unique surfactants.[2,6,11]

Commonly, IL-based surfactants are composed of cations derived from imidazolium (Im), phosphonium (P^+), pyridinium (Py^+), pyrrolidinium ($Pyrr^+$), piperidinium (Pip^+), and ammonium (N^+). The most common anions are bromide, chloride, tetrafluoroborate (BF_4^-), octyl sulfate ($C_8SO_4^-$), dodecyl sulfate ($C_{12}SO_4^-$), and dioctyl sulfosuccinate (AOT). Table 3.1 shows the main structures of IL-based surfactants reported in the literature during the last five years, together with their reported CMC values and main application fields.[10,12–79]

A quick search of the literature easily indicates the exponential increase in studies on IL-based surfactants, especially in the last five years. For this reason, this chapter focuses on the description of main IL-based surfactants including their classification, synthetic routes, characterization and properties, and the most relevant applications reported between 2012 and 2016, to highlight trends in the field.

3.2 IL-based Surfactants: Synthesis and Classification

One of the most important characteristics of the above-mentioned ILs is their synthetic tuneability, which leads to an impressive structural versatility. In addition, the reactions involved in their synthesis are quite simple, which also justifies their success. Due to the large number of possible combinations of cations and anions and the ease of tailoring ILs, a huge amount of IL-based surfactants can be designed.[80] As a simple classification, IL-based surfactants can be grouped mainly by attending to the cation structure, since most reported IL-based surfactants exhibit characteristics of cationic surfactants. Thus, monocationic, multicationic, and functionalized IL-based surfactants constitute a classification of the main IL-based surfactants types. However, it is important to take into account that not only the cation but also the anion plays an important role in the resulting physicochemical properties of the IL-based surfactant. Moreover, anionic IL-based surfactants have also been described.[19,81,82]

The synthetic route is quite simple for monocationic IL-based surfactants. It normally requires an initial nucleophilic substitution reaction (S_N) followed by a metathesis reaction.[83] The nucleophilic substitution is accomplished to functionalize the cation moiety of the resulting IL (using as starting materials imidazole, alkylimidazole, pyrrolidine, alkylpyrrolidine, pyridine, piperidine, triphenylphosphine, or trialkylamine), while the metathesis reaction allows the introduction of more complex anions in the IL structure. Figure 3.1(A) shows a scheme of the general route employed to synthetize ILs, using imidazolium cation as a representative example. The approach is valid for many other ILs containing cations of different nature. In the specific case of monoalkylguanidinium ILs, the use of 1H-pyrazol-1-carboxamidine hydrochloride as guanylating agent of the alkylamine containing the desired chain is required.[78]

In the case of dicationics (normally imidazolium-based), also known as Gemini surfactants, a general procedure has been reported by Anderson *et al.*[8,84] Two different synthetic routes have been described, depending on

Table 3.1 Main IL-based surfactants reported in the literature with their corresponding critical micelle concentration and most relevant applications.

Cation	Anion	CMC (mmol L^{-1})	Relevant applications (excluding synthesis and characterization studies)
C_4MIm^+	BF_4^-	800,[a,10] 820[b,10]	Analytical extraction[12] MEKC[13,14]
	$C_8SO_4^-$	34.9[a,15]	Interaction with proteins[16] Fabrication of materials[15]
	$C_{12}SO_4^-$	2.4[a,15]	IL-in-oil microemulsions[17] Fabrication of materials[15,18]
	AOT^-	1.78[a,19]	Fabrication of materials[19]
C_6MIm^+	Br^-	470,[a,8] 400,[b,20] 600[a,20]	Sensitizer for spectrophotometric determination of zinc[21]
	$C_{12}SO_4^-$	1.1[a,15]	Fabrication of materials[15]
C_8MIm^+	Br^-	150[a,b,20]	Analytical extraction[22,23] Sensitizer for spectrophotometric determination of zinc[21]
	Cl^-	90,[b,10] 100[a,10]	IL-in-oil microemulsions[24,25]
	$C_{12}SO_4^-$	0.4[a,15]	Fabrication of materials[15]
$C_{10}MIm^+$	Br^-	30,[b,20] 40[a,20]	Water-in-IL microemulsions[26] Sensitizer for spectrophotometric determination of zinc[21] Fabrication of materials[27]
	Cl^-	39.90,[a,28] 40.47,[b,28] 40.52[c,28]	
$C_{12}MIm^+$	Br^-	9.8[b,29]	Water-in-oil microemulsions[30,31] Analytical extraction[32] Interaction with DNA[33] Interaction with proteins[34,35] Fabrication of materials[27,36]
	Cl^-	13.17,[a,28] 13.47,[b,28] 14.0[c,28]	Water-in-oil microemulsions as reaction media[37–40] Analytical extraction[32] Pseudostationary phase in LC[41] Interaction with proteins[42]
	BF_4^-	7.6,[b,43] 9.2[a,43]	MEKC[44]
$C_{14}MIm^+$	Br^-	2.5[b,29]	Water-in-oil microemulsions[45] MEKC[46,47] Interaction with proteins[35]
	Cl^-	2.98,[a,28] 3.65,[c,28] 3.68[b,28]	Water-in-oil microemulsions[48] IL-in-oil microemulsions[49] MEKC[50]

$C_{16}MIm^+$	Br^-	0.61[b,29]	Analytical extraction[51,52]
			Interaction with proteins[35,53,54]
			Fabrication of materials[55,56]
	Cl^-	0.87,[a,28] 0.86,[b,28] 1.30[c,28]	IL-in-oil microemulsions[49]
			Analytical extraction[57]
			Fabrication of materials[55,58,59]
$C_{18}MIm^+$	Cl^-	0.40,[b,60] 0.45[a,60]	IL-in-oil microemulsions[49]
$C_{10}C_{10}Im^+$	Cl^-	0.49[b,61]	Analytical extraction[52]
$C_{16}C_4Im^+$	Br^-	0.1[a,8]	Analytical extraction[63]
$C_{16}C_9Im^+$	Br^-	0.9[a,64]	
$(C_{12}Im)_2C_4{}^{2+}$	$2Br^-$	72[a,8]	Interaction with proteins[65,66]
$(C_{16}Im)_2C_{12}{}^{2+}$	$2Br^-$	0.19,[b,67] 0.21[c,67]	Interaction with proteins[68]
$C_{14}MPyrr^+$	Br^-	3.30[b,69]	
$C_{16}MPyrr^+$	Br^-	0.83[b,69]	
$C_{12}Py^+$	Cl^-	1.76[a,70]	Water-in-oil microemulsions[70]
$C_{12}Ph_3P^+$	Br^-	1.82[b,71]	Fabrication of materials[72]
			Dispersion of multi-walled CNTs in water[73]
$C_{14}Ph_3P^+$	Br^-	0.77[b,71]	Dispersion of multi-walled CNTs in water[73]
$C_{16}Ph_3P^+$	Br^-	0.14[b,71]	Interaction with proteins[74]
$C_{12}C_1C_1C_1N^+$	Cl^-	30[b,41]	Pseudostationary phase for LC[41]
$(C_{12}C_{12}C_{12}N^+)_2C_2{}^{2+}$	$2Br^-$	1.99[a,75]	Corrosion inhibitors[76]
C_8Gu^+	Cl^-	43.2,[b,77] 43.8[c,77]	Analytical extraction[77]
$C_{10}Gu^+$	Cl^-	26[b,78]	
$C_{12}MPip^+$	Br^-	11.83[a,79]	
$C_{14}MPip^+$	Br^-	3.22[a,79]	
$C_{16}MPip^+$	Br^-	0.73[a,79]	

[a]Determined by surface tension measurements.
[b]Determined by conductivity measurements.
[c]Determined by fluorescence measurements. Im: imidazolium; C_n: alkyl chain with n carbon atoms; M: methyl; Ph: phenyl; AOT: dioctyl sulfosuccinate; $C_{12}SO_4{}^-$: dodecyl sulfate; $C_8SO_4{}^-$: octyl sulfate; Py: pyridinium; Pyrr: pyrrolidinium; Gu: guanidinium; Pip: piperidinium.

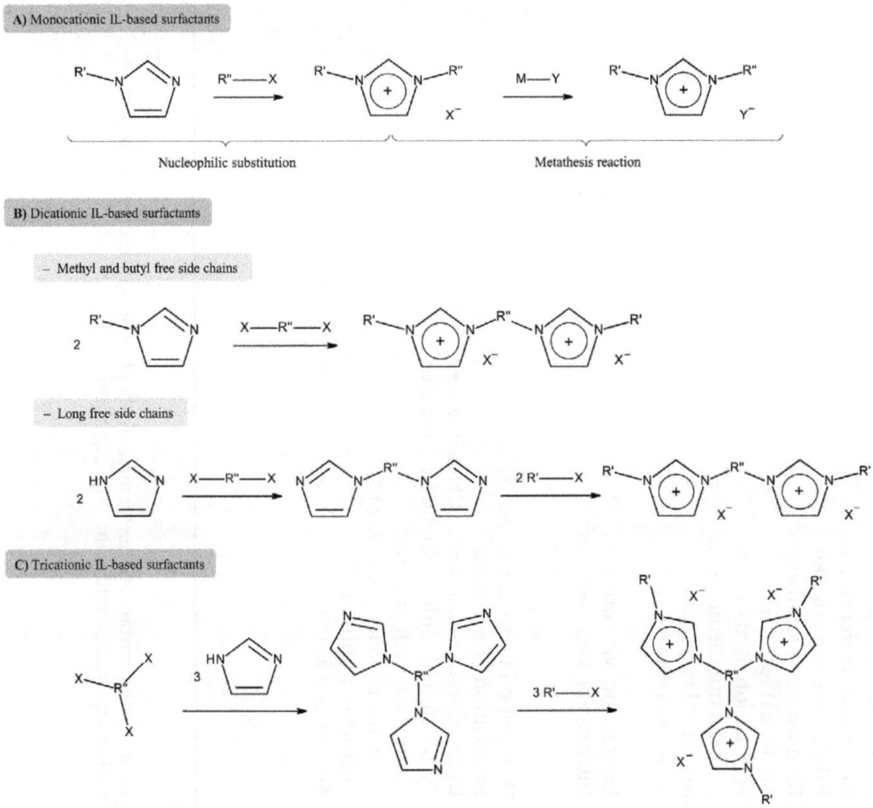

Figure 3.1 Scheme of general synthetic routes to prepare (A) monocationic IL-based surfactants, (B) dicationic IL-based surfactants, and (C) tricationic IL-based surfactants based on the procedures reported by Baltazar *et al.*[8]

the desired length of the free side chain of the imidazolium cations in the resulting IL. When the cations include methyl and butyl substituents, the corresponding alkylimidazole or alkylpyrrolidine reacts with the dibromoalkane containing the required spacer group. Two consecutive S_N reactions are required in the route when longer alkyl chains are desired in the dicationic IL. In a first step, imidazole or pyrrolidine reacts with the dibromoalkane to obtain the geminal structure with the proper spacer. In the second step, the free side chains (with more than four carbon atoms) are incorporated onto the imidazolium or pyrrolidinium cation. Figure 3.1(B) summarizes these two general procedures for dicationics (using imidazolium-based ionic liquids as examples).

Regarding tricationic IL-based surfactants (all imidazolium-based to date), their synthesis also require two consecutive S_N reactions, as for dicationic ILs. First, the group used as spacer reacts with imidazole to acquire the trigonal arrangement. Then, the free chain is included in the final structure of the tricationic IL-based surfactant using the corresponding haloalkane. Figure 3.1(C) shows an example of this synthetic method.

3.2.1 Monocationic IL-based Surfactants

Monocationic IL-based surfactants are those presenting a single cation in their structure. 1-Alkyl-3-methylimidazolium IL-based surfactants are the most widely studied within this group.[20,28,35,85,86] As shown in Table 3.1, this type represents around 50% of all IL-based surfactants described in the recent literature. C_4MIm-BF_4 is the shortest IL-based surfactant described, while C_{12}MIm-Br, C_{14}MIm-Br, and C_{16}MIm-Br are those more investigated in a variety of research fields.[27,29,30,35,47,51,56] 1-Alkyl-3-methylimidazolium surfactants with alkyl chains containing from 4 to 18 carbon atoms have been characterized as micellar systems.[8,11,15,20–22,49,60] Other imidazolium IL-based surfactants containing alkyl groups longer than methyl in the third position of the imidazolium cation have also been synthetized.[87–89] In several cases, symmetrical IL-based substituents (with identical alkyl chains in positions 1 and 3 of the imidazolium ring) have been described.[8,61,88]

Despite the success of imidazolium IL-based surfactants, the aggregation behaviour of other IL cations have also been investigated. Thereby, pyrrolidinium cation containing chains larger than dodecyl,[90,91] phosphonium cation containing three phenyl groups in their structure,[73,88,92] and monosubstituted pyridinium,[70] guanidinium,[78] and piperidinium[79] cations have been used to design IL-based surfactants.

Since most IL-based surfactants have cations with hydrophobic tails, and anions such as bromide, chloride, iodide, and tetrafluoroborate, they have been labelled as cationic surfactants. However, ILs containing cations with short substituents and anions with longer alkyl chains also constitute a new group of anionic IL-based surfactants.[19] The terms catanionic or biamphiphilic (BAILs) are employed to describe this type of IL-based surfactants, due to the amphiphilic character of both constituent ions.[15] To obtain these BAIL-based surfactants, anions such as octyl sulfate,[15,16] dodecyl sulfate,[15,82,93] and AOT[19,81,94] are required. These hydrophobic anions have mainly been combined with 1-akyl-3-methylimidazolium cations using alkyl chains with ten or fewer carbon atoms.[15,19,93] Kumar *et al.* have also synthetized BAIL-based surfactants pairing the AOT anion with more biocompatible moieties such as prolinium, cholinium, and guanidinium cations.[81] A new class of anionic IL-based surfactants using dodecylbenzenesulfonate anion linked to imidazolium[95] or phosphonium[96] cations have been prepared recently.

Zwitterionic IL-based surfactants (ZIL-based surfactants) were first described by Liu *et al.* in 2011.[97] These compounds include a hydrophilic part with both positively and negatively charged group, and a hydrophobic chain. The ZIL-based surfactants reported to date are composed of imidazolium cations including carboxylic groups as anionic sites in their structure: $CO_2^-C_{10}Im^+$,[98] $CO_2^-C_{12}Im^+$,[97–99] and $CO_2^-C_{14}Im^+$.[98] Despite the complexity of their chemical composition compared to cationic and anionic IL-based surfactants, their synthesis is quite simple: alkylimidazole reacts with bromoacetic acid to form the bromide salt, which is then neutralized with NaOH.[98] ZIL-based surfactants are characterized by higher surface activity compared to conventional zwitterionic surfactants.[98] Moreover, their micellar properties are less dependent on the pH and ionic strength of the media in comparison with cationic IL-based surfactants with similar side chains.[97]

3.2.2 Multicationic IL-based Surfactants

Among multicationic ILs, dicationic and tricationic IL-based surfactants have been those reported to date. They present two (for dicationics) or three (for tricationics) cations linked by a spacer. The influence of the spacer length on the micellar behaviour is more important than the length of the alkyl side chains in these multicationic IL-based surfactants.[100] Thus, the micellar behaviour is mainly due to the length of the spacer groups, even with IL cations with methyl as side chains. The main advantage of these multicationic IL-based surfactants compared to their monocationic analogues is their lower CMC value,[67,101] which leads to a reduction of the amount of IL required to exploit its aggregation properties.

As mentioned before for monocationic IL-based surfactants, Gemini IL-based surfactants containing imidazolium cations are the most common, combined in all cases with bromide anions.[8,67,68,100,101] Frizzo *et al.* have studied the aggregation behaviour of di(methylimidazolium)alkane ($(MIm)_2C_n^{2+}$) IL-based surfactants with spacers containing alkyl chains from 4 to 10 carbon atoms.[100] Dicationic imidazolium IL-based surfactants containing butyl, decyl, dodecyl, and hexadecyl as side free chains, mainly in combination with butane and dodecane as spacer groups, have also been widely reported.[8,67,68] Among them, $(C_{12}Im)_2C_4$–2Br is the Gemini IL-based surfactant with the largest number of applications.[65,66,102] Gemini surfactants based on pyrrolidinium cations containing butane groups as spacers have also been synthetized, and their micellar characteristics have been studied.[103]

More recently, some ammonium Gemini IL-based surfactants with antimicrobial properties have been synthetized and characterized.[75,104] Some of them with hydroxyl groups in the spacers turned out to be room temperature IL-based surfactants, due to the intermolecular hydrogen bonds formed.[104]

Regarding tricationic IL-based surfactants, only a few of them (all with imidazolium cations) have been reported.[67,105] Tris-octylimidazolium bromide IL-based surfactants with trimethylamine or 1,3,5-trimethylbenzene as spacers have been synthesized, and their surface activities have been properly investigated.[67] Trigonal surfactants with a tri-ester group as spacer have been prepared to obtain IL-based surfactants with lower toxicity.[105] These tricationic IL-based surfactants composed of chloride anion and tris-imidazolium or tris-benzimidazolium cations with alkyl chains (from 4 to 12 carbon atoms) or phenyl groups as free side chains/groups, respectively, present much better biodegradation than conventional surfactants.

3.2.3 Functionalized IL-based Surfactants

Based on the inherent structural versatility of ILs and environmental concerns around surfactants, several investigations are focused on the design of more biodegradable ILs.[106,107] In general, the toxicity of ILs is attributed to the cation nature and the side chain length.[106] Recent toxicological studies have pointed out that ILs possessing oxygenated groups in their cationic moiety exhibit lower toxicity.[108,109] In this sense, the incorporation of hydroxyl,[110,111] ester,[42,112,113] and ether[110,114] groups in the cationic moiety of IL-based surfactants has been performed to decrease the IL toxicity.

In other cases, functionalization of the ILs has been carried out to incorporate certain groups to ensure specific analyte–IL interactions. Thus, imidazolium IL-based surfactants tailored with aromatic groups such as vinyl,[115] pyrrole,[116] and benzyl[117] have been reported.

Metal-containing ILs have been revealed as promising catalysts or as novel reagents within organic chemistry.[118] Thus, a ZIL-based surfactant consisting of two dodecylimidazolium chains and the group Cu–$CuCl_2$ as spacer,[119] or ammonium cations combined with polyoxometalate anions to form IL-based surfactants,[120] have been described.

Amino acid-derived IL-based surfactants have also been prepared.[121,122] Trivedi *et al.*[122] synthetized several anionic IL-based surfactants using dodecyl sulfate as anion and natural amino acids as cations. Liu and Shamsi[121] reported the synthesis of IL-based surfactants containing ammonium cations with amino acid-functionalized chains, and bromide anion. An interesting highlight is the chiral component of these IL-based surfactants, which constitute an attractive feature for applications in separation science.

3.3 Characterization of IL-based Surfactants Properties: Micellar Behaviour

Since the first evidence of the surface active properties of ILs containing long substituents,[7,10] an increasing number of studies have focused on the synthesis and micellar characterization of novel IL-based surfactants. Knowledge of the aggregation behaviour of IL-based surfactants is crucial to develop and improve their applications in separation science, chemical synthesis, and/or in industrial technologies.

The CMC is the main parameter used in colloid science to evaluate monomer–micelle equilibrium in surfactant solutions. This value indicates the concentration range over which the surfactant molecules start building up aggregates in water. The CMC is estimated by the maximum change in a physical property of the solution with respect to the concentration of surfactant.

Table 3.1 includes the CMC values reported for most IL-based surfactants known to date. It can be observed from the table that an increase in the number of carbons in the alkylic side chain of IL-based surfactants composed of the same cation and anion is accompanied by an important decrease of the CMC. For example, the CMC of C_8MIm-Br is 150 mmol L^{-1},[20] for C_{10}MIm-Br it is around 30–40 mmol L^{-1},[20] while $C_{12}MIm$-Br presents a CMC value of 9.8 mmol L^{-1}.[29] Furthermore, micelles formed by ILs exhibit low CMC values when compared to conventional surfactants with similar structure. Thus, the anionic IL-based surfactant C_4MIm-$C_{12}SO_4$ presents a CMC value of 2.4 mmol L^{-1}, and C_8MIm-$C_{12}SO_4$ has a CMC value of 0.4 mmol L^{-1},[15] while the conventional structural analogue anionic surfactant sodium dodecyl sulfate (SDS) has a CMC of 8.2 mmol L^{-1}.[123] A similar situation is observed when comparing the cationic IL-based surfactant $C_{16}MIm$-Br, with a CMC of 0.61 mmol L^{-1},[29] and the conventional structurally analogue cationic surfactant hexadecyl(trimethyl)ammonium bromide (CTAB), with a CMC of 1 mmol L^{-1}.[71]

The CMC is strongly affected by the experimental conditions, such as the ionic strength, temperature, and the presence of organic solvents in the aqueous surfactant solution. The effect that organic solvents exert on the CMC has been exhaustively studied, particularly with methanol, ethanol, 1-propanol, 1-butanol, 1-pentanol, and acetonitrile for imidazolium IL-based surfactants.[67,124–126] In general, the addition of polar solvents to IL-based surfactant solutions leads to an increase of the CMC. The presence of ethylene glycol and its derivatives in IL-based surfactant solutions has been studied,[124,126] also exhibiting an increase in the resulting CMC. The influence of sodium salts on the CMC has also been assessed for imidazolium IL-based surfactants.[67,127] In this case, the CMC drastically decreases since the anions from the inorganic salt facilitate the aggregation by suppressing the electrostatic repulsion among the head groups of the IL-based surfactant. This behaviour was also observed for monoalkylguanidinium chloride IL-based surfactants.[77,78]

The most common conformation adopted by surfactants in aqueous solutions (at concentrations above the CMC) is that of spherical micelles. Recently, some authors have pointed out the spontaneous formation of vesicle structures at high concentrations of IL-based surfactants (well above the CMC).[15,81,128] The concentration over which IL-based surfactants form this self-assembly structure is designated as the critical vesicle concentration (CVC). IL-based surfactants can also form admicelles and hemimicelles, and these structures appear at slightly lower concentration than their CMCs. This aggregation takes place when IL-based surfactants interact with negatively charged solid supports.[6] Hemimicelles refer to monolayers of surfactants while admicelles involve bilayers arrangements. Figure 3.2 includes tentative shapes of all these micellar aggregations, using as an example the IL-based surfactant $C_{12}MIm$-Br.

The CMC can be experimentally determined using different techniques. Surface tension,[67,79,82,89,91,115,124,128,129,131] fluorescence,[61,67,79,82,89,115,128,129,131] and conductivity[67,79,82,91,100,124,129,130,132] measurements are the most common tools employed. In a few cases, small angle neutron scattering (SANS)[11,19,20,91] or turbidity measurements[133] have been used to estimate the CMC for IL-based surfactants. It is important to highlight that the CMC is not a concentration itself but a narrow range of concentrations.

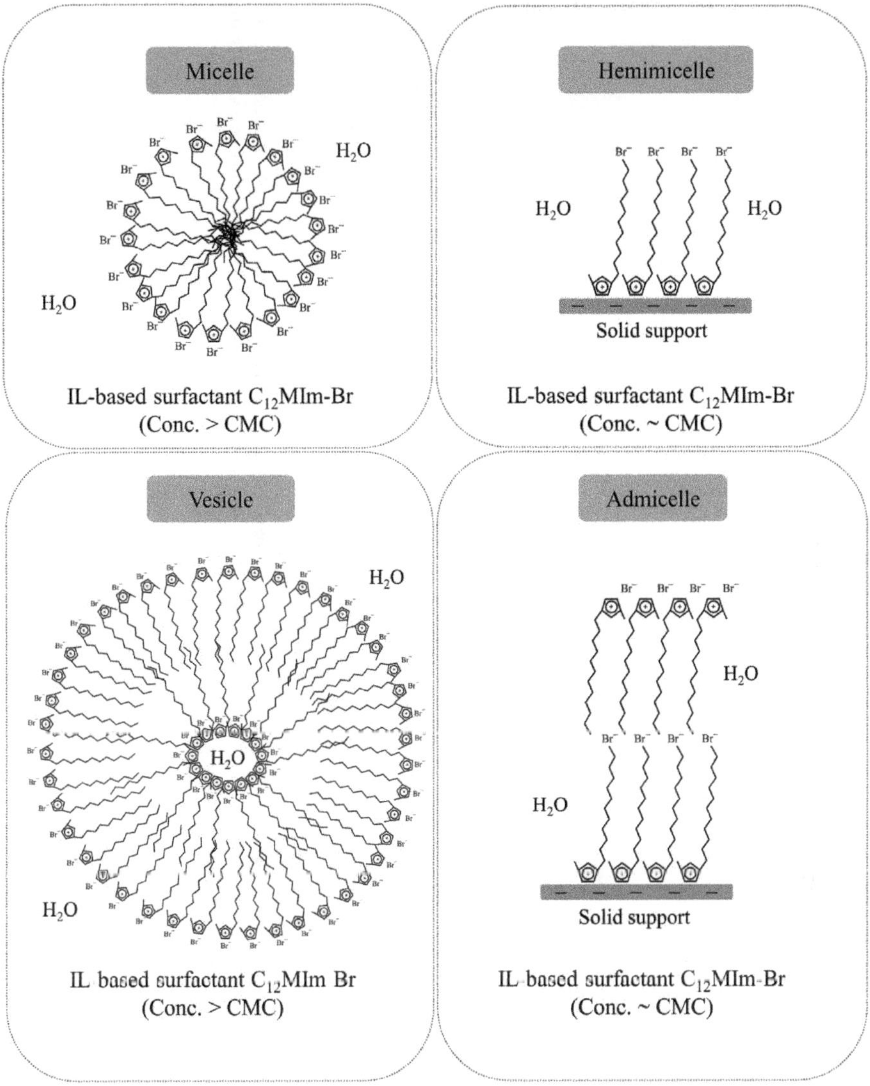

Figure 3.2 Main types of IL-based surfactant aggregations found in water: conventional micelles, vesicles, hemimicelles and admicelles.

The surface tension (γ) of an aqueous solution decreases with the addition of increasing amounts of surfactant, because surfactants tend to occupy first the surface of water. The aggregation of surfactants in the bulk solution takes place once the entire water surface is covered by monomers. Thus, the surface tension levels off and remains constant once the CMC is reached, as can be observed in Figure 3.3(A). Therefore, the CMC can be estimated from the break point observed in the surface tension *versus* surfactant concentration plots.[79,85,128] Many authors plot the variation of the surface

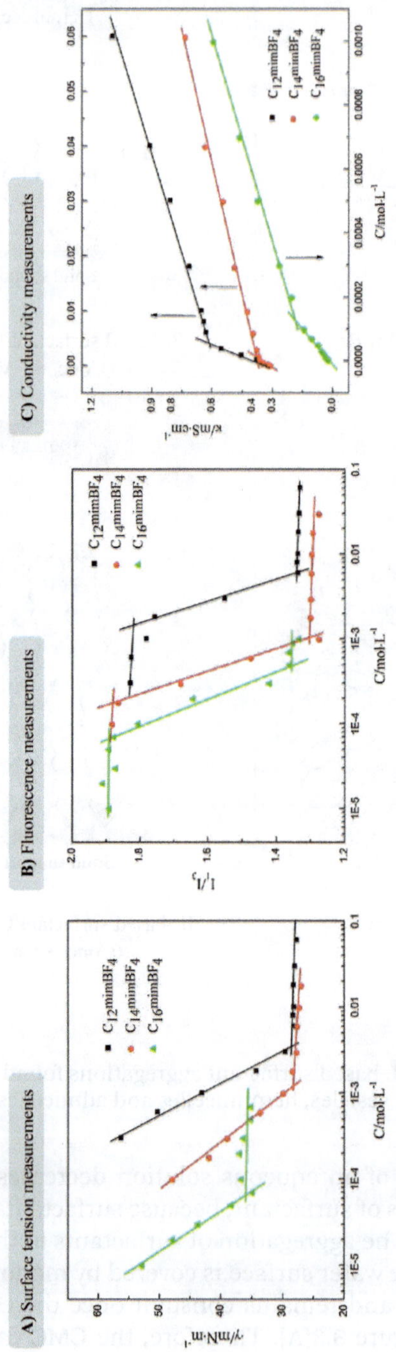

Figure 3.3 Plots obtained for (A) surface tension measurements, (B) fluorescence measurements, and (C) conductivity measurements for the CMC determination of $C_{12}MIm-BF_4$, $C_{14}MIm-BF_4$ and $C_{16}MIm-BF_4$ IL-based surfactants. Reprinted with permission from ref. 85. Copyright (2014) American Chemical Society.

tension *versus* the logarithm of the surfactant concentration to determine the CMC.[67,82,89,115,124,129–131]

There are other interesting interfacial characteristics that have been determined for IL-based surfactants by surface tensiometry, such as the surface tension at the CMC (γ_{cmc}),[15,67,75,85,89,91,104,124,125,134,135] the saturation adsorption (Γ_{max}),[15,67,75,89,91,104,114,124,125,134,135] the area occupied by a single surfactant molecule at the air–solution interface (A_{min}),[15,67,75,89,91,104,114,124,125,134,135] the adsorption efficiency (pC_{20}),[15,67,75,89,91,104,124,125,134,135] and the effectiveness of the surface tension reduction (Π_{cmc}).[15,75,85,125,135] Table 3.2 summarizes the values of these interfacial parameters reported for several imidazolium IL-based surfactants, included in the table as representative examples.

The saturation adsorption (Γ_{max}) is normally used to evaluate the effectiveness of adsorption of a surfactant at the air–aqueous interface since it represents its maximum adsorption. This parameter is calculated using the Gibbs adsorption isotherm equation:[136]

$$\Gamma_{max} = -\frac{1}{nRT} \times \frac{d\gamma}{d\ln(c)} \tag{3.1}$$

where γ is in mN m^{-1}, Γ_{max} is in μmol m^{-2}, R is the gas constant (8.314 J mol^{-1} K^{-1}), T is the temperature, c is the surfactant concentration, and $d\gamma/d\ln(c)$ is the slope in the surface tension isotherm when the concentration is near the CMC. As can be observed in Table 3.2, Γ_{max} values increase with increasing length of the alkyl chain of 1-alkyl-3-methylimidazolium type IL-based surfactants. In addition, dicationic IL-based surfactants provide the lowest values.

Table 3.2 Interfacial parameters (γ_{cmc}, Γ_{max}, A_{min}, pC_{20}, and Π_{cmc}), aggregation number (N_{agg}), degree of counterion dissociation (α_{mic}), and fractions of counterions bound to the micelle (β) for imidazolium IL-based surfactants in the recent literature.

IL-based surfactant	γ_{cmc} (mN m^{-1})	Γ_{max} (μmol m^{-2})	A_{min} (Å^2)	pC_{20}	Π_{cmc} (mN m^{-1})	N_{agg}	α_{mic}	β	Ref.
C_{10}MIm-Cl	27.3	1.9	85.0	2.5	44.5	-	0.510	0.490	91
C_{12}MIm-Cl	38.7	2.3	72.0	2.4	33.6	-	0.440	0.560	91
C_{16}MIm-Cl	37.0	3.4	49.0	3.2	34.8	-	0.550	0.450	9
C_{18}MIm-Cl	42.0	3.7	45.0	3.6	29.8	-	0.601	0.399	91
C_{12}MIm-Br	39.4	-	-	-	33.6	37	-	-	85
C_{14}MIm-Br	39.2	-	-	-	33.8	48	-	-	85
C_{16}MIm-Br	39.1	-	-	-	33.9	64	-	-	85
$(C_{12}Im)_2C_2$-2Br	33.6	1.2	135	4.5	38.4	21	0.230	0.770	135
$(C_{12}Im)_2C_4$-2Br	35.7	1.2	140	3.9	36.3	16	0.300	0.700	135
$(C_{12}Im)_2C_6$-2Br	39.5	1.2	143	3.7	32.5	14	0.400	0.600	135

The A_{min} values can be obtained from the saturated adsorption:

$$A_{min} = (N_A \Gamma_{max})^{-1} \times 10^{24} \quad (3.2)$$

where N_A is Avogadro's number and A_{min} is in nm^2. The A_{min} values reported in the literature, which are included in Table 3.2, are much higher for Gemini IL-based surfactant compared to monocationic types. This behaviour agrees with the greater size of dicationic IL-based surfactants, which makes them fill bigger areas.

The pC_{20} is also used to assess the efficiency of the adsorption of surfactant molecules at the interface. C_{20} refers to the surfactant concentration at which the initial surface tension of the solution is reduced by 20 mN m^{-1}.[137] The larger the value of pC_{20}, the higher the adsorption efficiency of the surfactant. This parameter is easily calculated as:

$$pC_{20} = -\log C_{20} \quad (3.3)$$

The values of pC_{20} found for alkyl(methyl)imidazolium IL-based surfactants increase with the length of the alkyl chain, being even higher for the dicationic ones (Table 3.2). Therefore, the adsorption efficiency is better the bigger/larger is the IL-based surfactant.

The parameter Π_{cmc} corresponds to the maximum reduction of surface tension caused by the dissolution of surfactant molecules.[137] It is given by the following equation:

$$\Pi_{cmc} = \gamma_0 - \gamma_{cmc} \quad (3.4)$$

where γ_0 is the initial surface tension of the solution and γ_{cmc} is the surface tension value at the CMC. Regarding the Π_{cmc} reported in the literature and included in Table 3.2, they decrease as the length of the alkyl chain increases. All these parameters are quite useful in order to test the colloid performance of these materials, particularly concerning further industrial applications.

The determination of the CMC by fluorescence measurements requires the utilization of a probe, which must present hydrophobic and fluorescence properties. Pyrene is the most common probe molecule used in such studies since its fluorescence spectrum is well known. This probe can be dissolved in the inner region of the micelles, which leads to a change in its two band intensities I_1 (~375 nm) and I_3 (~385 nm) of the emission spectrum.[123] If the I_1 to I_3 ratio is plotted *versus* the logarithm of surfactant concentration, a sigmoid curve is obtained (Figure 3.3(B)). The inflection point of this curve, which can be easily estimated by the second derivative method, determines the CMC.

The aggregation number (N_{agg}) represents the number of monomers of a surfactant required to form a micelle. N_{agg} values are not widely determined for IL-based surfactants since more reliable methodologies devoted to it (such as SANS) are tedious. A simpler method based on fluorescence quenching measurements has been developed by Turro and Yekta,[138] and used for the characterization of IL-based surfactants.[67,82,89,98,100,129]

This technique requires the utilization of a quenching agent of pyrene fluorescence activity that meets the following requirements: (1) both the probe and the quencher are dissolved in the micelles, (2) their distribution over the micelles is described by a Poisson distribution, and (3) they remain confined in the micelles longer than the lifetime of the excited probe. Under these conditions, the band intensities of pyrene at a constant concentration and a fixed wavelength in the presence (I) and absence (I_0) of quenching agent, and keeping constant the concentration of surfactant, are related to the quencher concentration by the following expression:

$$\ln\left(\frac{I_0}{I}\right) = \frac{N_{\text{agg}}}{C_{\text{surfactant}} - \text{CMC}} \times C_{\text{quencher}} \tag{3.5}$$

Thus, N_{agg} can be estimated from the slope of the Napierian logarithm (I_0/I) *versus* the concentration of the quenching agent. Benzophenone[82,85,89,94,97,98,100,135,139] and benzophenone-3[67] have been used as quenchers in this steady-state fluorescence quenching method for the characterization of IL-based surfactants. As can be expected taking into account their structure, the reported N_{agg} values reported for Gemini IL-based surfactants are quite lower than those reported for monocationic IL-based surfactants (Table 3.2). These results are in line with the fact that dicationic IL-based surfactants present lower CMCs than their monocationic analogues.

Several micellar parameters of IL-based surfactants can also be determined by conductivity measurements. The specific conductivity of a solution increases with the addition of an ionic surfactant. When the CMC is reached, the slope in a plot of specific conductivity *versus* surfactant concentration changes due to the formation of micelles in the media (Figure 3.3(C)). Therefore, the CMC of IL-based surfactants can be estimated as the concentration associated with this break point.[140] From Table 3.1, it is clear that despite differences in techniques there is an adequate agreement in CMC values determined by surface tension, fluorescence, and conductivity.

The degree of counterion dissociation (α_{mic}) is used to evaluate the ionization of micellar aggregates, and it is determined by conductivity measurements using the equation introduced by Evans:[140]

$$1000\ S_2 = \frac{\alpha_{\text{mic}}^2}{\left(N_{\text{agg}}\right)^{-2/3}}\left(1000\ S_1 - \Lambda_{X^-}\right) + \alpha_{\text{mic}}\ \Lambda_{X^-} \tag{3.6}$$

where S_1 and S_2 are the slopes of premicellar and post-micellar linear regions, respectively, and Λ_{X^-} refers to the equivalent conductivity of the surfactant counterion at infinite dilution. However, the approximation introduced by Hoffmann and Ulbricht[141] to calculate α_{mic} has become very popular in the IL-based surfactants research field because of its simplicity:[61,67,82,90,94,100,131,132,134]

$$\alpha_{\text{mic}} = \frac{S_2}{S_1} \tag{3.7}$$

Once α_{mic} is estimated, it is possible to calculate the fractions of counterions bound to the micelle (β) as $\beta = 1 - \alpha_{mic}$. As it is shown in Table 3.2, in general, the α_{mic} values decrease when shorter alkyl chains are used, and consequently β values increase.

3.4 Applications of IL-based Surfactants

Most studies reported in the literature dealing with IL-based surfactants are related to their synthesis and micellar characterization.[61,67,79,82,89,91,100,115,124,128–132] Many IL-based surfactants have been synthetized for the first time quite recently, and so it is very important to make a thorough characterization of their micellar behaviour. Given the outstanding performance of these novel micellar media, the number of characterization studies in the recent literature is understandable.

Figure 3.4 shows the most relevant application fields of IL-based surfactants from literature in the last five years, excluding the synthesis and characterization studies of micellar behaviour mentioned above.

Microemulsions are isotropic liquid mixtures formed by two or more immiscible liquids thermodynamically stabilized by a surfactant. Generally, these ternary systems are composed of oil, water, and a surfactant (water-in-oil microemulsions). However, ILs have been widely used instead of water or the organic phase, generating new types of microemulsion systems: IL-in-oil microemulsions (where an IL acts as polar phase), and water-in-IL microemulsions (in which the IL is the organic/oil phase). One of the most relevant applications of IL-based surfactants in recent years is their use as surfactants in microemulsions, constituting the most prolific field. Once dissolved in water, the long chain of the IL-based surfactant shows high affinity towards the apolar phase, while the head group presents high affinity for the polar phase. Figure 3.5 shows a scheme of a water-in-oil microemulsion using an IL as the surfactant.

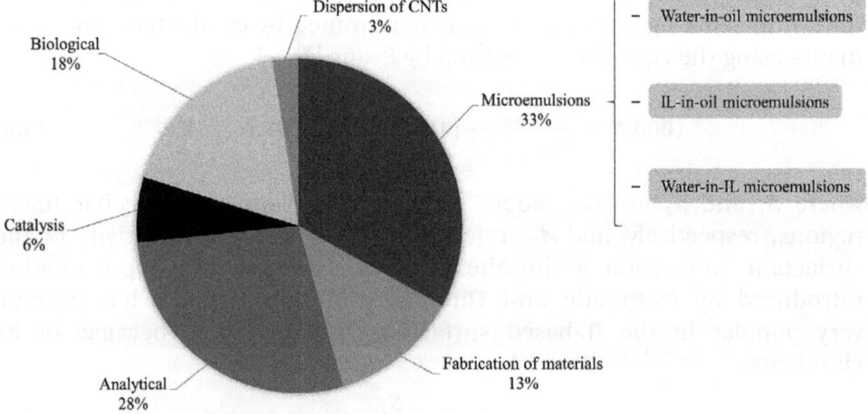

Figure 3.4 Main applications of IL-based surfactants, reported as a function of percentage of studies published in the recent literature.

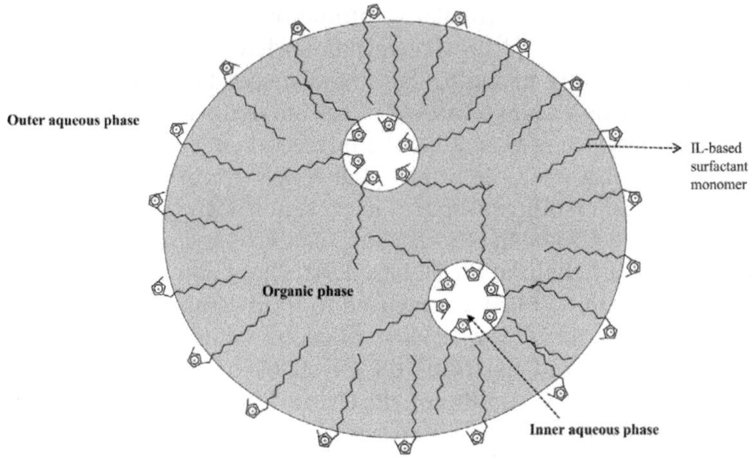

Figure 3.5 Scheme of a water-in-oil microemulsion using an IL-based surfactant (containing $C_{12}MIm^+$ in the example included).

IL-based surfactants forming part of water-in-oil microemulsions have been used in a great variety of applications, being very useful in enhanced oil recovery.[114,142–144] Furthermore, C_{12}MIm-Cl has been used as surfactant in water-in-oil microemulsions needed in polymerization reactions[37] and in the synthesis of AgCl nanoparticles.[38,39] The dicationic IL-based surfactant $(C_{14}Im)_2C_4$-2Br has been used as surfactant in microemulsion systems employed as solvent for the extraction of Au(III).[145]

Most studies that use IL-based surfactants in IL-in-oil microemulsion systems are devoted to the study of their phase behaviour and thermodynamic properties.[17,24,49,146,147]

Regarding water-in-IL microemulsions, C_{10}MIm-Br has been successfully used for the selective isolation of haemoglobin,[26] and the environmentally-friendly C_{12}MIm-HSO_3 has been used for the extraction of hydrophilic components from plants.[148]

Applications of IL-based surfactants in analytical chemistry have also been quite successful. The surfactants have been utilized as novel solvents in sample preparation,[12,22,23,32,63,77,148–151] combined with magnetic nanoparticles in analytical extraction approaches,[51,52,57,62] as pseudostationary phases in micellar electrokinetic chromatography (MEKC),[44,46,47,50] as pseudostationary phases in liquid chromatography,[41,125] as micellar media in capillary electrophoresis online preconcentration techniques such as sweeping MEKC or micelle to solvent stacking,[152–154] and as sensitizer of spectroscopic methods;[21] in all cases exploiting their micellar behaviour. In addition, they have been utilized as buffer modifiers in MEKC,[13,14] in these cases not taking advantage of their surface-active properties. Amino-acid based IL-based surfactants have also demonstrated successful enantio-separation of neutral compounds in capillary electrophoresis, by forming complexes with cyclodextrins.[154]

The biological activity, particularly antimicrobial, of some IL-based surfactants has been reported.[104,155,156] Several studies have focused on the influence of IL-based surfactants in biological systems, specifically their interactions with certain proteins such as β-casein,[34,35] bovine serum albumin,[53,113] and lysozyme.[16,65] Likewise, He *et al.* studied the interaction of the IL-based surfactant C_{12}MIm-Br and DNA in dilute brine.[33] Banerjee's group studied the changes in the activity, and stability of lipases in presence of IL-based surfactants.[54,68] Other effects have also been studied, such as those arising from the increase in size of the alkylic chain of the IL-based surfactants on biological molecular structures.[66,74] Thus, $C_{16}Ph_3$P-Br was used to form a supramolecular complex with β-cyclodextrin, forming vesicles in aqueous solution.[74] Another example is the study of Patel *et al.* regarding the effects of the symmetric Gemini type IL-based surfactant, $C_{12}C_{12}$Im-Br_2, on the channel form of gramicidin vesicles,[66] also demonstrating that the IL formed aggregates at the surface of such vesicles.

One of the most interesting applications of IL-based surfactants is their use as modifiers to control the size and morphology in the fabrication of novel nanomaterials. Thus, IL-based surfactants have been employed in the fabrication of gold nanocrystals and microplates,[15] ferrite nanoparticles,[58] calcium carbonate nanocrystals,[18] or in the control of the crystallographic orientation of WO_3 films.[55] C_{16}MIm-Cl and C_{16}MPy-Cl have been employed as templates to modify the mesoporous of zeolites,[59] while C_{16}MIm-Br has been employed as Br source in the successful fabrication of Ag/AgBr/BiOBr microspheres.[127] Other applications reported include their use in the synthesis of hybrid materials since IL-based surfactants enable the fabrication of inorganic–organic hybrid compounds due to their structural properties.[27,36,72,96]

IL-based surfactants have been satisfactorily used in chemical reactions due to their catalytic properties. This way, Zhu *et al.* synthesized several IL-based surfactants combining polyoxometalate anions and ammonium cations, and demonstrated their application in the oxidative desulfurization of organosulfur compounds.[120] Heidarizadeh and Majdi-nasab reported a reusable copper/IL-based surfactant catalytic system, $(C_{12}Im)_2Cu$-$CuCl_2$, in the N-arylation of heterocycles. The catalyst was an excellent alternative to the expensive palladium system since the formed micelles allowed an increase in the contact between the organic reactants and water.[119] Aghabarari *et al.* demonstrated the excellent catalytic activity of two imidazolium IL-based surfactants for the synthesis of biodiesel.[117] IL-based surfactants have also been employed in combination with other materials to improve their catalytic properties. Hence, Hu *et al.* reported the excellent catalytic activity of a system composed of metal nanoparticles and IL-based surfactants for the hydrochlorination of acetylene, which makes this system an excellent substitute for conventional mercury catalysts.[157] Furthermore, Lu *et al.* developed a catalytic epoxidation system based on $KMnO_4$ in an IL-based surfactant medium. The study showed an unusual oxidation behaviour of $KMnO_4$ under the crucial effects of the IL, not found in conventional solvents.[158]

Finally, interestingly, the application of IL-based surfactants improves the solubility of carbon nanotubes (CNTs) in water. Raiah *et al.* used imidazolium type IL-based surfactants for the dispersion and stabilization of double-walled

CNTs in water, which was maintained for long periods of time.[159] Another example was reported by Lu *et al.*, dispersing multi-walled CNTs in aqueous solution with the aid of phosphonium IL-based surfactants.[73] Furthermore, very recently, Tawfik has demonstrated the use of ammonium dicationic IL-based surfactants as corrosion inhibitors for carbon steel.[76]

3.5 Conclusions and Trends

Ionic-liquid-based surfactants constitute an interesting field of novel materials, with improved micellar behaviour over conventional surfactant systems. Their relatively simple synthetic tuneability together with the high variety of applications in which these smart materials have demonstrated outstanding performance ensure that a higher number of type and applications will continue to flourish.

Abbreviations

α_{mic}	counterion dissociation
γ	surface tension
γ_{cmc}	surface tension at the CMC
Γ_{max}	saturation adsorption
A_{min}	air–solution interface
pC_{20}	adsorption efficiency
Π_{cmc}	surface tension reduction
AOT	dioctyl sulfosuccinate
BAILs	biamphiphilic ionic liquids
BF_4^-	tetrafluoroborate
$C_8SO_4^-$	octyl sulfate
$C_{12}SO_4^-$	dodecyl sulfate
CMC	critical micelle concentration
CTAB	hexadecyl(trimethyl)ammonium bromide
CVC	critical vesicle concentration
ILs	ionic liquids
Im^+	imidazolium
MEKC	micellar electrokinetic chromatography
N^+	ammonium
N_{agg}	aggregation number of the micelle
P^+	phosphonium
Pip^+	piperidinium
Py^+	pyridinium
$Pyrr^+$	pyrrolidinium
SAILs	surface active ionic liquids
SANS	small-angle neutron scattering
SDS	sodium dodecyl sulfate
S_N	nucleophilic substitution reaction
ZILs	zwitterionic ionic liquids

Acknowledgements

V. P. thanks the Spanish Ministry of Economy and Competitiveness (MINECO) for the project MAT2014-57465-R. I. P.-F. and P. G.-H. give thanks for their PhD research contracts from ULL-La Caixa fellowship.

References

1. J. H. Davis Jr, *Chem. Lett.*, 2004, **33**, 1072.
2. M. J. Trujillo-Rodríguez, P. González-Hernández and V. Pino, in *Ionic Liquid-Based Surfactant Science: Formulation, Characterization, and Applications*, ed. B. K. Paul and S. P. Moulik, John Wiley & Sons Inc., Hoboken NJ, 2015, ch. 23, pp. 475–501.
3. M. J. Earle and K. R. Seddon, *Pure Appl. Chem.*, 2000, **72**, 1391.
4. N. V. Plechkova and K. R. Seddon, *Chem. Soc. Rev.*, 2008, **37**, 123.
5. M. Freemantle, *Chem. Eng. News*, 1998, **76**, 32.
6. V. Pino, M. Germán-Hernández, A. Martín-Pérez and J. L. Anderson, *Sep. Sci. Technol.*, 2012, **47**, 264.
7. J. Sirieix-Plénet, L. Gaillon and P. Letellier, *Talanta*, 2004, **63**, 979.
8. Q. Q. Baltazar, J. Chandawalla, K. Sawyerand and J. L. Anderson, *Colloids Surf., A*, 2007, **302**, 150.
9. P. Walden, *Bull. Acad. Imp. Sci. St.-Petersbourg*, 1914, **8**, 405.
10. J. Bowers, C. P. Butts, P. J. Martin and M. C. Vergara-Gutierrez, *Langmuir*, 2004, **20**(6), 2191.
11. N. A. Smirnova and E. A. Safonova, *Colloid J.*, 2012, **74**, 254.
12. W. Yu, Z. Liu, Q. Li, H. Zhang and Y. Yu, *Food Chem.*, 2015, **173**, 815.
13. W. Xiao, C. Chen, Q. Zhang, Q. H. Zhang, Y. J. Hu, Z. N. Xia and F. Q. Yang, *Chromatographia*, 2015, **78**, 1385.
14. J. Cao, J. Hu, J. Wei, B. Li, M. Zhang, C. Xiang and P. Li, *Molecules*, 2015, **20**, 15304.
15. K. S. Rao, T. J. Trivedi and A. Kumar, *J. Phys. Chem. B*, 2012, **116**, 14363.
16. B. Mandal, S. Mondal, A. Pan, S. P. Moulik and S. Ghosh, *Colloids Surf., A*, 2015, **484**, 345.
17. O. Rojas, B. Tiersch, C. Rabe, R. Stehle, A. Hoell, B. Arlt and J. Koetz, *Langmuir*, 2013, **29**, 6833.
18. Y. Zhao, W. Du, L. Sun, L. Yu, J. Jiao and R. Wang, *Colloid Polym. Sci.*, 2013, **291**, 2191.
19. P. Brown, C. P. Butts, J. Eastoe, D. Fermin, I. Grillo, H. C. Lee, D. Parker, D. Plana and R. M. Richardson, *Langmuir*, 2012, **28**, 2502.
20. I. Goodchild, L. Collier, S. L. Millar, I. Prokeš, J. C. D. Lord, C. P. Butts, J. Bowers, J. R. P. Webster and R. K. Heenan, *J. Colloid Interface Sci.*, 2007, **307**, 455.
21. T. Hou and X. Zhu, *J. Mol. Liq.*, 2012, **166**, 17.
22. X. Yang, S. Zhang, W. Yu, Z. Liu, L. Lei, N. Li, H. Zhang and Y. Yu, *Talanta*, 2014, **124**, 1.

23. X. J. Li, H. M. Yu, C. Gao, Y. G. Zu, W. Wang, M. Luo, C. B. Gu, C. J. Zhao and Y. J. Fu, *J. Sep. Sci.*, 2012, **35**, 3600.
24. A. Wang, L. Chen, F. Xu and Z. Yan, *J. Chem. Eng. Data*, 2015, **60**, 519.
25. A. Wang, L. Chen, D. Jiang, H. Zeng and Z. Yan, *Ind. Crops Prod.*, 2014, **62**, 515.
26. Q. X. Mao, H. Wang, Y. Shu, X. W. Chen and J. H. Wang, *RSC Adv.*, 2014, **4**, 8177.
27. T. Ito, K. Mikurube, K. Hasegawa, T. Matsumoto, K. Kosaka, H. Naruke and S. Koguchi, *Crystals*, 2014, **4**, 42.
28. O. A. El Seoud, P. A. R. Pires, T. Abdel-Moghny and E. L. Bastos, *J. Colloid Interface Sci.*, 2007, **313**, 296.
29. R. Vanyúr, L. Biczók and Z. Miskolczy, *Colloids Surf., A*, 2007, **299**, 256.
30. M. Zhu, J. Chai, L. Chen, L. Xu, W. Liu, S. Shang and J. Lu, *Fluid Phase Equilib.*, 2012, **314**, 90.
31. J. Chai, L. Xu, W. Liu and M. Zhu, *J. Chem. Eng. Data*, 2012, **57**, 2394.
32. S. S. Hu, L. Yi, X. Y. Li, J. Cao, L. H. Ye, W. Cao, J. H. Da, H. B. Dai and X. J. Liu, *J. Agric. Food Chem.*, 2014, **62**, 5275.
33. Y. He, Y. Shang, Z. Liu, S. Shao, H. Liu and Y. Hua, *Colloids Surf., B*, 2013, **101**, 398.
34. Y. Liu, L. Yang and R. Guo, *Soft Matter*, 2013, **9**, 3671.
35. C. Cao, T. Huang and F. P. Du, *Food Hydrocolloids*, 2014, **36**, 348.
36. L. Wang, X. Xin, M. Yang, J. Shen and S. Yuan, *RSC Adv.*, 2015, **5**, 68404.
37. C. Costa, V. H. S. Santos, C. Dariva, A. F. Santos, M. Fortuny, P. H. H. Araujo and C. Sayer, *J. Appl. Polym. Sci.*, 2013, **127**, 448.
38. L. Zhou, X. Q. Dai, J. J. Du, T. Wang, L. G. Wu, Y. C. Tang and J. Shen, *Ind. Eng. Chem. Res.*, 2015, **54**, 3326.
39. L. G. Wu, J. N. Shen, C. H. Du, T. Wang, Y. Teng and B. V. der Bruggen, *Sep. Purif. Technol.*, 2013, **114**, 117.
40. A. Z. Hezave, S. Dorostkar, S. Ayatollahi, M. Nabipour and B. Hemmateenejad, *Colloids Surf., A*, 2013, **421**, 63.
41. J. Flieger, A. Siwek, M. Pizon and A. Czajkowska-Zelazko, *J. Sep. Sci.*, 2013, **36**, 1530.
42. G. Singh and T. S. Kang, *J. Phys. Chem. B*, 2015, **119**, 10573.
43. B. Dong, N. Li, L. Zheng, L. Yu and T. Inoue, *Langmuir*, 2007, **23**, 4178.
44. C. Tejada-Casado, D. Moreno-González, A. M. García-Campaña and M. del Olmo-Iruela, *Electrophoresis*, 2015, **36**, 955.
45. Y. Tong, L. Han and Y. Yang, *Ind. Eng. Chem. Res.*, 2012, **51**, 16438.
46. A. H. Rageh and U. Pyell, *Electrophoresis*, 2015, **36**, 784.
47. A. H. Rageh and U. Pyell, *J. Chromatogr. A*, 2013, **1316**, 135.
48. T. H. Kang, Y. Jeon and M. W. Kim, *Phys. Chem. Chem. Phys.*, 2015, **17**, 27833.
49. A. Klee, S. Prevost, U. Gasser and M. Gradzielski, *J. Phys. Chem. B*, 2015, **119**, 4133.
50. J. Kazarjan, R. Mahlapuu, M. Hansen, U. Soomets, M. Kaljurand and M. Vaher, *J. Sep. Sci.*, 2015, **38**, 3461.

51. M. J. Trujillo-Rodríguez, V. Pino, J. L. Anderson, J. H. Ayala and A. M. Afonso, *Anal. Bioanal. Chem.*, 2015, **407**, 8753.
52. J. Wu, H. Zhao, D. Xiao, C. Pham-Huy, J. He and H. He, *J. Chromatogr. A*, 2016, **1454**, 1.
53. C. Cao, J. Lei, T. Huang and F. P. Du, *Soft Matter*, 2014, **10**, 8896.
54. S. Adak, S. Datta, S. Bhattacharya and R. Banerjee, *J. Mol. Catal. B: Enzym.*, 2015, **119**, 12.
55. H. Kaper, I. Djerdj, S. Gross, H. Amenitsch, M. Antonietti and B. M. Smarsly, *Phys. Chem. Chem. Phys.*, 2015, **17**, 18138.
56. T. Li, S. Luo and L. Yang, *J. Solid State Chem.*, 2013, **206**, 308.
57. E. Aliyari, M. Alvand and F. Shemirani, *RSC Adv.*, 2016, **6**, 64193.
58. P. Shikha, T. S. Kang and B. S. Randhawa, *RSC Adv.*, 2015, **5**, 96799.
59. A. Sachse, C. Wuttke, U. Díaz and M. O. de Souza, *Microporous Mesoporous Mater.*, 2015, **217**, 81.
60. C. Jungnickel, J. Łuczak, J. Ranke, J. F. Fernández, A. Müller and J. Thöming, *Colloids Surf., A*, 2008, **316**, 278.
61. M. Figueira-González, V. Francisco, L. García-Río, E. F. Marques, M. Parajó and P. Rodríguez-Dafonte, *J. Phys. Chem. B*, 2013, **117**, 2926.
62. L. Farzin, M. Shamsipur and S. Sheibani, *Microchim. Acta*, 2016, **183**, 2623.
63. J. F. Ayala-Cabrera, M. J. Trujillo-Rodríguez, V. Pino, O. M. Hernández-Torres, A. M. Afonso and J. Sirieix-Plénet, *Int. J. Environ. Anal. Chem.*, 2016, **96**(2), 101.
64. Y. Hu, L. Ge, J. Han and R. Guo, *Soft Matter*, 2015, **11**, 5624.
65. J. K. Maurya, M. U. H. Mir, U. K. Singh, N. Maurya, N. Dohare, S. Patel, A. Ali and R. Patel, *Biopolymers*, 2015, **103**, 406.
66. R. Patel, M. ud din Parray, U. K. Singh, A. Islam, P. Venkatesu, S. Singh and H. B. Bohidard, *Colloids Surf., A*, 2016, **508**, 150.
67. O. Nacham, A. Martín-Pérez, D. J. Steyer, M. J. Trujillo-Rodríguez, J. L. Anderson, V. Pino and A. M. Afonso, *Colloids Surf., A*, 2015, **469**, 224.
68. S. Adaka, S. Datta, S. Bhattacharya and R. Banerjee, *Int. J. Biol. Macromol.*, 2015, **81**, 560.
69. G. A. Baker, S. Pandey, S. Pandey and S. N. Baker, *Analyst*, 2004, **129**, 890–892.
70. A. Z. Hezave, S. Dorostkar, S. Ayatollahi, M. Nabipour and B. Hemmateenejad, *Fluid Phase Equilib.*, 2013, **360**, 139.
71. M. Prasad, S. P. Moulik, A. MacDonald and R. Palepu, *J. Phys. Chem. B*, 2004, **108**, 355.
72. R. C. Zhang, S. M. Hong and C. M. Koo, *J. Appl. Polym. Sci.*, 2014, **131**, 40648.
73. F. Lu, S. Zhang and L. Zheng, *J. Mol. Liq.*, 2012, **173**, 42.
74. S. Li, P. Xing, Y. Hou, J. Yang, X. Yang, B. Wang and A. Hao, *J. Mol. Liq.*, 2013, **188**, 74.
75. S. M. Tawfik, *J. Mol. Liq.*, 2015, **209**, 320.
76. S. M. Tawfik, *J. Mol. Liq.*, 2016, **216**, 624.

77. I. Pacheco-Fernández, V. Pino, J. H. Ayala and A. M. Afonso, *J. Chromatogr. A*, 2017, DOI: 10.1016/j.chroma.2017.04.061.
78. R. Bouchal, A. Hamel, P. Hesemann, M. In, B. Prelot and J. Zajac, *Int. J. Mol. Sci.*, 2016, **17**, 223.
79. Y. Zhao, X. Yue, X. Wang, D. Huang and X. Chen, *Colloids Surf., A*, 2012, **412**, 90.
80. C. Chiappe and C. S. Pomelli, *Eur. J. Org. Chem.*, 2014, 6120.
81. K. S. Rao, P. S. Gehlot, T. J. Trivedi and A. Kumar, *J. Colloid Interface Sci.*, 2014, **428**, 267.
82. J. Jiao, B. Dong, H. Zhang, Y. Zhao, X. Wang, R. Wang and L. Yu, *J. Phys. Chem. B*, 2012, **116**, 958.
83. B. Clare, A. Sirwardana and D. R. MacFarlane, *Top. Curr. Chem.*, 2009, **290**, 1.
84. J. L. Anderson, R. Ding, A. Ellern and D. W. Armstrong, *J. Am. Chem. Soc.*, 2005, **127**, 593.
85. Y. Wei, F. Wang, Z. Zhang, C. Ren and Y. Lin, *J. Chem. Eng. Data*, 2014, **59**, 1120.
86. J. Saien and S. Asadabadi, *Fluid Phase Equilib.*, 2015, **386**, 134.
87. Y. Hu, J. Han, L. Ge and R. Guo, *Langmuir*, 2015, **31**, 12618.
88. S. Livi, J. Duchet-Rumeau, T. N. Pham and J. F. Gérard, *J. Colloid Interface Sci.*, 2011, **354**, 555.
89. X. Liu, J. Hu, Y. Huang and Y. Fang, *J. Surfactants Deterg.*, 2013, **16**, 539.
90. J. Liu, Q. Zhang, Y. Huo, M. Zhao, D. Sun, X. Wei, S. Liu and L. Zheng, *Colloid Polym. Sci.*, 2012, **290**, 1721.
91. N. V. Sastry, N. M. Vaghela and V. K. Aswal, *Fluid Phase Equilib.*, 2012, **327**, 22.
92. K. Thakkar, B. Bharatiya, D. O. Shah, D. Ray, V. K. Aswal and P. Bahadur, *Colloids Surf., A*, 2015, **484**, 547.
93. J. Jiao, B. Han, M. Lin, N. Cheng, L. Yua and M. Liu, *J. Colloid Interface Sci.*, 2013, **412**, 24.
94. N. Cheng, X. Ma, X. Sheng, T. Wang, R. Wang, J. Jiao and L. Yu, *Colloids Surf., A*, 2014, **453**, 53.
95. K. S. Rao, P. S. Gehlot, H. Gupta, M. Drechsler and A. Kumar, *J. Phys. Chem. B*, 2015, **119**, 4263.
96. S. Livi, G. Sar, V. Bugatti, E. Espuche and J. Duchet-Rumeau, *RSC Adv.*, 2014, **4**, 26452.
97. X. F. Liu, L. L. Dong and Y. Fang, *J. Surfactants Deterg.*, 2011, **14**, 497.
98. X. Wang, J. Liu, L. Yua, J. Jiao, R. Wang and L. Sun, *J. Colloid Interface Sci.*, 2013, **391**, 103.
99. X. Wang, R. Wang, Y. Zheng, L. Sun, L. Yu, J. Jiao and R. Wang, *J. Phys. Chem. B*, 2013, **117**, 1886.
100. C. P. Frizzo, I. M. Gindri, C. R. Bender, A. Z. Tier, M. A. Villetti, D. C. Rodrigues, G. Machado and M. A. P. Martins, *Colloids Surf., A*, 2015, **468**, 285.

101. M. Ao, G. Xu, Y. Zhu and Y. Bai, *J. Colloid Interface Sci.*, 2008, **326**, 490–495.
102. Y. Liu, L. Yu, S. Zhang, J. Yuan, L. Shi and L. Zheng, *Colloids Surf., A*, 2010, **359**, 66.
103. S. Zhang, F. Lu and L. Zheng, *Colloid Polym. Sci.*, 2011, **289**, 1815.
104. H. Li, C. Yu, R. Chen, J. Li and J. Li, *Colloids Surf., A*, 2012, **395**, 116.
105. N. N. Al-Mohammed, R. S. D. Hussen, Y. Alias and Z. Abdullah, *RSC Adv.*, 2015, **5**, 2869.
106. M. Amde, J. F. Liu and L. Pang, *Environ. Sci. Technol.*, 2015, **49**, 12611.
107. T. P. T. Pham, C. W. Cho and Y. S. Yun, *Water Res.*, 2010, **44**, 352.
108. C. Samor, D. Malferrari, P. Valbonesi, A. Montecavalli, F. Moretti, P. Galletti, G. Sartor, E. Tagliavini, E. Fabbri and A. Pasteris, *Ecotoxicol. Environ. Saf.*, 2010, **73**, 1456.
109. R. F. M. Frade, A. A. Rosatella, C. S. Marques, L. C. Branco, P. S. Kulkarni, N. M. M. Mateus, C. A. M. Afonso and C. M. M. Duarte, *Green Chem.*, 2009, **11**, 1660.
110. J. H. Porada, D. Zauser, B. Feucht and C. Stubenrauch, *Soft Matter*, 2016, **12**, 6352.
111. V. G. Rao, C. Banerjee, S. Ghosh, S. Mandal, J. Kuchlyan and N. Sarkar, *J. Phys. Chem. B*, 2013, **117**, 7472.
112. W. Ding, K. Liu, X. Liu, H. Luan, C. Lv, T. Yu and G. Qu, *J. Appl. Polym. Sci.*, 2013, **129**, 2057.
113. X. Wang, J. Liu, L. Sun, L. Yu, J. Jiao and R. Wang, *J. Phys. Chem. B*, 2012, **116**, 12479.
114. A. M. Atta, H. A. Al-Lohedan, M. M. S. Abdullah and S. M. ElSaeed, *J. Ind. Eng. Chem.*, 2016, **33**, 122.
115. N. I. Malek, Z. S. Vaid, U. U. More and O. A. El Seoud, *Colloid Polym. Sci.*, 2015, **293**, 3213.
116. X. Fan and K. Zhao, *Soft Matter*, 2014, **10**, 3259.
117. B. Aghabarari, N. Dorostkar and M. V. Martinez-Huerta, *Fuel Process. Technol.*, 2014, **118**, 296.
118. I. J. B. Lin and C. S. Vasam, *J. Organomet. Chem.*, 2005, **690**, 3498.
119. F. Heidarizadeh and A. Majdi-nasab, *Tetrahedron Lett.*, 2015, **56**, 6360.
120. W. Zhu, G. Zhu, H. Li, Y. Chao, M. Zhang, D. Du, Q. Wang and Z. Zhao, *Fuel Process. Technol.*, 2013, **106**, 70.
121. Y. Liu and S. A. Shamsi, *J. Chromatogr. A*, 2014, **1360**, 296.
122. T. J. Trivedi, P. Bharmoria, T. Singh and A. Kumar, *J. Chem. Eng. Data*, 2012, **57**, 317.
123. J. Aguiar, P. Carpena, J. A. Molina-Bolívar and C. C. Ruiz, *J. Colloid Interface Sci.*, 2003, **258**, 116.
124. T. Nazemi and R. Sadeghi, *Colloids Surf., A*, 2014, **462**, 271.
125. V. Pino, C. Yao and J. L. Anderson, *J. Colloid Interface Sci.*, 2009, **333**, 548.
126. S. M. Rajput, U. U. More, Z. S. Vaid, K. D. Prajapati and N. I. Malek, *Colloids Surf., A*, 2016, **507**, 182.
127. G. Luo, X. Qi, C. Han, C. Liu and J. Gui, *J. Surfactants Deterg.*, 2013, **16**, 531.

128. H. Wang, L. Zhang, J. Wang, Z. Li and S. Zhang, *Chem. Commun.*, 2013, **49**, 5222.
129. R. Sharma, S. Mahajan and R. K. Mahajan, *Colloids Surf., A*, 2013, **427**, 62.
130. S. Mahajan, R. Sharma and R. K. Mahajan, *Langmuir*, 2012, **28**, 17238.
131. T. Singh, K. S. Rao and A. Kumar, *J. Phys. Chem. B*, 2012, **116**, 1612.
132. I. Rodríguez-Escontrela, I. Rodríguez-Palmeiro, O. Rodríguez, A. Arce and A. Soto, *Fluid Phase Equilib.*, 2016, **417**, 87.
133. Z. Miskolczy, K. Sebők-Nagy, L. Biczók and S. Göktürk, *Chem. Phys. Lett.*, 2004, **400**, 296.
134. S. Mahajan, R. Sharma and R. K. Mahajan, *Colloids Surf., A*, 2013, **424**, 96.
135. M. Ao, P. Huang, G. Xu, X. Yang and Y. Wang, *Colloid Polym. Sci.*, 2009, **287**, 395.
136. M. J. Jaycock and G. D. Parfitt, *Chemistry of Interfaces*, John Wiley & Sons, Inc., New York, 1981.
137. M. J. Rosen, *Surfactants and Interfacial Phenomena*, Wiley-Interscience, New York, 1989.
138. N. J. Turro and A. Yekta, *J. Am. Chem. Soc.*, 1978, **100**, 5951.
139. Y. Jin, L. Wang, T. Wang, P. Chen, Y. Bi and L. Yu, *J. Mol. Liq.*, 2015, **212**, 23.
140. H. C. Evans, *J. Chem. Soc.*, 1956, 579.
141. H. Hoffmann and W. Ulbricht, *Z. Phys. Chem. N. F.*, 1977, **106**, 167.
142. I. Rodríguez-Palmeiro, I. Rodríguez-Escontrela, O. Rodríguez, A. Arce and A. Soto, *RSC Adv.*, 2015, **5**, 37392.
143. S. Lago, H. Rodríguez, M. K. Khoshkbarchi, A. Soto and A. Arce, *RSC Adv.*, 2012, **2**, 9392.
144. S. Lago, M. Francisco, A. Arce and A. Soto, *Energy Fuels*, 2013, **27**, 5806.
145. S. Wang, Y. Zheng, H. Zhang, Y. Yan, X. Xin and Y. Yang, *Ind. Eng. Chem. Res.*, 2016, **55**, 2790.
146. C. Banerjee, S. Mandal, S. Ghosh, J. Kuchlyan, N. Kundu and N. Sarkar, *J. Phys. Chem. B*, 2013, **117**, 3927.
147. C. Banerjee, N. Kundu, S. Ghosh, S. Mandal, J. Kuchlyan and N. Sarkar, *J. Phys. Chem. B*, 2013, **117**, 9508.
148. J. Chen, J. Cao, W. Gao, L. W. Qi and P. Li, *Analyst*, 2013, **138**, 5933.
149. M. Germán-Hernández, V. Pino, J. L. Anderson and A. M. Afonso, *J. Chromatogr. A*, 2012, **1227**, 29.
150. B. Delgado, V. Pino, J. L. Anderson, J. H. Ayala, A. M. Afonso and V. González, *Talanta*, 2012, **99**, 972.
151. F. Huang, P. Berton, C. Lu, N. Siraj, C. Wang, P. K. S. Magut and I. M. Warner, *Electrophoresis*, 2014, **35**, 2463.
152. H. L. Su, M. T. Lan and Y. Z. Hsieh, *J. Chromatogr. A*, 2009, **1216**, 5313.
153. J. P. Quirino, P. Anres, J. Sirieix- Plénet, N. Delaunay and P. Gareil, *J. Chromatogr. A*, 2011, **1218**, 5718.
154. Q. Wang, H. Qiu, H. Han, X. Liu and S. Jiang, *J. Sep. Sci.*, 2012, **35**, 589.

155. Y. V. Nancharaiah, G. K. K. Reddy, P. Lalithamanasa and V. P. Venugopalan, *Biofouling*, 2012, **28**(10), 1141.
156. P. C. A. G. Pinto, D. M. G. P. Ribeiro, A. M. O. Azevedo, V. Dela Justina, E. Cunha, K. Bica, M. Vasiloiu, S. Reisa and M. L. M. F. S. Saraiva, *New J. Chem.*, 2013, **37**, 4095.
157. J. Hu, Q. Yang, L. Yang, Z. Zhang, B. Su, Z. Bao, Q. Ren, H. Xing and S. Dai, *ACS Catal.*, 2015, **5**, 6724.
158. Y. J. Lu, W. L. Wong and C. F. Chow, *Catal. Commun.*, 2015, **69**, 25.
159. K. Raiah, A. Djalab, A. Hadj-Ziane-Zafoura, B. Soula, A. M. Galibert and E. Flahaut, *Colloids Surf., A*, 2015, **469**, 107.

CHAPTER 4

Surfactant Fluorinated Ionic Liquids

A. B. PEREIRO*[a,b], J. M. M. ARAÚJO[a,b], J. M. S. S. ESPERANÇA[a,b] AND L. P. N. REBELO[a,b]

[a]LAQV, REQUIMTE, Departamento de Química, Faculdade de Ciências e Tecnologia, Universidade Nova de Lisboa, 2829-516 Caparica, Portugal; [b]Instituto de Tecnologia Química e Biológica António Xavier, Universidade Nova de Lisboa (ITQB NOVA), Avenida da República, 2780-157 Oeiras, Portugal
*E-mail: anab@fct.unl.pt

4.1 Introduction

Over recent decades, the capacity to design ionic liquids (ILs) for a wide range of applications has driven this field to become one of the most popular areas of research.[1–3] Due to their interesting and sometimes unexpected physicochemical properties, these compounds can represent a greener solution for engineering problems. Despite the large number of studies already available, there are still some almost unexplored themes, for instance the case of the fluorinated ionic liquids (FILs) family. Studies with FILs, defined as ionic liquids with fluorinated alkyl chain lengths equal to or greater than four carbon atoms, are scarce in contrast with some conventional fluorinated anions, such as bis(trifluoromethylsulfonyl)imide, hexafluorophosphate or tetrafluoroborate.[4–13] This chapter details the behaviour of FILs, with long fluorinated chains, in contrast with conventional fluorinated ionic liquids,

Smart Materials No. 28
Ionic Liquid Devices
Edited by Ali Eftekhari

Published by the Royal Society of Chemistry, www.rsc.org

herein denoted simply as fluoro-containing ionic liquids, which have less than four carbons in the fluorous alkyl chain. The carbon–fluorine bond present in FILs, the strongest and most stable, single covalent bond in organic chemistry,[4] deeply influences the molecular structure of FILs by increasing their rigidity and decreasing their polarity.[14] These particular ILs can have fluorinated alkyl chains in either the anion or the cation presenting fluorinated phases that have a quite distinct behaviour from polar/hydrophilic and apolar (hydrogenated or fluorinated)/hydrophobic phases.[15] The presence of these unusual, though useful, fluorinated groups in the cation and/or anion will give the ionic liquids attractive properties not observed in more conventional ionic liquids. To the best of our knowledge, FILs research has focused solely on the synthesis and characterization of these compounds, as well as on their application as materials and as reaction media.[6–13]

By combining the best characteristics of highly fluorinated compounds (exceptional low surface tension, chemical and biological inertness and as unique components and tools for the engineering of stable, self-assembled supramolecular and colloidal systems) with those of many ionic liquids (almost null vapour pressure, high thermal stability, tuneable solvent quality, non-flammability and greener potential), FILs have opened new windows of opportunity in a wide variety of areas where fluorocarbon compounds are used. The industrial production of fluorocarbon compounds has greatly increased in recent decades as a result of widespread applications as fire retardants, surfactants, polymers, lubricants, oxygen therapeutics and insecticides.[16,17] Fluorocarbon compounds might present very distinct behaviour depending on the characteristics of each compound. On the one hand, there are inert compounds, such as perfluorocarbons (PFCs) used in biomedical applications.[18] On the other hand, there are environmental contaminants, such as greenhouse PFCs gases and perfluoroalkyl acids that have been found to be toxic to various species, bioaccumulative and persistent in food chains.[19,20] Moreover, the gas-philic behaviour of these fluorinated compounds in general is well known.[21,22]

Fully or partially fluorinated media are of special interest in fluorous phase organic synthesis, in separation methods, as substitutes for chlorinated solvents, as surfactants in supercritical solvents, in uranium enrichment, and in environmental probes as dielectric solvents.[18] Inert fluorocarbon compounds have been widely used in different biomedical applications, *e.g.* as imaging agents,[23–25] and as fluorocarbon gels, nanocompartmented supramolecular assemblies and colloids.[26,27] In addition, they have been employed in the control and stabilization of emulsions, microbubbles and other colloids[28,29] as well as in the pulmonary delivery of drugs and genes[30] and oxygen therapeutics[18,31,32] where these compounds are used as gas carriers in liquid ventilation and intravenous formulations. However, in most cases, fluorocarbon compounds present a very low solubility in water, which constitutes a handicap in several biological applications. Interestingly, some novel FILs show total miscibility in water and could be used in these biological applications.[33]

This work provides a critical review and analysis of our most important publications related to these novel fluorinated ionic liquids. Guidelines for understanding the complex behaviour of these neoteric biomaterials are also provided. Finally, important applications of these compounds are summarized. The FILs studied in this work are detailed in Table 4.1 together with their structures and abbreviations.

4.2 Nanosegregation in Fluorinated Ionic Liquids

Ionic liquids are currently regarded as nanosegregated fluids, with a polar network permeated by non-polar domains. It has been realized that, for longer alkyl chain lengths, ILs show a structured liquid phase with the formation of apolar and polar nanodomains corresponding to the apolar part and the charged, polar part of the ionic liquid.[34–37] This structural characteristic

Table 4.1 Chemical structures and acronyms of the fluorinated ionic liquids.

Cation designation	Chemical structure	Abbreviation
1-Alkyl-3-methylimidazolium $n = 2, 4, 6, 8.$ 10 and 12	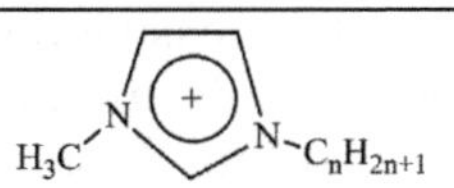	$[C_nC_1Im]^+$
1-Alkyl-1-methylpyrrolidinium $n = 2$ and 4	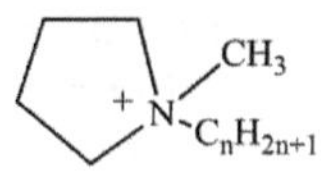	$[C_nC_1pyr]^+$
1-Ethyl-3-methylpyridinium	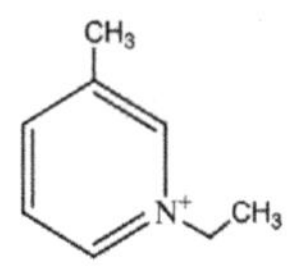	$[C_2C_1py]^+$
(2-Hydroxyethyl) trimethylammonium	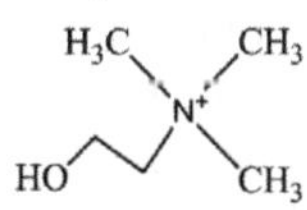	$[N_{1112OH}]^+$
Tetrabutylammonium	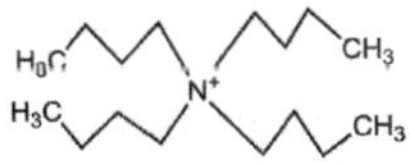	$[N_{4444}]^+$
Anion designation	**Chemical structure**	**Abbreviation**
Perfluoromethanesulfonate	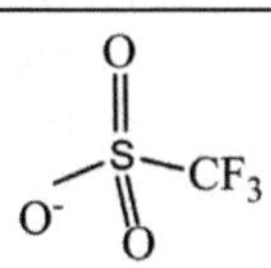	$[CF_3SO_3]^-$
Perfluorobutanesulfonate	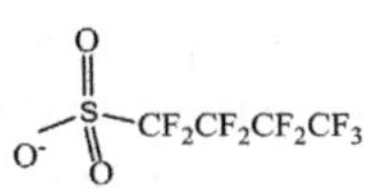	$[C_4F_9SO_3]^-$
Perfluorooctanesulfonate	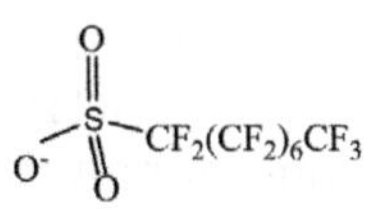	$[C_8F_{17}SO_3]^-$
Bis(nonafluorobutylsulfonyl) imide	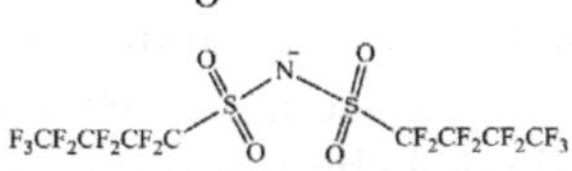	$[N(C_4F_9SO_2)_2]^-$

of ionic liquids was first reported using all-atom simulations[37] and afterwards confirmed by experimental techniques. The first experimental evidence of nanoscale heterogeneities in neat liquid and supercooled RTILs was attained by Triolo and co-workers[34] using X-ray diffraction. In more recent years, several research groups confirmed the nanostructured nature of many ILs[34,36,38–41] which increased insight into the peculiar solvent behaviour of ILs at a molecular level.[42] In addition, the formation of distinct hydrophobic and hydrophilic solvation fields has already been observed in binary mixtures of small chain fluorinated alcohols (up to three carbons) and water, as a consequence of the strong formation of fluorinated alcohol clusters and water clusters.[43]

The incorporation of fluorinated chains (with at least four carbons) in ILs, can induce the formation of one polar and two apolar, fluorinated and hydrogenated, nanosegregated domains. The new nanosegregated fluorinated domain of fluorinated ionic liquids can modify their mesomorphic properties and molecular geometry,[4,44–47] easing the accommodation of fluorinated solutes (gases or liquids) and increasing the tuneability power as compared to that of conventional ionic liquids. In a previous work,[47] we have characterized two FILs based on the perfluorobutanesulfonate anion, demonstrating for the first time the formation of a third nanostructured domain in aprotic ionic liquids. In addition, it was shown that these three mesoscopic regions may enable the formation of multiple types of crystals. The solid–fluid transitions were thoroughly studied using scanning calorimetry, rheology and molecular modelling techniques in the referred work,[47] with the goal of understanding their dynamics and the thermodynamics and developing potential applications. The results point out that the three types of mesoscopic domains – polar, non-polar and fluorinated nanosegregated domains – can act as interchangeable jigsaw pieces enabling the formation of multiple types of crystals (Figure 4.1). The perfluorinated and aliphatic chains tend to group separately and could be the latent reason for the reported rich phase scenario. Additionally, the hydrogenated and fluorinated alkyl chain length play a role in the nanosegregation and in the final characteristics of FILs.[48]

The formation of these three types of nanosegregated domains (polar, non-polar and fluorinated), also corroborated by the work of distinct research groups, has a profound impact on ionic liquid research. Shen *et al.*[4] studied the self-assembly nanostructure of protic ILs with perfluorinated anions, using small- and wide-angle X-ray scattering showing the segregation of the polar, hydrocarbon and fluorocarbon domains in the liquid state. These authors[4] established a structure–property relationship between the chemical structure and their liquid nanostructure, demonstrating that the ions of the protic FILs can be tailored to achieve specific intermediate range order length scales, modified to increase or decrease the intensity of the ordering. Russina and co-workers[46] employed RTILs with medium length fluorinated tails to verify the phenomenon of mutual self-assembly into domains containing polar, hydrophobic and fluorophilic moieties at the mesoscopic level, using NMR and SWAXS measurements. Hollóczki *et al.*[45] studied the effect of inserting fluorinated alkyl chains on the cations of aprotic FIL-mixtures,

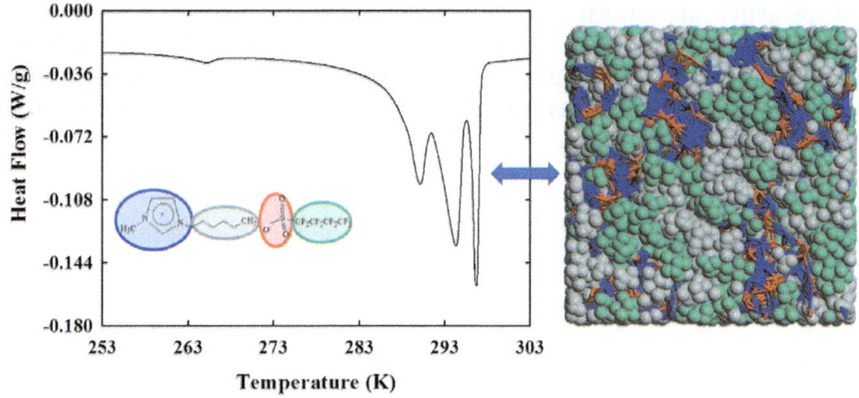

Figure 4.1 Differential scanning calorimeter curve (at a scan rate of 1 K min^{-1}) and molecular dynamic simulation snapshots for $[C_6C_1Im][C_4F_9SO_3]$ FIL. Red (negative charges) and blue (positive charges) areas represent the interactions from the polar network of the two ions, grey space-filled areas represent the domains caused by the hydrogenated side chains while the groups of green spheres illustrate the perfluoroalkyl apolar domains. They are also represented in the structure of ionic liquid with the corresponding colour.

detecting the formation of three stable microphases (polar, lipophilic and fluorous). The aggregation behaviour of the fluorinated chains, producing a fluorous microphase in the liquid state, was detected by domain analysis. Likewise, evidence of microsegregation of fluoro chains of ionic liquids anions at a molecular level was reported by Castiglione *et al.*,[44] using intermolecular anion–cation nuclear Overhauser enhancements (NOEs).

4.3 Influence of Nanosegregation on the Surface, Phase and Aggregation Behaviour

The strong C–F bond dictates the molecular structure of fluorinated ionic liquids, causing the formation of a rigid structure and decreasing its polarity[49] which has an impact on their melting point, thermal stability, liquidus range, density, viscosity, conductivity, surface tension, solubility and hydrophobicity.[33,48,50–55] The characterization of surface properties, phase and aggregation behaviour coupled with the properties–chemical structures relationship is de rigueur for an attentive selection of the adequate fluorinated ionic liquid for a particular application.

4.3.1 Surface Properties

One of the most important properties for the chemical industry is the surface tension, which is imperative in mass-transfer operations, such as distillation, absorption, extraction and adsorption. Despite coherent and accurate measurements of ionic liquids, surface tension data are difficult to obtain due to the presence of impurities.[56–64] The surface tension of ILs show general trends whereby this property decreases with increasing alkyl chain length up to an aliphatic alkyl chain no longer than hexyl, at which point this surface property attains an almost constant value.[65] Even so, some of us[54] have recently demonstrated that the nanosegregation has an important role in the surface tension of FILs. In this work, fluorinated ionic liquids composed of 1-alkyl-3-methylimidazolium cations ($[C_nC_1Im]$ with n = 2, 4, 6, 8, 10 or 12) combined with the perfluorobutanesulfonate anion showed the lowest surface tension values reported to date for ionic liquids. The studied FILs also present low surface entropies and surface enthalpies, lower than those of non-fluorinated ILs. All these observations denote a very high surface organization as well as a markedly structured liquid phase. However, the most noteworthy innovation of the referred study[54] was the identification, for the first time, of a minimum at n = 8 in the surface tension *versus* the length of the IL cation alkyl side chain (Figure 4.2).

Surface tension depends on the liquid structure and orientation of the molecules and is a measure of the cohesion of liquids. For most conventional ILs, it is well established that both the cation and the anion occupy the same plane

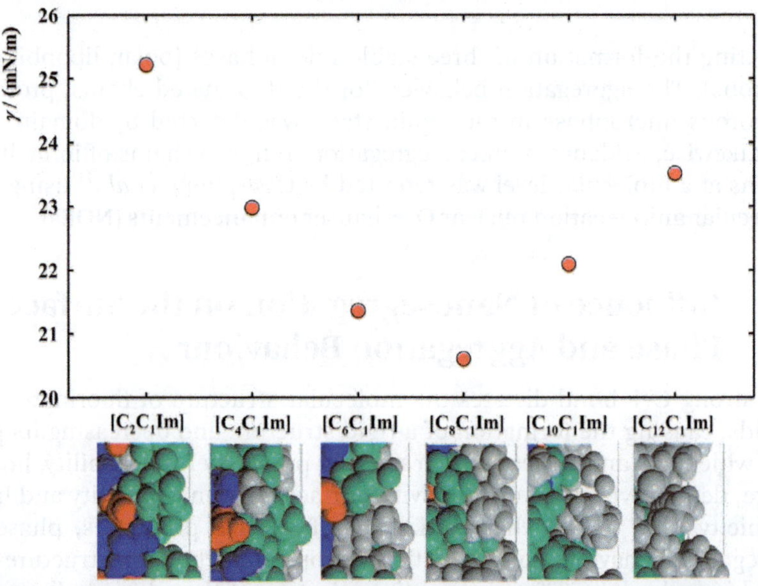

Figure 4.2 Surface tension values as a function of the imidazolium alkyl chain size at 313 K and simulation snapshots of the liquid–vacuum surface of six ionic liquid slabs for the $[C_nC_1Im][C_4F_9SO_3]$ family. In the simulation snapshots, green atoms: apolar moiety ($-CF_2-CF_2-CF_3$) of the perfluorobutyl chains; blue atoms: imidazolium rings and adjoining atoms of the cations (CH_3–Im–CH_2–); red atoms: sulfonyl groups and adjoining atoms (SO_3CF_2–); grey atoms: apolar moiety ($-(CH_2)_{n-2}-CH_3$) of the alkyl side chains.

at the liquid–gas interface, typically the cations' and/or the anions' aliphatic moieties projecting towards the interface forming a top layer, while the cations' polar head groups and those of the anions form the second layer.[66] The FILs herein studied, which are more complex than conventional ILs, feature the ability to form three nanosegregated domains together with the fluorinated chains presenting lower energy than the aliphatic and surface tensions values in the same range as for perfluorocarbons. This fact allows one to envision both the anion and the cation occupying the same plane at the interface (gas–liquid); however, with the perfluorinated alkyl chains, instead of the aliphatic, protruding towards the interface or the gas phase. As the aliphatic imidazolium chain size increases, the surface arrangement changes; above an octyl cation alkyl chain size the aliphatic moieties present lower energy than the perfluorobutyl chain and both the aliphatic and perfluorinated chains protrude towards the gas phase. A further increase of the aliphatic chains length, although both structure arrangements may occur at the same time, prompts more aliphatic and less perfluorinated chains to position at the interface, leading to an increment of the surface tension towards non-FILs values. The competition between fluorinated and aliphatic alkyl chains towards the interface (gas–liquid) is responsible for the unexpected behaviour found in these fluorinated ionic liquids (Figure 4.2).[54] This behaviour was also demonstrated by molecular dynamic simulations results, together with estimations based on the Langmuir principle, which reveal the structure and composition of the liquid–gas interface.

4.3.2 Phase Behaviour

The development of novel biomaterials based on fluorinated ionic liquids, ILs with improved tuneable properties, that could replace partially or totally the perfluorocarbons (PFCs) used nowadays in biomedical applications requires knowledge about their phase behaviour. This goal prompted some of the present authors[51,52] to determine the phase equilibria (liquid–liquid equilibria and solid–liquid equilibria) of FILs containing anionic fluorinated chains equal to or longer than four carbons (between four and eight) with PFCs and water over a wide range of temperatures. Two typical perfluorocarbons used in biomedical applications, perfluorooctane and perfluorodecalin, were selected.[67–69] Furthermore, in an attempt to map several ionic liquid families, FILs based on the pyridinium, ammonium, imidazolium and pyrrolidinium cation were included.[33,51,52]

The largest solubility of the studied PFCs in the FIL rich-phase was achieved with the FIL based on ammonium cation, presenting solubility values in molar fraction of ≈0.68 for perfluorodecalin and of ≈0.71 for perfluorooctane (Figure 4.3).[51] Additionally, the results show that the solubility of each perfluorocarbon in the rich phase of FIL rises proportionately to the following classification: pyridinium < imidazolium < tetraalkylammonium.

In addition, the solubility of water in the FIL-rich phase rises with the increase of hydrogenated alkyl chain in the imidazolium cation for fluorinated ILs based on perfluorobutanesulfonate anion (Figure 4.4). This trend can be related to the increment of surfactant power of the studied FILs when the length of hydrogenated alkyl chain on the imidazolium cation is increased (see Section 4.3.3).

Moreover, perfluorosulfonate-functionalized ILs with short alkyl chains in the cation ($[C_2C_1Im][C_8F_{17}SO_3]$, $[C_2C_1Im][C_4F_9SO_3]$, $[C_2C_1py][C_4F_9SO_3]$, $[N_{1112OH}]$ $[C_4F_9SO_3]$) surprisingly show a total miscibility in water. This fact could allow their use in biological applications, where PFCs present a handicap because

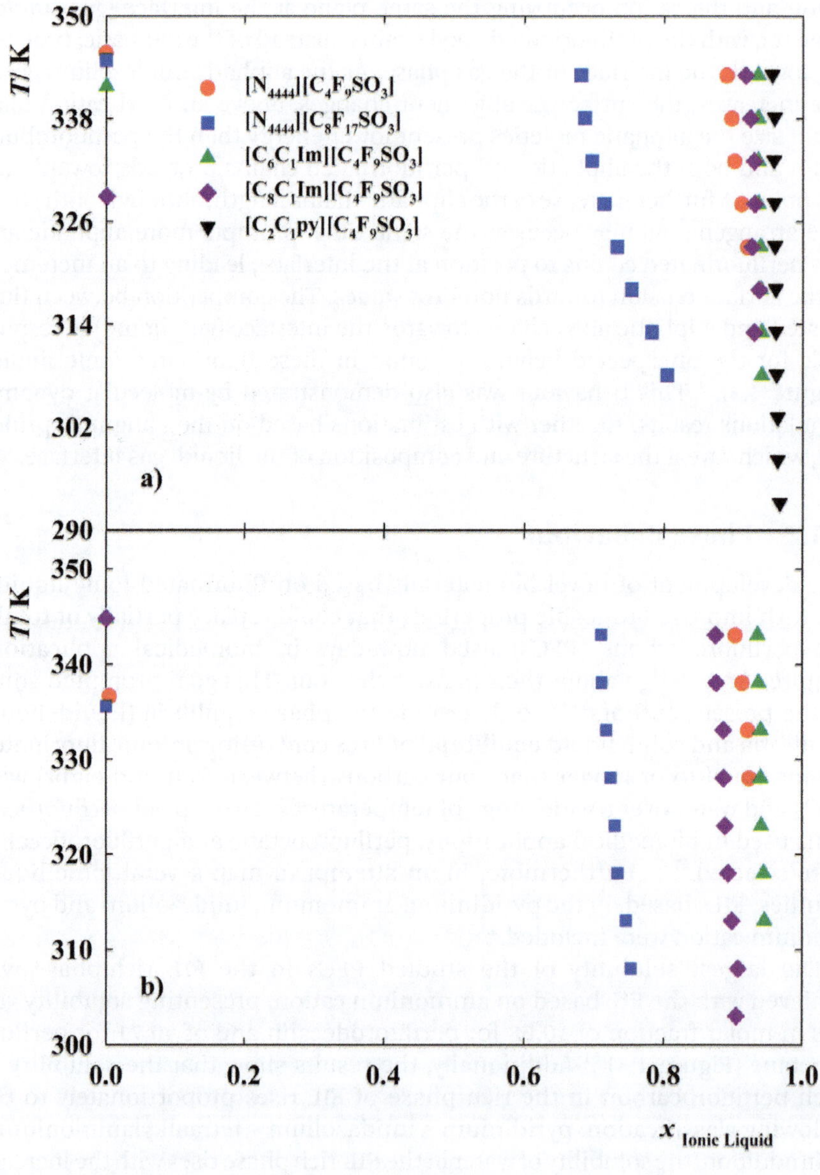

Figure 4.3 Liquid–liquid phase diagrams for binary systems: (a) perfluorodecalin + fluorinated ionic liquids; and (b) perfluorooctane + fluorinated ionic liquids.

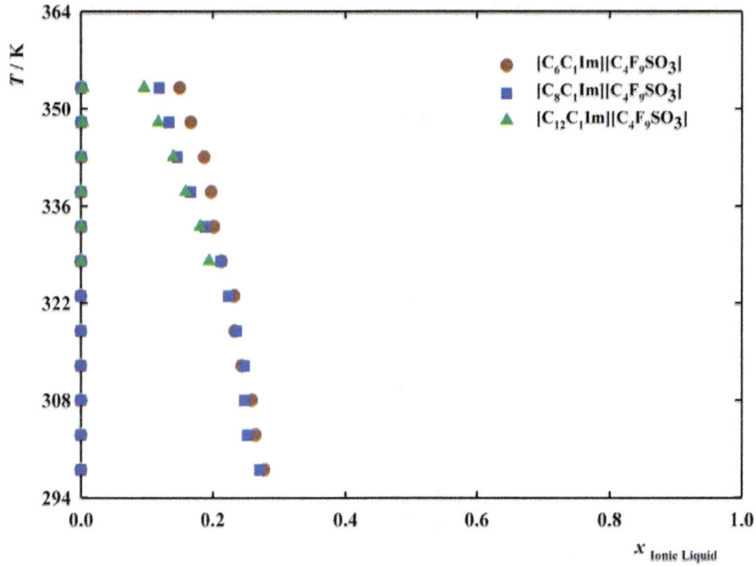

Figure 4.4 Liquid–liquid phase diagrams for binary mixtures of water + fluorinated ionic liquids.

their aqueous solubility (water and biological fluids) is in most cases too low.[33] This complete water miscibility may be related to the stable self-assembled structures of these FILs in aqueous medium (see Section 4.3.3).[33]

Likewise, the solubility of solid FILs in water was also studied at 298.15 K (Figure 4.5) and their behaviour was largely dependent on the cation nature.[52] The results show that the FIL presenting the lowest solubility is based on ammonium cation, $[N_{4444}][C_4F_9SO_3]$. Even so, this FIL shows high surfactant behaviour, and a low critical micelle concentration value (see Section 4.3.3). Finally, the asymmetry of $[C_4C_1pyr][C_4F_9SO_3]$ could explain the increase in solubility in water.

4.3.3 Aggregation Behaviour

The FILs studied herein involve cations and/or anions with high tensioactive properties in aqueous solution.[33,52] Thence, the critical micelle concentrations (CMCs) of these FILs must be determined to establish how their structure influences their use, toxicity and distribution. The CMCs at 298.15 K were determined, followed by evaluation of the self-assembled structures of FILs in aqueous solution.[33,52] Both analyses are essential, allowing an accurate assessment of the self-aggregation behaviour of these novel surfactants based on fluorinated ionic liquids in aqueous solutions.

The self-aggregation behaviour of these functionalized FIL surfactants in aqueous solution indicates that the increase of the hydrogenated alkyl chain length in the imidazolium cation decreases the CMC value (see Figure 4.6a, where the lowest CMC values correspond to the highest surfactant power). In addition, the FIL based on the ammonium cation, $[N_{4444}][C_4F_9SO_3]$, showed

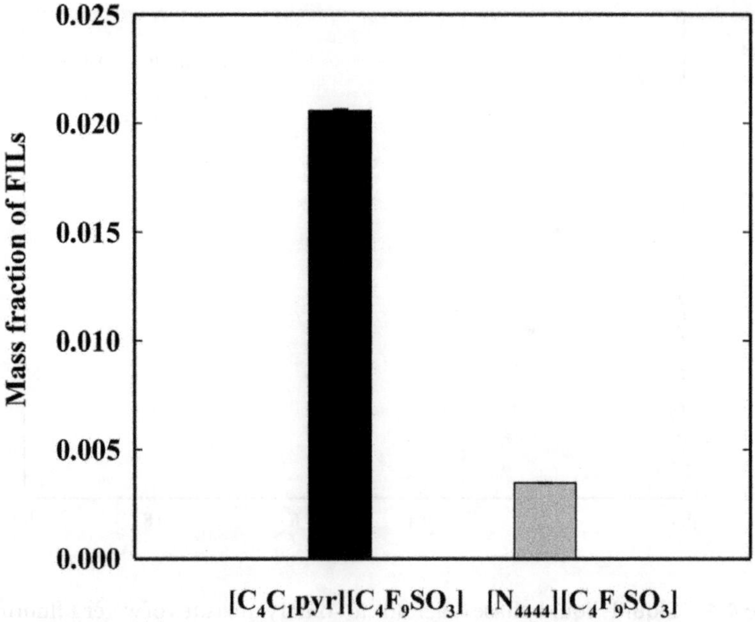

Figure 4.5 Solubilities of solid fluorinated ionic liquids ($[C_4C_1pyr][C_4F_9SO_3]$ and $[N_{4444}][C_4F_9SO_3]$) in water at 298.15 K.

the highest surfactant behaviour when compared with FILs with the same anion and cations with similar hydrogenated alkyl chain length, $[C_6C_1Im][C_4F_9SO_3]$, $[N_{4444}][C_4F_9SO_3]$ and $[C_4C_1pyr][C_4F_9SO_3]$ (Figure 4.6b).

One of the most relevant results is that FILs with only four carbon atoms, in the perfluorinated alkyl chain, improve the surfactant power of conventional perfluoro-surfactants with eight carbon atoms and the hydrogenated counterparts (Figure 4.7).[70–72] Another relevant result is the total miscibility and self-aggregation behaviour in water of these novel perfluorinated anionic surfactant ILs with short hydrogenated alkyl chains in the cation ($[C_2C_1Im][C_4F_9SO_3]$, $[C_2C_1py][C_4F_9SO_3]$, $[N_{1112OH}][C_4F_9SO_3]$).[33] In this specific case, the FIL based on ammonium cation presents lower surfactant power than the pyridinium- or imidazolium-based FILs (Figure 4.7).

Furthermore, these perfluoro-anionic amphiphilic ILs form distinct aggregated structures depending on the total FIL concentration in aqueous solution. The critical aggregation concentration (CAC) values obtained using different techniques are in corroboration and permit a rigorous identification of the phase transitions of the different aggregations of these FILs in aqueous solutions.[33] The possible mechanism for the phase transitions complies with a micellar shape change from monomers to spherical micelles (first transition), from spherical micelles to globular micelle (second transition), and from globular micelles to cylindrical or lamellar micelles (third transition) (Figure 4.8). The reported stable self-assembled structures in aqueous medium may explain the complete water miscibility of these novel fluorinated ionic liquids surfactants.

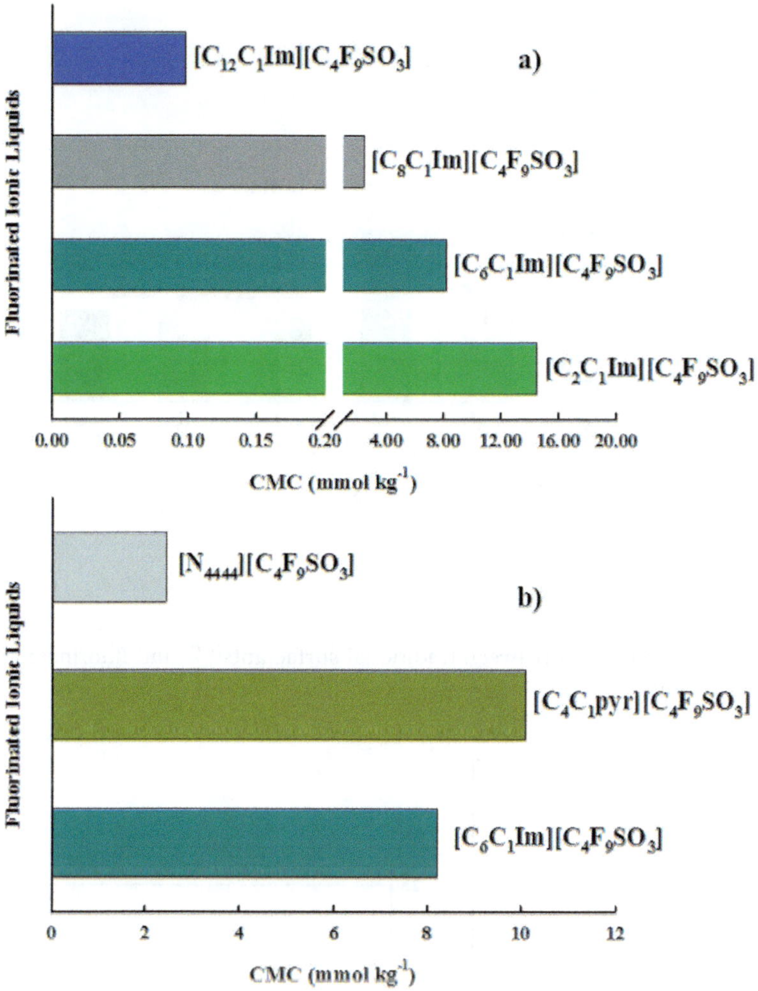

Figure 4.6 Comparison of critical micelle concentration values of the fluorinated ionic liquids at 298.15 K. a) Study of the increment of the hydrogenated alkyl chain length; b) comparison between different cations.

4.3.4 Binary Mixtures of Fluorinated Ionic Liquids

The melting temperature of fluorinated ionic liquids is commonly higher due to the presence of fluorine atoms. Consequently, FILs liquidus range decreases with the increasing incorporation of fluorine atoms, *e.g.* for ILs based on bis(perfluoroalkylsulfonyl)imide, perfluoroalkanoate or perfluoroalkylsulfonate anions (Table 4.2).[48,55,73] The preparation of eutectic mixtures, by mixing different FILs, may overcome this handicap. Additionally, a boost to the tuneability of neat FILs can be attained. These binary mixtures form eutectic mixtures promoting a depression in their melting points as compared to the neat FILs. Thus, this fact permits us to obtain a wider range of solvents that could be liquid at temperatures close or below room temperature.

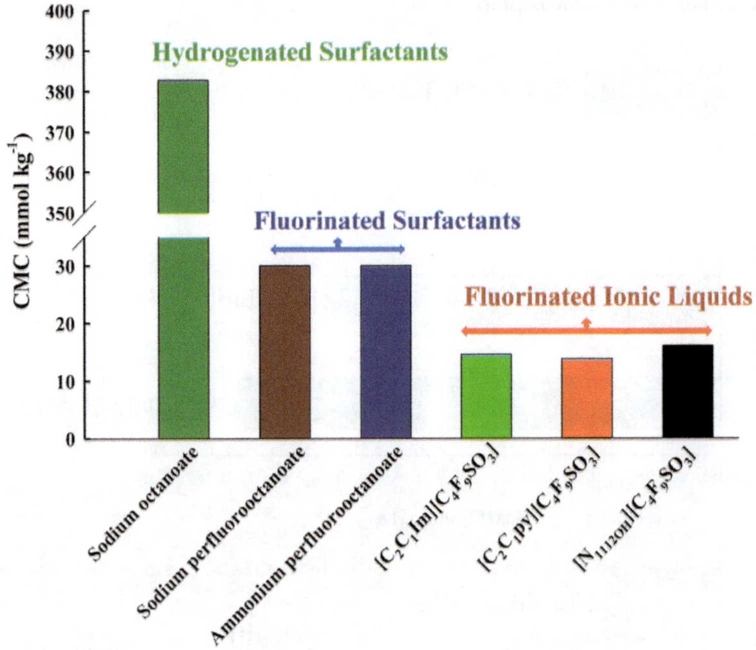

Figure 4.7 Comparison between traditional surfactants[70–72] and fluorinated ionic liquids.

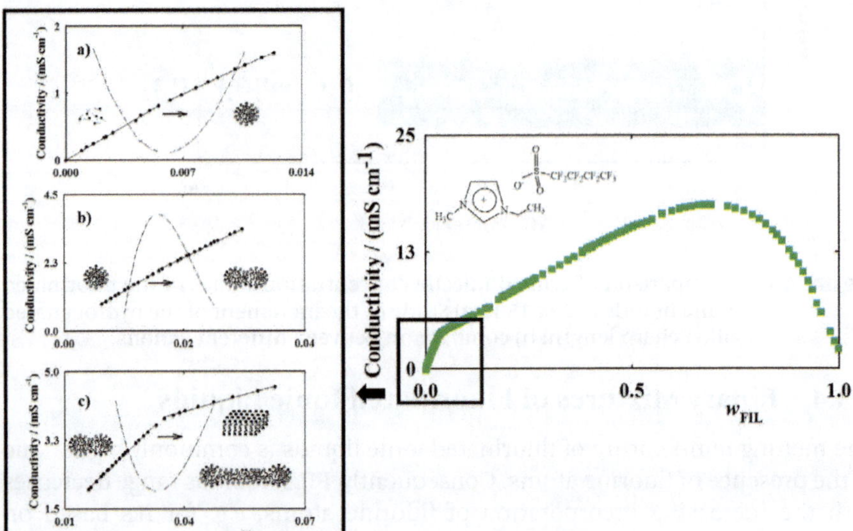

Figure 4.8 Conductivity profile of $[C_2C_1Im][C_4F_9SO_3]$ in aqueous solution at 298.15 K: (a) first critical aggregation concentration, (b) second critical aggregation concentration, and (c) third critical aggregation concentration. Dashed lines show maximum change in the gradient of ionic conductivity as a function of FIL concentration, second derivative.

Table 4.2 Melting point, T_m, of fluorinated ionic liquids and fluoro-containing ionic liquids.

Name/abbreviation	Chemical structure	Ref.	T_m (K)
1-Ethyl-3-methylimidazolium perfluoromethanesulfonate $[C_2C_1Im][CF_3SO_3]$	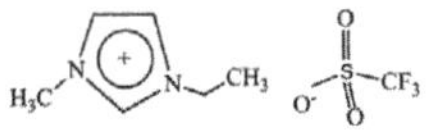	48	260
1-Ethyl-3-methylimidazolium perfluorobutanesulfonate $[C_2C_1Im][C_4F_9SO_3]$	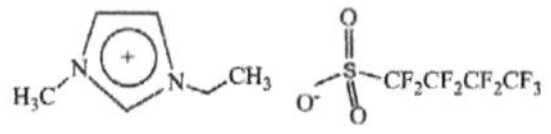	48	293
1-Ethyl-3-methylimidazolium perfluorooctanesulfonate $[C_2C_1Im][C_8F_{17}SO_3]$	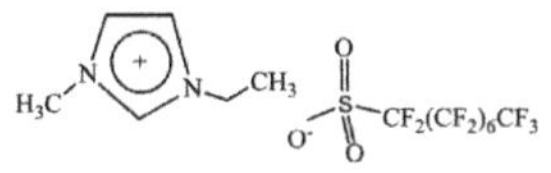	48	368
1-Ethyl-3-methylimidazolium perfluoroethanoate $[C_2C_1Im][CF_3CO_2]$	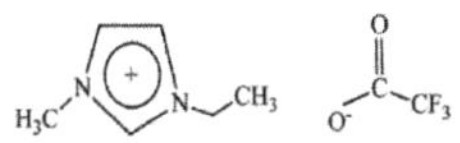	48	255
1-Ethyl-3-methylimidazolium perfluoropentanoate $[C_2C_1Im][C_4F_9CO_2]$	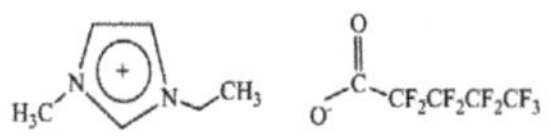	48	278
1-Methyl-3-octylimidazolium perfluoroethanoate $[C_8C_1Im][CF_3CO_2]$	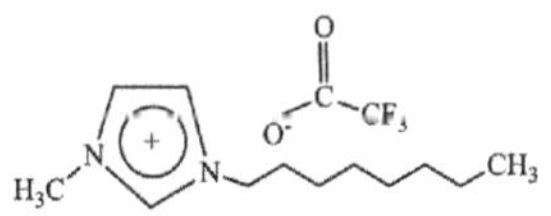	48	279
1-Methyl-3-octylimidazolium perfluoropentanoate $[C_8C_1Im][C_4F_9CO_2]$	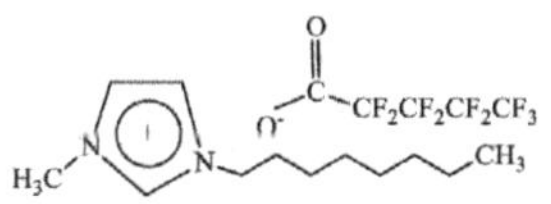	48	297
1-Butyl-1-methylpyrrolidinium bis(fluorosulfonyl)imide $[C_4C_1pyr][N(FSO_2)_2]$	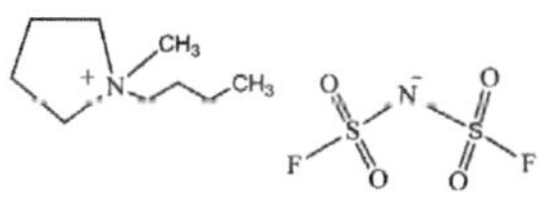	73	255
1-Butyl-1-methylpyrrolidinium bis(trifluoromethylsulfonyl) imide $[C_4C_1pyr][N(CF_3SO_2)_2]$	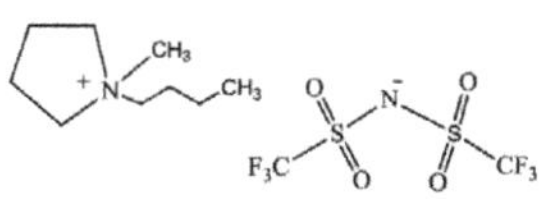	73	266
1-Butyl-1-methylpyrrolidinium bis(pentafluoroethylsulfonyl) imide $[C_4C_1pyr][N(C_2F_5SO_2)_2]$	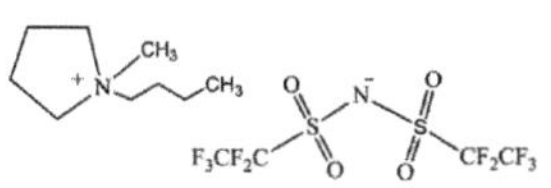	73	280
1-Butyl-1-methylpyrrolidinium bis(nonafluorobutylsulfonyl) imide $[C_4C_1pyr][N(C_4F_9SO_2)_2]$	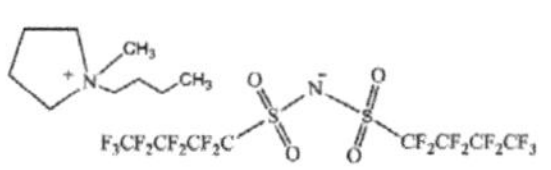	55	371

Hence, some of the present authors,[74] to take advantage of the potential of FILs, studied different binary mixtures of FILs to increase their range of applications by decreasing their melting temperatures by the formation of eutectic mixtures. Binary mixtures of fluoro-containing and fluorinated ILs were used with the aim of decreasing their melting points as well as to understand and characterize these mixtures and their phase transitions.[74] An investigation of four binary mixtures, each of them with a common ion, using five different FILs was carried out. A wide range of solid–liquid equilibria with mixtures exhibiting (i) polymorphs with solid–solid phase transitions and the formation of intermediate compounds; (ii) quasi-ideal eutectic behaviour; and (iii) solid solutions were unusually observed (Figure 4.9). Alongside these interesting behaviours, neoteric fluorinate ILs with lower melting points were developed by the formation of binary mixtures, expanding thus the application range of fluorinated ionic liquids.

Figure 4.9 illustrates the phase diagrams of these binary mixtures, indicating that the systems $[C_2C_1pyr][C_4F_9SO_3]$ + $[C_2C_1pyr][CF_3SO_3]$ and $[C_4C_1pyr][N(C_4F_9SO_2)_2]$ + $[C_4C_1pyr][C_4F_9SO_3]$ present quasi-ideal eutectic behaviour. Nevertheless, a solid solution was identified for the mixture constituted by $[C_4C_1pyr][N(C_4F_9SO_2)_2]$ + $[C_2C_1pyr][N(C_4F_9SO_2)_2]$ because this mixture did not show a eutectic point. Conversely, an intermediate compound was found

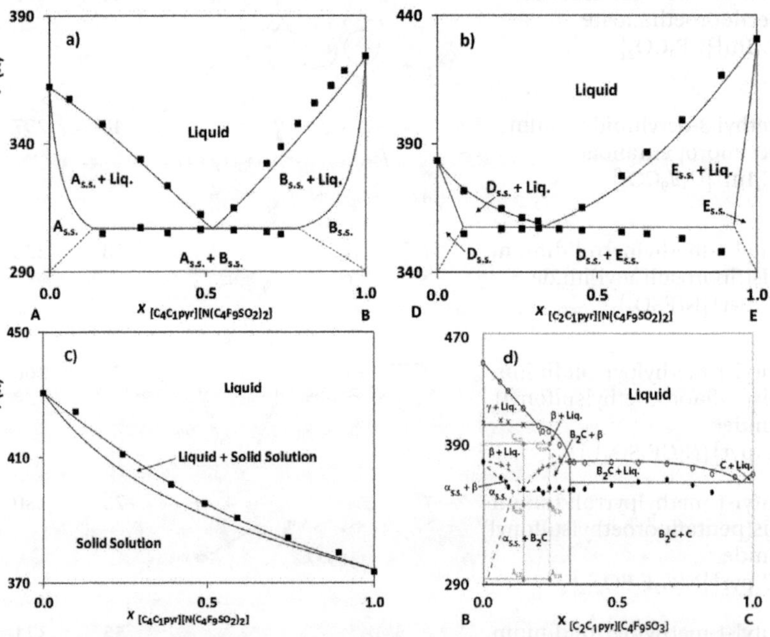

Figure 4.9 Solid–liquid phase diagrams for the binary mixtures: (a) $[C_4C_1pyr][C_4F_9SO_3]$ + $[C_4C_1pyr][N(C_4F_9SO_2)_2]$; (b) $[C_2C_1pyr][N(C_4F_9SO_2)_2]$ + $[C_2C_1pyr][CF_3SO_3]$; (c) $[C_4C_1pyr][N(C_4F_9SO_2)_2]$ + $[C_2C_1pyr][N(C_4F_9SO_2)_2]$; and (d) $[C_4C_1pyr][C_4F_9SO_3]$ + $[N_{1112(OH)}][C_4F_9SO_3]$.

for the mixture $[C_4C_1pyr][C_4F_9SO_3]$ + $[N_{1112(OH)}][C_4F_9SO_3]$ at about $x = 0.33$ of $[C_4C_1pyr][C_4F_9SO_3]$. Further, this last mixture shows a eutectoid point at around $x = 0.15$ of $[C_4C_1pyr][C_4F_9SO_3]$ with several solid–solid transitions.[74] Aside these intriguing behaviours, the range of FILs that can be used at lower temperatures for target applications was broadened. Then, novel fluorinated ionic liquids with lower melting points were developed by mixing high-melting temperature ILs.

4.4 Applications of Fluorinated Ionic Liquids

This work launches the characterization and evaluation of FILs, aiming to develop new and improved fluorinated media to be used in areas where enhancing the solubility in fluorinated media is the keystone. Fully or partially fluorinated ionic liquids are of interest to replace fluorinated organic compounds in areas such as components of pharmaceuticals, fire retardants, insecticides, lubricants, polymers, refrigerants and surfactants.[75] Two main areas can be clearly identified for the application of the new approaches and concepts using FILs: (i) biomedical applications (such as artificial blood substitutes and protein drug delivery systems) and (ii) environmental impact mitigation of fluorinated contaminants (such as greenhouse PFCs gases and perfluoroalkyl acids).

The unique properties of FILs, namely their high capacity for dissolving gases and low surface tension, low-intensity interactions with normal organic compounds[76] and outstanding chemical and biological inertness, extend the already enormous list of potential applications. The molecular structure of FILs leads to high solubility and low energy requirements for regenerating the solvent and expelling the molecules either upon an increase in temperature or/and decrease in pressure.[49]

4.4.1 Artificial Blood Substitutes

The end of the twentieth century was marked by the appearance of different generations of artificial blood substitutes. These oxygen carriers, often referred to as oxygen therapeutics, are nowadays in clinical trials. Their importance is tremendous, especially in biomedical procedures where there is a risk of extensive blood loss like in surgeries. In fact, the delivery and transport of respiratory gases *in vivo* by novel methods has become one of the most challenging research topics and PFC-based emulsions are among the main candidates for a reliable and safe oxygen therapeutics.[17,77] PFCs are inert organic compounds, totally fluorinated, that can dissolve large volumes of respiratory gases.[17,77] For intravascular administration they need to be emulsified due to their immiscibility in water.

The studies performed to date explore the possibility of PFCs replacement (either partially or totally) with selected FILs in oxygen therapeutic emulsions.[33,50,51,78] The results indicate that FILs solubilize higher amounts of respiratory gases (oxygen and carbon dioxide) than PFC emulsions. The potential use of FILs as oxygen therapeutics has been previously proposed in the literature.[79] FILs have also shown a good surfactant behaviour, easing the stabilization of dispersions of perfluorocarbons in a conventional IL and water solutions.[33,80] The use of FILs as surfactants in PFC-based emulsions will enhance stability effect promoting a higher capacity to solubilize respiratory gases than PFC-based emulsions alone. This gas delivery has been proposed based on the high stability of the C–F bonds where the repulsive forces of fluorine atoms lead not only to high gas solubilities but also to their easier regeneration.[81] The use of FILs for improving the available oxygen therapeutics is justified by their designer nature, their negligible vapour pressure at atmospheric conditions, their easy recovery, and, therefore, their recyclability. Nonetheless, physiological safety and health studies among others will have to be carried out prior to any possible application. Cytotoxicity studies of FILs, which are of vital importance for this application, have already been carried out for the first time (see Section 4.4.2).[50] The research carried out constitutes the first steps to establish the use of a new and improved generation of oxygen therapeutics containing fluorinated ionic liquids.[33,50,51,78]

4.4.2 Drug Delivery Systems (DDSs)

The high rates of drug development failure in clinical trials, generally around 90%,[82] the increasing competition from generic drugs and the reduction of healthcare costs are intensifying the economic pressure in the pharmaceutical market. This situation, together with the current societal challenges, namely the search for safer and more efficient drugs, has contributed to the revitalised interest in protein drug candidates. Interest in natural biological products, such as therapeutic proteins, started in the early 1980s and promised to be the solution for all kind of diseases.[83] More than 200 therapeutic proteins have been approved for clinical use in the European Union and the USA since the approval of insulin in 1982.[82] Clinical trials are already underway using therapeutic proteins for the treatment of Alzheimer's, autoimmune diseases, AIDS/HIV, cancer, immune disorders, infectious diseases and Parkinson's, among others.[84–86] However, therapeutic proteins still have some handicaps: (i) their three-dimensional structure, spatial occupation and hydrophilic/hydrophobic nature can hinder *in vivo* delivery; (ii) sensitivity to environmental conditions (*e.g.* threat of degradation in the gastrointestinal tract preclude their oral delivery); (iii) high dosage needed to obtain therapeutic efficacy, with the potential side effects and toxicity; (iv) poor bioavailability due to their short half-lives in the blood stream resulting from their degradation by enzymes such as proteases; and (v) common administration by intravenous injections, which is not always well tolerated by the pacient.[87] Drug delivery systems (DDSs) have been seen as an adequate solution to these problems because they can be time-controlled or present site-specific delivery, reduce the drug dosage needed, minimize adverse side effects and improve the action of a drug through faster onset.[84]

Recently, ILs have been recognized for their biological and pharmaceutical potential. Specifically, their ability to solubilize and stabilize proteins *in vitro* for extended times has been studied.[88,89] Other previous investigations have shown the adequacy of ILs to solubilize poorly soluble active pharmaceutical ingredients as well as biochemical compounds, such as nucleic acid bases.[90,91]

Literature shows that DDSs for therapeutic proteins have been developed using liposomes, microparticles, nanoparticles, fluorocarbons surfactants and microencapsulation.[86,87,92] Furthermore, the general effects on the essential properties of peptides and proteins arising from the incorporation of fluorinated amino acids have been studied by protein engineering researchers.[93] Nevertheless, more than a decade of research in protein DDSs did not produce an efficient system. Studies tackling FILs surfactants as drug delivery systems for therapeutic proteins are not yet available.

Novel surfactants based on FILs might increase the safety and effectiveness of therapeutic proteins, by increasing the pre-intake protection until reaching the target receptor, while decreasing their immunogenicity. Moreover, controlling the balance of forces present, namely van der Waals, coulombic and hydrogen bonding, the size of the fluorinated domains formed,[47] the surfactant behaviour, and the solubility in water and biological fluids[33] will provide the key features to use FILs in biological applications where fluorocarbon compounds in general present a significant handicap (overly low solubility in water and biological fluids). There are already available studies that can be used in FILs selection, where it is shown that some FILs are totally water miscible with improved surfactant behaviour[33] and present low cytotoxicity in two different human cell lines (Caco-2 and HepG2).[50] It has been shown that only the perfluorooctanesulfonate anion presents some cytotoxicity in one of the human cell cultures at high FIL concentrations (Caco-2 cell with log EC_{50} = 3.65, 4467 μM; but HepG2 with log $EC_{50} > 4$, >10 000 μM).[50] However, the corresponding acid, perfluorooctane sulfonic acid (log EC_{50} = 1.48, 30 μM in PC12 cells)[94] belongs to a new class of environmental contaminants, globally distributed and persistent, bioaccumulative and toxic to various species.[19] It is thus important to become aware of the expansion of production and use of these perfluorinated compounds in forthcoming years due to their utility, economic value and industrial application.[19] In fact, the perfluorobutanesulfonate acid has recently been commercialized as an alternative to the perfluorooctanesulfonate acid in a variety of applications (3M, Technical Data Bulletin). This is relevant since the perfluorobutanesulfonate anion, $[C_4F_9SO_3]^-$, shares some of the oleophobic and hydrophobic properties of perfluorooctanesulfonate, $[C_8F_{17}SO_3]^-$, but it is not considered to be as bioaccumulative or toxic.[95] The FILs based on the perfluorobutanesulfonate anion might be the proper choice as neoteric alternative compounds, being more environmentally friendly and presenting tuneable physicochemical properties as compared to other fluorinated compounds. Furthermore, the critical aggregation concentration values of fluorinated surfactants are comparable to those of hydrogenated surfactants with a chain length approximately 1.5 times longer.[33,70,96] On the other hand, octanol–water partition coefficients and cytotoxicity in two other human cell cultures (HaCaT and EA.hy926) are currently under study in our laboratory. Preliminary results show that the FILs presented here are non-toxic in four different human cell culture types (Caco-2, HepG2, HaCaT and EA.hy926) and present small octanol–water partition coefficients when compared with traditional solvents. Nonetheless, the physiological safety, the biodegradability, and several health studies, among others, will have to be carried out to fully evaluate the real application of these compounds.

4.4.3 Separation Processes

The use of fluorinated compounds has risen in recent decades, but some of them are actually recognized as fluorinated contaminants and are particularly harmful for human health and the environment. They accumulate in living organisms, are transported by air, water and migratory species and accumulate in terrestrial and aquatic ecosystems.[19] Consequently, the environmental impact mitigation of fluorinated contaminants is essential.

Fluorinated ionic liquids can be used as alternatives to currently used volatile organic solvents to efficiently separate perfluoroalkyl acid contaminants from industrial effluents through the use of extraction processes. Fluorinated ionic liquids will be used as solvents to design more efficient and environmentally friendly liquid–liquid extraction processes, significantly reducing the hazard of aquatic industrial effluents. FILs can also be used in gas separation processes, as supported or composite membranes, reducing emissions of greenhouse gases that contribute to climate change, a major goal of the Kyoto Protocol.[78] FILs have emerged recently in the recovery of contaminants either in gaseous mixtures or in liquid effluents due to their remarkable performances.[50,97] The carbon dioxide-phylic behaviour of fluorinated compounds in general[21,98] and ILs in particular[99] is well known. Gas solubility data for the most common ILs are already available in large amounts in the open literature.[100,101] The use of conventional ionic liquids with functional groups increased significantly their absorption capacity. Nevertheless, the mechanism of dissolution of gases in ILs is still not fully understood.[101] However, only a few works have been devoted to the FILs,[14,102–106] and thus considerable additional research is still required for understanding the full potential of this IL family. The studies available have been focused only on the solubility of carbon dioxide,[14,102,104–106] nitrogen, oxygen and methane[106] and difluoromethane[103] with little attention given to the effects of fluorination on the behaviour of other gases or gas pair selectivity. One paper has also studied the solubility of perfluoromethane, perfluoroethane and perfluoropropane in a conventional ionic liquid.[107] Some of the present authors studied the gas solubility or permeation data of fluorinated gases in FILs.[78]

4.5 Conclusion

An emerging research field of interest in the ionic liquids area is the increment of their intrinsic tuneability. The fluorinated ionic liquids studied in this work can be used as "3 in 1" solvents, due to their nanosegregated domains, increasing the solubilization power and enlarging the tuneability of traditional ionic liquids. Their physicochemical properties can be fine-tuned through the careful selection of cations and anions leading to the desired physicochemical and toxicological properties for each application. In addition, the FILs herein studied might be used as neoteric alternative compounds, being more environmentally friendly and presenting more tuneable physicochemical properties than other fluorinated compounds. However, a wide collection of rigorous physicochemical characterizations for these novel compounds is required to understand the relationships between the ionic liquid intermolecular forces and each macroscopic property. Therefore, this chapter provides a critical review and analysis of the characteristic and complex behaviour of these neoteric biomaterials so that guidelines for selecting the most suitable FILs for each application can be established. This work allows the development of environmentally benign selective fluorinated systems and clearly demonstrates the benefits of using FILs, which can be enormous to implement truly real processes.

Acknowledgements

This work was supported by the Associate Laboratory for Green Chemistry LAQV which is financed by national funds from FCT/MEC (UID/QUI/50006/2013) and co-financed by the ERDF under the PT2020 Partnership Agreement (POCI-01-0145-FEDER - 007265). The authors would like to thank the financial support from FCT/MCTES (Portugal), through Investigador FCT (A. B. Pereiro, J. M. M. Araújo and J. M. S. S. Esperança), and projects PTDC/QEQ-EPR/5841/2014, PTDC/QEQ-FTT/3289/2014, IF/00210/2014/CP1244/CT0003, PTDC/EQU-FTT/118800/2010 and UID/Multi/04551/2013. Financial support from Marie Curie Actions Intra-European Fellowships (FP7 – PEOPLE – 2009 - IEF - 252355 – HALOGENILS) and a Transbio SUDOE collaborative project (TRANSBIO-BCN-GT2-A1-03) is also gratefully acknowledged. A. B. Pereiro acknowledges the Short Term Scientific Mission Grants (STSM-CM1206-020315-056728 and STSM-CM1206-010316-071781), within the framework of the COST Action EXIL - Exchange on Ionic Liquids, CM1206.

References

1. M. Deetlefs, M. Fanselow and K. R. Seddon, *RSC Adv.*, 2016, **6**, 4280.
2. N. V. Plechkova and K. R. Seddon, *Chem. Soc. Rev.*, 2008, **37**, 123.
3. T. Welton, *Chem. Rev.*, 1999, **99**, 2071.
4. Y. Shen, D. F. Kennedy, T. L. Greaves, A. Weerawardena, R. J. Mulder, N. Kirby, G. H. Song and C. J. Drummond, *Phys. Chem. Chem. Phys.*, 2012, **14**, 7981.

5. T. L. Greaves, D. F. Kennedy, Y. Shen, A. Hawley, G. H. Song and C. J. Drummond, *Phys. Chem. Chem. Phys.*, 2013, **15**, 7592.
6. H. Xue and J. M. Shreeve, *Eur. J. Inorg. Chem.*, 2005, 2573.
7. Y. Tsukada, K. Iwamoto, H. Furutani, Y. Matsushita, Y. Abe, K. Matsumoto, K. Monda, S. Hayase, M. Kawatsura and T. Itoh, *Tetrahedron Lett.*, 2006, **47**, 1801.
8. R. Arvai, F. Toulgoat, M. Medebielle, B. Langlois, F. Alloin, C. Iojoiu and J. Y. J. Sanchez, *Fluorine Chem.*, 2008, **129**, 1029.
9. T. Linder and J. Sundermeyer, *Chem. Commun.*, 2009, 2914.
10. M. Skalicky, M. Rybackova, O. Kysilka, M. Kvicalova, J. Cvacka, J. Cejka and J. Kvicala, *J. Fluorine Chem.*, 2009, **130**, 966.
11. J. J. Tindale, K. L. Mouland and P. J. Ragogna, *J. Mol. Liq.*, 2010, **152**, 14.
12. S. Livi, J. Duchet-Rumeau and J. F. Gerard, *Eur. Polym. J.*, 2011, **47**, 1361.
13. L. V. N. R. Ganapatibhotla, L. Wu, J. P. Zheng, X. L. Jia, D. Roy, J. B. McLaughlin and S. Krishnan, *J. Mater. Chem.*, 2011, **21**, 19275.
14. D. Almantariotis, T. Gefflaut, A. A. H. Pádua, J.-Y. Coxam and M. F. C. Gomes, *J. Phys. Chem. B*, 2010, **114**, 3608.
15. J. J. Tindale, C. Na, M. C. Jennings and P. J. Ragogna, *Can. J. Chem.*, 2007, **85**(10), 660.
16. A. B. Lindstrom, M. J. Strynar and E. L. Libelo, *Environ. Sci. Technol.*, 2011, **45**, 7954.
17. K. C. Lowe, *Blood Rev.*, 1999, **13**, 171.
18. J. G. Riess, *Chem. Rev.*, 2001, **101**, 2797.
19. A. B. Lindstrom, M. J. Strynar, E. L. Libelo and J. A. Field, *Environ. Sci. Technol.*, 2011, **45**, 7951.
20. W. T. Tsai, H. P. Chen and W. Y. Hsien, *J. Loss Prev. Process Ind.*, 2002, **15**, 65.
21. A. M. A. Dias, R. P. Bonifacio, I. M. Marrucho, A. A. H. Padua and M. F. C. Gomes, *Phys. Chem. Chem. Phys.*, 2003, **5**, 543.
22. A. M. A. Dias, J. C. Pamies, J. A. P. Coutinho, I. M. Marrucho and L. F. Vega, *J. Phys. Chem. B*, 2004, **108**, 1450.
23. S. Rossi, G. Waton and M. P. Krafft, *Langmuir*, 2010, **26**, 1649.
24. E. G. Schutt, D. H. Klein, R. M. Mattrey and J. G. Riess, *Angew. Chem.*, 2003, **42**, 3218.
25. E. Stride and M. Edirisinghe, *Med. Biol. Eng. Comput.*, 2009, **47**, 809.
26. M. P. Krafft, *J. Polym. Sci., Part A: Polym. Chem.*, 2006, **44**, 4251.
27. M. P. Krafft and J. G. Riess, *Angew. Chem.*, 1994, **33**, 1100.
28. S. M. Bertilla, J. Thomas, P. Marie and M. P. Krafft, *Langmuir*, 2004, **20**, 3920.
29. M. C. Garnett, S. Stolnik, S. E. Dunn, I. Armstrong, W. Lin, E. Schacht, P. Ferutti, M. Vert, M. C. Davies, L. Illum and S. S. Davis, *MRS Bull.*, 1999, **24**, 49.
30. H. A. Courrier, T. F. Vandamme and M. P. Krafft, *Colloids Surf., A*, 2004, **244**, 141.
31. J. G. Riess, *Artif. Cells, Blood Substitutes, Biotechnol.*, 2005, **33**, 47.

32. J. G. Riess, *Artif. Cells, Blood Substitutes, Biotechnol.*, 2006, **34**, 567.
33. A. B. Pereiro, J. M. M. Araujo, F. S. Teixeira, I. M. Marrucho, M. M. Pineiro and L. P. N. Rebelo, *Langmuir*, 2015, **31**, 1283.
34. A. Triolo, O. Russina, H.-J. Bleif and E. Di Cola, *J. Phys. Chem. B*, 2007, **111**, 4641.
35. K. Shimizu, M. F. C. Gomes, A. A. Pádua, L. P. Rebelo and J. N. C. Lopes, *J. Mol. Struct.: THEOCHEM*, 2010, **946**, 70.
36. T. L. Greaves, D. F. Kennedy, S. T. Mudie and C. J. Drummond, *J. Phys. Chem. B*, 2010, **114**, 10022.
37. J. N. C. Lopes and A. A. H. Pádua, *J. Phys. Chem. B*, 2006, **110**, 3330.
38. A. Triolo, O. Russina, B. Fazio, R. Triolo and E. Di Cola, *Chem. Phys. Lett.*, 2008, **457**, 362.
39. T. L. Greaves, D. F. Kennedy, A. Weerawardena, N. M. K. Tse, N. Kirby and C. J. Drummond, *J. Phys. Chem. B*, 2011, **115**, 2055.
40. T. Pott and P. Meleard, *Phys. Chem. Chem. Phys.*, 2009, **11**, 5469.
41. R. Hayes, S. Z. El Abedin and R. Atkin, *J. Phys. Chem. B*, 2009, **113**, 7049.
42. J. N. C. Lopes, M. F. C. Gomes and A. A. H. Pádua, *J. Phys. Chem. B*, 2006, **110**, 16816.
43. T. Takamuku, M. Tobiishi and H. Saito, *J. Solution Chem.*, 2011, **40**, 2046.
44. F. Castiglione, M. Moreno, G. Raos, A. Famulari, A. Mele, G. B. Appetecchi and S. Passerini, *J. Phys. Chem. B*, 2009, **113**, 10750.
45. O. Hollóczki, M. Macchiagodena, H. Weber, M. Thomas, M. Brehm, A. Stark, O. Russina, A. Triolo and B. Kirchner, *ChemPhysChem*, 2015, **16**, 3325.
46. O. Russina, F. L. Celso, M. Di Michiel, S. Passerini, G. B. Appetecchi, F. Castiglione, A. Mele, R. Caminiti and A. Triolo, *Faraday Discuss.*, 2013, **167**, 499.
47. A. B. Pereiro, M. J. Pastoriza-Gallego, K. Shimizu, I. M. Marrucho, J. N. C. Lopes, M. M. Piñeiro and L. P. N. Rebelo, *J. Phys. Chem. B*, 2013, **117**, 10826.
48. N. S. M. Vieira, P. M. Reis, K. Shimizu, O. A. Cortes, I. M. Marrucho, J. M. M. Araújo, J. M. S. S. Esperança, J. N. C. Lopes, A. B. Pereiro and L. P. N. Rebelo, *RSC Adv.*, 2015, **5**, 65337.
49. M. F. C. Gomes and A. A. H. Padua, *J. Phys. Chem. B*, 2003, **107**, 14020.
50. A. B. Pereiro, J. M. M. Araújo, S. Martinho, F. Alves, S. Nunes, A. Matias, C. M. M. Duarte, L. P. N. Rebelo and I. M. Marrucho, *ACS Sustainable Chem. Eng.*, 2013, **1**, 427.
51. S. Martinho, J. M. M. Araújo, L. P. N. Rebelo, A. B. Pereiro and I. M. Marrucho, *J. Chem. Thermodyn.*, 2013, **64**, 71.
52. F. S. Teixeira, N. S. M. Vieira, O. A. Cortes, J. M. M. Araújo, I. M. Marrucho, L. P. N. Rebelo and A. B. Pereiro, *J. Chem. Thermodyn.*, 2015, **82**, 99.
53. N. S. M. Vieira, A. Luís, P. M. Reis, P. J. Carvalho, J. A. Lopes-da-Silva, J. M. S. S. Esperança, J. M. M. Araújo, L. P. N. Rebelo, M. G. Freire and A. B. Pereiro, *J. Chem. Thermodyn.*, 2015, **97**, 354.

54. A. Luís, K. Shimizu, J. M. M. Araújo, P. J. Carvalho, J. A. Lopes-da-Silva, J. N. C. Lopes, L. P. N. Rebelo, J. A. P. Coutinho, M. G. Freire and A. B. Pereiro, *Langmuir*, 2016, **32**, 6130.
55. N. S. M. Vieira, A. Luís, P. M. Reis, P. J. Carvalho, J. A. Lopes-da-Silva, J. M. S. S. Esperança, J. M. M. Araújo, L. P. N. Rebelo, M. G. Freire and A. B. Pereiro, *J. Chem. Thermodyn.*, 2016, **97**, 354.
56. A. Miguel and G. C. Rutledge, *J. Chem. Phys.*, 2010, **132**, 114704.
57. I. A. McLure, V. A. M. Soares and B. Edmonds, *J. Chem. Soc., Faraday Trans.*, 1982, **78**, 2251.
58. N. Nishikido, W. Mahler and P. Mukerjee, *Langmuir*, 1989, **5**, 227.
59. M. G. Freire, P. J. Carvalho, A. J. Queimada, I. M. Marrucho and J. A. P. Coutinho, *J. Chem. Eng. Data*, 2006, **51**, 1820.
60. R. N. Haszeldine and F. Smith, *J. Chem. Soc.*, 1951, **16**, 603.
61. J. K. Yoon and D. J. Burgess, *J. Colloid Interface Sci.*, 1992, **151**, 402.
62. V. E. Stiles and G. H. Cady, *J. Am. Chem. Soc.*, 1952, **337**, 3771.
63. G. D. Oliver, S. Blumkin and C. W. Cunningham, *J. Am. Chem. Soc.*, 1951, **73**, 5722.
64. T. Sakka and Y. H. Ogata, *J. Fluorine Chem.*, 2005, **126**, 371.
65. H. F. D. Almeida, M. G. Freire, A. M. Fernandes, J. A. Lopes-da-Silva, P. Morgado, K. Shimizu, E. J. M. Filipe, J. N. C. Lopes and M. N. B. F. Santos, *Langmuir*, 2014, **30**, 6408.
66. C. S. Santos and S. Baldelli, *Chem. Soc. Rev.*, 2010, **39**, 2136.
67. L. C. Clark, F. Becattini, S. Kaplan, V. Obrock, D. Cohen and C. Becker, *Science*, 1973, **181**, 680.
68. H. Okamoto, K. Yamanouchi and K. Yokoyama, *Chem. Pharm. Bull.*, 1975, **23**, 1452.
69. R. Naito and K. Yokoyama, *Perfluorochemicals Blood Substitutes*, Technical Information Series nº 7, Green Cross Corporation, Osaka, Japan, 1981.
70. E. Szajdzinska-Pietek and M. Wolszczak, *Langmuir*, 2000, **16**, 1675.
71. A. González-Pérez, J. M. Ruso, G. Prieto and F. Sarmiento, *Colloid Polym. Sci.*, 2004, **282**, 1133.
72. J. L. López-Fontán, F. Sarmiento and P. C. Schulz, *Colloid Polym. Sci.*, 2004, **283**, 862.
73. M. Kunze, S. Jeong, E. Paillard, M. Winter and S. Passerini, *J. Phys. Chem. C*, 2010, **14**, 12364.
74. A. R. R. Teles, H. Correia, G. J. Maximo, L. P. N. Rebelo, M. G. Freire, A. B. Pereiro and J. A. P. Coutinho, *Phys. Chem. Chem. Phys.*, 2016, **18**, 25741.
75. R. E. Banks, B. E. Smart and J. C. Tatlow, *Organofluorine Chemistry–Principles and Commercial Applications*, Plenum Press, New York, 1994.
76. M. J. P. de Melo, A. M. A. Dias, M. Blesic, L. P. N. Rebelo, L. F. Vega, J. A. P. Coutinho and I. M. Marrucho, *Fluid Phase Equilib.*, 2006, **242**, 210.
77. M. Postel, J. G. Riess and J. G. Weers, *Artif. Cells, Blood Substitutes, Biotechnol.*, 1994, **22**, 991.
78. A. B. Pereiro, L. C. Tomé, S. Martinho, L. P. N. Rebelo and I. M. Marrucho, *Ind. Eng. Chem. Res.*, 2013, **52**, 4994.

79. J. H. J. Davis, Ionic Liquids Derived from Natural Products and Other Novel Chemistries, in *Green Industrial Applications of Ionic Liquids*, ed. R. D. Rogers, K. R. Seddon and S. Volkov, Kluwer Academic, 2003, p. 385.
80. T. L. Merrigan, E. D. Bates, S. C. Dorman and J. H. Davis, *Chem. Commun.*, 2000, 2051.
81. H. W. Pennline, D. R. Luebke, K. L. Jones, C. R. Myers, B. I. Morsi, Y. J. Heintz and J. B. Ilconich, *Fuel Process. Technol.*, 2008, **89**(9), 897.
82. D. S. Dimitrov, Therapeutic Proteins, *Methods in Molecular Biology*, Humana Press, 2012, pp. 1–26.
83. D. Ibraheem, A. Elaissari and H. Fessi, *Int. J. Pharm.*, 2014, **477**, 578.
84. I. Brasnjevic, H. W. M. Steinbusch, C. Schmitz and P. Martinez-Martinez, *Prog. Neurobiol.*, 2009, **87**, 212.
85. B. Leader, Q. J. Baca and D. E. Golan, *Nature*, 2008, **7**, 21.
86. M. L. Tan, P. F. M. Choong and C. R. Dass, *Peptides*, 2010, **31**, 184.
87. C. Dai, B. Wanga and H. Zhao, *Colloids Surf., B*, 2005, **41**, 117.
88. K. Fujita, D. R. MacFarlane, M. Forsyth, M. Yoshizawa-Fujita, K. Murata, N. Nakamura and H. Ohno, *Biomacromolecules*, 2007, **8**, 2080.
89. H. Ohno, C. Suzuki, K. Fukumoto, M. Yoshizawa and K. Fujita, *Chem. Lett.*, 2003, **32**, 450.
90. J. M. M. Araújo, A. B. Pereiro, F. Alves, I. M. Marrucho and L. P. N. Rebelo, *J. Chem. Thermodyn.*, 2013, **57**, 1.
91. J. M. M. Araújo, C. Florindo, A. B. Pereiro, N. S. M. Vieira, A. A. Matias, C. M. M. Duarte, L. P. N. Rebelo and I. M. Marrucho, *RSC Adv.*, 2014, **4**, 28126.
92. M. P. Krafft and J. G. Riess, *Biochimie*, 1998, **80**, 489.
93. M. Salwiczek, E. K. Nyakatura, U. I. M. Gerling, S. Ye and B. Koksch, *Chem. Soc. Rev.*, 2012, **41**, 2135.
94. E. D. Oldham, W. Xie, A. M. Farnoud, J. Fiegel and H.-J. Lehmler, *J. Phys. Chem. B*, 2012, **116**, 9999.
95. J. L. Newsted, S. A. Beach, S. P. Gallagher and J. P. Giesy, *Arch. Environ. Contam. Toxicol.*, 2008, **54**, 535.
96. M. N. Wadekar, J. Boekhoven, W. F. Jager, G. J. M. Koper and S. J. Picken, *Langmuir*, 2012, **28**, 3397.
97. H. Xue and J. M. Shreeve, *Eur. J. Inorg. Chem.*, 2005, **2005**(13), 2573.
98. A. M. A. Dias, H. Carrier, J. L. Daridon, J. C. Pamies, L. F. Vega, J. A. P. Coutinho and I. M. Marrucho, *Ind. Eng. Chem. Res.*, 2006, **45**(7), 2341.
99. Z.-Z. Yang, Y.-N. Zhao and L.-N. He, *RSC Adv.*, 2011, **1**(4), 545.
100. J. E. Bara, D. E. Camper, D. L. Gin and R. D. Noble, *Acc. Chem. Res.*, 2010, **43**(1), 152.
101. M. Ramdin, T. W. de Loos and T. J. H. Vlugt, *Ind. Eng. Chem. Res.*, 2012, **51**(24), 8149.
102. R. E. Baltus, B. H. Culbertson, S. Dai, H. Luo and D. W. DePaoli, *J. Phys. Chem. B*, 2004, **108**, 721.
103. M. B. Shiflett, M. A. Harmer, C. P. Junk and A. Yokozeki, *J. Chem. Eng. Data*, 2006, **51**, 483.

104. M. J. Muldoon, S. N. V. K. Aki, J. L. Anderson, J. K. Dixon and J. F. Brennecke, *J. Phys. Chem. B*, 2007, **111**, 9001.
105. Y. Hou and R. E. Baltus, *Ind. Eng. Chem. Res.*, 2007, **46**, 8166.
106. J. E. Bara, C. J. Gabriel, T. K. Carlisle, D. E. Camper, A. Finotello, D. L. Gin and R. D. Noble, *Chem. Eng. J.*, 2009, **147**, 43.
107. L. Pison, J. N. C. Lopes, L. P. N. Rebelo, A. A. H. Padua and M. F. C. Gomes, *J. Phys. Chem. B*, 2008, **112**, 12394.

CHAPTER 5

Ion Solvation and Transport in Ionic Liquids and Ionogels

L. M. VARELA*[a], E. LÓPEZ-LAGO[a] AND O. CABEZA[b]

[a]Grupo de Nanomateriais, Fotónica e Materia Branda, Departamentos de Física de Partículas e de Física Aplicada, Universidade de Santiago de Compostela, Campus Vida s/n E-15782, Santiago de Compostela, Spain; [b]Facultade de Ciencias, Universidade da Coruña, Campus A Zapateira s/n E-15071, A Coruña, Spain
*E-mail: luismiguel.varela@usc.es

5.1 Introduction

Ionic liquids or room-temperature molten salts are well-reputed green designer solvents, with very singular tunable properties that make them ideal candidates for a plethora of task-specific applications.[1–5] Among these properties, low vapour pressures, high electrical conductivities and very large electrochemical windows (redox stability) are probably the most prominent, although thermal stability, non-flammability and low nucleophilicity and capability of providing weakly coordinating or non-coordinating environments are also of acute importance, in many practical applications. Most of these require tailoring the properties of the specific ionic liquid by fine-tuning the structure with an appropriate combination of cations and anions, a way in which, theoretically, millions of binary and trillions of ternary combinations could be produced.[6] This overpopulation of ionic liquids has literally revolutionized the field of solvents and is the basis of their outstanding role

Smart Materials No. 28
Ionic Liquid Devices
Edited by Ali Eftekhari

Published by the Royal Society of Chemistry, www.rsc.org

in contemporary academic research and, increasingly, in industrial applications. Among the latter, their use as electrolytes in high-voltage electrochemical devices (batteries, supercapacitors, solar cells, electroactuators, advanced electrodeposition...), synthesis and catalysis, nanochemistry, as well as lubricants or thermal fluids in energy applications deserve specific mention.[7–11]

In many applications of ionic liquids, mechanical rigidity is of fundamental importance for designing layered devices and/or for avoiding leaks. In this case, ionogels emerge as a real alternative since they preserve most of the properties of the ionic liquid, despite being immobilized in a solid matrix. The formation of the gel phase demands the presence of some additional organic, inorganic or hybrid organic–inorganic component with which the ionic liquid is hybridized. As can be seen in the review of Vioux and co-workers[12] usually low molecular weight polymers and biopolymers, carbon nanotubes, silica, and mixtures of polymers and inorganic fillers are used as gelators. Moreover, water is also frequently used for jellifying ionic liquids forming hydrogels,[13] which is a route for the formation of cheap and biocompatible gels of great potential interest in photonic and biomedical applications, among others. However, despite some quite recent studies (see ref. 12 and references therein) showing an increasingly frequent route for the practical application of ionic liquids, the properties of ionogels (phase diagrams, structure, transport of different substances in the ionic gel phase) are still scarcely known.

Many properties of these systems are controllable by external stimuli, including temperature, pressure, moisture, pH, and electric or magnetic fields, to which ionic liquids respond with a "useful" reversible effect, so they can be considered text-book smart materials. Indeed, according to Spillman *et al.*,[14] a smart material/structure is a: "physical material/structure having: (i) a definite purpose, (ii) means and imperative to achieve that purpose, (iii) the pattern of functioning of a universal computer or Turing machine."

These capabilities mimic those usually attributed to smart materials, which can "sense environment and internal conditions, process sensed data according to internal algorithm, act/don't act on environment and/or internal condition, repeat sequence".[14]

This category of materials, whose properties vary significantly in the presence of external perturbations, is opening a new era in materials science and technology.[15] They are capable of executing specific functions by responding to external stimuli or embedding sensory capabilities, which obviously contribute notably to opening up new trends in material science, biotechnology, nanotechnology, micro- and nanoelectronics, photonics, and so forth. These materials can exhibit very different properties (electrical, magnetic, rheological, thermal, mechanical...) and, in the presence of external perturbations (electric or magnetic fields, electrical currents, pH changes, temperature...), can reversibly transform energy in several ways (photoelectric, thermoelectric, piezoelectric...).[16]

Among these materials (see Table I in the recent ref. 16 for a classification of smart materials), field responsive fluids stand out. These fluids undergo profound changes in their rheological behaviour with the application of external electric (electrorheological) or magnetic (magnetorheological) fields. These fluids are usually prepared by dispersing some active agent (metals, micron- or nano-sized particles...) in a basis fluid and, when necessary, adding a stabilizer. Meanwhile, ionic liquids meet all the requirements in order to be classified as smart materials, since they usually exhibit large variations of their properties associated with the modification of environmental variables (water content, temperature, electric field...), not least because of their usefulness as carrier fluids in field responsive fluids.[17] For instance, their (usually very high) hygroscopicity makes them ideal candidates for the detection of moisture, and this is mainly done by controlling the electrical conductivity of the ionic liquid, which can dramatically increase with very small amounts of added water upon mixing.[18] Moreover, the sensing capability of ionic liquids is also reflected in the viscosity of the mixtures, which is greatly reduced by water absorption. Electrorheological properties of pure ionic liquids have been well known since the work of Bicak *et al.*,[19] who proved the vibration damping of 2-hydroxyethylammonium formate under the effect of an applied electric field. Ionic liquids have also been used as dispersion media for the preparation of electrorheological dispersions, as reported by Marins *et al.*[20] among others. These authors reported the preparation of ionogels by confinement of phosphonium-based ionic liquids inside silica networks, which show good electrorheological behaviour. As mentioned above, these dense ionic fluids have also been reported as promising carrier fluids for the preparation of magnetorheological fluids. For instance, Guerrero-Sanchez *et al.*[21] reported improved colloidal stability of magnetite-based magnetorheological fluids using ionic liquids of the imidazolium and phosphonium families. Ionic liquids have also been used for preparing ferrofluids, as reported by Jain *et al.*,[22] who prepared dispersions of both bare and sterically stabilized 8–12 nm diameter superparamagnetic iron oxide nanoparticles in protic ethylammonium and aprotic imidazolium ionic liquids showing the occurrence of spiking in static magnetic fields similar to conventional aqueous and non-aqueous ferrofluids.

Ionic liquids are also members of the family of chromogenic compounds,[23] materials whose colour changes by the action of an external stimulus such as light (photochromism),[24] electric/magnetic field (electrochromism)[25] or temperature (thermochromism).[26] In some compounds these effects are reversible, allowing their use in switchable devices as smart windows in buildings or vehicles. Besides UV-VIS absorption, refractive index is also thermoresponsive, making these materials good candidates to act as optical limiters by thermal refraction[27] or by multiphoton absorption, especially in those incorporating a transition metal in their structures.[28] The nonlinear response can be enhanced by the addition of nanoparticles.[29] Second harmonic generation with important applications in imaging microscopy, laser technology, sensing has been observed, among others, in $[Sb_7S_8Br_2](AlCl_4)_3$, the first ionic liquid with a chalcogenide cation.[30,31] On the other hand, *N*,*N*-diethyl-*N*-methyl-*N*-(2-methoxyethyl) ammonium bis(trifluoromethylsulphonyl)imide (DEME-TFSI) has been also successfully used as polarization layer in an organic photocell due to the large capacitance associated with the formation of electric double layers and high ionic conductivities. With these fluids, charge displacement currents due to the electric double layer formed at the solid/liquid interfaces can be produced even for large separations between the electrodes.[32]

As can be deduced from the previous paragraphs, the properties that allow the characterization of ionic liquids as smart materials usually arise in mixtures of these solvents with other substances, and they are mostly determined by the specific form of solvation of the different substances in the ionic environment. Indeed, most of the current applications of ionic liquids require some level of doping of these liquids in order to produce the desired effect (*e.g.* electrochemical devices require the presence of some active salt in the ionic liquid matrix; ionogels require the presence of some jellyfier, and metal or nanoparticle doping is used to induce thermochromism or plasmonic properties), so mixtures of ionic liquids with both molecular cosolvents, inorganic and organic salts or even micron- or nanosized particles are, indeed, usually far more interesting than the pure ionic liquids. Nonetheless, the essential features of the mechanisms of solvation of different molecules, ions or nanoparticles in these materials are still scarcely known.

In the present chapter we review some recent activity of solvation of these substances in ionic liquids and ionogels, as well as their effect on mass and charge transport in these dense ionic media. We will review the essential mechanisms of solvation in bulk and nanoconfined ionic liquids, paying attention to the formation of metal complexes in these media as the basic mechanism behind optical and magnetic properties. In the last part of the review, the transport properties and their relationship to the specific structure of the mixtures will be considered.

5.2 Results and Discussion

As previously stated, ionic liquids are room-temperature molten salts that are extremely versatile as far as the dissolution of solutes is concerned. Indeed, the actual way of solvating different solutes in these dense ionic environments is the main factor responsible for the properties of the mixtures of ionic liquids with molecular cosolvent, salts, nanoparticles, and so on. These mechanisms have been thoroughly studied in the past both with simulations and with experiments, and excellent reviews exist, as can be seen for instance in ref. 33, and so we will not further review the matter here. That the dielectric constant of the cosolvent controls the miscibility with ionic liquids, which are only miscible with molecular compounds of medium or high dielectric constants, is well known, but further insight into the molecular mechanism responsible for the solvation of these substances is still needed. In the remainder of this section, we will provide a microscopic picture of the solvation of water, alcohols and inorganic salts in ionic liquids, as well as its influence on the formation and properties of ionogels and on the optical properties of these mixtures.

5.2.1 Solvation of Molecular Cosolvents and Inorganic Salts in Ionic Liquids

In this work we will mainly focus our attention on the solvation of water and alkanol molecules, as well as in mixtures with mono- and divalent salts. Solvation is the term that summarizes the – normally intricate – attraction and association of molecules of a solvent with molecules or ions of a solute, giving rise to solvate complexes as the solute particles spread out in the bulk solvent where they are stabilized. The detailed features of this process are frequently very complex and they are determined by the specific type of interactions between the solvent and solute particles. The main physical interactions involved in complexation are, of course, electrostatic multipolar interactions (especially important are dipole–dipole or van der Waals interactions and, in ionic systems, ion–ion and ion–dipole interactions), hydrogen bonding, and excluded volume or steric interactions. These interactions determine that compounds of the same type are frequently dissolvable in one another (*similia similibus solvuntur*). Indeed, polar components – those containing dipoles such as water – of high dielectric constants are good at dissolving ions, since they are capable of forming solvation shells of oriented solvent molecules in response to the ion field effect, and other polar solutes, while apolar solvents (such as organic ones) are efficient for dissolving apolar moieties.

Nonetheless, this paradigm has been challenged by the solvents forming the new category of ionic liquids, which are well known to be amphiphilically nanostructured solvents[34,35] capable of dissolving a huge amount of substances. Indeed, theoretically, their molecular structure can be fine-tuned for dissolving any given substance. Although the specific details of the solvation of different molecular and ionic species in such dense ionic environments are not yet well characterized, considerable experimental and computational activity has been registered in the field in the last decade or so, most of which has been recently reviewed by some of us in ref. 33. Here we provide some details of this so-called nanostructured solvation, which can adequately rationalize the way in which substances mix with ionic liquids. The latter are intrinsically organized into polar and apolar nanodomains where selective solvation can take place, which gives rise to interlaced networks in the bulk mixtures. This is at the origin of the very high solvation ability of the solvents, capable of dissolving both polar and apolar solutes.

Starting with water/ionic liquid mixtures – especially important due to the inevitable presence of moisture in these solvents, as well as for the importance of water-based systems for the formation of hydrogels and other relevant mixtures – water seems to be miscible with many ionic liquids depending mostly on the anion, and it is apparently solvated in these solvents, forming water clusters, whose size is essentially controlled by the hydrophobicity of the anions, which also controls the details of the water network in the bulk mixture. Experimentally, water mixtures have been thoroughly considered (see, for example, ref. 36–45), but here we will focus mainly on the microscopic picture of these mixtures arising from molecular dynamics simulations.

In 2011,[46] some of us reported the radial distribution functions (rdfs) of water with all the species in mixtures of water and imidazolium-based aprotic ionic liquids. In these liquids, unable to form hydrogen bonds with the solute, we can see that water tends to extensively form clusters at low water contents, as shown by the relatively large height of the first peaks of the water–water rdfs of these mixtures. These heights increase notably with the chain length (hydrophobicity) of the cation, as was also reported by Feng and Voth in ref. 47. The first peaks of the water–water rdfs of the ionic liquids with the hydrophobic PF_6^- reported in ref. 46 are much higher than those of their halogenated homologues with the same cationic chain lengths. Moreover, as water concentration increases, the peaks are progressively lower, reflecting more homogeneous mixing of water in the bulk mixtures. The situation is somewhat different in the case of mixtures with protic ionic liquids in which the mixing is more homogeneous (Figure 5.1). There we can see that the peaks of the water–water rdfs are notably lower than those previously reported in the apolar mixtures.[46] This is probably because water molecules are forming hydrogen bonds both with the anions and the cations of the ionic liquid,[48] leading to mixtures more intermingled at the molecular level. This picture of clusterised water is also confirmed for the single-particle dynamics (Figure 5.2), where oscillations in the velocity autocorrelation functions of water molecules (Figure 5.2a) reveal that water molecules undergo a "rattling" motion, with successive rebounds and velocity inversions inside their quite resilient molecular cages formed by their nearest neighbour molecules in the polar nanoregions of the bulk mixtures. Similar functions have been reported for mixtures with aprotic ionic liquids.[46] As we can also see in Figure 5.2, these correlations are almost completely damped out within *ca* 0.4 ps.[48] This motion is represented in Fourier space by the vibrational density of states (vDOS), plotted in Figure 5.2b, where we can see that water vibrations take place in the usual THz region with two basic modes, once centred at 5 THz and the other at 10 THz.

However, the situation is notably different in mixtures with alcohols. Varela and co-workers[49,50] previously reported MD studies of the structure and single particle dynamics of mixtures of aprotic ionic liquids and alkanols

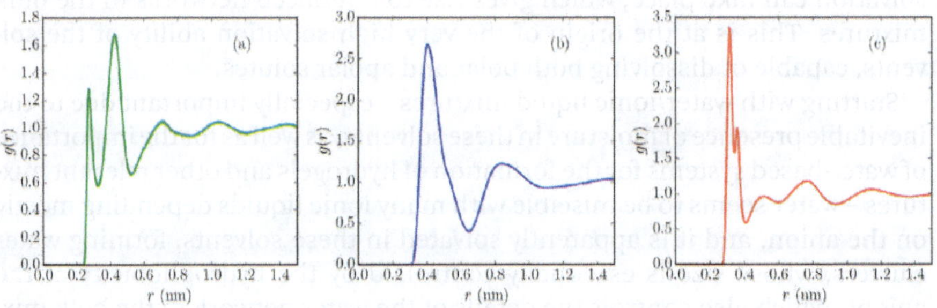

Figure 5.1 (a) Water–water, (b) water–anion and (c) water–cation radial distribution functions in mixtures of EAN and water (data taken from ref. 48).

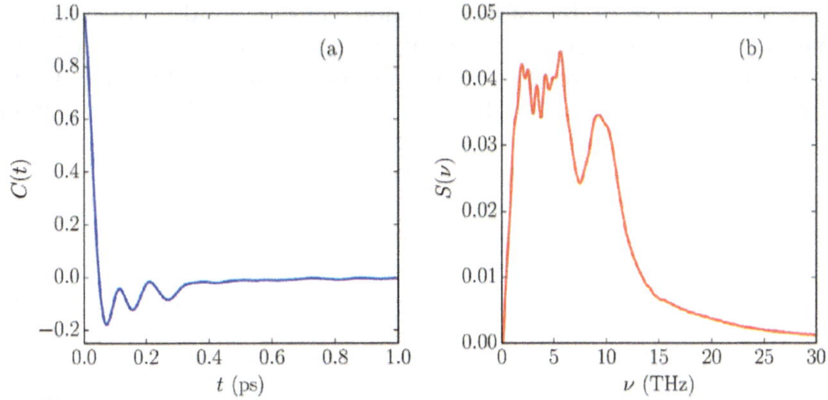

Figure 5.2 (a) Velocity autocorrelation function and (b) vibrational density of states of water molecules in mixtures of EAN and water (data taken from ref. 48).

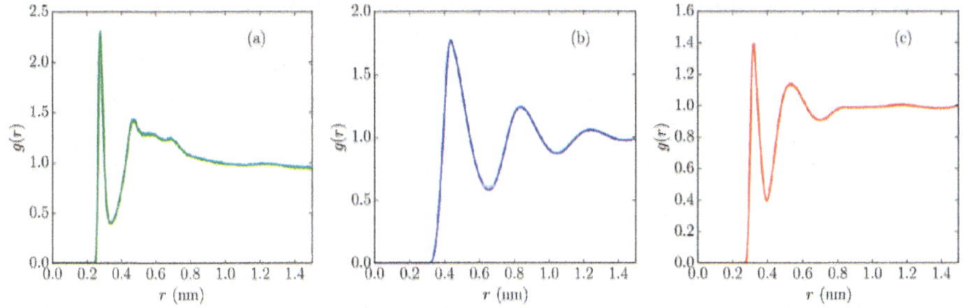

Figure 5.3 (a) Ethanol–ethanol, (b) ethanol–anion and (c) ethanol–cation radial distribution functions in mixtures of EAN and ethanol (data taken from ref. 48).

of different chain lengths (ethanol and methanol). The same group reported a few years later a computational analysis of the structure and dynamics of mixtures with protic ionic liquids.[48] The microscopic picture emerging from these studies is that of alcohol being more homogeneously dispersed in the mixtures, contrarily to mixtures with water, where the formation of a water network breaks the polar network of the ionic liquid. However, in the more solvophobic mixtures studied by Méndez-Morales *et al.*,[49] those with fluorinated anions, some evidence of the formation of a network of alcohol molecules was found. Moreover, we have very recently reported[51] that alcohol molecules are accommodated in the borders of the polar and apolar nanoregions, since these molecules have both types of regions in their molecular moieties. The situation is similar in the case of mixtures with protic ionic liquids, where the rdfs (Figure 5.3) do not reveal any significant clustering of alcohol molecules. Meanwhile, the single particle dynamics of these molecules in mixtures with aprotic[50] and protic[48] ionic liquids is notably different to that of water. No caging of alcohol molecules is found in mixtures with ethanol or methanol neither in the mixtures with the aprotic ionic liquids nor in those with the protic ones, and no oscillations in the velocity autocorrelation functions are seen (Figure 5.4a). The less tightly bound states of alcohol with respect to those of water are reflected also in the vDOS, since the central frequency of the only observed mode is notably lower than those of the modes of water in the same environment. Alcohol molecules are, of course, not so intensely hydrogen bonded to the ions of the ionic liquid as their water homologues, which explains this different behaviour. Docampo-Álvarez *et al.* also reported spatial distribution functions that revealed other differences of these mixtures with aqueous mixtures: while water mainly coordinates anions of the ionic liquid in a bidentate fashion (in the bisector of the molecular equatorial plane), alcohols coordinate the oxygen atoms of NO_3^- in a monodentate way as shown by spherical rings in the spatial distribution functions.[48]

Finally, we will briefly review the solvation of inorganic salts in ionic liquids. These mixtures have also been intensely studied due to interest in their electrochemical applications (alkali ions, Mg and Ca, essentially), and also due to the importance of transition metals and rare earths in extraction, recycling and optical applications. Starting with mixtures with lithium salts, many studies have been reported in the past for mixtures of Li^+ cations in ionic liquids including some from our own groups, and all of them seem to indicate that ions are solvated in the polar nanoregions of the ionic liquid, forming pseudo-solid structures. In 2015, Russina *et al.*,[52] studying experimentally and computationally mixtures of EAN and $LiNO_3$, summarized the main features of what could be a universal ion solvation mechanism in amphiphilically nanostructured ionic liquids. First, the authors found that the morphology of the ionic liquid is scarcely affected by the addition of

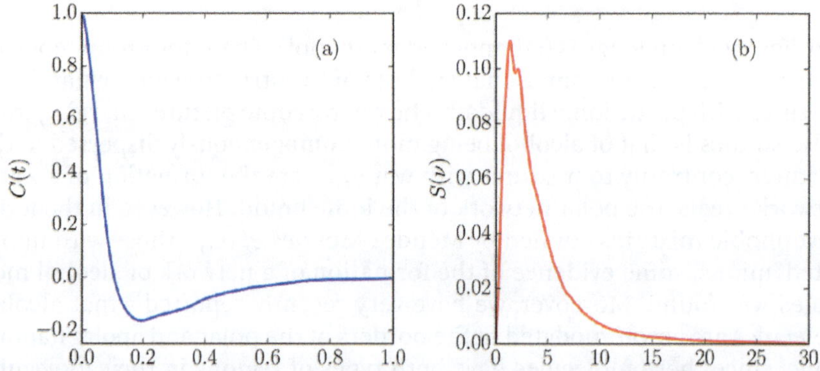

Figure 5.4 (a) Velocity autocorrelation function and (b) vibrational density of states of ethanol molecules in mixtures of EAN and ethanol (data taken from ref. 48).

the salt, specifically preserving the nanosegregation of EAN persistently. Secondly, $LiNO_3$ was found to accommodate into the polar nanodomains, with a very low probability of presence in the apolar portions of the mesoscopically structured mixture. Finally, the added lithium salt does not adopt the structure of its neat, molten state, but, instead, a further level of segregation of Li ions in the polar domains seems to exist which suggests crystalline-like order that eventually will lead to $LiNO_3$ separation. Indeed, these structures could also be the clusters reported by Méndez-Morales *et al.*[53] in mixtures of [Bmim][X] (X = BF_4^-, PF_6^-, $TFSI^-$) with Li and Na salts with a common anion, which are indeed ionic complexes of the form $[Li/Na(Anion)_n]^{1-n}$ that have mean lifetimes of the order of 10 ns with $n = 3$, 3, and 2 for BF_4^-, PF_6^- and $TFSI^-$, respectively. These complexes are actually observed in the rdfs reported in ref. 53 and in the cage correlation functions studied there. In the former, the existence of double first peaks in the Li-anion rdfs indicates the coexistence of two different populations, one coordinating anions in a monodentate way and the other in a bidentate fashion. The existence of these ionic aggregates has also been reported in mixtures of monovalent (Li) and divalent (Mg) salts with protic ionic liquids of the ethylammonium family.[54] In Figures 5.5 and 5.6 we present the rdfs of mixtures of these salts and EAN, and, as we can see, Li^+ and Mg^{2+} are essentially surrounded by anions of the ionic liquid. In the case of the monovalent anions, monodentate and bidentate configurations are found, while in the case of the divalent cation one of the populations is virtually suppressed. This confirms the structural picture of a pseudo-solid, complexes in the polar domains as the way of solvating added ions in ionic liquids, and it is completely compatible with the pictures that we described above for the molecular additives. As for the single-particle dynamics of ions in the bulk mixtures, velocity autocorrelation functions and vDOS of the monovalent and divalent cations are shown in Figure 5.7c and d, respectively. It is apparent in those representations that both cations

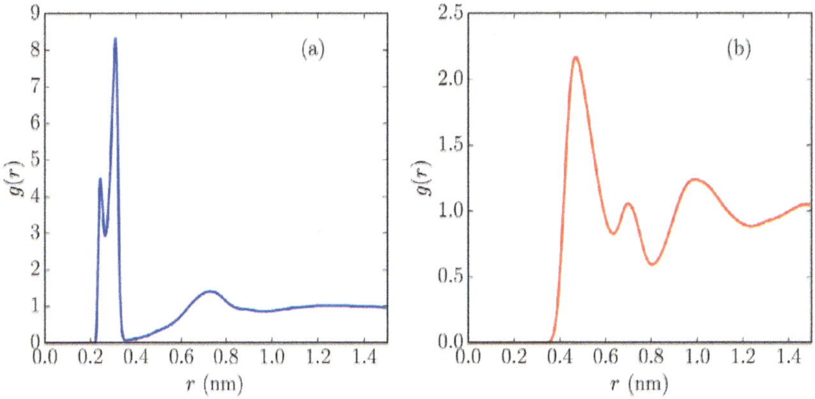

Figure 5.5 (a) Lithium–anion and (b) lithium–cation radial distribution functions in mixtures of EAN and $LiNO_3$ (data taken from ref. 54).

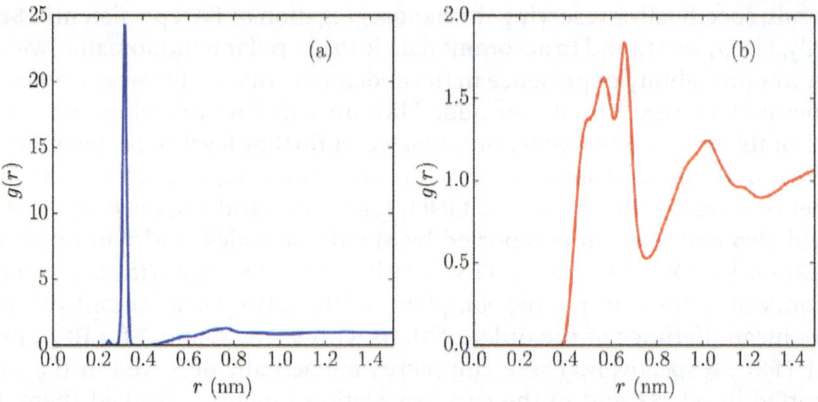

Figure 5.6 (a) Magnesium–anion and (b) magnesium–cation radial distribution functions in mixtures of EAN and $Mg(NO_3)_2$ (data taken from ref. 54).

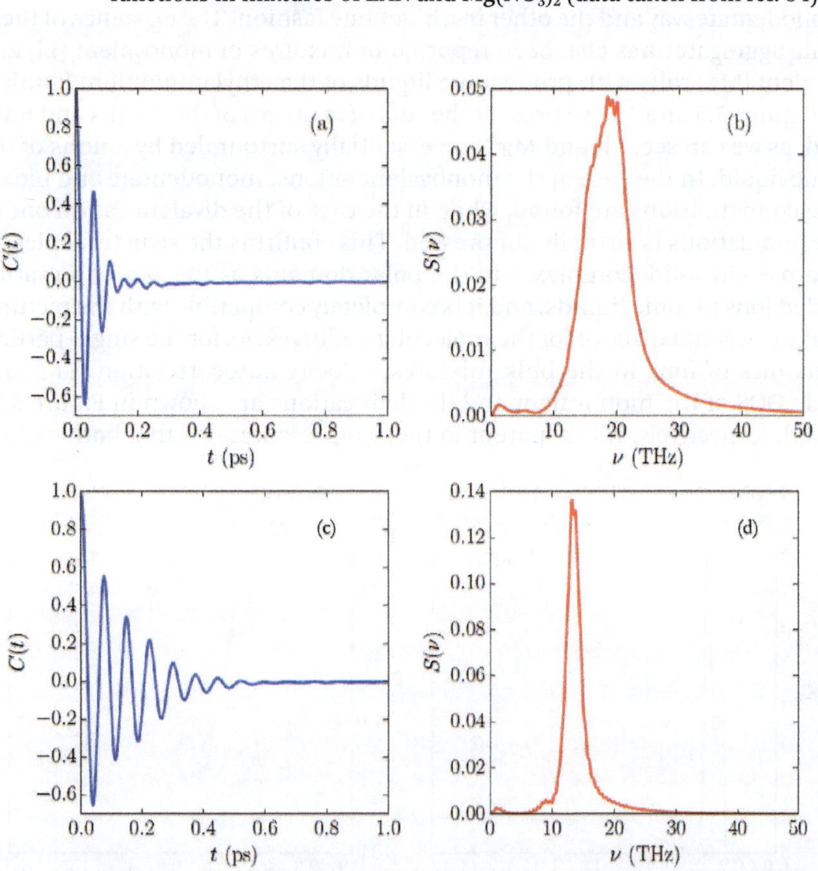

Figure 5.7 (a, c) Velocity autocorrelation functions and (b, d) vibrational density of states of salt cations in mixtures of EAN and $LiNO_3$ (a, b) and $Mg(NO_3)_2$ (c, d) (data taken from ref. 54).

exhibit virtually the same dynamics. However, Mg^{2+} is more scattered by collisions inside its cage than Li^{+}, although the oscillations of the latter are a little bit more energetic, as reflected by the blue shift of the vDOS relative to that of Mg^{2+}.

The solvation of multivalent ions in ionic environments has been also a matter of concern of other authors recently, and the complexation of multivalent cations in the ionic liquid has been confirmed. Matsumiya *et al.*[55] reported the stability and electrochemical behaviour of neutral, divalent and trivalent complexes of rare earths, noble metal (Eu, Sm, Pd, Fe) and ferrocene in phosphonium based ionic liquids, and compared the results to those in the protic environments of ammonium-type ionic liquids. The authors found that the diffusivity of these complexes depend on the charge of the centred metal and the circumstances around the ligand field in ionic liquids, and studied the role of d^5 electron transition and ligand–metal charge transfer (LMCT) transition in UV-Vis absorption spectra. The solvation/complexation of multivalent cations in ionic liquids has also been considered in order to study thermochromism in a nonaqueous environment with an eye to applications in smart optical devices (*e.g.* greenhouses, windows in smart buildings, *etc.*).[56] The thermochromic substances are, frequently, transition metals forming complexes with ligands, and whose optical properties respond (ideally, reversibly) to changes in temperature. These properties of the complexes are greatly influenced by the ligand field of the ionic liquid environment. Moreover, as proved by Babai and Mudring[57] and Arenz *et al.*[58] for Pr salts in 1-dodecyl-3-imidazolium bis(trifluoromethanesulphonyl) amide the replacement of conventional solvents by designer dense ionic solvents significantly increases the emission quantum yields and the lifetimes of the excited state of lanthanide compounds, increasing the luminescence with improved energy transfer mechanisms and allowing the tuning of emission colours with the appropriate choice of the anions. Moreover, Nóvoa-López *et al.*[28] reported comprehensive study of the nonlinear absorption in low-heat-capacity ionic liquids of 1-butyl-3-methyl-imidazolium cations with organometallic anions containing a transition metal (Co, Zn, Cu or Ni) and thiocyanate groups by an open aperture Z-scan, finding also strong laser induced thermal effects. In Co- and Ni-based ionic liquids the nonlinear absorption is associated to two-photon-like processes, while for Zn, Fe and Cu it seems to be associated to three-photon-like processes. These ionic liquids can act as optical limits since the absorption coefficient is light intensity dependent.

To summarize the contents of this section, all the evidence reported so far seems to indicate that molecular cosolvents and inorganic salts in mixtures with ionic liquids are selectively solvated in the nanoregions of the ionic liquid with similar polar/apolar characteristics, forming complexes with the ionic species in the bulk, which can be relatively long-lived. This specific form of solvation, termed nanostructured solvation, is behind most of the properties of these systems: phase behaviour, thermodynamics, optical, magnetic and transport properties, and so forth.

5.2.2 Ionogels: Preparation and Phase Diagrams

Traditionally, a substance is defined as a gel when there is a liquid phase dispersed in a continuous solid matrix. Thus, a gel is a colloid and then also a dispersion and a solution of, at least, two entities. The term gel is a contraction of gelatine and it is often attributed to Thomas Graham, one of the founders of colloid chemistry.[59] The impact of gels on daily life is ubiquitous, comprising almost any material sharing properties of liquids and solids, from hair gels, contact lens or jelly desserts. The main characteristic of a real gel is that it has solid-like rheology (it does not flow when in steady state), although the main component is a liquid. This is due to the existence of a three dimensional cross-linked network within the liquid component, which confers this elastic behaviour to the mixture. According to the specific nature of the bonds of this network, these heterogeneous systems can be physical or chemical. The latter usually form permanent bonds and so they are irreversible, while physical gels are reversible due to the formation of non-permanent bonds, hence being more appropriate for smart applications. The most classical classification of gels is due to Flory in 1974[60] and consists of four categories: (i) well-ordered lamellar structures (as phospholipids); (ii) disordered covalent polymeric networks (as vulcanized rubber or phenolic resins); (iii) polymer networks formed through physical aggregation (as gelatines); (iv) particulate, disordered structures (as precipitates consisting of reticular networks of fibres). Recent developments have led to a new transversal classification, such as ionogels (those that are electrical conductors) or low molecular weight gelators (LMWGs), which consists of organogels and hydrogels (depending on whether they are composed by organic compounds or by water, respectively). As can be deduced, gels form a very broad class of materials, both natural and artificial, which are quite difficult to classify properly. Many interesting review papers and book chapters have been recently published devoted to gels.[61–65]

Recent developments in ionic liquids have opened up the opportunity to obtain many new ionogels based on those materials.[12,66] These are intended to exhibit the physicochemical properties of ionic liquids and the rheological properties of gels, being very suitable for electrochemical applications.[67] In this sense many different ways to prepare gels using ionic liquids have been reported and characterized. The usual way to do so is to mix monomers with an ionic liquid and polymerize the mixture.[68–70] Another way is to include a gelator in an aqueous mixture of a given ionic liquid.[71,72] The most recent method is to synthesize poly-ionic liquids which self-polymerize.[73] These materials have many potential applications not only in the fields of electrochemistry, but also for tribology or fuel cell membranes as reported in the above-mentioned references.

In the rest of this section we will focus on the study of ionic hydrogels, which are solely formed by an appropriate ionic liquid and water, often that naturally adsorbed from the atmospheric moisture in a reversible fashion. The mesophase is due to the formation of physical H-bonds among water molecules and the anions or cations of the ionic liquids, one of them having a surfactant like character. Thus, the alkyl tails point to the centre of the aggregate while charged heads point to the bulk forming micelles. The quantity of water adsorbed by the hygroscopic ionic liquids depends on the humidity degree.[74] As an example of this behaviour, Figure 5.8 shows the evolution of the mass content of water of four samples with time, two of 1-ethyl-3-methylimidazolium ethyl sulphate ([EMIM][$EtSO_4$]) (open circles) and two of 1-ethyl-3-methylimidazolium tetrafluoroborate ([EMIM][BF_4]) (solid circles). The large symbols represent the samples placed in an Eppendorf bin open to the atmosphere while the small ones correspond to two closed Eppendorf bins, which were used as references. As can be observed, water content varies continuously in the open tube depending on the humidity grade, and their maxima correspond to the dampest days while their minima are reached during the driest days. Water adsorbed in an atmosphere saturated with moisture can represent up to 70% of the total weight of the sample, depending on the ionic liquid used, but it is released back to the atmosphere when the sample is left in a dry chamber (with less than a 10% humidity grade). This adsorption equilibrium of water in ionic liquids has been demonstrated by doping an ionic liquid with heavy water and analysing it using infrared spectroscopy. Figure 5.9 shows 32 continuous FTIR spectra of a 1-ethyl-3-methylimidazolium octyl sulphate ([EMIM][OSO_4]), known commercially as ECOENG 218, mixed with heavy water (with a concentration of molar fraction of the ionic liquid about of 0.2). Each scan takes about 6 s and so the complete figure was obtained in less than 5 min (arrows indicate the time

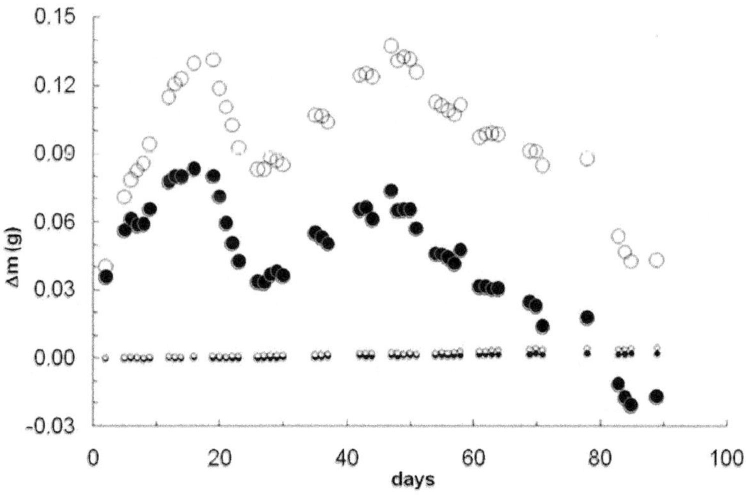

Figure 5.8 Variation of mass suffered by [EMIM] ethyl sulphate (O) and [EMIM][BF_4] (•) after being left for 3 months outside the window of our laboratory. The larger symbols represent the Eppendorf bins opened to atmosphere, while the small symbols represent those closed and used as reference.

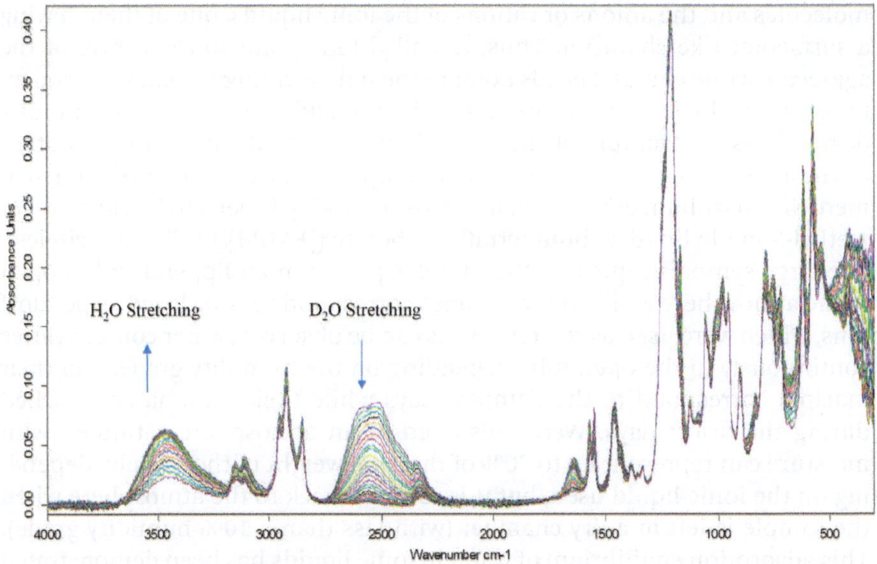

Figure 5.9 Thirty two FTIR spectra for a [EMIM][OSO$_4$] sample doped originally with heavy water ($x_{\text{ionic liquid}}$ = 0.2) with the sample open to the atmosphere during the measurement. Arrows indicate the order of the scans, opposite for the D_2O stretching peak (at about 2500 cm^{-1} and decreasing with time) than that for H_2O (at about 3400 cm^{-1} and increasing).

order of the scans). As can be seen, the peaks corresponding to stretching of D_2O molecules (at about 2500 cm^{-1}) decreased with time (so D_2O molecules were leaving the sample) while those corresponding to H_2O molecules (at about 3400 cm^{-1}) increased in a similar way due to the absorption of atmospheric water. At the end of the measurement the presence of heavy water in the sample was negligible and the net absorbed molecules is null (the peak height corresponding to water does not vary for the last scans). Such exchange of the two kind of water molecules can be stopped if the sample is closed during the measurement, but it restarts when the sample is opened to the atmosphere again. Obviously, the speed of the exchange process and the relaxation time to the equilibrium depends on the surface/volume ratio of the sample exposed to the atmosphere. The ionic liquid [EMIM][OSO$_4$] was the first one to form an ionic hydrogel during our investigations of the hygroscopic degree of different ionic liquids. We observed that [EMIM][OSO$_4$] was in a rigid-gel state after some hours left in a humidity saturated atmosphere. Later, studying the influence of cosolvents on the physical properties of alkyl sulphate ionic liquids with water and ethanol, we found that the octyl sulphate formed a gel state at about room temperature for some aqueous mixtures but not for ethanol.[75,76] Then, we focused on studying the gel state for this ionic liquid, building its phase diagram, where we observed that the rigid gel state appears for water concentrations between roughly 10 and 25 wt%.

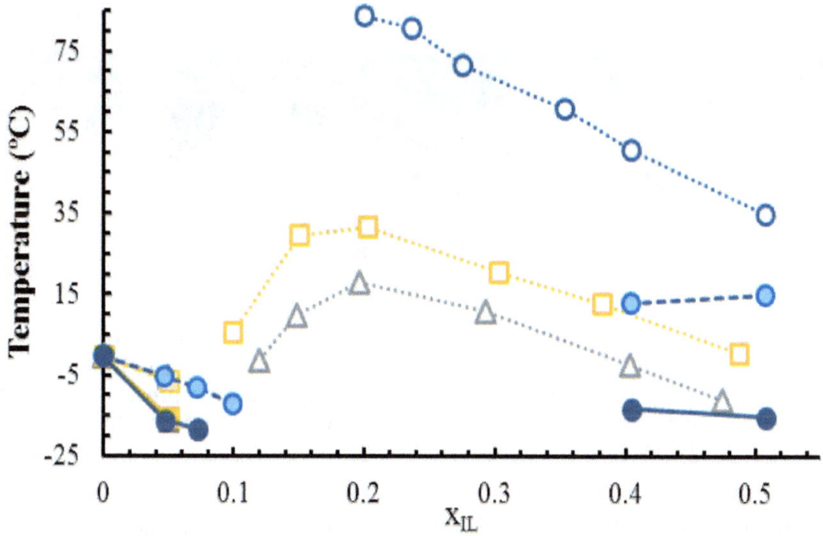

Figure 5.10 Phase diagram for the studied [EMIM] alkyl sulphates. Dotted line with open symbols represents liquid to gel transition, dashed line with clear symbols is for the transition from crystalline solid to liquid (or gel), while the continuous line with solid symbols represents the liquid (or gel) to crystal solid transition temperature. In addition, blue dot symbols are for [EMIM][DSO_4], yellow square symbols are for [EMIM][NSO_4] and grey triangle symbols are for [EMIM][OSO_4].

Moreover, the ionic conductivities, densities and viscosities for many different concentrations and over a broad range of temperatures were reported.[77] Recently, we have observed (but not yet reported) that jellification happens at room temperature for some other ionic liquids of the same family, as [EMIM] nonyl sulphate, [NSO_4]; [EMIM] decyl sulphate, [DSO_4]; 1-ethyl-1-metyl pyrrolidinium [DSO_4] or 1-ethyl-2,3 dimethyl imidazolium [DSO_4]. In addition, we observed that some protic ionic liquids also are ionic hydrogel formers (such as hexylammonium nitrate, HAN, or octylammonium nitrate, OAN). In Figure 5.10 we present the phase diagram for the three [EMIM] alkyl sulphates mentioned above for temperatures higher than −20 °C. The transition from crystalline solid to liquid (or gel) presents thermal hysteresis, as was observed for many other ionic liquids.[78] In Figure 5.10, the liquid to gel transitions (which do not show any hysteresis) for several ionic liquid mixtures are shown. As can be seen in this figure, the longer the alkyl chain, the higher the liquid–gel transition temperature, reaching about 85 °C for [EMIM][DSO_4]. At high enough water content, both solid transition temperatures are independent of the alkyl chain length. Interestingly, the three ionic liquids need a minimum of equimolar composition to become rigid gel (which represents only 5 wt% of water), while

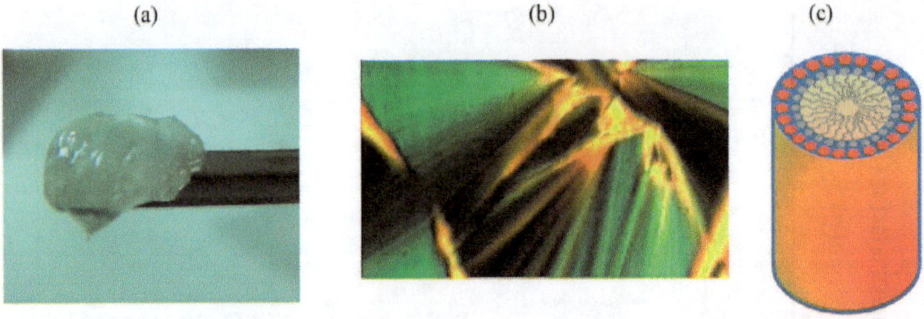

Figure 5.11 (a) Photograph of the rigid-gel state of [EMIM][DSO_4]. (b) White polarized micrograph of the same sample. (c) Artistic representation of a tubular micelle formed, where red pentagons represent the [EMIM] cations, grey tailed heads the alkyl sulphate anions and blue the water molecules.

the optimum concentration of the gel state is about $x_{\text{ionic liquid}} = 0.2$ (which represents one H_2O for each oxygen atom of the sulphate anion to form all possible H-bonds between both). The maximum water content in the gel state and its transition temperature increase with alkyl chain length. To explain gel formation, it must be taken into account that long alkyl chained sulphate anions have a surfactant like behaviour, forming micelles in the presence of water, and also have to form those of nonyl or decyl sulphates.[79] This surfactant character is not enough to form a quasi-solid mesomorphic state, because it is present in many other ionic liquids.[80] The chromonic character of the imidazolium, due to its aromatic ring, tends to align the imidazolium molecules, forming lamellar stacks glued by π-cloud interaction between neighbouring cations.[81] Both facts, the surfactant character of the anion and the chromonic character of the cation, contribute to generate tubular micelles in the aqueous mixtures, giving rise to a gel at the macroscopic level (Figure 5.11(a)), while at the microscopic level producing lyotropic mesophases as shown in Figure 5.11(b), obtained from polarized white light microscopy. Finally, an artistic representation of a micelle with such a microstructure is given in Figure 5.11(c), where red pentagons stand for [EMIM] cations, grey tailed heads represent alkyl sulphate anions and blue represent water molecules. Direct evidence of a nanostructure in these compounds has been observed by small-angle X-ray spectroscopy, SAXS.[82] Indeed, in Figure 5.9 it is clearly observed that the peak corresponding to H_2O stretching moves to higher wavenumbers with time, which indicates that H-bonds are weakened. As the sample was cool in a rigid gel state when we began that FTIR experiment, and it finished in a liquid state in the final runs, so H-bonds are stronger in the mesomorphic state than in the isotropic one, as expected. A very interesting aspect of the described ionic hydrogel is that its ionic conductivity does not appreciably change when it is in liquid or rigid-gel states, still being high enough (>1 S m^{-1}) to be used as electrolyte. Finally, we have discovered that salt addition changes the properties of this hydrogel, decreasing its electrical conductivity and increasing its viscosity, and also the liquid to rigid gel temperature transition can be tuned.[83]

5.2.3 Optical Properties

Optical limiters are essential components to protect eyes, sensors, detectors or any device from high intensity radiation. Negative nonlinear refraction and multiphotonic absorption are efficient mechanisms for implementing this function; they make the nonlinear medium itself regulate the irradiance of a beam traversing it by either increasing its divergence (as in the case of thermal negative nonlinear refraction) or by attenuating it (in the case of nonlinear absorption). The strength of the nonlinear effect depends on the input irradiance. On the other hand, saturable absorbers become transparent when the input irradiance increases above a reference value, finding application in laser technology.[84,85] One of the strategies followed by researchers for maximizing the nonlinear response[86] consists of inserting in the molecular structure elements with delocalized electrons, *e.g.* second and third row transition metals. These elements are usually combined with ions that supply electrons forming meta-ligand complexes. In addition, the anion needs to be paired with a large nonpolarizing cation to minimize the electrostatic interaction with the anion, which prevents delocalization of the electrons. The use of metal–ligand complexes determines also the colour of the material; consequently, many materials synthesized in this way are not transparent over the entire visible spectral region.

The first reports on the influence of transition metal-containing ionic liquids on the third-order nonlinear refraction and absorption of ionic liquids were published by del Sesto *et al.*[86,87] The ionic liquids under study were formed by tetradecyl(trihexyl)phosphonium cations paired with metal–ligand systems as anions. Tetraisothiocyanate cobaltate ($[Co(SCN)_4]^{2-}$), cobalt dicabollide ($[CoCB]^-$), cobalt phthalocyanine tetrasulphonate ($[CoPc]^{4-}$) and nickel dithiolate ($[Ni(dimit)_2]^{2-}$) have been used as anions. At a molecular level, the strength of cubic nonlinear effects is related to the second hyperpolarizability, and at the macroscopic level it is characterized in terms of the cubic susceptibility. Both are wavelength dependent. The relation between them is not evident in liquids and solids due to molecular interactions. Second hyperpolarizability is usually predicted by computational methods and susceptibility experimentally measured. By means of *ab initio* calculations, del Sesto *et al.*[86,87] predicted the second hyperpolarizability of the anions and the influence of the cation on the third-order nonlinear response. Besides, they measured the nonlinear refractive index and the nonlinear absorption (they are related to the third-order susceptibility) in the 700–1000 nm spectral region by the *z*-scan method, taking care to minimize thermal effects on the refractive index. They found some discrepancies between the measured and the predicted magnitudes (sign, strength) although qualitatively, allowing for the strength of the nonlinearity, the reported behaviour found for different anions follows the trends predicted numerically.

Recently, Nóvoa-López *et al.*[28] reported the existence of high nonlinear absorption at 810 nm in metal isothiocyanates of 1-butyl-3-methylimidazolium ionic liquids ($[BMIM]_m[X(SCN)_n]_q$ (X = Cu, Ni, Zn, Fe and Co; *m*, *n*, *q* are integers)). All the ionic liquids are coloured. Almost all are very weak absorbers at 810 nm except the one based on iron isothiocyanate. The observed nonlinear absorption is caused by a multiphotonic process, two or three photons being involved depending on the ionic liquid, and thermal effects. Those containing Fe, Co or Ni exhibit two-photon absorption-like behaviour and the ones with Zn or Cu, three-photon absorption. Notably, $[BMIM_4][Ni(SCN)_6]$ is the only solid sample at room temperature, melting at 333 K according to DSC experiments. Below the melting point it presents a crystalline structure. The use of a femtosecond laser source with a high repetition rate favours the development of thermo-optical effects, which reduces the refractive index and increases the divergence of the beam. Thermal effects are much stronger in these ionic liquids than those observed in others such as those studied in ref. 27, 88–93. The excess heating leads to effects like cavitation in liquid samples or to photodarkening in solid samples.[28]

5.2.3.1 Thermochromism

Many metal-complex-based ionic liquids are chromogenic materials,[94] which means that their colour can be altered by an external stimuli (temperature, moisture, pressure...). When temperature causes the colour change, the material is classified as thermochromic. Thermochromism may be continuous or discontinuous and reversible or irreversible. Discontinuous thermochromism finds applications in switchable devices. Continuous and reversible thermochromic materials are used as temperature indicators or smart glazing, for example. If they are sensitive to solar heat, they may be used as smart windows contributing to energy saving. Achievable temperatures by solar non-concentrated power are between 30 and 80 °C, so the use of ionic liquids can overcome the problems derived from evaporation of other liquids after different heating–cooling cycles.

The thermochromism of metal complexes in solvents is often attributed to tetrahedral–octahedral configuration changes in the metal. Wei *et al.*[95] prepared thermochromic compounds based on 1-butyl-3-methylimidazolium nickel tetrachloride ($[BMIM][NiCl_4]$) in 1-(3 hydroxypropyl)-3-methylimidazolium tetrafluoroborate and hexafluorophosphate ($[BMIM][BF_4]$ and $[BMIM][PF_6]$, respectively). They created a thermochromic composite formed by $[BMIM][NiCl_4]/[BMIM][BF_4]$ immobilized inside a film of PDVF that exhibits reversible thermochromism induced by solar heat.[96] Zhu *et al.* created a composite, by combining vanadium dioxide nanoparticles with an ionic liquid, of very high thermochromic performance.[97] Solutions of cobalt(II) isothiocyanate in [EMIM][SCN] or [BMIM][SCN] exhibit discontinuous and reversible thermochromism at low temperatures (233 K) due to reversible switching of cobalt coordination from tetrahedral to octahedral. The ionic liquid acts as a reservoir of SCN^- anions, which have quite large mobilities at low temperatures. The effect is cation dependent since it has not been observed with phosphonium cations. The alkyl chain length is also known to have a great influence.[26] Kohno *et al.*[98] prepared a new thermochromic ionic liquid formed by cobalt(II) bis(salicylate) anion and two tetra-*n*-butylphosphonium cations ($[P_{4444}]_2[Co(Sal)_2]$). In this ionic liquid, the colour change takes place when the material is exposed to liquid or gas water, because each cobalt coordinates with two water molecules forming the $[Co(Sal)_2(H_2O)_2]^{2-}$ anion. The water can be released by heating at moderate temperature or by vacuum procedures, so the ionic liquid recovers its original composition. Addition of water has been demonstrated to be crucial for favouring the thermochromism of cobalt in cobalt nitrate and cobalt chloride solutions in 1-(2-hydroxyethyl)-3-methylimizazolium nitrate ($[EOHMIM][NO_3]$).[99] Thermochromism has also been observed in ionic liquids containing uranyl isothiocyanate complexes: $[BMIM]_3[UO_2(SCN)_5]$, $[BMIM]_4[UO_2(SCN)_6]$ and $[EMIM]_3[UO_2(SCN)_5]$.[100] The authors attribute the origin of the effect to cation–anion interactions, as well as to changes in coordination number in uranyl ions. They noted that thermochromism is not observed after dilution in [BMIM][CSN] or $[BMIM][Tf_2N]$, but only in pure ionic liquids indicating that the anion–anion distance has a great influence in the process.

Meanwhile, nickel compounds have also been introduced in a cation to generate thermochromic ionic liquids with $[Tf_2N]^-$ as anion, as reported by Lan *et al.*[101] These authors studied three ionic liquids based on Ni(acac) $(Me_2NC_2H_4NR_1R_2)]$ (where acac = acetylacetonate) as cation and Tf_2N as anion. The chains R_1 and R_2 marked the difference among them ([A] R_1 = C_3H_6OEt, R_2 = Me; [B] R_1 = C_3H_6OMe, R_2 = Me). In the liquid state, the colour depends on the cationic open–closed forms ratio (open: diamagnetic species with square-planar four-coordination; closed: paramagnetic species with intramolecular side chain coordination), which can be related to the magnetic susceptibility values at each temperature. Neither ionic liquids exhibit thermochromism in the solid state, since they adopt the closed form. The ionic liquid [B] presents solvatochromism, as can be observed when it is diluted in different solvents (dichloromethane, acetone and DMSO), the colour of the solution being dependent on the solvent coordination. In addition, solutions in acetone and dichloromethane exhibit also thermochromism due to changes in coordination.

On our side, we have also observed a continuous colour change in pure $[BMIM_4][Ni(SCN)_6]$ from a green/blue at room temperature to a yellow/green at 333 K turning to grey and black at higher temperatures. Whenever the sample reaches a temperature lower than 525 K, the colour change is reversible as can be seen in Figure 5.12.

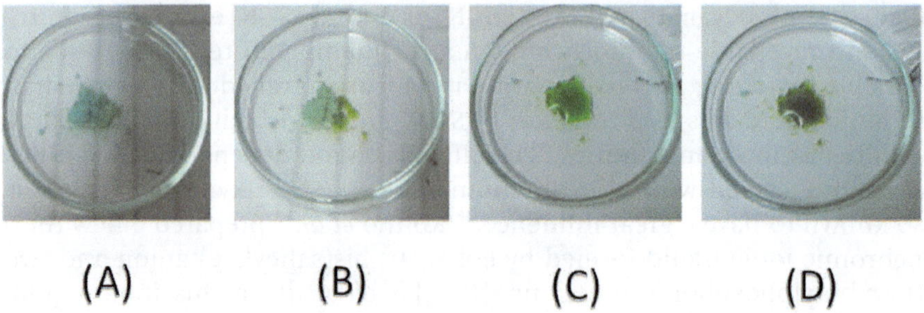

Figure 5.12 Colour change observed in $[BMIM]_4[Ni(NCS)_6]$ with increasing temperature: (A) 303, (B) 332, (C) 343 and (D) 368 K.

5.2.3.2 Electrochromism

Displays,[102] anti-glare mirrors,[103] smart windows[104] and so on are examples of advanced devices based on electrochromic materials.[105] An electrochromic device consists of three basic components: conducting surfaces, electrochromic compounds and electrolytes. Ionic liquids containing transition metal-complexes can act simultaneously as electrochromic material and electrolyte whenever they have anions or cations capable of reversible oxidation/reduction processes involving strong changes in colour, and/or their magnetic properties.[106]

Branco *et al.*[106] synthesized electrochromic ionic liquids based on cobalt(III) ethylenediaminetetraacetic complexes, paired with cations like 1-ethyl-3-methylimidazolium [EMIM], 1-butyl-3-methylimidazolium [BMIM], 1-octyl-3-methylimidazolium [OMIM], tri-octylmethylammonium [ALIQUAT] and trihexyltetradecylphosphonium $[P_{6,6,6,14}]$. The authors found that the choice of the transition metal is relevant since other compounds of chromium(III) and iron(III) ethylenediaminetetraacetic complexes combined with alkylimidazoliums do not exhibit enough contrast in their colour change to be proposed as electrochromic materials. Besides, they showed that the alkyl chain length allows fine tuning of the redox electrochromic behaviour. Apart from their electrochromic behaviour these liquids present a reversible diamagnetic/paramagnetic switch accompanying the colour change, and may find applications as paramagnetic materials for medical diagnosis since these metal–ligand complexes, combined with adequate cations, may reduce the toxicity of the traditional contrast agents based on lanthanides and transition metals.[106] Branco *et al.*[106] also reported the preparation of liquid electrochromic cells with trioctylmethylammonium vanadate ($[ALIQUAT][VO_3]$), trihexyltetradecylphosphonium vanadate ($[P_{6,6,6,14}][VO_3]$), 1-ethyl-3-methylimidazolium vanadate ($[EMIM][VO_3]$), 1-octyl-3-methylimidazolium vanadate ($[C_8MIM][VO_3]$) and 1-ethanol-3-methylimidazolium vanadate ($[C_2OHMIM][VO_3]$).

5.2.4 Magnetic Properties

Some of the earliest room temperature ionic liquids contained transition metal complexes in their anions (see for example ref. 1 in ref. 107). However, the first studies on magnetic ionic liquids are more recent.[108–110] The presence of transition metals in the cation or anion confers the ionic liquids with magnetic properties, with some of them being magneto-responsive in such a way that properties like conductivity or viscosity may be altered by the application of an external magnetic field.[111,112] In this line, Hayashi and Hamaguchi[108] discovered the paramagnetic behaviour of [BMIM][$FeCl_4$]. Over a wide range of temperatures (5–300 K), Yoshida *et al.*[109] studied the thermal and physical properties of ionic liquids formed by the cation 1-ethyl-3-methylimidazolium and metal chloride anions such as $(FeCl_4)^-$, $(Fe_{0.5}Ga_{0.5}Cl_4)^-$ and ($GaCl_4$). García-Saiz *et al.*[113] reported the influence of pressure in 1-ethyl-3-methylimidazolium tetrachloroferrate, [Emim][$FeCl_4$], material that shows a long-range antiferromagnetic ordering below the Néel temperature (T_N) ≈ 3.8, while most magnetic ionic liquids show a paramagnetic behaviour, evidencing the absence of long-range interaction between anions. The authors detected a ferromagnetic–antiferromagnetic three-dimensional pressure-driven order–disorder transition. At low pressures, the magnetic susceptibility presents a small peak at 3.8 K indicating the onset of magnetic ordering. As pressure increases up to 1 GPa, the susceptibility grows significantly when the temperature falls from 5–6 to 2 K, behaviour that is compatible with ferromagnetic ordering. For pressures as high as 1 GPa the susceptibility of the ionic liquids recovers its zero pressure value. García-Saiz *et al.* also studied the influence of magnetic fields and pressure on the reversibility of ZFC-FC magnetic susceptibility curves for temperatures between 2 and 7 K, showing that the irreversibility degree decreases as the magnetic field increases. The higher the pressure, the lower the magnetic field strength needed for hysteresis to disappear. Using Raman spectroscopy, they reported structural changes in the ionic liquids, and they found that throughout the full range of studied pressures $FeCl_4$ keeps the tetrahedral coordination but they recognized variations in the stretching and bending modes. Raman spectroscopy also revealed that the cation environment is more sensitive to pressure than that of the anion. Magnetic hysteresis is found to be coupled with a solid–liquid transition. Experiments reveal that it may be controlled with low magnetic fields and weak pressures.

On the other hand, del Sesto *et al.*[107] studied ionic liquids including tetraalkylphosphonium cation (PR_4) combined with $[FeCl_4]^{2-}$, $[MnCl_4]^{2-}$, $[CoCl_4]^{2-}$, $[GdCl_4]^{2-}$, and $[Co(NCS)_4]^{2-}$ and also C_{10}Mim $FeCl_4$ and C_4Mim $MnBr_4$. These ionic liquids show weak antiferromagnetic interaction discarding the formation of cluster and paramagnetic behaviour over a temperature interval of 20–350 K, responding strongly to applied magnetic fields. The advantage of using the tetraalkylphosphonium cations is the lack of acidic protons, in such a way that hydrogen bonding networks are prevented, contrary to what happens with imidazolium cations. In addition, phosphonium cations are hydrophobic, allowing magnetic transport in aqueous systems. Strongly diamagnetic particles of nitrogen gas as well as silica gel particles and carbon tetrachloride drops have also been transported in 1-butyl-3-methylimidazolium tetrachloroferrate.[114] The response in both cases obeys a similar mechanism, but obviously diamagnetic particles are repelled by the field and the paramagnetic are attracted.

Funasako *et al.*[115] showed that the magnetic susceptibility of the solid phase of a paramagnetic ionic liquid, butyl(octamethyl)-ferrocenium bis(trifluoromethanesulphonyl)amide, can be controlled applying weak magnetic fields during the solidification process, between 0.1 T (minimum to detect changes) and 2 T (when susceptibility reaches saturation). They observed hysteresis with the solidification temperature of 309 K and the solid–liquid transition of 299 K. The fact that the phase transition occurs at temperatures close to room temperature makes this ionic liquid a potential component in information recording devices. The change in the magnetic susceptibility is due to a change in the orientation of the material during the crystallization process, which can be seen through polarizing microscopy. The requirement for magnetic order to take place is simply that the structure of the crystal be magnetically anisotropic.

The magnetic properties of transition metal complexes in ionic liquids can be modified by external stimuli such as temperature or pressure, for instance. In fact, temperature-driven paramagnetic switching[26] has been observed in solutions of $[Co(SCN)]^-$ in [EMIM][SCN] and [BMIM][SCN] due to reversible switching of the coordination of cobalt(II) from a tetrahedral $[Co(SCN)_4]^{2-}$ environment at ambient temperature to an octahedral $[Co(SCN)_6]^{4-}$ coordination at temperatures below 230 K. The liquid acts as a SCN^- anion reservoir. This effect seems to be cation-dependent because larger phosphonium cations such as $[P_{6,6,6,14}]^+$ do not exhibit this behaviour, which is, besides, highly dependent on the length of the chain on the cation. Reversible coordination changes in cobalt isothiocyanate complexes have been induced by temperature and pressure in solutions of $Co(SCN)_2$ in $[P_{6,6,6,14}]$ [SCN].[111] The coordination changes are accompanied by an abrupt decrease of the dc conductivity in such a way that this magnitude may be used to monitor the coordination switching under different conditions of pressure and temperature.

5.2.5 Ionic Liquids Doped with Nanoparticles

Metal nanoparticles are of great interest in many technological areas such as biomedicine, energy, environment protection or information technologies, among many others. They can have properties different from those of typical of bulk media that depend on the size, shape, surrounding medium, composition and concentration. The optical absorption spectra of metal nanoparticles are characterized by surface plasmon resonance (SPR), which shifts to longer wavelengths with increasing particle size.[116] Thus, for instance, spherical silver nanoparticles show a plasmon band around 420 nm; the resonant wavelength shifts from 400 nm to 500 nm when the size goes from 10 to 100 nm, including a secondary peak at a lower wavelength for sizes larger than 50 nm, whose origin lies in the contribution of the quadrupole resonance. Copper nanoparticles of 30–50 nm diameter at 579 nm, gold nanoparticles at 530 nm, whereas ceria nanoparticles exhibit a strong absorption band below 400 nm, in the UV region.

The shape of nanoparticles also determines the peak of the SPR. For example, spherical nanoparticles present one absorption band (due to transversal oscillation) whereas rod like nanoparticles present two absorption bands corresponding to two oscillating modes: transversal and longitudinal. The electric field in the longitudinal oscillation mode is aligned to the axis of the rod, and the electric field of the transversal oscillation to the orthogonal direction. The SPR associated with the longitudinal mode shifts to low frequencies the more elongated the rods are.[117]

One of the challenges during the preparation of small nanoparticles is preventing agglomeration processes (flocculation or coagulation), which would lead to the growth of larger particles at the expense of the smaller ones. For that, a protective layer must be formed around the nanoparticle, adding some stabilizing agent. As mentioned previously, it is well known that ionic liquids are nanostructured fluids due to their particular electrostatic, hydrogen bond and van der Waals interactions. When a nanoparticle is immersed in an ionic liquid an adapted ionic layer is formed around it, allowing easy preparation of nanoparticle dispersions in ionic liquids without any additional stabilizer (see, for example, ref. 118–121 for further details).

The first paper dealing with the stabilization of nanoparticles in ionic liquids dates back to 2001 when Deshmukh *et al.* stabilized palladium nanoparticles in imidazolium based ionic liquids.[122] Since then, research dealing with gold NPs, and also silver, copper, platinum, palladium, iron and nickel nanoparticles, has developed.[123–127] Silver nanoparticles are particularly interesting due to low cost, bactericide activity, low toxicity and for being environmentally friendly, which confer them with potential applications in physics, chemistry, medicine, materials science and so forth. Wang and Yang[128] prepared dispersions of silver nanoparticles using an ionic liquid as solvent, oleic acid as capping agent and silver trifluoroacetate as reagent. Redel *et al.*[129] studied the influence of anionic volume of the ionic liquid on the nanoparticle size using hydrogen as reducing agent. They found that the nanoparticle diameter increases as the anionic volume increases. Kim *et al.*[130] reported the reduction of silver nitrate with ethylene glycol in presence of ionic liquid. Lazarus *et al.*[131] prepared silver nanoparticles by reduction of silver tetrafluoroborate in ionic liquids, considering the role of impurities in the formation of polydispersed nanoparticles with irregular shaping. Recently, Corrêa *et al.*[132] reported the synthesis of spherical silver nanoparticles in pure ionic liquids. The best results were obtained by using 1-butyl-3-methylimidazolium bis(trifluoromethanesulphonyl)imide and citrate tetrabutylammonium ionic liquids. In addition, the latter does not need PVP stabilized agent, since no changes were observeed in the plasmon band in UV-VIS spectra after several months. Typically, the surface plasmon absorption of silver nanoparticles exhibits a short wavelength band in the visible region around 420 nm, which is due to the transverse electronic oscillation. Spherical silver nanoparticles with surface plasmon bands in the range 400–430 nm were produced in all those studied ionic liquids, with the only exception being for the 1-octyl-3-methylimidazolium bis(trifluoromethanesulphonyl)imide.

On the other hand, gold nanoparticles have a wide range of applications, from charge storage systems to sensing of organic and biomolecules to photonic device fabrications. Applications include chemical sensors, ferrofluids, quantum dots, quantum computers, lithography, light-emitting diodes and flat-panel displays. Gold nanoparticles are highly customizable in size and shape, stable in a wide variety of environments, they are inert, non-toxic, and they have controllable optical-electronic properties. Gold NPs, due to heat dissipation of the absorbed energy by the SPR to the surrounding environment, can locally destroy, for example, a tumour cell, while weakly disturbing neighbouring ones. The wavelength of a surface plasmon resonance for gold spherical nanoparticles in water is around 530 nm. Variations in the nanoparticle size shift the peak of the plasmon. For example, Link and El-Sayed reported that spherical nanoparticles of 9, 22, 48 and 99 nm diameters have their SPR peaks at 517, 521, 533 and 575 nm, respectively.[133] The shape of the nanoparticle has a strong influence on the peak of the absorption band. It can vary from 527 to 1100 nm, depending on the shape and the dimensions ratio (see for example Table 1 in ref. 134). The growth and stability of gold nanoparticles has been investigated in ionic liquids such as [BMIM][N(CN)$_2$], [BMIM][Tf$_2$N], [BMIM][BF$_4$] and[BMIM][PF$_6$] by Vanecht *et al.*[135,136] Nanoparticle growth in these ionic liquids has been carried out by sputtering. It has been found that the growth is faster in [BMIM][N(CN)$_2$] than in the other ionic liquids. Viscosity is identified as the probable cause. The influence of the different ionic liquids on the stability of gold nanoparticle has been also investigated. The time elapsed until coagulation starts is proportional to the viscosity of the ionic liquid. Coagulation starts 2 hr, 12 h, 2 days, and 5 days after sputtering for [BMIM][N(CN)$_2$], [BMIM][Tf$_2$N], [BMIM][BF$_4$] and [BMIM][PF$_6$], respectively. In relation to sedimentation, it starts after 1 day for [BMIM][N(CN)$_2$] and [BMIM][Tf2N], after 5 days for [BMIM][BF$_4$] and after 9 days for [BMIM][PF$_6$].

Dispersions of nanoparticles of other metals have also received attention. For instance, hexagonal nanoparticles of CdS of 3–13 nm diameter have been prepared in [BMIM][MeSO$_4$], [BMIM][BF$_4$] and [BMIM][PF$_6$], respectively, showing an absorption band at 460 nm that shifts towards higher wavelengths with increasing particle size of the nanoparticles.[137] Finally, we mention that bimetallic nanoparticles have also been of interest due to their potential applications in biosensors and optoelectronic devices because their properties, as well as the frequency of surface plasmon resonance, are different from those observed in the corresponding monometallic nanoparticles.

Okazaki *et al.* reported the synthesis of spherical Au–Ag nanoparticles in [BMIM][PF_6].[138] Au and Ag nanoparticles have their SPR frequencies in the visible spectral region at different wavelengths. The addition of silver nanoparticles to [BMIM][PF_6] turns the ionic liquid yellow, the SPR wavelength now being 404 nm. The simultaneous addition of Au shifts linearly the plasmon frequency towards 504 nm, the SPR wavelength of pure Au nanoparticles. Contrarily, the optical absorption spectra of a mixture of Au nanoparticles with another of Ag nanoparticles solution presents two peaks, at 404 and 504 nm, whose intensity depend on the proportions of the mixture. Ag–Cu nanoparticles dispersed in [EMIM][BF_4] have also been prepared, by Suzuki and co-workers.[139]

5.3 Mass and Charge Transport in Mixtures of Ionic Liquids

Electrical conductivity of ionic liquids is one of their most characteristic properties, and it is a fundamental one in order to apply these compounds in electrochemical devices. Ionic liquids are among the best conducting liquids, with electrical conductivities only rivalled by mercury at ambient temperature. Although the electrical conductivity of pure ionic liquids is itself very important in many applications, in this work we will focus our attention on mixtures with molecular cosolvents and inorganic salts. The conductivity of these mixtures is indeed very much related to the solvation details of ions and molecules, since it is known that the integral solvation time of a dipolar solute is inversely proportional to the electrical conductivity of the ionic solvent or solution.[140]

As regards mixtures with molecular cosolvents, the addition of even small amounts to ionic liquids induces drastic changes in the conductivities of the mixtures,[18,141] which confer on these solvents a clear smart-material character. Moreover, conductivity maxima are usually registered in these mixtures (Figure 5.13), so they can be fine-tuned for optimizing their practical applications. As can be seen in Figure 5.13, the electrical conductivity of aqueous electrolyte solutions and mixtures of ionic liquids with water and alcohols follows a very simple parabolic trend (conducting dome) up to very high ionic concentrations, which has been theoretically rationalized. Probably the most promising theoretical approach to explain this experimental behaviour of the ionic conductivity values for mixtures of ionic liquids, saturated salts or deep eutectic solvents (DES) with molecular cosolvents was reported by Varela *et al.*,[142] who used a lattice model of the mixture – made up of low and high mobility cells – to calculate the conductivity of the mixture normalized by its maximum value, reporting:

$$\frac{\sigma}{\sigma_\mathrm{m}} = 2\frac{\phi}{\phi_\mathrm{m}}\left(1 - \frac{\phi}{2\phi_\mathrm{m}}\right) \tag{5.1}$$

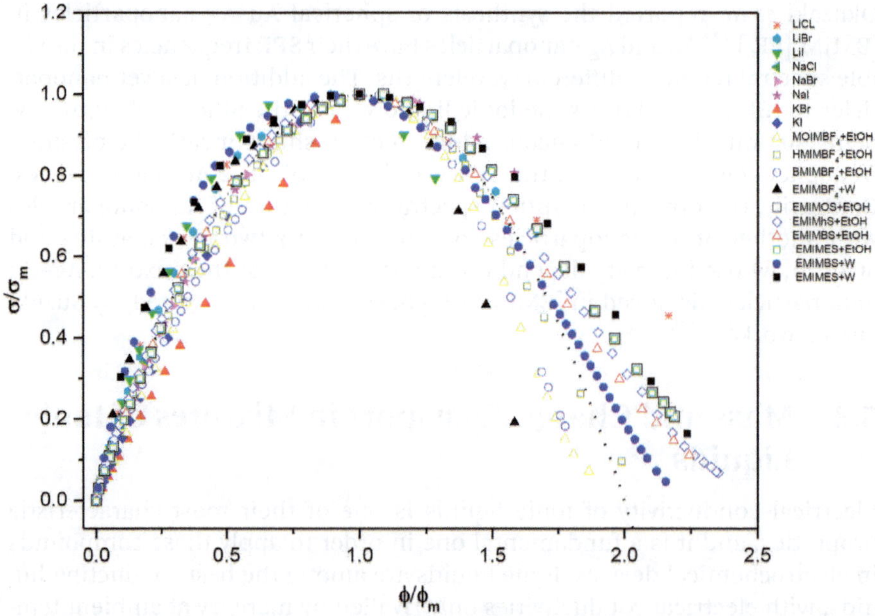

Figure 5.13 Normalized electrical conductivity *vs* normalized concentration for different aqueous electrolyte solutions and ionic liquid mixtures with water and ethanol at 298.15 K (data taken from ref. 142 and 146). Dotted line corresponds to the predictions of eqn (5.1).

where ϕ represents the volume fraction of ionic ions in the mixture and ϕ_m its value at the maximum conductivity. The predictions of the above equation are shown in Figure 5.13, and they are surprisingly accurate for a plethora of mixtures up to fairly high concentrations. Notwithstanding this, deviations are registered for very highly concentrated mixtures, which are probably related to the specific details of the solvation of water and alcohols in the dense ionic environment. Work is currently being done in order to generalize the theoretical treatment in ref. 142 to include the tendency of the solvent to selectively form clusters as a potential way to account for the deviations in the highly concentrated region.

Going back to [EMIM][C_nSO_4] ionic liquids family, some of which are ionic hydrogel formers, the electrical conductivity of pure compounds is not very high (about 0.05 S m^{-1} at 25 °C for [EMIM][OSO_4]) but its value increases when mixed with water up to 2.2 S m^{-1} at the same temperature. This giant increase in ionic conductivity with water happens in the majority of ionic liquids.[18] Other cosolvents will also increase the ionic conductivity, but more moderately. Thus, with ethanol a maximum of *ca* 1 S m^{-1} for the same compound at the same temperature can be reached.[141] Although it is not yet clear, the reason for this seems to be the different structure of the mixtures of these cosolvents, as revealed in Section 5.2. The ionic conductivity of most members of this family has been recently reported without added water.[143] The conductivity of ionic liquids decreases exponentially with the number of carbon atoms of the alkyl chain, with this number between 1 and 8. The hydrogen sulphate (n = 0) presents a much lower value than it must have (which is explained as due to its large increase in viscosity) while for $n > 8$ the ionic conductivity is similar for all long-chained alkyl sulphates (due to the very low mobility, if any, of the anion forming the tubular micelles). A Walden plot, which relates molar conductivity with fluidity,[144] shows that the pure [EMIM][HSO_4] and the aqueous solution (in rigid gel state) of [EMIM][OSO_4] are in the superionic regime, contrarily to the majority of ionic liquids. The behaviour of the former was recently attributed to the existence of proton transport of charge (Grotthuss mechanism), while the properties of the latter may be explained assuming a one-dimensional charge transport mechanism in the nanostructured columnar environment.[145] In the alkyl sulphate rigid gel the electrical charge would be transported by the [EMIM] cations, which have more mobility in the crust of the tubular micelles than the alkyl sulphate anions, which will be held in a fixed place by the H-bonds between water molecules and the alkyl sulphate head (as depicted in Figure 5.11c).

5.4 Conclusions

We have presented a critical review and some novel results on the importance of solvation mechanisms on the structure, optical, magnetic and transport properties of mixtures of ionic liquids and molecular cosolvents. The conclusion is that the solvation of substances in these dense ionic environments takes place in a selective way, with their polar and apolar regions being accommodated in the corresponding nanoregions of the bulk ionic liquid with the same character, in what has been termed nanostructured solvation. This specific solvation is critical for the structure and transport properties of the mixtures, as well as for their response to external stimuli, and hence is very usually behind the smart character of ionic liquids.

Acknowledgements

The authors wish to thank E. Tojo and J. A. Seijas for the synthesis of some of the ionic liquids presented here, and also M. Cabanas, technician at UDC, for some of the new measurements hereby reported. The financial support of the Spanish Ministry of Economy and Competitiveness (Projects MAT2014-57943-C3-1-P, MAT2014-57943-C3-2-P, MAT2014-57943-C3-3-P) is gratefully acknowledged. Moreover, this work was funded by the Xunta de Galicia (AGRUP2015/11 and GRC ED431C 2016/001 both partially supported by FEDER). Facilities provided by the Galician Supercomputing Centre (CESGA) are also acknowledged.

References

1. J. S. Wilkes and M. J. Zaworotko, *J. Chem. Soc., Chem. Commun.*, 1992, 965.
2. R. Hagiwara and Y. Ito, *J. Fluorine Chem.*, 2000, **105**, 221.
3. M. J. Earle and K. R. Seddon, *Pure Appl. Chem.*, 2000, **72**, 1391.
4. M. J. Earle, J. M. S. S. Esperança, M. A. Gilea, J. N. C. Lopes, L. P. N. Rebelo, J. W. Magee, K. R. Seddon and J. A. Widegren, *Nature (London)*, 2006, **439**, 831.
5. C. Wakai, A. Oleinikova, M. Ott and H. Weingärtner, *J. Phys. Chem. B*, 2005, **109**, 17028.
6. A. J. Carmichael and K. R. Seddon, *J. Phys. Org. Chem.*, 2000, **13**, 591.
7. R. D. Rogers and K. R. Seddon, *Science*, 2003, **302**, 792.
8. R. D. Rogers and K. R. Seddon, *Ionic Liquids: Industrial Applications for Green Chemistry*, American Chemical Society, Washington, DC, 2002.
9. V. Borgel, E. Markevich, D. Aurbach, G. Semrau and M. Schmidt, *J. Power Sources*, 2009, **189**, 331.
10. J. L. Anderson, D. W. Armstrong and G. T. Wei, *Anal. Chem.*, 2006, **78**, 2892–2902.
11. T. Welton, *Chem. Rev.*, 1999, **99**, 2071.
12. J. Le Bideau, L. Viau and A. Vioux, *Chem. Soc. Rev.*, 2011, **40**, 907.
13. O. Cabeza, J. Vila, E. Rilo, M. Domínguez-Pérez, L. Otero-Cernadas, E. López-Lago, T. Méndez-Morales and L. M. Varela, *J. Chem. Thermodyn.*, 2014, **75**, 52.
14. W. B. Spillman Jr, J. S. Sirkis and P. T. Gardiner, *Smart Mater. Struct.*, 1996, **5**, 247.
15. M. V. Gandhi and B. D. Thompson, *Smart Materials and Structures*, Springer, The Netherlands, 1992.
16. A. Hajalilou, S. Amri Mazlan, H. Lavvafi and K. Shameli, *Field Responsive Fluids as Smart Materials*, Springer, Singapore, 2016.
17. L. Rodríguez-Arco, A. Gómez-Ramírez, J. D. G. Durán and M. T. López-López, *New Perspectives for Magnetic Fluid-based Devices Using Novel Ionic Liquids as Carriers*, INTECH Open Access Publisher, 2012.
18. J. Vila, P. Ginés, E. Rilo, O. Cabeza and L. M. Varela, *Fluid Phase Equilib.*, 2006, **247**, 32.
19. M. M. A. Bicak, H. T. Belek and A. Göksenli, *Acta Polytechni. Czech Tech. Univ. Prague*, 2005, **45**, 68.
20. J. A. Marins, B. G. Soares, A. A. Silva, M. G. Hurtado and S. Livi, *J. Colloid Interface Sci.*, 2013, **405**, 64.
21. C. Guerrero-Sanchez, T. Lara-Ceniceros, E. Jimenez-Regalado, M. Ras and U. S. Schubert, *Adv. Mater.*, 2007, **19**, 1740.
22. N. Jain, X. Zhang, B. S. Hawkett and G. G. Warr, *ACS Appl. Mater. Interfaces*, 2011, **3**, 662.
23. C. M. Lamper, *Mater. Today*, 2004, **7**, 28.
24. L. C. Branco and F. Pina, *Chem. Commun.*, 2009, **41**, 6204.
25. A. Kavanagh, K. J. Fraser, R. Byrne and D. Diamond, *ACS Appl. Mater. Interfaces*, 2013, **5**, 55.

26. S. J. Osborne, S. Wellens, C. Ward, S. Felton, R. M. Bowman, K. Binnemans, M. Swadźba-Kwaśny, H. Q. Gunaratne and P. Nockemann, *Dalton Trans.*, 2015, **44**, 11286.
27. J. A. Nóvoa-López, E. López Lago, M. Domínguez, J. Troncoso, L. M. Varela, R. de la Fuente, O. Cabeza, H. Michinel and J. R. Rodríguez, *Opt. Laser Technol.*, 2014, **61**, 1.
28. J. A. Nóvoa-López, E. López Lago, J. A. Seijas, M. P. Vázquez-Tato, J. Troncoso, R. de la Fuente, J. R. Salgueiro and H. Michinel, *Opt. Mat.*, 2016, **52**, 144.
29. N. F. Corrêa, C. E. Santos, D. R. Valadão, L. F. de Oliveira, J. Dupont, M. A. Alencar and J. M. Hickmann, *Opt. Mater. Express*, 2016, **6**(1), 244.
30. Q. Zhang, I. Chung, J. I. Jang, J. B. Ketterson and M. G. Kanatzidis, *J. Am. Chem. Soc.*, 2009, **131**(29), 9896.
31. A. H. Reshak, H. Kamarudin, S. Auluck and I. V. Kityk, *J. Phys. Chem. B*, 2011, **115**, 11763.
32. B. Li, Y. Noda, L. Hu, H. Yoshikawa, M. M. Matsushita and K. Awaga, *Appl. Phys. Lett.*, 2012, **100**, 163304.
33. L. M. Varela, T. Méndez-Morales, J. Carrete, V. Gómez-González, B. Docampo-Álvarez, L. J. Gallego, O. Cabeza and O. Russina, *J. Mol. Liq.*, 2015, **210**, 178.
34. J. N. A. Canongia Lopes and A. A. H. Pádua, *J. Phys. Chem. B*, 2006, **110**, 3330.
35. A. Triolo, O. Russina, H.-J. Bleif and E. Di Cola, *J. Phys. Chem. B*, 2007, **111**, 4641.
36. J. Carrete, M. García, J. Rodríguez, O. Cabeza and L. M. Varela, *Fluid Phase Equilib.*, 2011, **301**, 118.
37. S. Cuadrado-Prado, M. Domínguez-Pérez, E. Rilo, S. García-Garabal, L. Segade, C. Franjo and O. Cabeza, *Fluid Phase Equilib*, 2009, **278**, 36.
38. J. Jacquemin, P. Husson, A. A. H. Padua and V. Majer, *Green Chem.*, 2006, **8**, 172.
39. B. Mokhtarani, A. Sharifi, H. R. Mortaheb, M. Mirzaei, M. Mafi and F. Sadeghian, *J. Chem. Thermodyn.*, 2009, **41**, 323.
40. I. B. Malham, P. Letellier and M. Turmine, *J. Phys. Chem. B*, 2006, **110**, 14212.
41. E. Rilo, J. Vila, M. García, L. M. Varela and O. Cabeza, *J. Chem. Eng. Data*, 2010, **55**, 5156.
42. E. Rilo, J. Vila, S. García-Garabal, L. M. Varela and O. Cabeza, *J. Phys. Chem. B*, 2013, **117**, 1411.
43. S. Fendt, S. Padmanabhan, H. W. Blanch and J. M. Prausnitz, *J. Chem. Eng. Data*, 2011, **56**, 31.
44. T. Murphy, L. M. Varela, G. B. Webber, G. G. Warr and R. Atkin, *J. Phys. Chem. B*, 2014, **118**, 12017.
45. S. Cha, M. Ao, W. Sung, B. Moon, B. Ahlstrom, P. Johansson, Y. Ouchi and D. Kim, *Phys. Chem. Chem. Phys.*, 2014, **16**, 9591.
46. T. Méndez-Morales, J. Carrete, O. Cabeza, L. J. Gallego and L. M. Varela, *J. Phys. Chem. B*, 2011, **115**, 6995.

47. S. Feng and G. A. Voth, *Fluid Phase Equilib.*, 2010, **294**, 148.
48. B. Docampo-Álvarez, V. Gómez-González, T. Méndez-Morales, J. Carrete, J. R. Rodríguez, O. Cabeza, L. J. Gallego and L. M. Varela, *J. Chem. Phys.*, 2014, **140**, 214502.
49. T. Méndez-Morales, J. Carrete, O. Cabeza, L. J. Gallego and L. M. Varela, *J. Phys. Chem. B*, 2011, **115**, 11170.
50. T. Méndez-Morales, J. Carrete, M. García, O. Cabeza, L. J. Gallego and L. M. Varela, *J. Phys. Chem. B*, 2011, **115**, 15313.
51. H. Montes-Campos, J. M. Otero-Mato, T. Méndez-Morales, E. López-Lago, O. Russina, O. Cabeza, L. J. Gallego and L. M. Varela, *J. Chem. Phys.*, 2017, **146**(12), 124503.
52. O. Russina, R. Caminiti, T. Méndez-Morales, J. Carrete, O. Cabeza, L. J. Gallego, L. M. Varela and A. Triolo, *J. Mol. Liq.*, 2015, **205**, 16.
53. T. Méndez-Morales, J. Carrete, S. Bouzón-Capelo, M. Pérez-Rodríguez, O. Cabeza, L. J. Gallego and L. M. Varela, *J. Phys. Chem. B*, 2013, **117**, 3207.
54. V. Gómez-González, B. Docampo-Álvarez, O. Cabeza, M. V. Fedorov, R. M. Lynden-Bell, L. J. Gallego and L. M. Varela, *J. Chem. Phys.*, 2015, **143**, 124507.
55. M. Matsumiya, S. Suda, T. Katsuhiko, M. Sugiya, S.-y. Kishioka and H. Matsuura, *J. Electroanal. Chem.*, 2008, **622**, 129.
56. N. Banić, M. Vraneš, B. Abramović, J. Csanádi and S. Gadžurić, *Dalton Trans.*, 2014, **43**, 15515.
57. A. Babai and A. V. Mudring, *Chem. Mater.*, 2005, **17**, 6230.
58. S. Arenz, A. Babai, K. Binnemans, K. Driesen, R. Giernoth, A. V. Mudring and P. Nockemann, *Chem. Phys. Lett.*, 2005, **402**, 75.
59. T. Graham, *Philos. Trans.*, 1861, **151**, 183.
60. P. J. Flory, *Faraday Discuss. Chem. Soc.*, 1974, **57**, 7.
61. R. Koynova and B. Tenchov, *OA Biochem.*, 2013, **1**, 1.
62. K. Trickett and J. Eastoe, *Adv. Colloid Interface Sci.*, 2008, **144**, 66.
63. J. W. Steed, *Chem. Commun.*, 2011, **47**, 1379.
64. M. O. M. Piepenbrock, G. O. Lloyd, N. Clarke and J. L. Steed, *Chem. Rev.*, 2010, **110**, 1960.
65. L. A. Estroff and A. D. Hamilton, *Chem. Rev.*, 2003, **104**, 1201.
66. P. C. Marr and A. C. Marr, *Green Chem.*, 2016, **18**, 105.
67. Y.-S. Ye, J. Rick and B.-J. Hwang, *J. Mater. Chem. A*, 2013, **1**, 2719.
68. A. Noda and M. Watanabe, *Electrochim. Acta*, 2000, **45**, 1265.
69. P. Vidinha, N. M. T. Lourenço, C. Pinheiro, A. R. Brás, T. Carvalho, T. Santos-Silva, A. Mukhopadhyay, M. J. Romaô, J. Parola, M. Dionisio, J. M. S. Cabral, C. A. M. Afonso and S. Barreiros, *Chem. Commun.*, 2008, 5842.
70. S. Doherty, J. G. Knight, J. R. Ellison, P. Goodrich, L. Hall, C. Hardacre, M. J. Muldoon, S. Park, A. Ribeiro, C. A. Nieto de Castro, M. J. Lourenço and P. Davey, *Green Chem.*, 2014, **16**, 1470.
71. N. Yoshimoto, T. Shira and M. Morita, *Electrochim. Acta*, 2005, **50**, 3866.
72. M. Cai, Y. Liang, F. Zhou and W. Liu, *J. Mater. Chem.*, 2011, **21**, 13399.
73. A. S. Shaplov, R. Marcilla and D. Mecerreyes, *Electrochim. Acta*, 2015, **175**, 18.

74. S. Cuadrado-Prado, M. Domínguez-Pérez, E. Rilo, S. García-Garabal, L. Segade, C. Franjo and O. Cabeza, *Fluid Phase Equilib.*, 2009, **278**, 36.
75. E. Rilo, J. Vila, S. García-Garabal, L. M. Varela and O. Cabeza, *J. Phys. Chem. B*, 2013, **117**, 1411.
76. E. Rilo, L. M. Varela and O. Cabeza, *J. Chem. Eng. Data*, 2012, **57**, 2136.
77. O. Cabeza, J. Vila, E. Rilo, M. Domínguez-Pérez, L. Otero-Cernadas, E. López-Lago, T. Méndez-Morales and L. M. Varela, *J. Chem. Thermodyn.*, 2014, **75**, 52.
78. J. Vila, B. Fernández-Castro, E. Rilo, J. Carrete, M. Domínguez-Pérez, J. R. Rodríguez, M. García, L. M. Varela and O. Cabeza, *Fluid Phase Equilib.*, 2012, **320**, 1.
79. T. Singh, M. Drechsler, A. H. E. Müeller, I. Mukhopadhayay and A. Kumar, *Phys. Chem. Chem. Phys.*, 2010, **12**, 11728.
80. E. Rilo, J. Pico, S. García-Garabal, L. M. Varela and O. Cabeza, *Fluid Phase Equilib.*, 2009, **285**, 83.
81. J. E. Lydon, *Curr. Opin. Colloid Interface Sci.*, 2004, **8**, 480–490.
82. O. Russina, l. Gontrani, B. Fazio, D. Lombardo, A. Triolo and R. Caminiti, *Chem. Phys. Lett.*, 2010, **493**, 259.
83. O. Cabeza, L. Segade, M. Domínguez-Pérez, E. Rilo, S. García-Garabal, D. Ausín, A. Martinelli, E. López-Lago and L. M. Varela, *J. Chem. Thermodyn.*, 2017, **112**, 267.
84. J. A. Morris and C. R. Pollock, *Opt. Lett.*, 1990, **15**, 440.
85. A. J. DeMaria, D. A. Stetser and H. Heynau, *Appl. Phys. Lett.*, 1966, **8**, 174.
86. R. E. del Sesto, D. S. Dudis, F. Ghebremichael, N. E. Heimer, T. K. C. Low, J. S. Wilkes and A. T. Yeates, in *Nonlinear Optical Ionic Liquids, Ionic Liquids IIIB: Fundamentals, Progress, Challenges, and Opportunities*, ed. R. D. Rogers and K. R. Seddon, ACS Symposium Series, 2005, ch. 12, vol. 902, pp. 144–158.
87. R. E. del Sesto, D. S. Dudis, F. Ghebremichael, N. E. Heimer, T. K. C. Low, J. S. Wilkes and A. T. Yeates, Modeling, synthesis and characterization of third order nonlinear optical salts, *Proc. SPIE*, 2003, **5292**, 292–299.
88. C. D. Tran, S. Challa and M. Franko, *Anal. Chem.*, 2005, **77**, 7442.
89. R. F. Souza, M. A. R. C. Alencar, M. R. Meneghetti, J. Dupont and J. M. Hickmann, *J. Phys.: Condens. Matter*, 2008, **20**, 155102.
90. C. E. Valencia-Loredo, K. A. Barrera-Rivera, M. Trejo-Durán, E. Alvarado-Méndez, A. Martínez Richa and J. A. Andrade-Lucio, Nonlinear optical characterization of ionics liquids, *Proc. SPIE*, 2009, **7386**, 738610.
91. M. Trejo-Duran, E. Alvarado-Méndez, E. Vargas-Rodríguez, J. M. Estudillo-Ayala and R. I. Mata-Chavez, Nonlinear optical characterization of ionics liquids of 1-methylpyrrolidine family, *Proc. SPIE*, 2011, **8412**, 84121X.
92. E. Santos, M. A. Alencar, P. Migowski, J. Dupont and J. M. Hickmann, *Chem. Phys.*, 2012, **403**, 33.
93. E. A. Cássio, E. A. Santos, M. A. R. C. Alencar, P. Migowski, J. Dupont and J. M. Hickmann, *Adv. Mater. Sci. Eng.*, 2013, 104914.
94. D. S. Lamber, *Mater. Today*, 2004, **7**, 28.
95. X. Wei, L. Yu, D. Wang, X. Jin and G. Z. Chen, *Green Chem.*, 2008, **10**, 296.

96. X. Wei, L. Yu, X. Jin, D. Wang and G. Z. Chen, *Adv. Mater.*, 2009, **21**, 776.
97. J. Zhu, A. Huang, H. Ma, Y. Ma, K. Tong, S. Ji, S. Bao, X. Cao and P. Jin, *ACS Appl. Mater. Interfaces*, 2016, **8**, 29742.
98. Y. Kohno, M. G. Cowan, M. Masuda, I. Bhowmick, I. M. P. Shores, D. L. Gin and R. D. Noble, *Chem. Commun.*, 2014, **50**, 6633.
99. N. Banić, M. Vraneš, B. Abramović, J. Csanádi and J. Gadžurić, *Dalton Trans.*, 2014, **43**, 15515.
100. N. Aoyagi, K. Shimojo, N. R. Brooks, R. Nagaishi, H. Naganawa, K. Van Hecke, L. Van Meervelt, K. Binnemans and T. Kimura, *Chem. Commun.*, 2011, **47**, 4490.
101. X. Lan, T. Mochida, Y. Funasako, K. Takahashi, T. Sakurai and H. Ohta, *Chemistry*, 2017, **23**, 823.
102. W. Lu, B. R. Mattes, A. G. Fadeev and B. Qi, Stable conjugated polymer electrochromic devices incorporating ionic liquids, US Pat. 6667825 B2, 2003.
103. A. Agrawal, J. C. Lopez Tonazzi, R. S. LeCompte and J. P. Cronin, Electrochromic mirrors and other electrooptic devices, US Pat. 7300166 B2, 2010.
104. G. C. Granqvist, A. Azens, J. Isidorsson, M. Kharrazi, L. Kullman, T. Lindström, G. A. Niklasson, C.-G. Ribbing, D. Rönnow and M. Veszelei, *J. Non-Cryst. Solids*, 1997, **218**, 273.
105. P. M. S. Monk, J. R. Mortimer and D. R. Rosseinsky, *Electrochromism and Electrochromic Devices*, Cambridge University Press, 2007.
106. A. Branco, L. C. Branco and F. Pina, *Chem. Commun.*, 2011, **47**, 2300–2302.
107. R. E. del Sesto, T. M. McCleskey, A. K. Burrell, G. A. Baker, J. D. Thompson, B. L. Scott, J. S. Wilkes and P. Williams, *Chem. Commun.*, 2008, 447.
108. S. Hayashi and H. Hamaguchi, *Chem. Lett.*, 2004, **33**, 1590.
109. Y. Yoshida, J. Fujii, K. Muroi, A. Otsuka, G. Saito, M. Takashasi and T. Yoko, *Synth. Met.*, 2005, **153**, 421.
110. R. E. del Sesto, C. Corley, A. Robertson and J. S. Wilkes, *J. Organomet. Chem.*, 2005, **690**, 2536.
111. E. McCourt, Z. Wojnarowska, J. Jacquemin, P. Nockemann, H. G. Manyar, L. Hawelek and M. Paluch, *J. Phys. Chem. C*, 2016, **120**, 10156.
112. S. H. Lee, S. H. Ha, S. S. Ha, H. B. Jin, C. Y. You and Y. M. Koo, *J. Appl. Phys.*, 2007, **101**, 09J102.
113. A. García-Saiz, I. de Pedro, J. A. Blanco, J. González and J. Rodríguez Fernández, *J. Phys. Chem. B*, 2013, **117**, 3198.
114. M. Okuno, H. Hamaguchi and S. Hayashi, *Appl. Phys. Lett.*, 2006, **89**, 132506.
115. Y. Funasako, T. Mochida, T. Inagaki, T. Sakurai, H. Ohta, K. Furukawa and T. Nakamura, *Chem. Commun.*, 2011, **47**, 4475.
116. M. A. García, *J. Phys. D: Appl. Phys.*, 2011, **44**, 283001.
117. B. M. van der Zande, M. R. Böhmer, L. G. Fokkink and C. Schönenberger, *Langmuir*, 2000, **16**, 451.
118. C. Janiak, *Z. Naturforsch. B*, 2013, **68b**, 1059.

119. J. Dupont, *Chem. Soc.*, 2004, **15**, 341.
120. J. Dupont and J. D. Scholten, *Chem. Soc. Rev.*, 2010, **39**, 1780.
121. J. Dupont and P. Suárez, *Phys. Chem. Chem. Phys.*, 2006, **8**, 2441.
122. R. R. Deshmukh, R. Rajagopal and K. V. Srinivasan, *Chem. Commun.*, 2001, 1544.
123. P. Singh, A. Katyal, R. Kalraa and R. Chandra, *Tetrahedron Lett.*, 2008, **49**, 727.
124. C. C. Cassol, A. P. Umpierre, G. Machado, S. I. Wolke and J. Dupont, *J. Am. Chem. Soc.*, 2005, **127**, 3298.
125. K. S. Kim, D. Demberelnyamba and H. Lee, *Langmuir*, 2004, **20**, 556.
126. J. Krämer, E. Redel, R. Thomann and C. Janiak, *Organometallics*, 2008, **27**, 1976.
127. P. Migowski, G. Machado, S. R. Texeira, M. C. Alves, J. Morais, A. Traverse and J. Dupont, *Phys. Chem. Chem. Phys.*, 2007, **9**, 4814.
128. Y. Wang and H. Yang, *Chem. Commun.*, 2006, **24**, 2545.
129. E. Redel, R. Thomann and C. Janiak, *Inorg. Chem.*, 2008, **47**, 14.
130. T. Y. Kim, J. H. Yeon, S. R. Kim, C. Y. Kim, J. P. Shim and K. S. Suh, *Phys. Chem. Chem. Phys.*, 2011, **13**, 16138.
131. L. L. Lazarus, C. T. Riche, N. Malmstadt and R. L. Brutchey, *Langmuir*, 2012, **28**, 5987.
132. C. M. Corrêa, M. A. Bizeto and F. F. Camilo, *J. Nanopart. Res.*, 2016, **18**, 132.
133. S. Link and M. A. El-Sayed, *J. Phys. Chem. B*, 1999, **103**, 4212.
134. H. Chen, X. Kou, Z. Yang, W. Ni and J. Wang, *Langmuir*, 2008, **24**, 5233.
135. E. Vanecht, K. Binnemans, J. W. Seo, L. Stappers and J. Fransaer, *Phys. Chem. Chem. Phys.*, 2011, **13**, 13565.
136. E. Vanecht, K. Binnemans, S. Patskovsky, M. Meunier, J. W. Seo, L. Stappers and J. Fransaer, *Phys. Chem. Chem. Phys.*, 2012, **14**, 5662.
137. K. Biswas and C. N. R. Rao, *Chemistry*, 2007, **13**, 6123.
138. K. Okazaki, T. Kiyama, K. Hirahara, N. Tanaka, S. Kuwabata and T. Torimoto, *Chem. Commun.*, 2008, 691.
139. S. Suzuki, Y. Tomita, S. Kuwabata and T. Torimoto, *Dalton Trans.*, 2015, **44**, 4186.
140. X. X. Zhang, M. Liang, N. P. Ernsting and M. Maroncelli, *J. Phys. Chem. Lett.*, 2013, **4**, 1205.
141. E. Rilo, J. Vila, S. García-Garabal, L. M. Varela and O. Cabeza, *J. Phys. Chem. B*, 2013, **117**, 1411.
142. L. M. Varela, J. Carrete, M. García, L. J. Gallego, M. Turmine, E. Rilo and O. Cabeza, *Fluid Phase Equilib.*, 2010, **298**, 280.
143. S. García-Garabal, J. Vila, E. Rilo, M. Domínguez-Pérez, L. Segade, E. Tojo, P. Verdía, L. M. Varela and O. Cabeza, *Electrochim. Acta*, 2017, **231**, 94.
144. W. Xu, E. I. Cooper and C. A. Angell, *J. Phys. Chem. B*, 2003, **107**, 6170.
145. H. Shimura, M. Yoshio, K. Hoshino, T. Mukai, H. Ohno and T. Kato, *J. Am. Chem. Soc.*, 2008, **130**, 1759.
146. V. M. M. Lobo and J. L. Quaresma, *Handbook of Electrolyte Solutions: Part A and Part B, Physical Sciences Data 41*, Elsevier, 1989.

CHAPTER 6

Laser Deposition of Nano-ionic Liquids and Their Process Applications in a Vacuum

YUJI MATSUMOTO* AND SHINGO MARUYAMA

Department of Applied Chemistry, School of Engineering, Tohoku University, 6-6-07, Aramaki-Aza-Aoba, Aoba-ku, Sendai City, Miyagi 980-8579, Japan
*E-mail: y-matsumoto@tohoku.ac.jp

6.1 Introduction

Vacuum technologies involved in material processing and analysis, as represented by the industrial semiconductor processes, have become indispensable today for modern electronics. Such vacuum technologies are now extending their application to liquid materials as well as solid materials. The belief that a "liquid cannot be stable in vacuum" is just a belief, probably due to our daily experience that water is ready to evaporate in vacuum. On the contrary, there are more materials than expected that are substantially stable as a liquid in vacuum, such as Ga, In and Al melts with a low melting point,[1–3] high-temperature oxide melts[4] and others including silicon-oils[5] and bis(2-ethylhexyl) sebacate (B2EHS).[6] These kinds of liquids that are stable in vacuum allow us not only to prepare a model liquid system subjected

Smart Materials No. 28
Ionic Liquid Devices
Edited by Ali Eftekhari

Published by the Royal Society of Chemistry, www.rsc.org

to precision high-vacuum analyses, but also to develop new material processes with liquid in a vacuum.

Recently, there has been growing interest in ionic liquids (ILs) that are available as a liquid in a vacuum. ILs are a family of molten salts of a pair of ions either or both of which is (are) molecular ions.[7] Most ILs are in a liquid state at room temperature (RT), transparent in a visible light region, and immiscible with water or oil, or both. They also have a high viscosity and an extremely low vapor pressure. The origin of the low vapor pressure is still controversial but is probably because of their long-range Coulomb interaction.[8,9] The strong molecular ion interaction results in a vaporization enthalpy typically ranging from 120 to 150 kJ mol^{-1},[10] almost three or four times larger than that of water (40.7 kJ mol^{-1}). In fact, the vapor pressure is extrapolated to be on the order of 10^{-10} Torr at RT,[11] indicative of a stable survival of ILs even in an ultra-high vacuum (UHV).

Owing to the extremely low vapor pressure of ILs, many researchers so far tried to introduce ILs into a vacuum for mainly two purposes, as already pointed out in the general cases. One is that the physical and chemical properties of ILs are investigated in a vacuum by using high-vacuum analysis techniques as a model liquid system. Examples include fundamental studies of ILs with X-ray-photoelectron spectroscopy (XPS),[12,13] thermal desorption spectroscopy (TDS)[14,15] and scanning electron microscopes (SEMs).[16] More recently, high-resolution transmission electron microscopy (HR-TEM)[17] and low-temperature scanning tunneling microscopes (STMs)[18] have become available for molecular level analyses of ILs. The second aim is to use ILs in vacuum processing. For example, ILs are used as a solvent into which metal is sputter-deposited for an efficient production of metal nanoparticles,[19] as a nonmetal liquid electrode in a plasma process[20] and as an electrolyte solution to realize electrochemical analysis[21,22] in vacuum.

Following these pioneering works with ILs in a vacuum, we also have developed some new vacuum processes with IL since 2010, especially elaborating a new vacuum deposition technique for ILs.[23] As already overviewed above, a bulky IL has been used in most vacuum analyses and process applications, which may become an obstacle to their further applications in nano-science and -technology. The vacuum deposition of ILs may be one of the promising techniques to prepare nanoscale ILs, "nano-ILs", which enable the nanoscale physics and chemistry of ILs, in analogy with solid-state physics and semiconductor technology.

Here, a question may arise: What are such "nano-ILs" useful for? One potential application from a chemical point of view is that such nano-ILs can be used as a nanoscale reactor as well as a solvent available in a vacuum (Figure 6.1(a)). In the nano-IL reactor, the low diffusion process of reactants and/or products in a high-viscous IL solvent will not become a serious problem, unlike the case of bulk ILs. The nanoscale IL reactor may not only simplify a chemical process, but also provide a new reaction field characteristic to the nanoscale. Another potential application arises from a more device physics–oriented viewpoint. If various forms of the nano-ILs come to be manufactured,

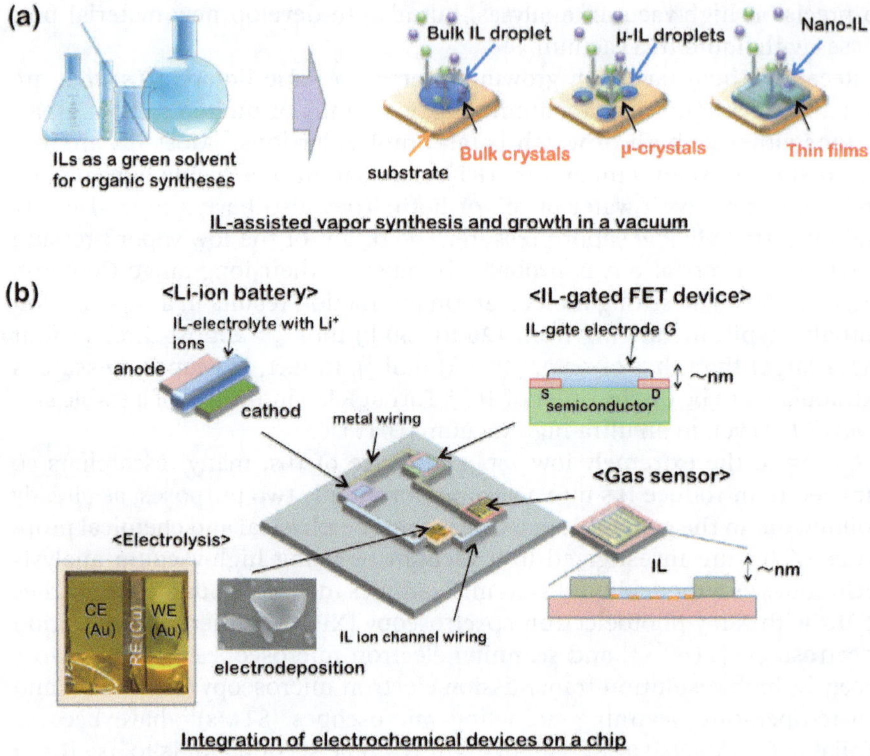

Figure 6.1 Applications of "nano-ILs". (a) Nano-IL reactors for IL-assisted vapor synthesis and growth in a vacuum. (b) Integration of electrochemical devices for IL-based ionics.

e.g. as thin film electrolytes and ion channel wires, such nano-ILs components allow us to downsize and integrate electrochemical devices for IL-based ionics operatable in a vacuum(!), in a similar manner to the way in which solid-state devices are integrated (Figure 6.1(b)). For example, the downsizing and integration of inorganic and organic field effect transistors (FETs) with IL as a gate electrode would be very interesting. IL-gated FET devices have recently attracted much attention because their higher gate voltage applicability than that in the all solid-state FET devices can realize new functionalities such as a reversible control of superconducting transition in semiconductors.[24,25]

These applications of nano-ILs require an understanding of fundamental behaviors and properties, such as ion conductivity, viscosity and glass transition temperatures as well as liquid structures, characteristic to the "nano-ILs. In fact, recent studies claim that when an IL comes into contact with a solid substrate a solid-like structure of the IL is formed within a nanoscale region just above the substrate surface[26] and its viscosity increases.[27]

With this background, in Section 6.2 we first introduce our originally developed laser deposition technique of ILs, along with a brief comparison of other deposition techniques of the IL, and then discuss the fundamental deposition process unique to the IL and demonstrate the fabrication of various nano-ILs. In Section 6.3, we propose a new concept of IL-assisted vapor synthesis and growth, in which precursors and reactants are supplied in some way from the gas phase in a vacuum into the micro- and nano-IL reactors, prepared on a substrate in advance, and verify its capability as a new synthetic route to high-quality organic and inorganic single crystals and films. Finally, in Section 6.4, we conclude this chapter with some future aspects of the laser deposition techniques of the nano-ILs.

6.2 Laser Deposition of Ionic Liquids

The year 2005, until when many reports on ILs investigated in a vacuum had been published with a common preconception among the research community that ILs would not evaporate even in vacuum, saw a breakthrough in IL-related research in a vacuum. L. P. N. Rebelo and co-workers first demonstrated that ILs could evaporate without decomposition in vacuum.[11,28] There had been few reports on the evaporation of ILs to measure their thermodynamic properties such as a vapor pressure and vaporization enthalpy,[10] the distillation of ILs in a vacuum[29] and the deposition processes of any ILs in a vacuum.[13] Vapor deposition of ILs and related studies that have been so far reported in the literature are listed in Table 6.1.

6.2.1 Continuous-wave Infrared Laser Deposition Method

A pulsed laser deposition (PLD) method with intense ultraviolet (UV) laser,[60] such as an excimer laser or a high-order harmonic generated Nd:YAG laser, has been used to deposit inorganic materials with a high melting temperature, *e.g.* oxides, nitrides, and carbides as well as metals. In contrast, for materials with a low melting temperature, no matter whether they are inorganic or organic materials, a thermal evaporation method with a resistive or electron bombardment heating device, *e.g.* known as a "Knudsen cell" or "E-gun evaporator", is used. For deposition of IL, the thermal evaporation method is usually used, as listed in Table 6.1. However, some ILs are thermally unstable even at their evaporation temperature, which may cause a gradual decomposition of such an IL because of its continuous, long heat exposure even in a vacuum.

To avoid such possible decomposition of ILs during deposition, we have developed a continuous wave infrared (CW-IR) laser deposition system as shown in Figure 6.2(a).[23] The deposition system is essentially the same as the PLD system, but a CW-IR laser (808 nm) is used instead of an intense UV laser in order to thermally evaporate the target material. Prior to the first attempt to deposit IL, we had verified how useful this IR laser deposition method was for depositing some low-molecular-weight organic semiconductor compounds such as pentacene[61,62] and C_{60}.[63] For example, a clear observation of the reflection high energy electron diffraction (RHEED) intensity oscillation,

Table 6.1 List of publications on vacuum deposition of ILs and characterization.

Ionic liquid	Substrate	Deposition technique	Observation technique	Reference (Author, Year)
[EMIM][Tf_2N]	Glass	Thermal evaporation	XPS	Cremer 2008[13]
[EMIM][Tf_2N]	Ni(111)	Thermal evaporation	TOF-SIMS	Souda 2008[30]
[EMIM][Tf_2N]	HOPG	Thermal evaporation	TOF-SIMS	Souda 2009[31]
[BMIM][Tf_2N]	Al_2O_3/NiAl(110)	Thermal evaporation	IRRAS	Sobota 2010[32]
[bmim][PF_6]	α-Al_2O_3(0001)	CW-IR laser deposition	OM, AFM	Kato 2010[33]
[emim][NTf_2], [bmim][NTf_2], [bmim][PF_6], [bmim][BF_4], [PC_{66614}][NTf_2], [hmim][OTf], [hmim][OTf], [emim][$EtSO_4$],	α-Al_2O_3(0001), Si(100), C_{60}/Si(100)	CW-IR laser deposition	OM, LM, AFM, SEM, NMR, Raman, ESI-TOF-MS	Maruyama 2010[23]
[C_1C_1Im][Tf_2N], [C_8C_1Im][Tf_2N]	Au(111)	Thermal evaporation	XPS	Cremer 2011[34]
[bmim][NTf_2]-[bmim][PF_6]	α-Al_2O_3(0001)	CW-IR laser deposition	Raman, LM	Maruyama 2011[35]
[Py1,4][FAP]	Au(111)	Thermal evaporation	STM	Waldmann 2011[36]
[C_1C_1Im][Tf_2N]	Au(110)	Thermal evaporation	UPS, LEED, STM	Foulston 2012[37]
[BMIM][Tf_2N]	Al_2O_3/NiAl(110)	Thermal evaporation	IRRAS	Sobota 2012[38]
[C_1C_1Im][Tf_2N], [C_4C_1Im][Tf_2N]	Mica	Thermal evaporation	XPS	Deyko 2013[39]
[5-oxo-C_6C_1Im][Tf_2N]	CeO_2(111)/Cu(111) CeO_{2-x}/Cu(111)	Thermal evaporation	IRRAS	Schernich 2013[40]
[BMP][TFSA]	Ag(111)	Thermal evaporation	STM, XPS	Buchner 2013[41]

$[EMIM][Tf_2N]$	Hydroxylated $SiO_2/Si(100)$ Porous $SiO_2/Si(100)$ $FAS/SiO_2/Si(100)$ $ODMS/SiO_2/Si(100)$ $PMMA/SiO_2/Si(100)$ $Li[Tf_2N]/SiO_2/Si(100)$	Thermal evaporation	TOF-SIMS	Souda 2013[42]
[BMP][TFSA]	Ag(111), Au(111)	Thermal evaporation	STM	Uhl 2013[43]
[BMP][TFSA]	Cu(111)	Thermal evaporation	STM	Uhl 2014[44]
$[C_8C_1im][NTf_2]$, $[C_8C_1im]Cl$	Au(111)	Electrospray ionization deposition	XPS	Rietzler 2014[45]
$[C_2C_1Im][OTf]$	Pd(111), $Al_2O_3/NiAl(110)$	Thermal evaporation	IRRAS	Schernich 2014[46]
$[C_2C_1Im][OTf]$	Pd(111)	Thermal evaporation	IRRAS	Schernich 2014[47]
$[EMIM][Tf_2N]$ $[BMIM][Tf_2N]$ $[HMIM][Tf_2N]$ $[OMIM][Tf_2N]$	HOPG, Au(111)	Thermal evaporation	MIES, XPS, UPS	Krischok 2014[48]
[BMIM][TFSA]	Au(111)/Mica	Vacuum spray deposition	XPS, AFM	Morino 2014[49]
[OMIM][TFSA]	PEDOT:PSS/ITO/glass	CW-IR laser deposition	OM	Takeyama 2014[50]
[EMIM][TFSA], [OMIM][TFSA]	Au(111), Ag(111)	Thermal evaporation	STM	Uhl 2015[51]
$[C_2C_1Im][OTf]$	Pd(111)	Thermal evaporation	IRRAS	Mehl 2015[52]
$[C_8C_1Im][BF_4]$	Cu(111)	Thermal evaporation	XPS	Syres 2015[53]
$[bmim][PF_6]$	α-$Al_2O_3(0001)$	CW-IR laser deposition	OM	Yamauchi 2016[54]
$[C_2C_1Im][OTf]$	Pd(111)	Thermal evaporation	IRRAS	Bauer 2016[55]
$[C_nC_1im][NTf_2]$ ($n = 1–8$)	ITO/glass	Thermal evaporation	SEM	Costa 2016[56]
[BMP][TFSA]	$TiO_2(110)$	Thermal evaporation	STM, XPS	Uhl 2016[57]
$[C_2C_1Im][OTf]$	Co/Pd(111)	Thermal evaporation	IRRAS	Mehl 2016[58]
$[C_2C_1Im][OTf]$	Au(111) Pd/Au(111)	Thermal evaporation	XPS	Rietzler 2016[59]

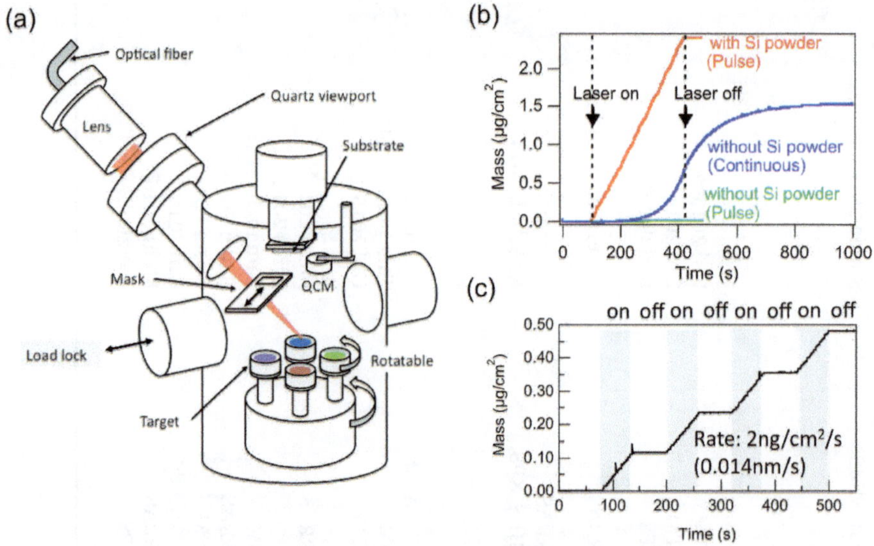

Figure 6.2 CW-IR laser deposition of [bmim][TFSA] IL. (a) Schematic of the CW-IR laser deposition system. (b) Comparison of stability in the CW-IR deposition of IL with three different set-up conditions: pulsed mode without Si powder, continuous mode without Si powder and pulsed mode with Si powder. (c) Digital controllability of deposition of IL with a constant deposition rate.

which resulted from a layer-by-layer growth of C_{60},[63] during the deposition of C_{60} on a mica substrate is evidence for its superiority to any other conventional thermal evaporation methods. For deposition of ILs, since most ILs are transparent to IR light the IL target is prepared with Si powder mixed together as an IR absorber for an efficient absorption of the IR laser light. In Figure 6.2(b), the deposition amounts of an IL (μg cm^{-2}), estimated by a quartz-crystal microbalance (QCM) sensor, are plotted as a function of deposition time for different set-up conditions. Stable deposition can be achieved only when the Si-containing IL target is irradiated by the CW-IR laser in a pulsed mode (*e.g.* 10 ms pulse width, 2Hz repetition frequency, and 260 W cm^{-2} power density). As shown in Figure 6.2(c), furthermore, the deposition is ready to start upon the laser irradiation and the deposition amount increases linearly with time, *i.e.* with a constant deposition rate, which can be controlled to be as small as 2 ng cm^{-2} s^{-1} (=1.3 fl cm^{-2} s^{-1}) or 0.014 nm s^{-1} in effective thickness (=the total volume of deposited IL divided by the deposited area), and the deposition stops quickly upon turning off the laser.

Figure 6.3 shows a comparison of spectral features of ^{1}H nuclear magnetic resonance (NMR) spectroscopy and electrospray ionization time-of-flight mass spectrometry (ESI-TOF-MS) for the IL [bmim][TFSA] deposited by the CW-IR laser deposition method, along with those of the corresponding as-supplied IL. There is almost no difference between their spectral features

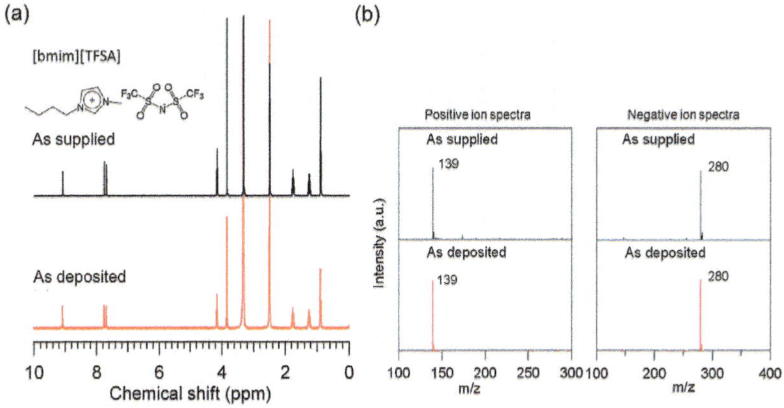

Figure 6.3 Comparison of spectral features of ^{1}H NMR (a) and ESI-TOF-MS (b) for [bmim][TFSA] IL before and after deposition.

in both spectroscopy data, indicative of no sign of its decomposition during the deposition. For various kinds of ILs, their stability during the deposition was also examined by ^{1}H, ^{19}F and ^{31}P NMR spectroscopy, as is summarized together with the results of fractional distillation of ILs by Seddon and co-workers[29] in Table 6.2.

All the ILs listed in Table 6.2, except for [emim][$EtSO_4$], show no significant decomposition, with a fraction of decomposition less than 1%, estimated by ^{1}H NMR. Notably, [PC_{66614}][TFSA] could be successfully deposited in UHV and [hmim][OTf] also even in 1 Torr O_2, although their fractions of decomposition are more than 50% when distilled by using a Kugelrohr oven.

While most ILs, as already mentioned in Section 6.1, are in a liquid state at RT, some ILs are in a solid state at RT with a relatively high melting point. The CW-IR laser deposition method can be used also to deposit such solid-state ILs. For example, ethyl(trimethyl)ammonium bis(trifluoromethylsulfonyl)amide [N1112][TFSA] is a solid at RT with a melting point of 109 °C. Figure 6.4 displays a set of grazing incidence X-ray diffraction (XRD) patterns of 10 nm- and 100 nm-thick [N1112][TFSA] IL samples deposited at RT by the CW-IR laser deposition method, measured at a beam line of BL46XU in SPring-8. Clear diffraction peaks for the 100 nm-thick IL are observed without any unidentified peaks, and also even for the 10 nm-thick IL, indicating that the [N1112][TFSA] IL can be deposited without decomposition and the as-deposited IL is in the solid state.

6.2.2 Fabrication of Various Micro- and Nano-ILs

The CW-IR laser deposition method, by which ILs can be deposited on the nanoscale in a vacuum, allows us to fabricate various micro- and nanoscale ILs.

Table 6.2 Deposition condition and fraction of decomposition in the deposited IL estimated by ^{1}H NMR.

IL	Ambient (Torr)	Frequency (Hz)	Pulse width (ms)	Laser power (W cm^{-2})	Deposition time (h)	Fraction of decomposition in deposited IL (%)	Fraction of decomposition in deposited IL[29]
[emim][NTf$_2$]	4×10^{-8}	4	14	340	11.5	<1	-
[bmim][NTf$_2$]	3×10^{-8}	4	10	260	39.5	<1	1% (8 h, 6 mbar)
[bmim][PF$_6$]	2×10^{-8}	4	14	340	30	<1	1% (4 h, 6 mbar)
[bmim][BF$_4$]	1×10^{-8}	4	10	260	72	<1	-
[PC$_{66614}$][NTf$_2$]	8×10^{-7}	4	14	260	6.5	<1	65% (4 h, 6 mbar)
[hmim][OTf]	1×10^{-8}	4	22	340	12	<1	50% (4 h, 6 mbar)
[hmim][OTf]	1 (O_2)	4	28	340	7.4	<1	
[emim][EtSO$_4$]	6×10^{-6}	4	28	390	6	<10	99% (5 h, 7 mbar)

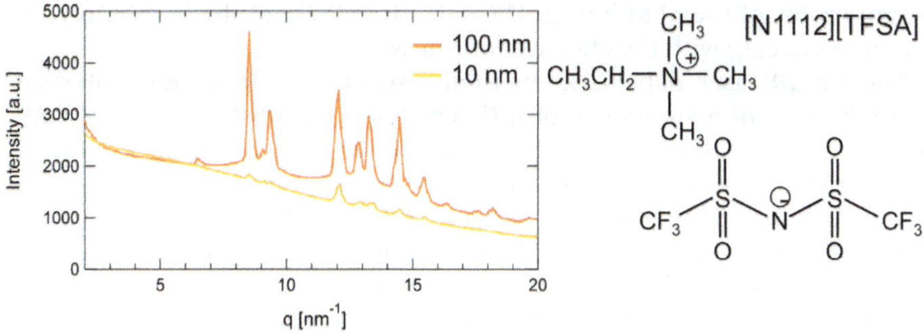

Figure 6.4 Set of grazing incidence X-ray diffraction (XRD) patterns of 10 nm- and 100 nm-thick [N1112][TFSA] IL deposited at room temperature by the CW-IR deposition method.

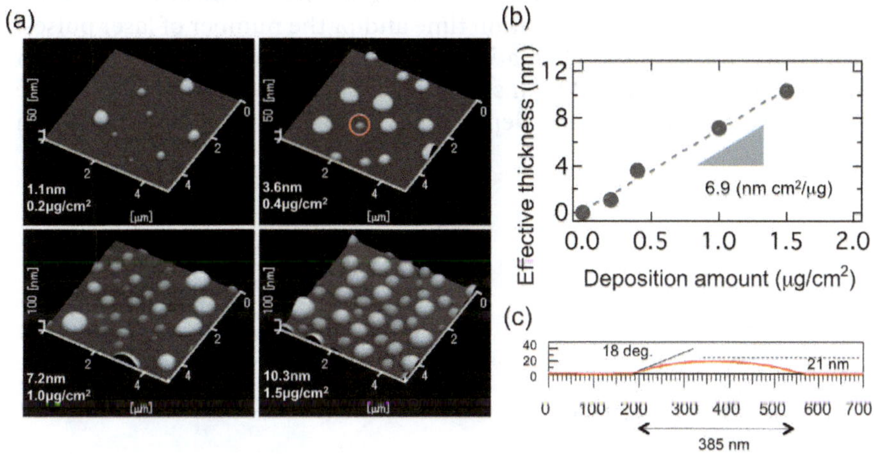

Figure 6.5 (a) Series of AFM images, recorded in a tapping mode, for IL deposited on an α-Al_2O_3(0001) substrate at room temperature with different deposition amount monitored by a QCM sensor. (b) Effective thickness plotted as a function of the deposition amount, giving a good linear relationship between them. (c) Contact angle of a single droplet of nano-IL indicated by a red circle in the AFM image of (a).

6.2.2.1 *Micro- and Nano-IL Droplets*[23,35]

Figure 6.5(a) shows a series of AFM images, recorded in a tapping mode, for [bmim][TFSA] IL deposited on an α-Al_2O_3(0001) substrate at RT with different deposition amount monitored by a QCM sensor. Since the α-Al_2O_3(0001) is not completely wettable to the IL, there are many IL droplets observed on the substrate. There is a good linear relationship between the effective thickness estimated by AFM and the deposition amount as displayed in Figure 6.5(b). The contact angle of a single nano-IL droplet, for example indicated by a red circle in the AFM image of Figure 6.5(a), is about 18°, measured from the height profile of the AFM image. The density and size of the IL droplets are found to increase with the effective thickness.

The CW-IR laser deposition method also enables fabrication of such nano-droplets of a composite of different ILs on a substrate. A most simple way may be to simultaneously deposit two or more ILs by using a target of their mixture with a given mixing ratio. However, if the vapor pressures of the ILs differ too much the distillation of ILs will occur rather than their co-deposition because the CW-IR laser deposition is essentially a thermal evaporation process; the deposited IL droplets would have mixing ratios substantially different from that of the target. In this case, a nanoscale alternate deposition of ILs with each target IL is one of the most powerful approaches to fabricate micro- and nano-droplets of a composite of different ILs. Furthermore, another advantage of the alternate deposition process is that the mixing ratio of the deposited micro- and nano-droplet of ILs can be easily controlled during the deposition according to their deposition ratio, for example by changing the deposition time and/or the number of laser pulses. Figure 6.6(a) illustrates a schematic of the nanoscale alternate deposition of [bmim][TFSA] and [bmim][PF_6] ILs on an α-Al_2O_3(0001) substrate at RT, where the effective thickness of the deposited ILs for each alternate deposition

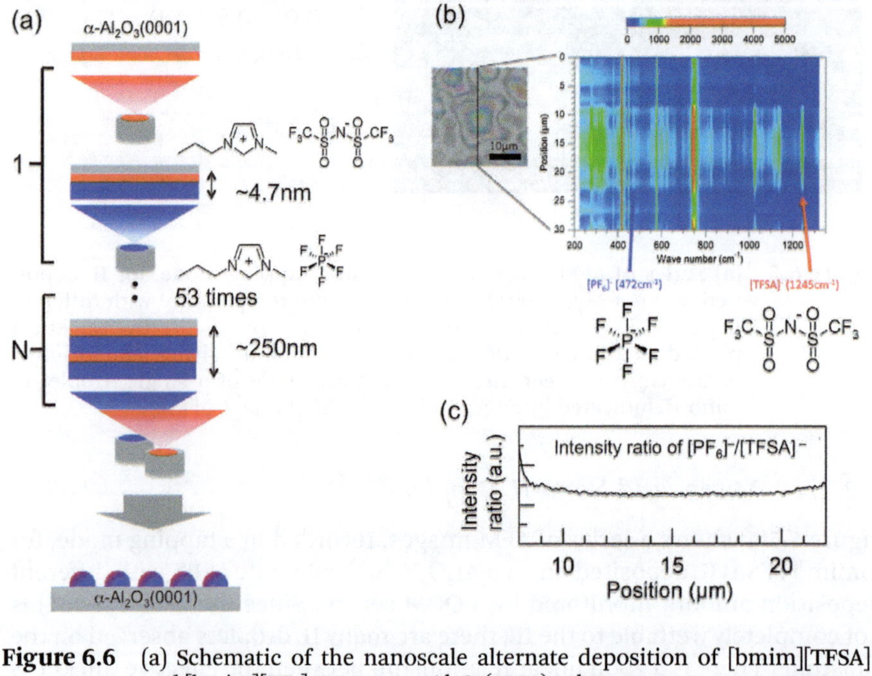

Figure 6.6 (a) Schematic of the nanoscale alternate deposition of [bmim][TFSA] and [bmim][PF_6] ILs on an α-Al_2O_3(0001) substrate at room temperature. (b) Mapping of the peak intensities in line scanned Raman spectra for an isolated droplet. (c) Peak intensity ratio of $[PF_6]^-$ to $[TFSA]^-$, which are found at 472 and 1245 cm^{-1}, respectively, plotted along the scanned line.

cycle is about 4.7 nm and the 53 times repetition of the cycle produces a *ca.* 250 nm-thick ILs composite in droplet form. It is known that [bmim][TFSA] and [bmim][PF_6] ILs are miscible at any mixing ratio. To confirm that the two ILs are homogeneously mixed on the substrate, the peak intensities at 472 and 1245 cm^{-1}, which are assigned, respectively, to [PF_6]$^-$ and [TFSA]$^-$,[64,65] in line scanned Raman spectra were mapped for an isolated droplet with a spatial resolution of less than 500 nm (Figure 6.6(b)). The peak intensity ratio of [PF_6]$^-$ to [TFSA]$^-$ is almost constant, as plotted along the scanned line in Figure 6.6(c), indicative of their homogeneous mixing within the resolution limit of the line scanned Raman spectra.

[bmim][TFSA] and [bmim][PF_6] ILs have a different viscosity and surface tension: 0.040 (Pa s)[66] and 31.8 (mN m^{-1})[10] for [bmim][TFSA], and 0.182 (Pa s)[66] and 45.9 (mN m^{-1})[10] for [bmim][PF_6], respectively. Such different physical properties may cause different behaviors of droplet formation between [bmim][TFSA] and [bmim][PF_6]. Figure 6.7(a) shows a set of laser microscope (LM) images of [bmim][TFSA] and [bmim][PF_6] IL composites with different mixing ratios.[35] The droplet density (cm^{-2}) increases with an increase of [bmim][PF_6] ratio, which can be seen more clearly in Figure 6.7(b). During droplet formation, smaller droplets coalesce to become larger droplets, thereby decreasing the droplet density as the deposition amount increases. In this case, the coalescence probability would be somewhat related to the mobility of IL droplets on a substrate, *i.e.* to the viscosity of the IL. If this is the case, the viscosity of IL droplets containing more [bmim][PF_6] becomes larger and thus their coalescence will be more prevented, resulting in a

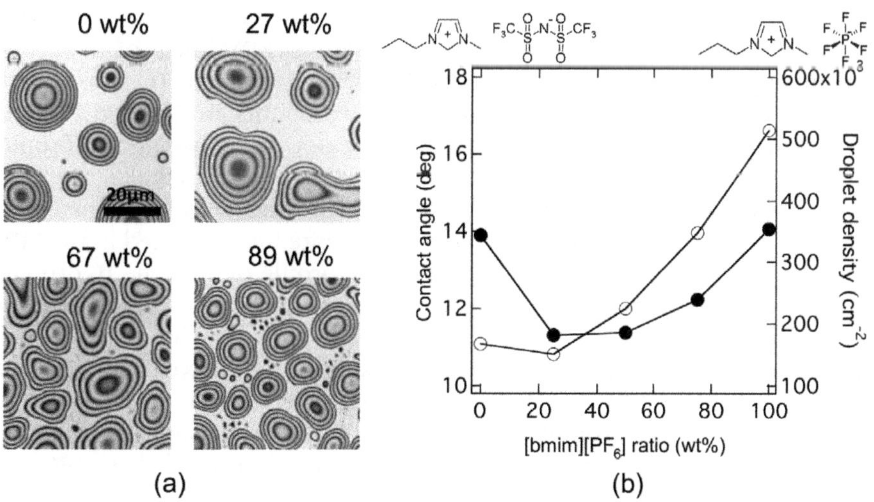

Figure 6.7 (a) Set of LM images of [bmim][TFSA] and [bmim][PF_6] IL composites with different mixing ratios. (b) The droplet density (o) and the contact angle (•) of [bmim][TFSA] and [bmim][PF_6] IL droplet composites, estimated from the LM images, plotted as a function of the [bmim][PF_6] ratio (wt%).

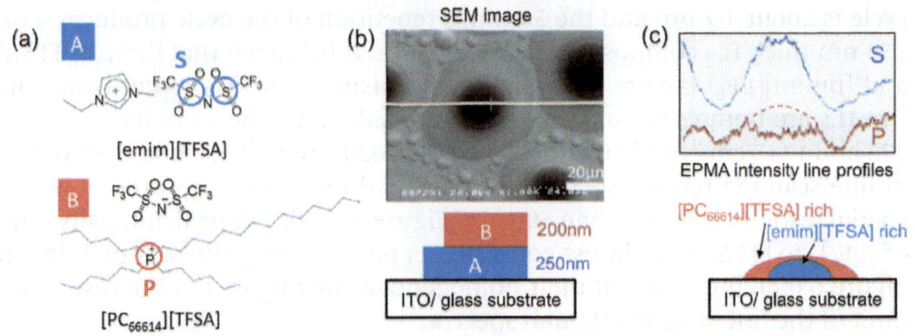

Figure 6.8 (a) Molecular structures of [emim][TFSA] and $[PC_{66614}]$[TFSA] ILs. (b) Typical SEM image of an IL composite of [emim][TFSA] and $[PC_{66614}]$[TFSA], whose total deposition amounts are 250 and 200 nm, respectively. (c) Set of S and P intensity line profiles scanned on the IL droplet along the white line indicated in the SEM, along with a phase separation model.

higher droplet density. On the other hand, although the surface tension is different between these ILs, the contact angles of [bmim][TFSA] and [bmim]$[PF_6]$ ILs are both about 14°; interestingly, the contact angle is a minimum at a certain mixing ratio of [bmim]$[PF_6]$ to [bmim][TFSA], around at 40 wt% (Figure 6.7(b)).

Unlike the completely miscible combination of [bmim][TFSA] and [bmim]$[PF_6]$, [emim][TFSA] and $[PC_{66614}]$[TFSA], whose molecular structures are shown in Figure 6.8(a), are partially immiscible: a phase separation into two different composition phases occurs in a mol fraction range of [emim][TFSA] between 0.3 and 1.0 at RT.[67] Figure 6.8(b) is a typical SEM image of an IL composite of [emim][TFSA] and $[PC_{66614}]$[TFSA] deposited on an ITO substrate by the CW-IR laser method; the total deposition amounts are 250 and 200 nm, respectively. A clear phase separation is observed in an isolated IL droplet with a dark region surrounded by a bright region like a fried egg! The electron probe micro-analysis equipped with SEM provides S and P intensity line profiles scanned on the IL droplet along the white line indicated in the SEM as shown in Figure 6.8(c). The intensity line profile of S on the IL droplet, which comes from [TFSA] anions included in both ILs, shows a curved line like an arch, reflected by the droplet shape. In contrast, the intensity of P, which comes only from $[PC_{66614}]$ cations, is substantially lower than expected from the droplet shape (dotted line) at the center region. From these different intensity line profiles between S and P, it is suggested that the dark region corresponds to an [emim][TFSA]-rich region, surrounded by a $[PC_{66614}]$[TFSA]-rich region as shown in the model in Figure 6.8(c).

6.2.2.2 *Nano-IL Thin Films*

Most oxide surfaces, such as α-$Al_2O_3(0001)$, $SiO_x/Si(100)$ and ITO glass that are often used as a substrate, are not completely wettable to IL, resulting in droplet formation when IL is deposited on the oxide substrates. Fabrication

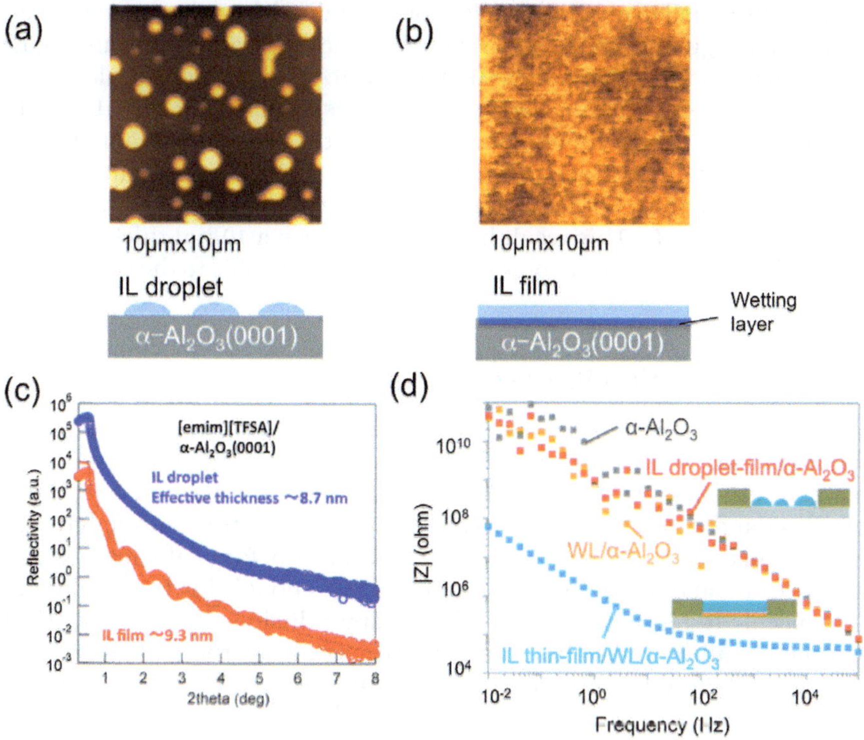

Figure 6.9 (a) and (b) AFM images of a 10 nm-thick [emim][TFSA] IL deposited on non-treated and WL-covered α-Al_2O_3(0001) substrates, respectively. (c) X-ray reflectivity measurements of the IL film and droplet samples. (d) Set of frequency-dependent absolute values of impedances for a 10 nm-thick IL film and IL droplet samples, along with that of the WL-covered insulating substrate for a reference.

of nano-IL thin films thus requires an improvement in the wettability of the oxide surfaces to the IL. One of the simplest methods is the "wetting layer (WL) treatment":[54] a droplet of 20 µl [emim][TFSA] was placed on the substrate and then evaporated by heating at 300 °C in a vacuum for 20 min. The resultant "evaporation trace" left on the substrate surface was found to work as a "wetting layer (WL)" to dramatically improve the wettability of ILs. Figure 6.9(a) and (b) show AFM images of a 10 nm-thick [emim][TFSA] IL deposited on non-treated and WL-covered α-Al_2O_3(0001) substrates, respectively. There are almost no structures of IL on the WL-covered substrate, while a clear droplet formation occurs on the non-treated substrate, suggesting that the IL on the WL-covered substrate should form a flat film. In fact, the IL film sample is so flat that its X-ray reflectivity measurement exhibits a clear fringe pattern with 2θ (θ is an incident angle of the X-rays) though the IL droplet sample has no fringe pattern (Figure 6.9(c)). From the fringe pattern the thickness of the IL film is estimated to be 9.3 nm, which is very close to the as-designed thickness of 10 nm. Furthermore, to confirm that the IL film is completely continuous even on the nanoscale, and not composed of IL droplets too small to observe by AFM, ionic conductivity measurements were performed. Figure 6.9(d) shows a set of frequency-dependent absolute values of impedances for a 10 nm-thick IL film and IL droplet samples, along with that of the WL-covered insulating substrate as a reference. The IL droplet sample shows an ionic conductivity as small as the substrate, an almost insulating behavior. In contrast, the IL film sample shows a significant decrease in the absolute values of impedances for a wide frequency region, between 0.01 Hz and 10 kHz. This result indicates that even a 10 nm-thick IL film is uniform enough to have an ionic conductivity comparable to the corresponding bulk IL.

6.2.3 Evaporation Process of IL in Vacuum

When IL is evaporated during deposition, the vapor pressure of IL, *e.g.* [bmim][TFSA], is on the order of 10^{-3} Torr, assuming that the evaporation temperature is above 200 °C.[11] In this case, since the IL molecule, a pair of cation and anion, is ~1 nm in diameter, the mean free path length, the average distance that the IL molecule can travel before collisions, is estimated to be about 0.5 cm.[68] The value is not enough for the evaporated IL molecule to reach a substrate without any collisions. As a result, the IL molecules will coalesce into droplet particles through their collisions with each other before reaching the substrate. When such IL droplet particles are adsorbed onto the substrate, their relatively large particle sizes may prevent their surface migration sufficiently and thereby their further coalescence on the substrate at RT. Thus, the density of IL droplets observed on the substrate is determined by how many IL droplet particles are formed in the vapor before they reach the substrate rather than by a degree of the supersaturation on the substrate according to the deposition rate, as is often found in the vapor deposition of solid thin films. To verify the present model for the evaporation process of IL, [emim][TFSA] IL was deposited with different deposition rates in the pulsed-mode on a non-treated α-Al_2O_3(0001) substrate and how the droplet density and size of IL change with the deposition amount are examined by AFM; the results are shown in Figure 6.10. For the samples with the same deposition amount but different deposition rates the droplet density seems to differ little under the present pulsed deposition conditions. This is more clearly found when the droplet density is plotted against the deposition rate for different deposition amounts in effective thickness as displayed in Figure 6.11(a). The droplet density is almost constant over a wide range of the deposition rate between 0.2 and 46 nm min^{-1}! Consequently, the droplet density, as shown in Figure 6.11(b), linearly increases with the effective thickness up to around 10 nm, at which deposition amount, however, the IL droplet density becomes large enough for the IL droplets to overlap and coalesce with each other on the substrate, decreasing the droplet density. As illustrated in Figure 6.11(c), if the molecular density in the vapor every laser pulse becomes low it is expected that the collision times will decrease and that the size of each IL droplet particle will, therefore, become smaller, while the droplet density increases. In fact, as shown in Figures 6.10 and 11(a) and (b), when IL is deposited in the CW-mode instead even at a deposition rate as low as 3 nm min^{-1}, at which the molecular density in the vapor is relatively low, the IL droplets observed on the substrate are much smaller than any other cases in the pulsed-mode, much increasing the droplet density. In this way, the smallest IL droplet that CW-mode can produce is 100 nm in diameter and 0.1 aL in volume.

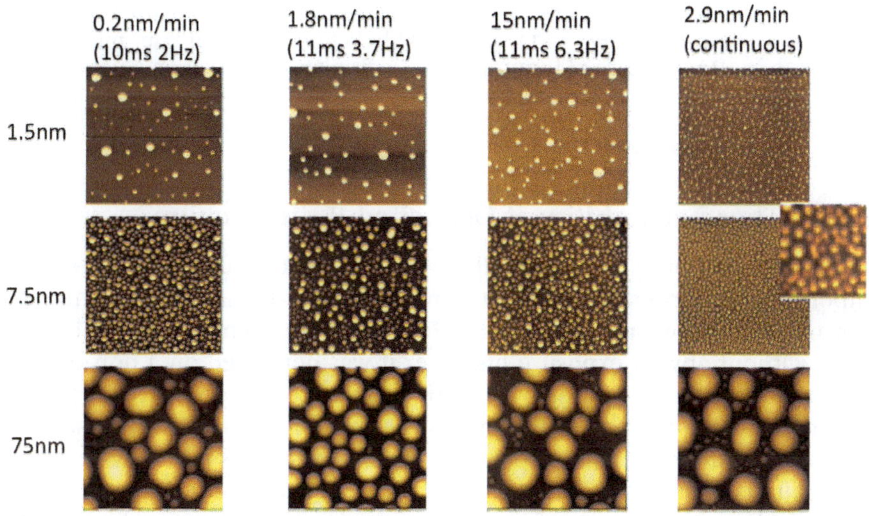

Figure 6.10 Set of AFM images (10 μm × 10 μm) of [emim][TFSA] IL deposited with different deposition rates in the pulsed- and CW-mode on a non-treated α-Al_2O_3(0001) substrate.

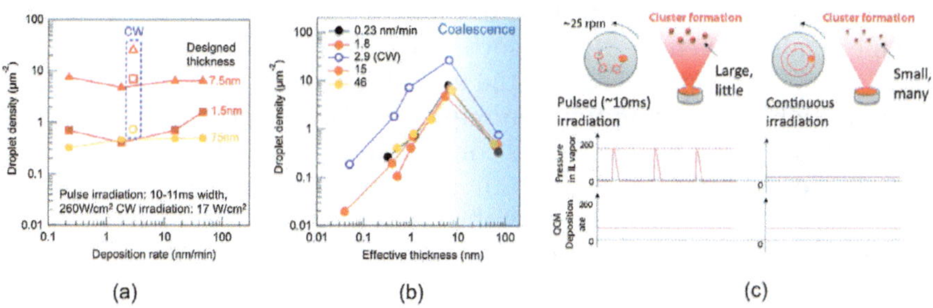

Figure 6.11 (a) Droplet density plotted against the deposition rate for different deposition amounts in effective thickness. (b) Droplet density plotted against the deposition amount in effective thickness for different deposition rates. (c) Schematic of the evaporation process of IL in vacuum by the IR-laser deposition in pulsed- and CW-modes.

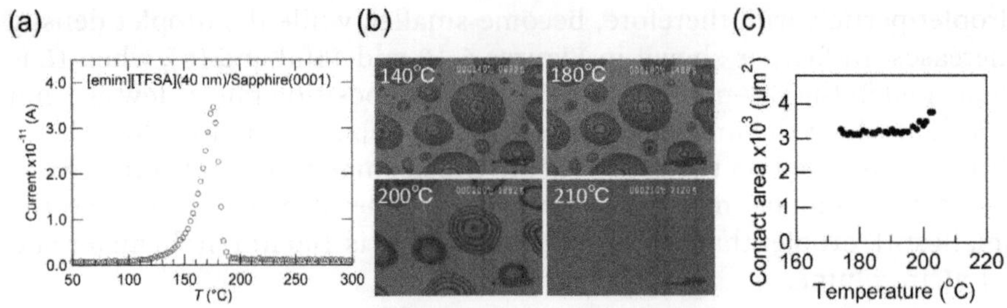

Figure 6.12 (a) TDS of a 40 nm-thick [emim][TFSA] IL measured without any air exposure just after it was deposited on a non-treated α-Al_2O_3(0001) substrate (m/z = 111, $[emim]^+$). (b) Set of sequential laser microscope (LM) images of [bmim][TFSA] IL droplets deposited on a non-treated α-Al_2O_3(0001) substrate when heated in a vacuum (scale bar = 25 µm). (c) Projected desorption area in total (contact area) plotted against temperature.

6.2.4 Thermal Stability of Deposited ILs in Vacuum

The thermal stability of, especially, nano ILs deposited on a substrate is important for process applications in a vacuum. Figure 6.12(a) shows a thermal desorption spectrum (TDS) of a 40 nm-thick [emim][TFSA] IL measured without any air exposure just after it was deposited on a non-treated α-Al_2O_3(0001) substrate, often by detecting a mass-to-charge ratio of m/z = 111 ($[emim]^+$) as one of its mass fragments. Desorption of the IL can be seen to start at around 120 °C; the temperature at which the mass intensity becomes maximum is about 180 °C in this case. The peak shape observed in the TDS is a typical one that resulted from the zero-order desorption mechanism. In fact, Figure 6.12(b) is a set of sequential laser microscope (LM) images of IL micro-droplets of [bmim][TFSA], a similar type to [emim][TFSA], deposited on a non-treated α-Al_2O_3(0001) substrate when heated in a vacuum. Notably, little change is found in the diameter of the IL droplets during the desorption process. The projected desorption area in total is almost constant, as clearly shown in Figure 6.12(c), indicative of the validity of the zero-order desorption mechanism when the IL is desorbed. If this is the case, the evaporation enthalpy can be estimated to be ~128 kJ mol^{-1}, which is well comparable to the corresponding evaporation enthalpy in bulk as reported in the literature.[14,15] From these results, it was confirmed that some ILs could be used above RT, for example up to 100 °C in a vacuum, though it depends on the kind of IL. Such thermal stability of the ILs in a liquid state is a great advantage of ILs for process applications in a vacuum even above RT.

6.3 Process Applications of ILs in Vacuum

Since some ILs are stable in a liquid state in vacuum at RT, there have been reports on ILs used for process applications in a vacuum. They include sputter deposition of nanoparticles (NPs) in an IL,[19] *in situ* electrochemistry[21,69,70] as well as some combination processes of plasma with IL.[20,22] However, in most cases the ILs are in bulk and used at RT; there have been few cases where the ILs are deposited and/or used above RT in a vacuum.

6.3.1 A New Concept of IL-assisted Vapor Synthesis and Growth in a Vacuum

As already pointed out in Section 6.1, the present IL deposition technique allows us to prepare nano-reactors of ILs, in either droplet or thin film form, on a solid substrate in a vacuum. If some precursors and reactants are supplied in some way into such IL reactors from the gas phase, we will be able to use ILs as a solvent not only for crystal growth but also for chemical synthesis like a polymerization in vacuum; this is termed IL assisted vapor synthesis and growth in a vacuum (Figure 6.13).

A similar vacuum process has historically been known as a vapor–liquid–solid (VLS) mechanism, proposed by R. S. Wagner and co-workers in 1964, who succeeded in the growth of single crystal Si nanowires with Au melt droplets in a chemical vapor deposition (CVD) process.[71] However, it was about 40 years after Wagner's first report that the applicability of this VLS mechanism to the PLD growth of single crystal oxide films was verified for high-*Tc*-superconducting (HTSC) films with a Ba–Cu–O liquid flux in 2002[72] and ferroelectric Bi-layered oxide films with a Bi–Cu–O liquid flux in 2006.[73] Owing to its similarity to the flux growth process for inorganic single crystals, though, our process is based on PLD most suitable for oxide epitaxy and it is, thereby, often called flux-mediated epitaxy.[74] This process, in principle, can be naturally extended to the growth of organic compounds; indeed, for example, organic single crystal growths were attempted by using B2EHS as a liquid solvent.[6,7] B2EHS is, however, stable in a liquid state only at RT, and cannot remain at higher temperature, in a vacuum. As a way around this thermal instability of conventional organic solvents in a vacuum, the use of ILs was proposed.

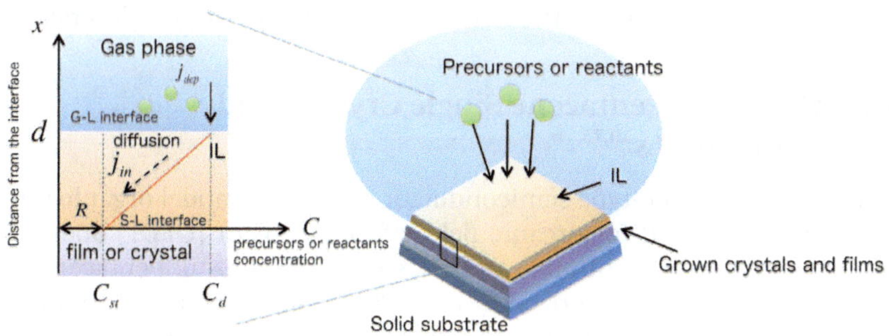

Figure 6.13 Schematic of IL-assisted vapor synthesis and growth in a vacuum, along with an ideal concentration profile of precursors and reactants in the IL at a steady-state condition.

These processes commonly offer some advantages over both conventional vapor deposition and bulk flux (solution) growth processes: (1) the process is close to being under the equilibrium condition, as in the bulk flux growth, but a new feature is that (2) the supersaturation condition can be controlled by the deposition rate of precursors from the gas phase at a constant growth temperature. In this case various solvents can be used, even if they have a low solubility and/or less temperature-dependent solubility, that are difficult to use in the bulk process.

Here, let us consider the case of crystal growth in an IL. Ideally, at the steady-state condition, as shown in Figure 6.13, the deposition rate of the precursors at the gas–liquid (G-L) interface (j_{dep}), their diffusion rate in the liquid (j_{in}) and the crystallization rate at the solid–liquid (S-L) interface (R) should all be the same, as shown in eqn (6.1):

$$j_{dep} = j_{in} = D\frac{\partial C(x)}{\partial x} \approx D\frac{C_d - C_{st}}{d} = R/\upsilon_c \tag{6.1}$$

where C_d and C_{st} are the concentrations at the G-L interface ($x = d$; d is the thickness of IL) and S-L interface ($x = 0$), respectively, and D is the diffusion constant, υ_c is the volume of a building unit in the crystal phase. In a rather simplified picture of crystal growth for a rough crystal face, the growth rate follows the first-order reaction, *i.e.* $R = \upsilon_c k(C_{st} - C_0)$, where k is the rate constant and C_0 is the equilibrium concentration in IL. In this case, the crystallization rate R is expressed by eqn (6.2):

$$R = \upsilon_c\left(C_d - C_0\right)\left(\frac{kD/d}{k + D/d}\right) \tag{6.2}$$

When the diffusion rate, D/d, is sufficiently greater than the crystallization rate, k, the latter is the rate-limiting step, resulting in $R \rightarrow \upsilon_c k(C_d - C_0)$. At the other extreme of the diffusion regime, $D/d \ll k$, the diffusion controls the crystallization rate R, resulting in $R \rightarrow \upsilon_c D/d(C_d - C_0)$. Notably, the thickness of IL d plays an important role in the control of the crystallization rate in the diffusion regime: the thinner the IL is, the higher the crystallization rate.

6.3.2 Growth of Pentacene Single Crystals and Films *via* ILs[50,75,76]

Pentacene is a typical p-type semiconductor material, a model organic semiconductor often tested for organic field effect transistor (OFET) operation in either polycrystalline thin film or bulk singe crystal form. Pentacene has several polymorphs characterized by their d(001)-spacings of 1.41, 1.45 and 1.50–1.54 nm termed "single-crystal" phase, "bulk" phase and "thin film" phase, respectively, among which the "single-crystal" phase pentacene is the most thermodynamically stable phase, but is often difficult to obtain. Single-crystal phase pentacene has been grown only by solution growth, though pentacene has low solubility in most organic solvents including ILs. Therefore, its solution growth processes should take a long time for large-scale single-crystal phase pentacene.

6.3.2.1 Growth of Bulk Pentacene Single Crystals in an IL Macro-droplet[75]

The process of IL-assisted vapor growth in a vacuum has been found to solve such a low solubility limit in the solution growth and many crystals of "single-crystal" phase pentacene can be obtained after just a short time, *e.g.* 30 min deposition. The process is very simple, as shown in Figure 6.14: First, a droplet of 1 µl IL [omim][TFSA] in volume is dropped onto a substrate (sapphire, glass, Si, *etc.*) (Figure 6.14(a)), then the substrate is introduced into the CW-IR laser deposition chamber, as already explained in Section 6.2.1, to deposit pentacene at a deposition rate of a few tens nm min^{-1} over the IL droplet above RT, *e.g.* at 110 °C. As shown in Figure 6.14(b), a 30 min deposition results in the growth of many platelet pentacene crystals of "single-crystal" phase in the IL droplet. This result indicates that the present IL-assisted vapor growth process is very close to the thermodynamic equilibrium even though the deposition rate is very high as compared with those in the conventional vacuum deposition of organic semiconductor molecules. Figure 6.14(c) shows a piece of the grown platelet crystal of pentacene; the size is as large as a few hundreds µm and its crystal surface exhibits a molecular steps-and-terraces structure with a unique step height of about 1.4 nm, corresponding to the *d*(001)-spacing for the single-crystal phase pentacene as shown in the inset.

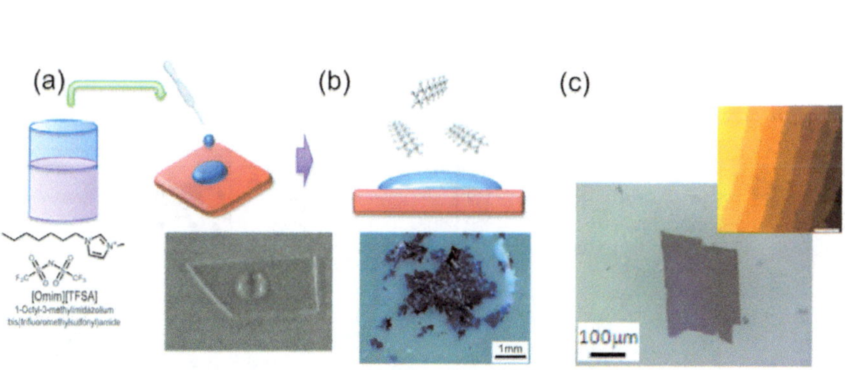

Figure 6.14 Process of IL-assisted vapor growth of pentacene crystals with an IL macro-droplet on a non-treated α-Al_2O_3(0001) substrate in a vacuum. (a) Preparation of an IL [omim][TFSA] pre-dropped substrate, (b) Many platelet pentacene single crystals grown in the IL droplet just after a 30 min deposition of pentacene at around 110 °C over the IL droplet. (c) Expanded photo image of a platelet pentacene crystal, along with a typical AFM image (5 µm × 5 µm) of a steps-and-terraces structure of the crystal surface.

When organic crystals are grown from organic solvents by a solution growth method, a possible inclusion of the solvent molecules and/or their decomposed derivatives into the crystal as impurities often becomes a problem, especially for their electronic device applications. However, this would not be the case with the use of ILs as an organic solvent in the solution growth of organic crystals. This is because most ILs are thermally stable and their molecular-sizes are relatively large compared with those of the organic solvents conventionally used, preventing the IL ions and/or their decomposed derivatives from being incorporated in the grown crystals. In fact, the transistor characteristics of single-crystal phase pentacene were investigated by fabricating their IL-gated FET devices. As shown in Figure 6.15(a), a single piece of crystal pentacene is picked up and transferred onto a patterned electrode substrate, and then another kind of IL, [P13][TFSA], used as a gate insulating layer, is poured into the space between the platelet crystal and the gate electrode; the inset is a picture of the fabricated device. A typical transistor action as a p-type semiconductor was found, as shown in Figure 6.15(b), in the gate voltage V_G dependence of source-drain current I_D (I_D–V_G curve). Figure 6.15(c) shows plots of the field effect mobility μ_{FE} values, estimated by

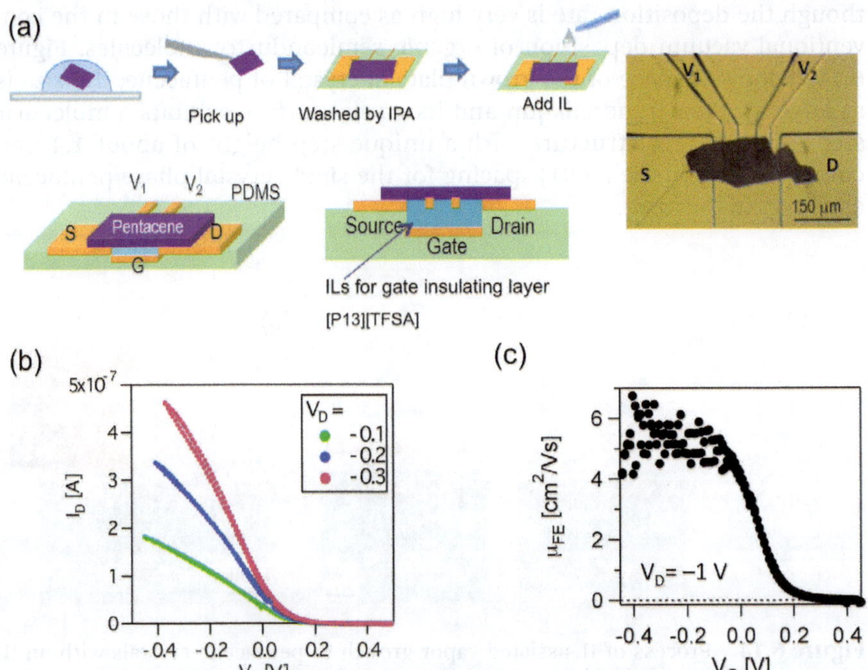

Figure 6.15 (a) Schematic process to fabricate the IL-gated FET device of the single-crystal phase pentacene, along with a picture of the fabricated device. (b) Transfer characteristics (I_D–V_G) curve at different V_D = −0.1, −0.2 and −0.3 V, respectively. (c) The μ_{FE} values plotted against V_G at V_D = −0.1 V.

a four-terminal method, against V_G at a constant source-drain voltage of V_D = −0.1 V. The observed mobility, as high as 5 cm^2 V^{-1} s^{-1}, is one of the largest values in pentacene FET devices so far reported in the literature, providing experimental evidence for the high-quality of pentacene crystals obtained by the IL-assisted vapor growth technique in terms of not only the crystallinity but also the semiconductor properties.[76] A similar result was obtained for a different solution-crystallization method *via* thermal conversion of pentacene precursor molecules soluble in the IL: the crystals, grown even in a different IL ([emim][TFSA]), are also of the single-crystal phase, showing high mobility values of 1–2 cm^2 V^{-1} s^{-1}.[77]

6.3.2.2 *Confined Growth of Pentacene within a Nano Thin Film-IL*[50]

As seen above, when pentacene molecules were deposited into an IL macro-droplet on a substrate they were nucleated in the IL before reaching the substrate, producing many platelet crystals floating in the IL (Figure 6.14(b)). This result suggests that the concentration of pentacene in the IL exceeds the critical concentration required for nucleation to occur during the deposition. This is probably because of the low solubility limit and slow diffusivity of pentacene in the IL. According to eqn (6.2), the quantity D/d represents an effective diffusivity of solute molecules in a solution, *i.e.* pentacene molecules in the IL. Therefore, if it is possible to decrease the thickness of the IL, d, the effective diffusivity can increase, even though the diffusion constant remains the same. Consequently, the nucleation of pentacene in the IL will be suppressed, and the nucleation and subsequent growth of pentacene will occur directly on the substrate.

In fact, the two-dimensional growth of pentacene directly on a substrate was found when pentacene was deposited into a nano thin film-IL. As discussed in Section 6.2.2, IL thin films can be fabricated on a substrate by using the CW-IR laser deposition technique if the substrate is properly treated so as to be wettable to the IL. In the present experiments, PEDOT : PSS, a polymer mixture, which is commonly used as a semitransparent, hole-transporting layer in organic solar cells and electroluminescence devices, was coated as a wetting layer on an ITO substrate; this was then followed by the deposition of a 100 nm-thick [omim][TFSA] IL film. Figure 6.16(a) is a set of sequential images of pentacene growing in the IL thin film taken by a digital optical microscope. After the induction period of about 10 s from the beginning of the deposition, the nucleation of pentacene and its subsequent two-dimensional growth were found to proceed confined within the IL thin film. As the solubility limit of pentacene in the IL is very small, it should be under a supercooling state in the induction period so that the lateral growth rate in the two-dimensional growth can reach as high as 16 μm s^{-1} in the IL. After the deposition of pentacene in the IL thin film, the obtained pentacene film was rinsed using 2-propanol (IPA) to test its adhesion to the substrate. The result is shown in Figure 6.16(b): the pentacene film, as appeared in purple,

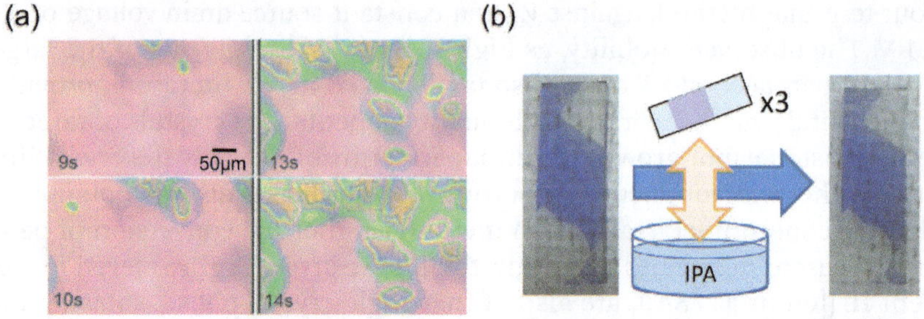

Figure 6.16 (a) Set of sequential images of pentacene growing in the IL thin film taken by a digital optical microscope. (b) Pictures of pentacene film samples before and after being rinsed with IPA.

remained adhered to the substrate even after three cycles of treatments with IPA. This result suggests that the nano thin film IL forces the pentacene molecules to nucleate directly on the substrate surface because the effective diffusivity became large enough in such a nano thin film IL.

6.3.3 Epitaxial Growth of High-quality C_{60} Films *via* IL[78]

Fullerene (C_{60}) is known as one of the few n-type π-conjugated semiconductors, with a spherical molecular structure. There have been many studies on the epitaxial growth of C_{60} films, particularly on van der Waals substrates. However, even in epitaxial films C_{60} tends to grow in microcrystalline or dendrite forms by the conventional vapor growth processes. To improve the crystal quality of epitaxial C_{60} films, the IL-assisted vapor growth technique is applicable. In this case, the key point is the preferential nucleation of C_{60} molecules on a substrate even *via* IL, unlike the case of pentacene. In fact, when C_{60} fullerene molecules are deposited into either a macro-droplet or thin layer of IL they are always nucleated directly on various substrates, and the epitaxial growth of C_{60} was confirmed on a basal plane of MoS_2, which is well known as a van der Waals substrate for epitaxial growth of C_{60}. In Figure 6.17(a) and (b) a good comparison is displayed between 100 nm-thick C_{60} films grown on MoS_2 with and without IL, where the growth temperature and deposition rate are 100 °C and about 2 nm min^{-1}, respectively. As shown in Figure 6.17(a), when grown with the IL, each C_{60} grain is visible even by LM and their surfaces clearly exhibit a molecular steps-and-terraces structure, as observed by AFM. In contrast, the C_{60} film grown without the IL is found to consist of very small grains (<1 μm) by AFM so that there is apparently a smooth film surface observed by LM, as shown in Figure 6.17(b).

As is often the case with solution growth processes, however, the control of nucleation sites may become a problem in the IL-assisted vapor growth of a C_{60} film. C_{60} seems to turn to grow preferentially along the out-of-plane direction rather than the in-plane direction before the grains become large enough to coalesce with each other. Consequently, the film does not become

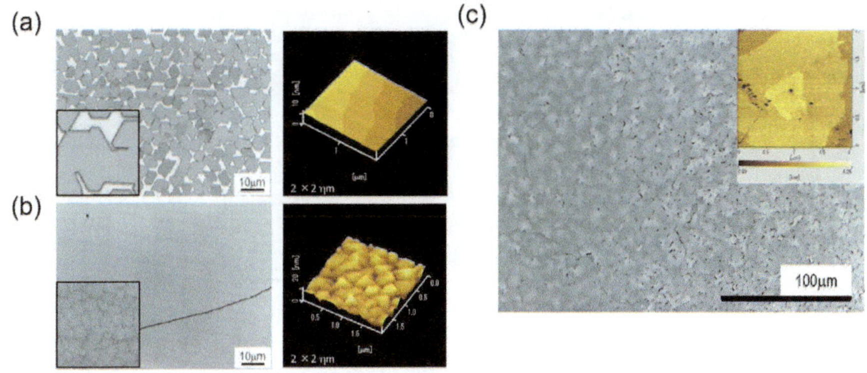

Figure 6.17 (a) and (b) Set of laser microscope and AFM images of 100 nm-thick C_{60} films grown on MoS_2, (a) with and (b) without IL, where the growth temperature and deposition rate are 100 °C and about 2 nm min^{-1}, respectively. (c) Laser microscope and AFM images of a 180 nm-thick C_{60} film grown on MoS_2 with IL at a deposition rate of 6 nm min^{-1}.

continuous, consisting of many single crystalline grains, though each crystal grain is of high crystallinity as discussed above. To obtain a continuous C_{60} film, it would be effective to increase the number of nuclei at the initial stage of the nucleation on the MoS_2 substrate by increasing the degree of supersaturation. According to eqn (6.2) in Section 6.3.1, the supersaturation, $C_d - C_0/C_0$, is proportional to the deposition rate j_{dep} (=R). In fact, when C_{60} molecules were deposited into the IL at a deposition rate of 6 nm min^{-1}, almost three times larger than that of the C_{60} film in Figure 6.17(a), at 100 °C for 30 min, a continuous C_{60} film could be obtained as shown in Figure 6.17(c), with each of C_{60} grain still visible even by LM, exhibiting a steps-and-terraces structure by AFM (inset).

6.3.4 Stabilization of a Flat (111) Polar Surface of KBr Films Grown *via* IL[33,54]

ILs are often used as a green solvent for various organic synthetic processes, but taking their strong polar nature into account the use of ILs should also be favored in the synthesis and growth of inorganic ionic compounds. For example, an ionic KBr crystal is a member of the alkali halides group, having a rock-salt crystal structure and its (111) surface is a polar layer composed solely of either K^+ or Br^- ions, resulting in its electrostatic instability. It has been thus generally accepted that any atomically flat rock-salt type (111) surfaces are difficult to obtain. If KBr(111) films are grown in the IL, the strong polar environment of the IL is expected to stabilize the flat KBr(111) polar surface.

To verify such a polar environment effect of the IL, the IL-assisted vapor growth technique was also applied to the growth of KBr(111) films. As with

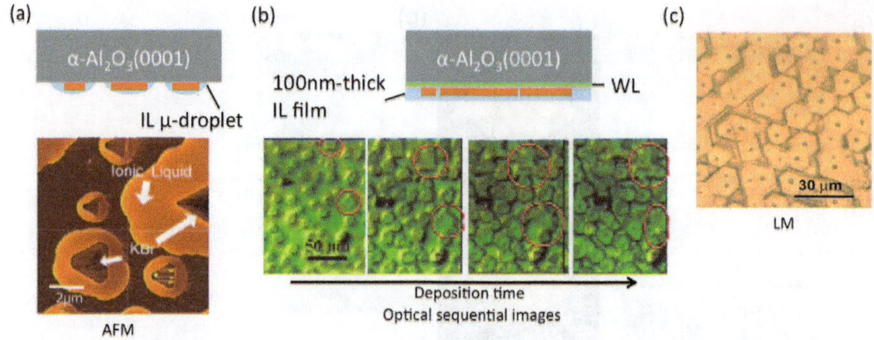

Figure 6.18 Growth of KBr(111) films *via* [bmim][PF_6] IL on (a) non-treated and (b) WL-covered α-Al_2O_3(0001) substrates, respectively. Optical sequential images show the growth of KBr film crystals in the IL thin film during the deposition of KBr at 110 °C. (c) LM image after the deposition of KBr(111) film.

other alkali halides, KBr is expected to be easily vacuum-evaporated in a neutral molecular form to safely come into the IL without any chemical reactions with the IL solvent. Figure 6.18 shows the growth of flat KBr(111) films *via* [bmim][PF_6] IL on (a) non-treated and (b) & (c) WL-covered α-Al_2O_3(0001) substrates. As shown in the AFM image of KBr(111) film crystals (35 nm thick) deposited on the non-treated substrate at 110 °C, the IL μ-droplets (50 nm in effective thickness) act as a reactor, only within which are the KBr(111) film crystals formed and therefore their crystal sizes are limited by the size of the corresponding IL droplets, to be, at most, a few μm. In contrast, when KBr was deposited into the 100 nm-thick IL film deposited on the WL-covered substrate under the same growth conditions, KBr film crystals appeared initially at random and their sizes were already large enough to be visible, as shown in a series of sequential optical images taken during the deposition (Figure 6.18(b)), and they gradually increased in lateral size until they coalesced with each other because of no size limitation in the IL film. Finally a continuous, flat KBr(111) film (1000 nm thick) was formed all over the substrate, as shown in the LM image after the deposition (Figure 6.18(c)).

6.3.5 Reactive Vapor Synthesis and Growth *via* IL

In all examples so far presented above, a single source of molecular or ionic compounds is simply deposited as a precursor into the IL to be crystallized through their van der Waals or Coulomb interactions, in which no chemical reaction, *i.e.* no activation process, is involved. As already pointed out, an IL is a solvent that can not only enhance the diffusion process of precursors but can also activate chemical process. To demonstrate such a reactive process in an IL in a vacuum, two different chemical sources were simultaneously deposited into the IL on a substrate to react with each other. In this section, two examples are shown: one is the growth of charge-transfer (CT) organic complex crystals, and the other is the synthesis of polymer films.

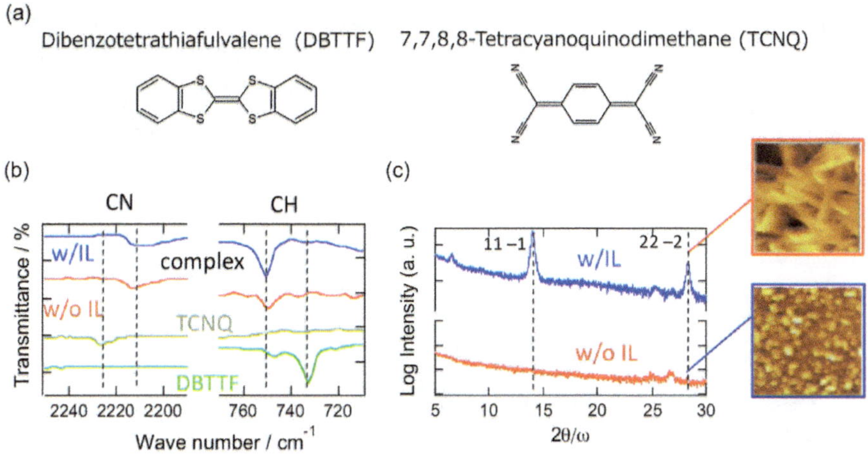

Figure 6.19 (a) Molecular structures of DBTTF and TCNQ. (b) Set of FTIR spectra for the film samples co-deposited w/ and w/o the IL layer at room temperature, along with those of vacuum-deposited DBTTF and TCNQ samples for a reference. (c) XRD patterns and AFM images (5 μm × 5 μm) of the DBTTF-TCNQ complexes samples obtained w/ and w/o the IL.

6.3.5.1 DBTTF-TCNQ Complex Organic Crystals[79]

The reactive crystal growth of organic CT complexes, composed of dibenzotetrathiafulvalene (DBTTF) (donor) and tetracyanoquinodimethane (TCNQ) (acceptor) molecules, whose molecular structures are shown in Figure 6.19(a), was attempted on a 1 μm-thick IL ([omim][TFSA]) layer-covered Si substrate. In this case, to improve the wettability of the IL on the Si substrate, a poly(vinyl alcohol) (PVA) layer was first spin-coated as a wetting layer. Figure 6.19(b) shows a set of FTIR spectra for the film samples co-deposited with and without the IL layer at RT, along with those of vacuum-deposited DBTTF and TCNQ samples for a reference. Irrespective of whether the IL was present, the peak at 2226 cm^{-1}, assigned to the C≡N stretching mode in TCNQ, was shifted to 2210 cm^{-1}. In addition, also, the peak at 732 cm^{-1}, assigned to the C–H bending mode in DBTTF, was shifted to 750 cm^{-1}. The concurrent peak shift indicates that the complex formation can occur for both samples even at RT. In contrast, the XRD patterns of these DBTTF-TCNQ complexes samples exhibited a remarkable difference in intensity as shown in Figure 6.19(c). The intensity of the XRD peaks as well as each grain size as shown in AFM images becomes much larger for the complex film grown with IL, confirming that the crystallinity is much improved by the effect of the IL.

6.3.5.2 Direct Synthesis of Porous Polyurea Films[80]

Network-structured, porous polymer films, exhibiting the merits of their micropores and the resultant large surface area, are often used as a filter in separation processes, a support material in heterogeneous catalysis and so on.

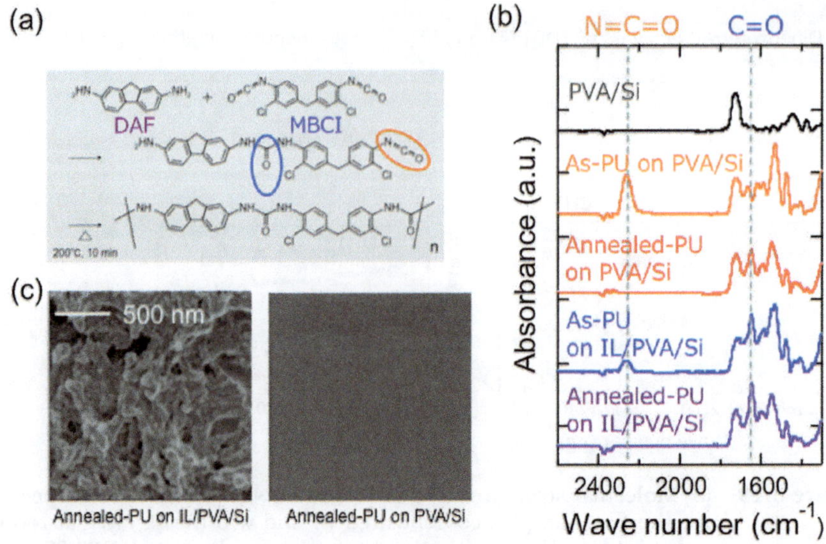

Figure 6.20 (a) Reaction scheme of PU. (b) Set of FTIR spectra for the PU film samples co-deposited w/ and w/o the IL layer at room temperature, before and after the post-annealing treatment. (c) SEM images of these PU films after the post-annealing treatment fabricated w/ and w/o the IL layer.

For example, Zhu *et al.* reported that a bulk polyurea (PU) exhibits a micropore structure when synthesized with ILs *via* interfacial full polymerization even at RT.[81] This result suggests that the PU film should be a good model system to demonstrate the capability of the IL-assisted vapor synthesis for polymer films. The reaction scheme of PU is shown in Figure 6.20(a); in the present experiment the PU films were synthesized by co-depositing the monomers, 2, 7-diaminofluorene (DAF) and 4,4′-methylenebis (2-chlorophenyl isocyanate) (MBCI) at RT into a 1 µm-thick IL ([omim][TFSA]) layer-covered Si substrate, as in the similar case for the growth of DBTTF-TCNQ complex crystals above. Figure 6.20(b) shows a comparison of FTIR spectra between PU films fabricated with and without the IL layer, before and after the post-annealing treatment. An absorption peak found at around 2270 cm^{-1} in both as-deposited films is from the N=C=O group, indicating incompletion of the polymerization reaction. After the post-annealing treatment, these N=C=O peaks completely disappeared, while an adsorption peak at 1650 cm^{-1}, which is from the urea C=O group, appeared instead, suggestive of further polymerization during the post-annealing treatment. It should be point out that the PU film fabricated with the IL layer before the annealing has a peak intensity of the remaining N=C=O group that is relatively much smaller than that without the IL, which is consistent with the ready appearance of the urea C=O group. This result indicates that the IL has a substantial effect in accelerating the polymerization reaction even at RT (as seen in the deposition process), as is expected from the result of Zhu *et al.* Furthermore, as shown in Figure 6.20(c), the PU film fabricated with the IL layer showed a network structure surface, while such a network structure was never found in PU films fabricated without the IL layer, regardless of being annealed.[82]

6.4 Conclusions

This chapter has outlined new applications of ILs in a vacuum. The CW-IR laser deposition, originally developed by our group, is one of the most promising techniques, potentially opening up new research fields of nanoscale physics and chemistry of ILs. This new deposition method allows us to fabricate various micro- and nano-ILs on a nanometer scale in a vacuum in a similar manner to the way in which micro- and nano-solid state materials have been fabricated in a vacuum. A new concept of IL-assisted vapor synthesis and growth in a vacuum has been introduced as one of the applications of such micro- and nano-ILs, in which they are used as micro- and nano-scale reactors of IL in chemical processes. The capability of this unique process has been verified, exemplified by the growth of "single-crystal" phase pentacene crystals and films together with their superior EDL-FET device performance, the epitaxial growth of a high-quality C_{60} film, the stabilization of a flat (111) polar film surface of KBr, the reactive synthesis and growth of DBTTF-TCNQ CT complex crystals and porous PU films, all of which have resulted from IL solvent-driven enhancements of crystal growth and/or chemical reaction. In fact, the IL-assisted vapor growth process is now being accepted as one of the new routes to the growth of various organic single crystals.[83,84] Furthermore, the applications of micro- and nano-ILs are not limited to such IL-assisted chemical processes, but open to a new micro- and nano-ionics operatable in a vacuum. For example, it will become possible to integrate various IL-based electrochemical devices to produce new functionalities that could never emerge from each single device, which is one of our next targets to attack in the near future.

Acknowledgements

This work was partially supported by a Grant-in-Aid for Scientific Research (No. 15K13613) from the Ministry of Education, Culture, Sports, Science, and Technology of Japan and by the Incorporated Administrative Agency New Energy and Industrial Technology Development Organization (NEDO) under the Ministry of Economy, Trade, and Industry (METI). The synchrotron radiation experiments were performed at the BL46XU of SPring-8 with the approval of the Japan Synchrotron Radiation Research Institute (JASRI) (Proposal No. 2016B1784, 2016A1672, and 2015A1845).

References

1. C.-D. Lee, C. Park, H. J. Lee, K.-S. Lee, S.-J. Park, C. G. Park, S. K. Noh and N. Koguchi, *Jpn. J. Appl. Phys.*, 1998, **37**, 7158–7160.
2. N. Koguchi, S. Takahashi and T. Chikyow, *J. Cryst. Growth*, 1991, **111**, 688–692.
3. H.-N. No and S.-T. Wu, *Jpn. J. Appl. Phys.*, 1998, **37**, 274–278.
4. R. Takahashi, Y. Tsuruta, Y. Yonezawa, T. Ohsawa, H. Koinuma and Y. Matsumoto, *J. Appl. Phys.*, 2007, **101**, 033511.
5. L. C. Bradley and M. Gupta, *Langmuir*, 2015, **31**, 7999–8005.
6. M. Voigt, S. Dorsfeld, A. Volz and M. Sokolowski, *Phys. Rev. Lett.*, 2002, **91**, 026103.
7. Y. Ishii, T. Shimada, N. Okazaki and T. Hasegawa, *Langmuir*, 2007, **23**, 6864–6869.
8. R. Hayes, G. G. Warr and R. Atkin, *Chem. Rev.*, 2015, **115**, 6357–6426.
9. N. M. Micaelo, A. M. Baptista and C. M. Soares, *J. Phys. Chem. B*, 2006, **110**, 14444–14451.
10. D. H. Zaitsau, G. J. Kabo, A. A. Strechan, Y. U. Paulechka, A. Tschersich, S. P. Verevkin and A. Heintz, *J. Phys. Chem. A*, 2006, **110**, 7303–7306.
11. Y. U. Paulechka, D. H. Zaitsau, G. J. Kabo and A. A. Strechan, *Thermochim. Acta*, 2005, **439**, 158–160.
12. O. Höfft, S. Bahr, M. Himmerlich, S. Krischok, J. A. Schaefer and V. Kempter, *Langmuir*, 2006, **22**, 7120–7123.
13. T. Cremer, M. Killian, J. M. Gottfried, N. Paape, P. Wasserscheid, F. Maier and H.-P. Steinrück, *ChemPhysChem*, 2008, **9**, 2185–2190.
14. J. P. Armstrong, C. Hurst, R. G. Jones, P. Licence, K. R. J. Lovelock, C. J. Satterley and I. J. Villar-Garcia, *Phys. Chem. Chem. Phys.*, 2007, **9**, 982–990.
15. A. W. Taylor, K. R. J. Lovelock, A. Deyko, P. Licence and R. G. Jones, *Phys. Chem. Chem. Phys.*, 2010, **12**, 1772–1793.
16. S. Kuwabata, A. Kongkanand, D. Oyamatsu and T. Torimoto, *Chem. Lett.*, 2006, **35**, 600–601.
17. S. Chen, K. Kobayashi, R. Kitaura, Y. Miyata and H. Shinohara, *ACS Nano*, 2011, **5**, 4902–4908.
18. F. Buchner, K. Forster-Tonigold, B. Uhl, D. Alwast, N. Wagner, H. Frankhondeh, A. Groß and R. J. Behm, *ACS Nano*, 2013, **9**, 7773.
19. T. Torimoto, K. Okazaki, T. Kiyama, K. Hirahara, N. Tanaka and S. Kuwabata, *Appl. Phys. Lett.*, 2006, **89**, 243117.
20. T. Kaneko, K. Baba and R. Hatakeyama, *J. Appl. Phys.*, 2009, **105**, 103306.
21. K. Fukui, Y. Yokota and A. Imanishi, *Chem. Rec.*, 2014, **14**, 964–973.
22. M. Brettholle, O. Höfft, L. Klarhöfer, S. Mathes, W. Macus-Friedrichs, S. Z. E. I. Abedin, S. Krischok, J. Janek and F. Endres, *Phys. Chem. Chem. Phys.*, 2010, **12**, 1750–1755.
23. S. Maruyama, Y. Takeyama, H. Taniguchi, H. Fukumoto, M. Itoh, H. Kumigashira, M. Oshima, T. Yamamoto and Y. Matsumoto, *ACS Nano*, 2010, **4**, 5946–5952.
24. J. T. Ye, Y. J. Zhang, R. Akashi, M. S. Bahramy, R. Arita and Y. Iwasa, *Science*, 2012, **338**, 1193–1196.

25. K. Ueno, S. Nakamura, H. Shimotani, H. T. Yuan, N. Kimura, T. Nojima, H. Aoki, Y. Iwasa and M. Kawasaki, *Nat. Nanotechnol.*, 2011, **6**, 408–412.
26. M. Mezger, H. Schröder, H. Reichert, S. Schramm, J. S. Okasinski, S. Schöder, V. Honkimäki, M. Deutsch, B. M. Ocko, J. Ralston, M. Rohwerder, M. Stratmann and H. Dosch, *Science*, 2008, **322**, 424–428.
27. K. Ueno, M. Kasuya, M. Watanabe, M. Mizukami and K. Kurihara, *Phys. Chem. Chem. Phys.*, 2010, **12**, 4066–4071.
28. L. P. N. Rebelo, J. N. Canongia Lopes, J. M. S. S. Esperanca and E. Fillpe, *J. Phys. Chem. B*, 2005, **109**, 6040–6043.
29. M. J. Earle, M. S. S. Esperanca, M. A. Gilea, J. N. Canongia Lopes, L. P. N. Rebelo, J. W. Magee, K. R. Seddon and J. A. Widegren, *Nature*, 2006, **439**, 831–834.
30. R. Souda, *J. Phys. Chem. B*, 2008, **112**, 15349–15354.
31. R. Souda, *J. Phys. Chem. B*, 2009, **113**, 12973–12977.
32. M. Sobota, I. Nikiforidis, W. Hieringer, N. Paape, M. Happel, H. P. Steinrück, A. Görling, P. Wasserscheid, M. Laurin and J. Libuda, *Langmuir*, 2010, **26**, 7199–7207.
33. S. Kato, Y. Takeyama, S. Maruyama and Y. Matsumoto, *Cryst. Growth Des.*, 2010, **10**, 3608–3611.
34. T. Cremer, M. Stark, A. Deyko, H.-P. Steinrück and F. Maier, *Langmuir*, 2011, **27**, 3662–3671.
35. S. Maruyama, H. Taniguchi, Y. Takeyama, M. Itoh and Y. Matsumoto, *Sci. Technol. Adv. Mater.*, 2011, **12**, 54204.
36. T. Waldmann, H. H. Huang, H. E. Hoster, O. Höfft, F. Endres and R. J. Behm, *ChemPhysChem*, 2011, **12**, 2565–2567.
37. R. Foulston, S. Gangopadhyay, C. Chiutu, P. Moriarty and R. G. Jones, *Phys. Chem. Chem. Phys.*, 2012, **14**, 6054–6066.
38. M. Sobota, S. Schernich, H. Schulz, W. Hieringer, N. Paape, P. Wasserscheid, A. Görling, M. Laurin and J. Libuda, *Phys. Chem. Chem. Phys.*, 2012, **14**, 10603–10612.
39. A. Deyko, T. Cremer, F. Rietzler, S. Perkin, L. Crowhurst, T. Welton, H.-P. Steinruck and F. Maier, *J. Phys. Chem. C*, 2013, **117**, 5101–5111.
40. S. Schernich, M. Laurin, Y. Lykhach, N. Tsud, M. Sobota, T. Skála, K. C. Prince, N. Taccardi, V. Wagner, H.-P. Steinrück, V. Matolín, P. Wasserscheid and J. Libuda, *ChemPhysChem*, 2013, **14**, 3673–3677.
41. F. Buchner, K. Forster-Tonigold, B. Uhl, D. Alwast, N. Wagner, H. Farkhondeh, A. Gross and R. J. Behm, *ACS Nano*, 2013, **7**, 7773–7784.
42. R. Souda, *J. Phys. Chem. C*, 2013, **117**, 21281–21287.
43. B. Uhl, F. Buchner, D. Alwast, N. Wagner and R. Jürgen Behm, *Beilstein J. Nanotechnol.*, 2013, **4**, 903–918.
44. B. Uhl, F. Buchner, S. Gabler, M. Bozorgchenani and R. Jürgen Behm, *Chem. Commun. (Cambridge, U. K.)*, 2014, **50**, 8601–8604.
45. F. Rietzler, M. Piermaier, A. Deyko, H.-P. Steinrück and F. Maier, *Langmuir*, 2014, **30**, 1063–1071.
46. S. Schernich, D. Kostyshyn, V. Wagner, N. Taccardi, M. Laurin, P. Wasserscheid and J. Libuda, *J. Phys. Chem. C*, 2014, **118**, 3188–3193.

47. S. Schernich, V. Wagner, N. Taccardi, P. Wasserscheid, M. Laurin and J. Libuda, *Langmuir*, 2014, **30**, 6846–6851.
48. S. Krischok, A. Ulbrich, T. Ikari, V. Kempter, M. Marschewski and O. Höfft, *Nucl. Instrum. Methods Phys. Res., Sect. B*, 2014, **340**, 51–57.
49. Y. Morino, Y. Kanai, A. Imanishi, Y. Yokota and K. I. Fukui, *Jpn. J. Appl. Phys.*, 2014, **53**, 05FY01.
50. Y. Takeyama, S. Mantoku, S. Maruyama and Y. Matsumoto, *CrystEngComm*, 2014, **16**, 684.
51. B. Uhl, H. Huang, D. Alwast, F. Buchner and R. J. Behm, *Phys. Chem. Chem. Phys.*, 2015, **17**, 23816–23832.
52. S. Mehl, A. Toghan, T. Bauer, O. Brummel, N. Taccardi, P. Wasserscheid and J. Libuda, *Langmuir*, 2015, **31**, 12126–12139.
53. K. L. Syres and R. G. Jones, *Langmuir*, 2015, **31**, 9799–9808.
54. M. Yamauchi, S. Maruyama, N. Ohashi, K. Toyabe and Y. Matsumoto, *CrystEngComm*, 2016, **18**, 3399–3403.
55. T. Bauer, S. Mehl, O. Brummel, K. Pohako-Esko, P. Wasserscheid and J. Libuda, *J. Phys. Chem. C*, 2016, **120**, 4453.
56. J. C. S. Costa, A. Mendes and L. M. N. B. F. Santos, *ChemPhysChem*, 2016, 2123–2127.
57. B. Uhl, M. Hekmatfar, F. Buchner and R. J. Behm, *Phys. Chem. Chem. Phys.*, 2016, **18**, 6618–6636.
58. S. Mehl, T. Bauer, O. Brummel, K. Pohako-Esko, P. Schulz, P. Wasserscheid and J. Libuda, *Langmuir*, 2016, **32**, 8613–8622.
59. F. Rietzler, B. May, H.-P. Steinrück and F. Maier, *Phys. Chem. Chem. Phys.*, 2016, **18**, 25143–25150.
60. *Pulsed Laser Deposition of Thin Films: Applications-led Growth of Functional Materials*, ed. R. Eason, John Wiley & Sons, Inc., Hoboken, New Jersey, 1st edn, 2007.
61. S. Yaginuma, J. Yamaguchi, M. Haemori, K. Itaka, Y. Matsumoto, M. Kondo and H. Koinuma, *J. Phys.: Conf. Ser.*, 2007, **59**, 520–525.
62. S. Yaginuma, K. Itaka, Y. Matsumoto, T. Ohnishi, M. Lippmaa, T. Nagata, T. Chikyow and H. Koinuma, *Appl. Surf. Sci.*, 2008, **254**, 2336–2341.
63. S. Yaginuma, K. Itaka, M. Haemori, M. Katayama, K. Ueno, T. Ohnishi, M. Lippmaa, Y. Matsumoto and H. Koinuma, *Appl. Phys. Express*, 2008, **1**, 015005.
64. I. Rey, P. Johansson, J. Lindgren, J. C. Lassègues, J. Grondin and L. Servant, *J. Phys. Chem. A*, 1998, **102**, 3249–3258.
65. E. R. Talaty, S. Raja, V. J. Storhaug, A. Dölle and W. R. Carper, *J. Phys. Chem. B*, 2004, **108**, 13177–13184.
66. H. Tokuda, S. Tsuzuki, Md. A. B. H. Susan, K. Hayamizu and M. Watanabe, *J. Phys. Chem. B*, 2006, **110**, 19593–19600.
67. A. Arce, M. J. Earle, S. Katdare, H. Rodriguez and K. R. Seddon, *Chem. Commun.*, 2006, 2548–2550.
68. G. M. Barrow, *Physical Chemistry*, McGraw-Hill, Inc., 4th edn, 1979.
69. S. Arimoto, M. Sugimura, H. Kageyama, T. Torimoto and S. Kuwabata, *Electrochim. Acta*, 2008, **53**, 6228–6234.

70. S. Maruyama, Y. Takeyama and Y. Matsumoto, *Appl. Phys. Express*, 2011, **4**, 051602.
71. R. S. Wagner and W. C. Ellis, *Appl. Phys. Lett.*, 1964, **4**, 89–91.
72. K. S. Yun, B. D. Choi, Y. Matsumoto, J. H. Song, N. Kand, T. Itoh, M. Kawasaki and H. Koinuma, *Appl. Phys. Lett.*, 2002, **80**, 61–63.
73. R. Takahashi, Y. Yonezawa, M. Ohtani, M. Kawasaki, K. Nakajima, T. Chikyow, H. Koinuma and Y. Matsumoto, *Adv. Funct. Mater.*, 2006, **16**, 485–491.
74. Y. Matsumoto, R. Takahashi and H. Koinuma, *J. Cryst. Growth*, 2005, **275**, 325–330.
75. Y. Takeyama, S. Maruyama and Y. Matsumoto, *Cryst. Growth Des.*, 2011, **11**, 2273–2278.
76. Y. Takeyama, S. Ono and Y. Matsumoto, *Appl. Phys. Lett.*, 2012, **101**, 083303.
77. J. Soeda, T. Okamoto, A. Hamaguchi, Y. Ikeda, H. Sato, A. Yamano and J. Takeya, *Org. Electron.*, 2013, **14**, 1211–1217.
78. Y. Takeyama, S. Maruyama, H. Taniguchi, M. Itoh, K. Ueno and Y. Matsumoto, *CrystEngComm*, 2012, **14**, 4939–4945.
79. K. Kuroishi, S. Maruyama, N. Ohashi, M. Watanabe, K. Naito and Y. Matsumoto, *J. Cryst. Growth*, 2016, **453**, 34–39.
80. Y. Ohsawa, R. Takahashi, S. Maruyama and Y. Matsumoto, *ACS Macro Lett.*, 2016, **5**, 1009–1013.
81. L. Zhu, C.-Y. Huang, Y. H. Patel, J. Wu and S. V. Malhotra, *Macromol. Rapid Commun.*, 2006, **27**, 1306–1311.
82. T. Yanase, T. Hasegawa, T. Nagahama and T. Shimada, *Jpn. J. Appl. Phys.*, 2012, **51**, 041603.
83. S. Horike, Y. Koshiba, M. Misaki and K. Ishida, *Jpn. J. Appl. Phys.*, 2014, **53**, 05FT03.
84. S. Horike, M. Misaki, Y. Koshiba, M. Morimoto and K. Ishida, *Chem. Lett.*, 2016, **45**, 1156–1158.

CHAPTER 7

Smart Design of Sustainable and Efficient ILs

GIUSEPPE MUSUMARRA*, ALESSIO PATERNÒ AND SALVATORE SCIRÈ

Dipartimento di Scienze Chimiche, Università di Catania, Viale A. Doria 6, 95125 Catania, Italy
*E-mail: gmusumarra@unict.it

7.1 Investigating the ILs Chemical Space: A Complex Task

The applications of ILs, which originally attracted much attention as green solvents due to their lower vapour pressure as compared to common volatile organic solvents, increased exponentially in the last decade.[1] Optimization of ILs properties is presently a major goal of scientific research, which has to also take into account ILs human hazard potential, their effects on aquatic ecosystems as well as economic considerations. In summary, optimization studies should address not only ILs efficiency for specific applications, but also their environmental and economic sustainability. Such an ambitious task is very difficult to achieve as the number of ILs is estimated to be one million and the number of ILs mixtures over ten million.

Each IL can be thought of as the intersection of two chemical spaces: the cation and the anion. Due to the large number of cation and anion combinations, the IL chemical space is so huge that it cannot be fully experimentally explored, therefore available data cover only a very small portion of it.

Smart Materials No. 28
Ionic Liquid Devices
Edited by Ali Eftekhari

Published by the Royal Society of Chemistry, www.rsc.org

Relatively small ILs subspaces can be investigated by means of enormous synthetic and experimental screening efforts resulting in extremely expensive and time-consuming experiments. Moreover, IL purification is often difficult and their purity can dramatically affect the experimental determination of their properties, as demonstrated for ILs containing halide impurities and water.[2] Hence the need to derive ILs *in silico* descriptors suitable for quantitative structure–property relationships (QSPRs) models allowing property prediction for unknown ILs before planning their synthesis and carrying out efficiency and toxicity tests.

7.2 The Multivariate Statistical Approach

7.2.1 Data Collection and Data Analysis

Multivariate statistical approaches can be adopted both for data collection and data analysis. Multivariate experimental design is aimed at the selection of a limited number of experiments providing maximum information useful to optimize experimental procedures. The optimization strategy usually involves two steps. The first one, considering all possible factors able to influence the process, is aimed at finding out which variables and which interactions are more important in the process. These are not many, as stated by the Pareto principle (the vital few, the trivial many) and the 80/20 rule (80% improvement in the process performance will come from changes in 20% of the process variables), concepts widely adopted in the economic field.[3] The identification of relevant variables, usually achieved by fractional factorial design,[4] is outside the scope of the present chapter focusing on the second step, *i.e.* the building of models for data analysis. Multivariate modelling may be adopted to achieve another important objective, *i.e.* extraction of important information present in a given data set by means of principal component analysis (PCA) and partial least squares (PLS).

7.2.2 Comparison Between MRA and PLS

Data are usually reported in a matrix (table) where a number M of variables (experimental values) is collected for each of N objects (*e.g.* chemical compounds). The geometrical representation of each object is one point in a M-dimensional space, where each variable defines an orthogonal axis. Multivariate methods are aimed at investigating the data structure using as similarity criterion either the Euclidean distance (linear discriminant analysis, kth nearest neighbour, cluster analysis and others) or the fit into a unique mathematical model based on least squares procedures.

Multiple regression analysis (MRA) was developed[5] to describe the values of a dependent variable y as a function of a number of independent variables x_k (eqn (7.1)):

$$y = \beta_0 x_0 + \beta_1 x_1 + \beta_2 x_2 + \ldots + \beta_k x_k + \varepsilon \tag{7.1}$$

where y = response variable, x = predictor variables, β_k = the regression coefficients, and ε = a random error. As this method is still the most popular multivariate approach, it is worth mentioning some of the statistical implications involved in its application. MRA assumes *a priori* that the x_k independent variables are all (and the only ones) relevant to the problem. Low regression coefficients β_k (depending on the variables numerical values) do not automatically mean that the variable can be neglected if the x values have not been normalized. Furthermore, the x_k variables are assumed to be "independent" from each other and error free. Correlated variables give rise to the problem of collinearity, which strongly affects correlation coefficients and the statistical tests providing overoptimstic modelling results. Last but not least, MRA is insensitive to grouping of the data points and cannot detect the presence of outliers. When using MRA it is strongly suggested not to rely only on statistical values such as correlation coefficient, slope, intercept, and standard deviation values provided by "black box" approaches, but to integrate the results with correlation plots. Figure 7.1 shows different correlation plots of ten samples providing the same slope and intercept.

The philosophy behind the PCA and PLS approaches, included in the SIMCA (soft independent modelling of class analogy)[6] strategy, allows both samples classification (PCA) and correlation (PLS) and consists in building separate class models after a preliminary investigation of the data structure called pattern recognition. To compare variables with different numerical values, each

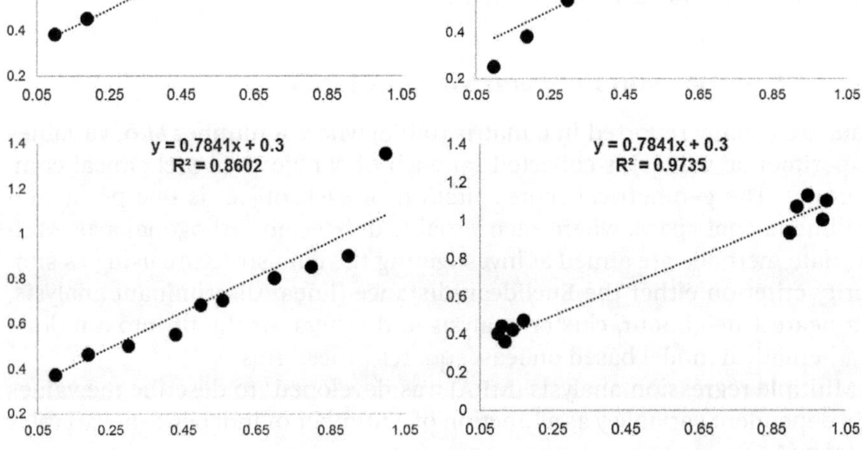

Figure 7.1 Examples of data distribution providing different plots with the same slope and intercept.

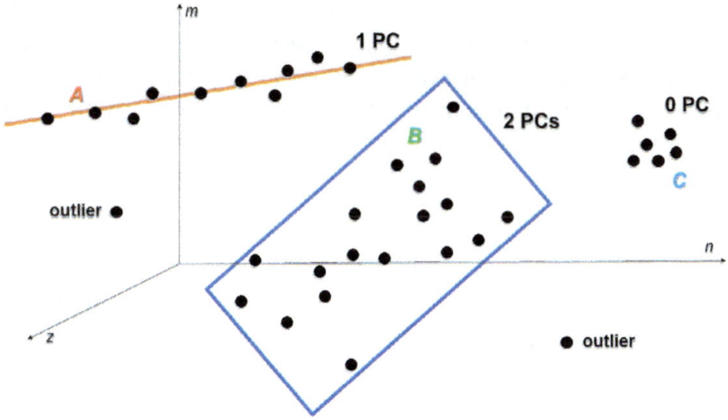

Figure 7.2 A possible distribution of experimental samples in a three-dimensional space evidencing the advantage of applying disjoint soft PC models with different dimensionalities as compared to a unique MRA model.

variable has to be normalized to the same variance by the so-called "scaling" procedure.[7] A large sample population may not be representative of a unique "*universum*" and the data may be distributed into different "classes". A graphical representation of a SIMCA pattern recognition reporting the data structure of 36 samples characterized by three variables ($M = 3$) is illustrated in Figure 7.2. Each sample either belongs to one of three different classes (A, B, C) or to none of them, *i.e.* is an outlier. This trend can be evidenced by a preliminary PCA including all the samples. Once the data structure has been elucidated, separate PCA models can be derived for each class.

PCA is an extremely useful technique to "summarise" all the information contained in the **X**-matrix in a more understandable form. PCA works by decomposing the **X**-matrix as the product of two smaller matrices (eqn (7.2)), which are called the loadings (**P**) and the scores (**T**) matrices:

$$\mathbf{X} = \mathbf{TP} + \mathbf{E} \tag{7.2}$$

The loadings matrix contains information about the variables. It is composed of a few vectors (principal components, PCs) which are linear combinations of the original **X**-variables. The scores matrix contains information about the objects. Each object is described in terms of its projections onto the PCs (instead of the original variables). The information not contained in these matrices remains as "unexplained **X**-variance" in a residual matrix (**E**) which has exactly the same dimensionality as the original **X**-matrix.

The PCs, among many others, have two interesting properties: (1) they are extracted in decreasing order of importance, and therefore the first PC always contains more information than the second, the second more than the third and so on; (2) PCs are orthogonal to each other: this means that there is no correlation between the information contained in different PCs.

In PCA, each new extracted PC increases further the amount of information (the variance, expressed by R^2 parameter indicating the goodness of fit) explained by the model. However, usually the first four or five PCs explain more than 90% of the **X**-variance. There is no simple or unique criterion to decide how many PCs should be considered significant. From a theoretical point of view, it is possible to use cross-validation techniques to decide the number of PCs to include evaluating the predictability of **X**-variation (Q^2 parameter) of each component and all together.[6]

Going back to Figure 7.2, for class A a model with a single PC is obtained and the data can be compressed by characterizing each sample using only one vector (the first PC "scores") which is the projection on the line corresponding to the maximum dispersion of data points. The second PC, exploring the dispersion of points along all directions perpendicular (*i.e.* statistically orthogonal) to the first PC, does not turn out to be statistically significant, as no significant dispersion of points can be observed. For class B the second PC turns out to be statistically significant and the dimensionality can be reduced from three to two by projecting the data points into a plane. For class C, no PC is found as the data are clustered with little variation. Examination of the two "outliers" can also be useful as it can suggest an explanation of why these samples behave differently from others.

In PCA no cause–effect relationship is assumed. The PLS analysis[8] is a regression technique whose goal is to explain one or more dependent variables (y's) in terms of a number of explanatory variables (x's) according to eqn (7.3):

$$\mathbf{Y} = \mathrm{f}(\mathbf{X}) + \mathbf{E} \tag{7.3}$$

It is possible to build many different models that fulfil the equation, each of them fitting the y's more or less accurately. Among them, the best one will fit the experimental y values even for objects not included in building the model. Consequently, this model becomes "predictive" and can be used to calculate reliable estimations of y values for new samples, prior to experimental measurement. It is important to note that the y variables, as any other experimental variable, contain errors. The models will try to fit the y as much as possible (increasing the R^2 parameter) but, when trying to improve too much the fitting, the model will explain also the noise. This phenomenon, called overfitting, is very dangerous because overfitted models appear to be very good (having high R^2) but often prove to be useless in prediction for y's of objects not included in the learning set (low Q^2) (see Figure 7.3).

PLS represents the ideal regression method to deal with **X**-matrices used in the 3D-QSAR, which usually contain much fewer objects (molecules) than variables. In these situations, MLR is almost useless for two main reasons: (1) MLR was developed to deal with situations in which the number of objects (N) is three times at least larger than the number of variables (K). This inconvenience can be overcome by using stepwise MLR, but then there is a high probability of obtaining relationships just by chance; (2) MLR assumes that

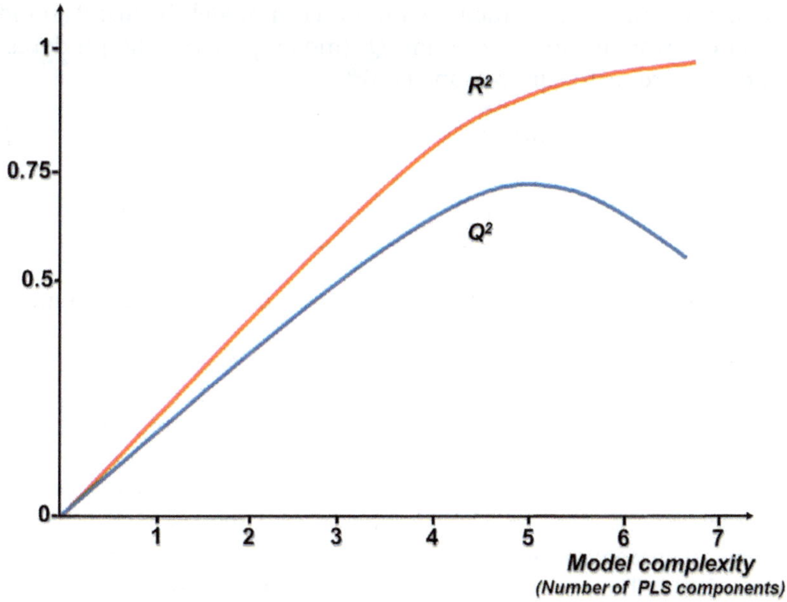

Figure 7.3 Example illustrating the variation of R^2 and Q^2 on increasing the complexity of a PLS model. At a given model complexity, an optimal balance between fit and predictability is achieved (5 PLS components).

the **X**-variables are "independent" and not correlated, but it is known that variables do not often fulfil these requirements.

Similarly to PCA, PLS decomposes the **X**-matrix into the product of two smaller matrices. The weights (a concept similar to the PCA loadings) matrix contains information about the variables, including a few vectors (latent variables, LVs) which are linear combinations of the original **X**-variables. The concept of LV is mostly equivalent to the PC in PCA. The scores matrix (**T**) contains information about the objects. Each object is described in terms of the LVs.

The main difference is that PCA obtains the PCs that represent at best the structure of the **X**-matrix and PLS obtains the LVs under the constraint that they have to represent the structure of the **X**- and **Y**-matrices and simultaneously maximize the fitting between the x's and the y's. The LVs share some important properties with the PCs, being extracted in decreasing order of importance and orthogonal to each other.

The selection of the correct dimensionality in PLS is of crucial importance. In fact, when too many PCs are included a serious overfitting could occur leading to a model with little or no validity with high R^2 but low or negative Q^2. The optimal number of PCs is usually evaluated by using cross-validation (CV) with the aim of maximizing the predictive ability of the model. CV works by building reduced models, where one or some of the objects are removed, and using them to predict the y variables of the held-out objects. Then the predicted and experimental y values are compared and for each model the SDEP (standard deviation of errors of prediction) and Q^2 (model predictability) indexes are computed according to eqn (7.4) and (7.5):[6]

$$\text{SDEP} = \sqrt{\left[\sum(y - y')^2 / N\right]} \quad (7.4)$$

$$Q^2 = 1 - \left[\sum(y - y')^2 / \sum(y - \acute{y})^2\right] \quad (7.5)$$

where y is the experimental value, y' the predicted value, $\acute{y}$ the average value and N the number of objects.

The CV technique is very valuable because it performs an "internal validation" of the model and obtains an estimation of the predictive ability without the help of external datasets. This is particularly important in quantitative structure–activity relationships (QSARs) studies, where the number of objects available is usually small, and it is not affordable to remove objects from the learning dataset. One of the main inconveniences of CV is that there is no general agreement on how to build the reduced groups and on the criterion to decide how many objects to keep. SIMCA, by default, divides the dataset into n (often 7) groups, each one containing an equal (or nearly equal) number of objects.[6] Then models are built keeping one of these groups out of the analysis until all of the objects are kept out once. The procedure should be repeated many times, in order to obtain stable results.

The use of an external prediction set is another approach to evaluate the predictive ability of the model. In this case, the objects in the original dataset are split up into two groups from the beginning of the analysis. The first one, the learning set, will be used to build the PLS model, the other one, the prediction set, allows verification of the correspondence between experimental and predicted **Y**-values. Undoubtedly this technique is more realistic to test the predictive ability, but the results depend critically upon how many and which objects are assigned to each group. In addition, datasets in QSAR often contain too few objects and it is not possible to remove objects from the analysis without a significant loss of information.

The statistical results obtained by the PLS method are able to detect what variables in the **X** block are relevant to determine the dependent variables (**Y** block) by means of the VIP (variables importance for the projection) values. The VIP values reflect, in fact, the importance of terms in the model both with respect to **Y**, *i.e.* its correlation to all the responses, and with respect to **X**, *i.e.* in explaining **X**-variation. The VIP values are calculated for each **X**-variable by summing the squares of the PLS loadings weighted by the amount of sum squares explained in each model component.

The predictive and interpretation potentialities of the PLS approach can be deduced by inspection of Figure 7.4. Once a reliable PLS model is obtained, predictions of the desired property (y) can be achieved from the easily accessible descriptor values in the **X** matrix. In addition to prediction, PLS parameters such as the loadings and VIP values provide an estimate of the variable

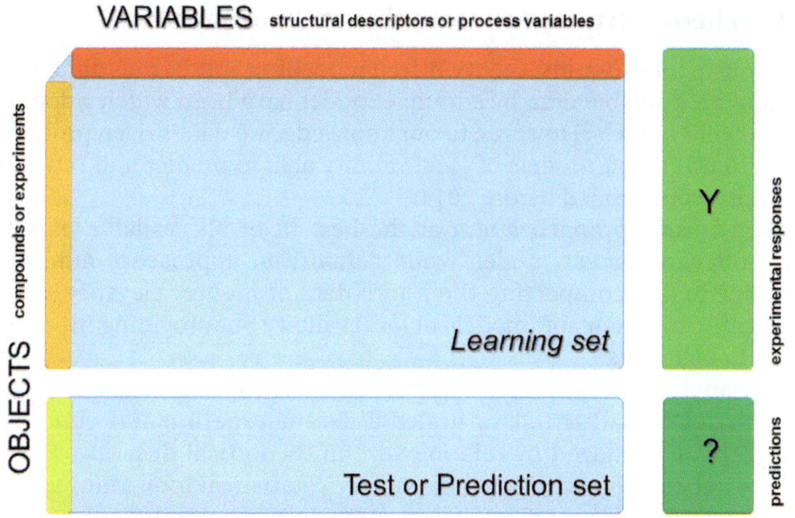

Figure 7.4 Example of a data matrix undergoing PLS. The predictive ability of the model is based on structural descriptors or process variables whose VIP values can be used for interpretation and design purposes.

importance in each component and in the overall model respectively. Interpretation of the significance of each variable (structural feature or process variable) in relation to the dependent variable (compound biological or physico-chemical property or process performance) is essential in optimization studies.

7.3 ILs QSPR

For a long time scientists have aimed to make predictions about physical and chemical properties. A QSAR can be defined as a mathematical relationship linking chemical structure to biological activity in a quantitative way for a series of compounds.

The development of QSARs was mainly due to Corwin Hansch[9] who developed MRA equations for the prediction of biological activities from structural and physico-chemical properties. Since then, QSAR modelling expanded including environmental toxicities due to the increasing attention to the eco-toxicological risk and to comply with legislation such as the European Union Regulation for the Registration, Evaluation, Authorisation and Restriction of Chemicals (REACH).[10]

The correlation between structural and compositional parameters and a physico-chemical property is called a quantitative structure–property relationship (QSPR), a more general term which presently also includes structure–biological activity relationships.[11] A QSPR equation has the form of eqn (7.6):[11]

$$y = ax_1 + bx_2 + cx_3 + \ldots k \tag{7.6}$$

7.3.1 Theory-driven and Data-driven Approaches

In the field of ionic liquids, theory-driven modelling approaches aimed at the best fit of all available data by a unique model have been widely adopted to develop QSPR models. However, to our knowledge no data-driven procedures leading to QSPR soft models of local validity able to predict and rationalize ILs toxicity were applied before 2014.

Theory-driven approaches aim at the best fit of all available data by a unique often nonlinear model, while data-driven approaches aim at raw data reduction by compacting them into data of higher relevance, eventually adopting different soft models of local validity but providing more easily interpretable results.

An advantage of theory-driven approaches is that a unique often nonlinear model fits all the data under different experimental conditions. Usually, this is achieved by relating several theoretical descriptors to the property under examination using MRA as a statistical tool. Many authors, to demonstrate the superiority of such models, report plots of predicted *vs* experimental data providing a better correlation as compared with that of previous models. However, a good correlation using many theoretical descriptors by means of the MRA approach has a high risk of descriptors collinearity and therefore may not be statistically significant. Unfortunately in many literature papers the numerical values of the descriptors are not reported, preventing other researchers from replicating the results and applying other approaches to the same data set, a desirable condition which often produces further achievements in scientific knowledge. Moreover, the model interpretation may be unclear and the results are not always usable in practice by experimentalists for the design of specific properties.

On the other hand, the data-driven approach presented here starts with an overview of the raw data and then compacts them into data of higher relevance eventually modelled by different soft models of local validity. Descriptors availability allows the readers to reproduce the results, providing a simple evaluation of their relevance. This approach is more flexible as it may adopt different data modelling techniques depending on the purpose of the investigation and on the data structure.

A drawback of data-driven modelling is that models are valid within a limited experimental domain, determined by the learning set samples. The selection of the samples to include in each class model is subjective, but the above-mentioned preliminary pattern recognition on the data structure can orient this choice leading to more accurate predictions. Moreover, once the model has been developed, the results are displayed in simple equations and easily interpretable plots. This allows experimentalists to predict and design themselves physico-chemical properties for a set of ILs for R&D studies taking into account not only ILs efficiency for specific applications, but also their environmental and economic sustainability.

7.3.2 New VolSurf+ *In silico* Descriptors for ILs Cationic and Anionic Counterparts

VolSurf+ is a powerful chemoinformatic tool aimed at producing and exploring the physico-chemical property space of a library of molecules starting from 3D maps of interaction energies between the molecule and chemical probes.

In VolSurf+, a molecular interaction field (MIF) approach previously applied to investigate antitumour drugs[12] identifies potential polar and hydrophobic interaction sites around target molecules by the water (OH2), the hydrophobic (DRY), the carbonyl oxygen (O) and amide nitrogen (N1) probes. The information contained in the MIFs is translated into a quantitative scale by calculating the volume or the surface of the interaction contours. In fact, in VolSurf+ 3D structure generation is automatically performed, generating a bunch of minimized 3D structures of molecules in water and apolar solvents from which the best energy minima conformers are selected. All the remaining structures are used to derive conformational dependent descriptors, such as flexibility parameters, Log*P*, partition coefficient in non-water solvent, amphiphilic moments, shape, electrostatic forces, H-bonds and hydrophobicity.

The VolSurf+ procedure is depicted in Figure 7.5. In the first step, a three-dimensional map (3D map), which is an image of the target–probe molecular interaction, is generated from the interactions of the OH2, DRY, O and N1 probes with a target molecule. In this 3D map each pixel contains information about a chemical property. The amount of information contained in a 3D map is related to the interacting molecular partners. The second step consists in the calculation of descriptors from the 3D maps obtained in the first step.

128 cationic *in silico* descriptors (Table 7.1) were calculated by importing the corresponding SMILES codes in VolSurf+ and using the aforementioned steps.[13] Derivation of the anions descriptors required a slightly different procedure as compared to that adopted for organic ILs cations.

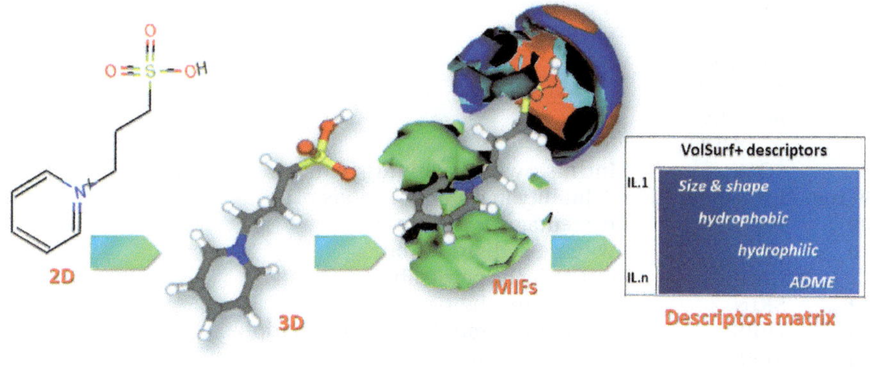

Figure 7.5 Derivation of descriptors in VolSurf+.

Table 7.1 The 128 VolSurf+ descriptors.

	Code	Name
1	V	Molecular volume
2	S	Molecular surface
3	R	Volume/surface ratio
4	G	Molecular globularity = S/S_{eq}, with S_{eq} = surface area of a sphere of volume V
5–12	W1–W8	Hydrophilic regions
13–20	D1–D8	Hydrophobic regions
21–26	WO1–WO6	H-bond donor volumes
27–32	WN1–WN6	H-bond acceptor volumes
33–36	IW1–IW4	Hydrophilic interaction energy: distance between hydrophilic areas and molecular barycentre
37–44	CW1–CW8	Capacity factor for OH2: hydrophilic regions/total molecular surface
45–48	ID1–ID4	Hydrophobic integy moment: distance between hydrophobic areas and molecular barycentre
49–56	CD1–CD8	Capacity factor for DRY: hydrophobic regions/total molecular surface
57–58	HL1–HL2	Hydrophilic–lipophilic balance
59	A	Amphiphilic moment
60	CP	Critical packing, predicting molecular packing such as in micelle formation
61	POL	Polarizability
62	MW	Molecular weight
63–64	FLEX, FLEX_RB	Flexibility parameters
65	NCC	Number of charged centres
66	DIFF	Diffusivity: dispersion of chemicals in water fluid at 25 °C
67	LOGP N-oCT	Log *P* octanol/water
68	LOGP c-Hex	Log *P* cyclohexane/water
69–72	PSA HSA PSAR PHSAR	Polar and hydrophobic surface areas
73–79	LgD5–LgD10	Log *D*: log*P* oct/water at different pH
80	AUS7.4	Available uncharged species
81–87	%FU4–%FU10	Percentage of unionized species at different pH
88–97	DRDRDR DRDRAC DRDRDO DRACAC DRACDO DRDODO ACACAC ACACDO ACDODO	Dry, H-bond donor, H-bond acceptor and mixed Dry, H-bond donor and acceptor 3D triplets pharmacophoric areas
98	SOLY	Intrinsic water solubility in mol L^{-1} at 25 °C
99–108	LgS3–LgS11	Solubility in water at various pH
109	PB	Percentage of protein binding
110	VD	Volume of distribution
111	CACO2	CACO2 cells permeability
112	SKIN	Skin permeability
113	LgBB	Log blood–brain barrier distribution
114	MetStab	Metabolic stability
115	HTSFlag	High-throughput screening flag
116–120	L0lgS–L4lgS	Solubility profiling coefficients
121–128	DD1–DD8	Differences of the hydrophobic volumes

It is well known that small negative ions like halides do not exist in solution as free compounds, but are complexed by water molecules producing hydrated complexes.[14] Therefore halides ($Cl^-/Br^-/I^-$) were initially imported in Sybyl[15] where a solvation shell was created within a 3 Å sphere by using three water molecules selected around the ions. Next, the hydrated ions were processed using the GRID force field to produce the corresponding MIFs. All the other ions, including ions containing metals or phosphorus, were directly imported by means of their SMILES codes in GRID force fields without hydration shell. Finally, VolSurf+ was used to convert the MIFs into molecular descriptors. Notably, due to the different information embedded in the file format, a large number of VolSurf+ anionic descriptors could not be generated. Thus, only 48 non-zero variables were extracted as anions descriptors.[13]

7.3.3 Multivariate Design in Principal Properties

The result of an experiment (*i.e.* chemical reaction or material manufacture) depends on experimental conditions (reaction or process variables). Although satisfactory performances may be achieved by intuition or innovative thought, it is rare that the first attempt provides optimal results. Hence the need to optimize experimental conditions. The one-variable-at-a-time (OVAT) approach, which consists of studying the "effect" of one variable while keeping all other variables constant, is nowadays widely adopted, despite the fact that almost one century ago R. A. Fisher pointed out that it is inappropriate when several factors are to be considered.[16] The OVAT approach implicitly assumes that the influence of the variable under examination does not depend on other variables kept constant. For example, it is evident that optimal temperature and reaction times with catalyst A may not be optimal reaction conditions if catalyst B is used. To take into account variable interactions statistical approaches considering all variables simultaneously have to be adopted. The reason for adopting the OVAT approach is that multivariate methods are still not widespread in the experimenters' community. The term multivariate design is wrongly associated to a very large number of experiments deriving from a combination of each single variable values. Experimenters often do not know how to use multivariate design and textbooks on statistics rarely explain how statistics may be useful to experimenters.

Multivariate strategies can be conveniently used to optimize desired properties (reaction yield, material performances, *etc.*) by simultaneous variation of experimental variables, which may be continuous (temperature, time, concentrations, *etc.*) or discrete (solvent, catalyst, *etc.*). However, parametrization of the latter variables is needed to use them in experimental design. Such a design requires the use of variables orthogonal to each other and this led to the derivation of principal properties (PPs), intrinsic properties representative of experimentally observable macroscopic descriptors. PPs, calculated as the principal component analysis (PCA) scores from several experimentally measured properties, were derived for solvents,[17,18] aldehydes and ketones,[19] amines,[20] Lewis acids,[21] lanthanide triflates,[22] aromatic substituents[23] and amino acids.[24–27] First and second generation principal properties for heteroaromatic moieties, based on aromaticity[28] and on 3D-GRID structural parameters,[29] respectively, have also been reported. Statistically orthogonal PPs are also less influenced by measurement errors and system-specific variations as compared to single descriptors.

A striking example of successful multivariate experimental design in PPs is provided by the Fischer indole synthesis,[30] a chemical reaction known for more than a century.[31] It is a two-step synthesis involving reaction of ketones with phenylhydrazines giving phenylhydrazones and a subsequent ring closure in the presence of a Lewis acid. When unsymmetrical ketones are used as starting materials, an indole regioisomeric mixture is obtained. A PLS analysis using the PPs of ketones, Lewis acid catalysts and solvents as descriptors, evidenced the effects of the above parameters on the indole isomeric distribution (the response). In addition to interpretation, PLS allowed the prediction of the regioisomeric ratio for new reactions. The approach achieved not only regiospecific synthesis of single indole regioisomers,[32] but almost quantitative yields in a single step (one-pot) reaction under milder conditions.[33]

PPs turned out to be suitable for design in QSARs,[23] especially for the design of biologically active peptides.[24,25] Dedicated PPs for amino acids were then derived for peptides QSARs[26] and for quantitative sequence–activity modelling[34] and more recently using MIF (molecular interaction fields) descriptors for GRID-based description of molecular similarity and ligand-based virtual screening.[27,35]

In the case of biologically active peptides, for a sequence of amino acids (AAs) building blocks, the adoption of experimental design in PPs considering two AAs would reduce the number of experiments investigating a chemical space covering 400 (20 × 20 natural AAs in different positions) peptides. The number of ILs resulting from the combination of cations and anions is over a million, therefore the experimental space is so huge that it cannot be fully explored and a rational selection of informative experiments is needed.

As a high number of descriptors cannot be easily handled, we recently revived interest in principal properties by compacting VolSurf+ physico-chemical parameters into ILs cationic and anionic PPs.[36] PP+ and PP− are respectively the scores of PCA carried out on data matrices including ILs cations and anions as "objects" and VolSurf+ descriptors as variables. PCA is an "open" statistical procedure depending on the analysed data matrix, chosen to provide an optimal balance between model (and therefore PPs) generality and prediction ability.

For the derivation of cationic PPs, structures including different heterocyclic aromatic and non-aromatic cationic cores (imidazolium, pyridinium, quinolinium, piperidinium, pyrrolidinium, morpholinium) with different side chains were considered as objects (Scheme 7.1 and 7.2). In the cationic data matrix the variables were the 128 VolSurf+ *in silico* descriptors, related to molecular

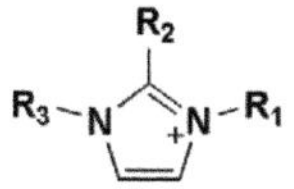

$R_1 = H \quad R_2 = H \quad R_3 = H$
$R_1 = CH_3 \quad R_2 = H \quad R_3 = H$
$R_1 = CH_3 \quad R_2 = CH_3 \quad R_3 = H$
$R_1 = CH_3 \quad R_2 = H \quad R_3 = CH_3$
$R_1 = C_2H_5 \quad R_2 = H \quad R_3 = H$
$R_1 = C_2H_5 \quad R_2 = H \quad R_3 = CH_3$
$R_1 = C_2H_5 \quad R_2 = H \quad R_3 = C_2H_5$
$R_1 = C_3H_7 \quad R_2 = H \quad R_3 = H$
$R_1 = C_3H_9 \quad R_2 = H \quad R_3 = CH_3$
$R_1 = C_3H_9 \quad R_2 = CH_3 \quad R_3 = CH_3$
$R_1 = C_3H_9 \quad R_2 = H \quad R_3 = C_2H_5$
$R_1 = C_3H_9 \quad R_2 = H \quad R_3 = C_3H_9$
$R_1 = C_4H_9 \quad R_2 = H \quad R_3 = H$
$R_1 = C_4H_9 \quad R_2 = H \quad R_3 = CH_3$
$R_1 = C_4H_9 \quad R_2 = CH_3 \quad R_3 = CH_3$
$R_1 = C_4H_9 \quad R_2 = H \quad R_3 = C_2H_5$
$R_1 = C_4H_9 \quad R_2 = H \quad R_3 = C_4H_9$
$R_1 = C_5H_{11} \quad R_2 = H \quad R_3 = H$
$R_1 = C_5H_{11} \quad R_2 = H \quad R_3 = CH_3$
$R_1 = C_5H_{11} \quad R_2 = H \quad R_3 = C_5H_{11}$
$R_1 = C_6H_{13} \quad R_2 = H \quad R_3 = H$
$R_1 = C_6H_{13} \quad R_2 = H \quad R_3 = CH_3$
$R_1 = C_6H_{13} \quad R_2 = CH_3 \quad R_3 = CH_3$
$R_1 = C_6H_{13} \quad R_2 = CH_3 \quad R_3 = CH_3$
$R_1 = C_6H_{13} \quad R_2 = H \quad R_3 = C_2H_5$
$R_1 = C_6H_{13} \quad R_2 = H \quad R_3 = C_6H_{13}$
$R_1 = C_7H_{15} \quad R_2 = H \quad R_3 = H$
$R_1 = C_7H_{15} \quad R_2 = H \quad R_3 = CH_3$
$R_1 = C_8H_{17} \quad R_2 = H \quad R_3 = H$
$R_1 = C_8H_{17} \quad R_2 = H \quad R_3 = CH_3$
$R_1 = C_9H_{19} \quad R_2 = H \quad R_3 = H$
$R_1 = C_9H_{19} \quad R_2 = H \quad R_3 = CH_3$
$R_1 = C_{10}H_{21} \quad R_2 = H \quad R_3 = H$
$R_1 = C_{10}H_{21} \quad R_2 = H \quad R_3 = CH_3$
$R_1 = C_{10}H_{21} \quad R_2 = H \quad R_3 = C_2H_5$
$R_1 = C_{10}H_{21} \quad R_2 = CH_3 \quad R_3 = C_{10}H_{21}$
$R_1 = C_{11}H_{23} \quad R_2 = H \quad R_3 = H$
$R_1 = C_{12}H_{25} \quad R_2 = H \quad R_3 = H$
$R_1 = C_{12}H_{25} \quad R_2 = H \quad R_3 = CH_3$
$R_1 = C_{14}H_{29} \quad R_2 = H \quad R_3 = CH_3$
$R_1 = C_{16}H_{33} \quad R_2 = H \quad R_3 = CH_3$
$R_1 = C_{18}H_{37} \quad R_2 = H \quad R_3 = CH_3$
$R_1 = CH_2CH{=}CH_2 \quad R_2 = H \quad R_3 = CH_3$
$R_1 = CH_2iPr \quad R_2 = H \quad R_3 = CH_3$
$R_1 = CH_2Ph \quad R_2 = H \quad R_3 = CH_3$
$R_1 = CH_2Ph \quad R_2 = H \quad R_3 = CH_3$
$R_1 = CH_2Ph \quad R_2 = H \quad R_3 = C_4H_9$
$R_1 = C_2H_4Ph \quad R_2 = H \quad R_3 = CH_3$
$R_1 = CH_2(4\text{-}MePh) \quad R_2 = H \quad R_3 = CH_3$
$R_1 = C_2H_4OH \quad R_2 = H \quad R_3 = CH_3$
$R_1 = C_2H_4OH \quad R_2 = CH_3 \quad R_3 = CH_3$
$R_1 = C_2H_4OH \quad R_2 = H \quad R_3 = C_4H_9$
$R_1 = C_2H_4OH \quad R_2 = H \quad R_3 = C_2H_2Ph$
$R_1 = C_2H_4OH \quad R_2 = CH_3 \quad R_3 = C_{14}H_{27}$
$R_1 = C_2H_4OH \quad R_2 = CH_3 \quad R_3 = C_{16}H_{33}$
$R_1 = C_2H_4OH \quad R_2 = H \quad R_3 = C_2H_4OH$
$R_1 = C_3H_6OH \quad R_2 = H \quad R_3 = CH_3$
$R_1 = CH_2CH(OH)CH_2OH \quad R_2 = H \quad R_3 = CH_3$
$R_1 = CH_2CH(OH)CH_2OH \quad R_2 = CH_3 \quad R_3 = CH_3$
$R_1 = C_4H_8OH \quad R_2 = H \quad R_3 = CH_3$
$R_1 = C_8H_{16}OH \quad R_2 = H \quad R_3 = CH_3$
$R_1 = CH_2Oet \quad R_2 = H \quad R_3 = CH_3$
$R_1 = CH_2Opr \quad R_2 = H \quad R_3 = CH_2Opr$
$R_1 = CH_2Obut \quad R_2 = H \quad R_3 = H$
$R_1 = CH_2Obut \quad R_2 = H \quad R_3 = CH_2Obut$
$R_1 = CH_2Opent \quad R_2 = H \quad R_3 = H$
$R_1 = CH_2Opent \quad R_2 = H \quad R_3 = CH_2Opent$
$R_1 = CH_2Ohex \quad R_2 = H \quad R_3 = H$
$R_1 = CH_2Ohex \quad R_2 = H \quad R_3 = CH_2Ohex$
$R_1 = CH_2Oept \quad R_2 = H \quad R_3 = H$
$R_1 = CH_2Oept \quad R_2 = H \quad R_3 = CH_2Oept$
$R_1 = CH_2Ooct \quad R_2 = H \quad R_3 = H$
$R_1 = CH_2Ooct \quad R_2 = H \quad R_3 = CH_2Ooct$
$R_1 = CH_2Onon \quad R_2 = H \quad R_3 = H$
$R_1 = CH_2Onon \quad R_2 = H \quad R_3 = CH_2Onon$
$R_1 = CH_2Odec \quad R_2 = H \quad R_3 = H$
$R_1 = CH_2Odec \quad R_2 = H \quad R_3 = CH_2Odec$
$R_1 = CH_2Oundec \quad R_2 = H \quad R_3 = H$
$R_1 = CH_2Oundec \quad R_2 = H \quad R_3 = CH_2Oundec$
$R_1 = CH_2Ododec \quad R_2 = H \quad R_3 = H$
$R_1 = CH_2Ododec \quad R_2 = H \quad R_3 = CH_2Ododec$
$R_1 = CH_2Otetradec \quad R_2 = H \quad R_3 = CH_2Otetradec$
$R_1 = CH_2Ohexadec \quad R_2 = H \quad R_3 = CH_2Ohexadec$
$R_1 = C_2H_4OCH_3 \quad R_2 = H \quad R_3 = CH_3$
$R_1 - C_3H_6OCH_3 \quad R_2 = H \quad R_3 = CH_3$
$R_1 = C_2H_4OEtOme \quad R_2 = H \quad R_3 = CH_3$
$R_1 = C_2H_4C(CH_3)O \quad R_2 = H \quad R_3 = CH_3$
$R_1 = C_3H_6COOH \quad R_2 = H \quad R_3 = CH_3$
$R_1 = C_7H_{14}COOH \quad R_2 = H \quad R_3 = CH_3$
$R_1 = CH_2COOCH_3 \quad R_2 = H \quad R_3 = CH_3$
$R_1 = CH_2COOet \quad R_2 = H \quad R_3 = CH_3$
$R_1 = CH_2COOpr \quad R_2 = H \quad R_3 = CH_3$
$R_1 = CH_2COOpr \quad R_2 = CH_3 \quad R_3 = CH_3$
$R_1 = CH_2COObut \quad R_2 = H \quad R_3 = CH_3$
$R_1 = CH_2COOpent \quad R_2 = H \quad R_3 = CH_3$
$R_1 = CH_2COOpent \quad R_2 = CH_3 \quad R_3 = CH_3$
$R_1 = CH_2COOhex \quad R_2 = H \quad R_3 = CH_3$
$R_1 = CH_2COOoct \quad R_2 = H \quad R_3 = CH_3$
$R_1 = C_2H_4COOet \quad R_2 = H \quad R_3 = CH_3$
$R_1 = C_{10}H_{20}COOet \quad R_2 = H \quad R_3 = CH_3$
$R_1 = CH_2CONHbut \quad R_2 = H \quad R_3 = CH_3$
$R_1 = CH_2CON(CH_3)but \quad R_2 = H \quad R_3 = CH_3$
$R_1 = CH_2CON(et)_2 \quad R_2 = H \quad R_3 = CH_3$
$R_1 = CH_2CN \quad R_2 = H \quad R_3 = CH_3$
$R_1 = C_2H_4C_6F_{13} \quad R_2 = H \quad R_3 = CH_3$
$R_1 = C_2H_4OH \quad R_2 = C_7H_{14}CH{=}CHoct \quad R_3 = C_2H_5$

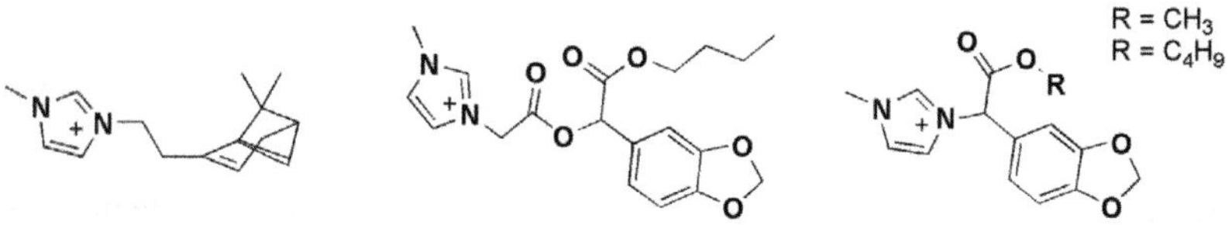

Scheme 7.1 Imidazolium cations for which PP+ values are available in ref. 36.

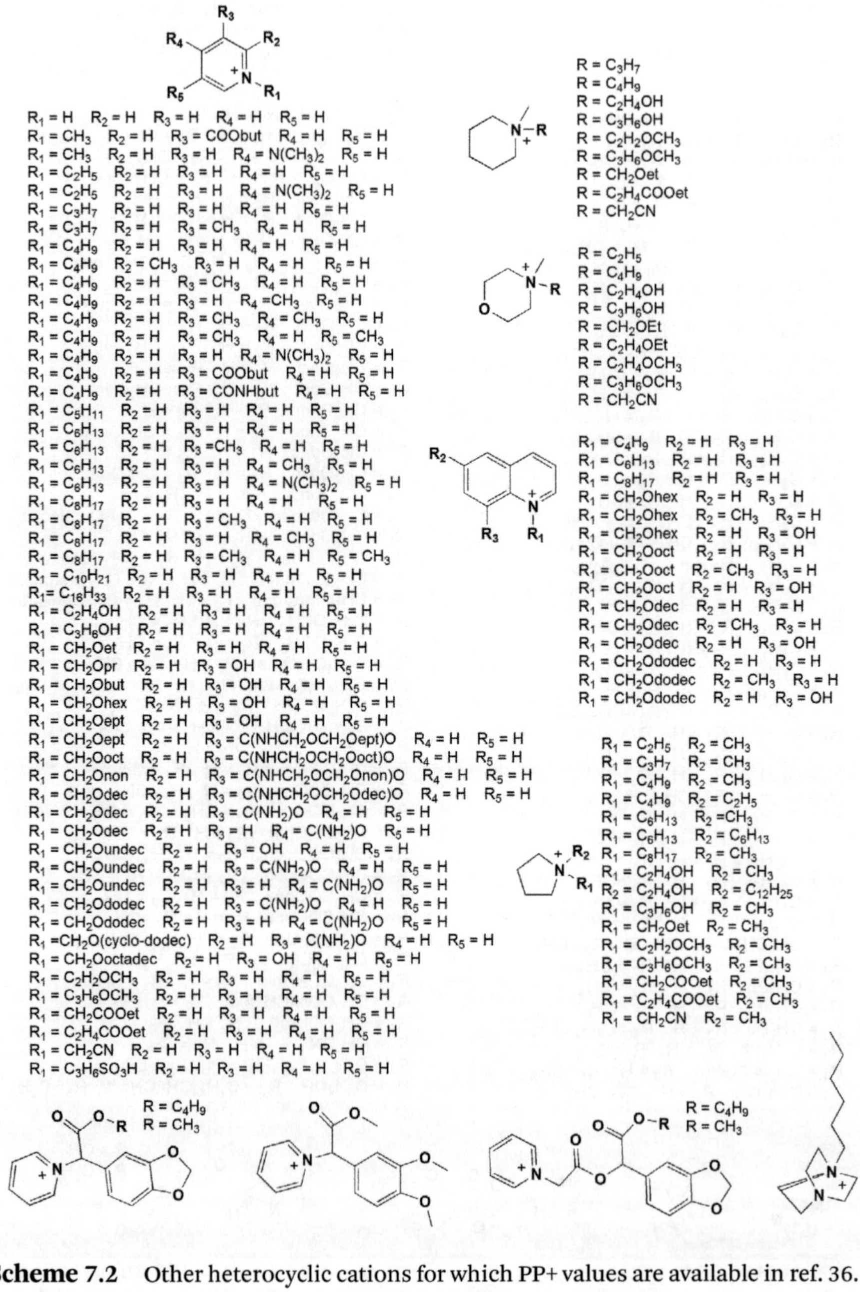

Scheme 7.2 Other heterocyclic cations for which PP+ values are available in ref. 36.

properties such as hydrophobic/hydrophilic balance, solubility, organic solvent/water partition coefficients, H-bonding ability both as donor and acceptor, as well as flexibility and skin permeability. A statistically significant 5 PC model explaining 77.5% of variance was obtained and the 5 PCA scores for 218 aromatic and non-aromatic heterocyclic cationic scaffolds were named PP+.[36]

The influence of cationic structural features on cations PPs can be interpreted by the analysis of the original VolSurf+ descriptors in their derivation.[36]

$PP+_1$ and $PP+_2$ account for the main cationic structural effects, the scaffold and the side chain effect. In particular six-membered non-aromatic rings are characterized by positive $PP+_1$ values while quinoliniums by negative $PP+_1$ values. Moreover, $PP+_1$ values decrease on increasing side chain length. The second cationic PP distinguishes cations according to their substituents features, with high positive values found for side chains bearing heteroatoms such as oxygen and nitrogen in both ester/amidic and ether groups, while high negative $PP+_2$ values are characteristic of alkyl side chains.[36] The above structural features influence ILs physico-chemical properties. Solubility is a property depending on the solute and the solvent features and, in the case of organic molecules, also on pH. VolSurf+ takes into account both these aspects, which are well explained by the first cationic PP+. Highly positive $PP+_1$ values indicate high solubility in water, whereas solubility in octanol is better described by high negative values. Moreover, negative $PP+_1$ is related to molecular size, shape, volume and surface, and expresses a diffused hydrophobic character onto the molecule surface.[36] In addition to solubility, the capability to give H-bonds is a fundamental property for organic molecules. The presence of groups able to behave as H-bonds acceptors and giving hydrophilic character to the whole molecule is expressed by $PP+_2$ positive values, whereas negative $PP+_2$ are associated with descriptors related to the dispersion in water and to the tendency to favour supermolecular aggregates.[36]

The interpretation of the third and fourth PP+ can be better elucidated by the geometrical representation of ILs in Figure 7.6 when superimposed to the corresponding loadings plot (Figure 7.7) reporting the original VolSurf+ descriptors. $PP+_3$ and $PP+_4$ evidence specific effects. In the upper left quadrant of Figure 7.6 we find morpholinium, pyrrolidinium, and piperidinium non-aromatic scaffolds well described by the HL1-HL2 (hydrophilic–lipophilic balance, blue square) and SKIN (skin permeability, yellow triangle) descriptors located in the corresponding quadrant of Figure 7.7.

Mono-*N*-alkyl and *N*-alkyloxymethyl imidazoliums and the unsubstituted pyridinium, bearing a basic unsubstituted ring nitrogen, exhibit negative $PP+_3$ and $PP+_4$ values in Figure 7.6. Accordingly, the percentage of unionized species at different pH (%FU4–%FU10, green diamond) and the blood–brain barrier distribution (LgBB, red star) are located in the same quadrant of Figure 7.7.

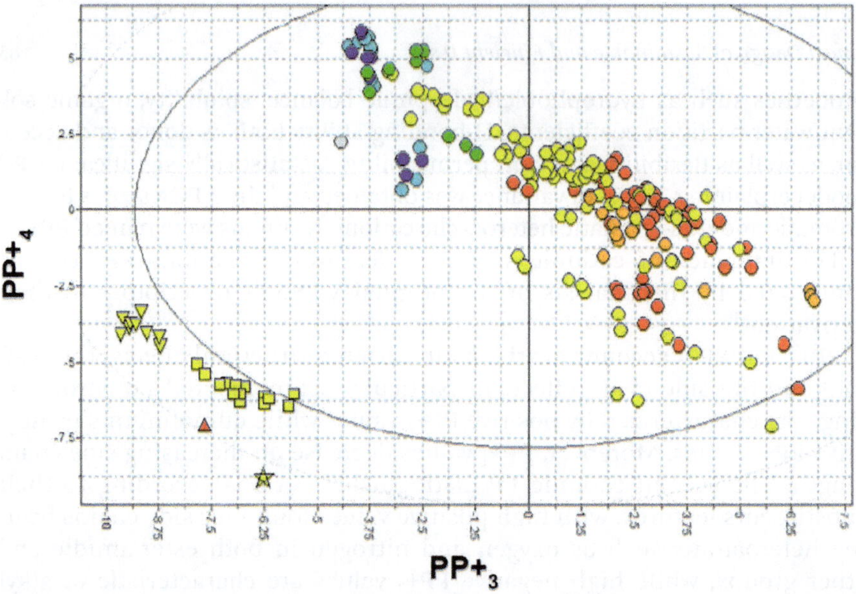

Figure 7.6 PP+$_3$ *vs* PP+$_4$ scores plot reporting the 218 cations. Symbols as follows: ● pyrrolidiniums; ● piperidiniums; ● morpholiniums; ● *N*-substituted pyridiniums; ▲ unsubstituted pyridinium; ● quinoliniums; ● *N*,*N*-substituted imidazoliums; ■ *N*-alkyl substituted imidazoliums; ▼ *N*-alkyloxymethyl substituted imidazoliums; ★ 1-ethyl-2-(8-heptadecenyl)-4,5-dihydro-3-(2-hydroxyethyl) imidazolium; ● diaza bicyclooctanium. (Data from ref. 36.)

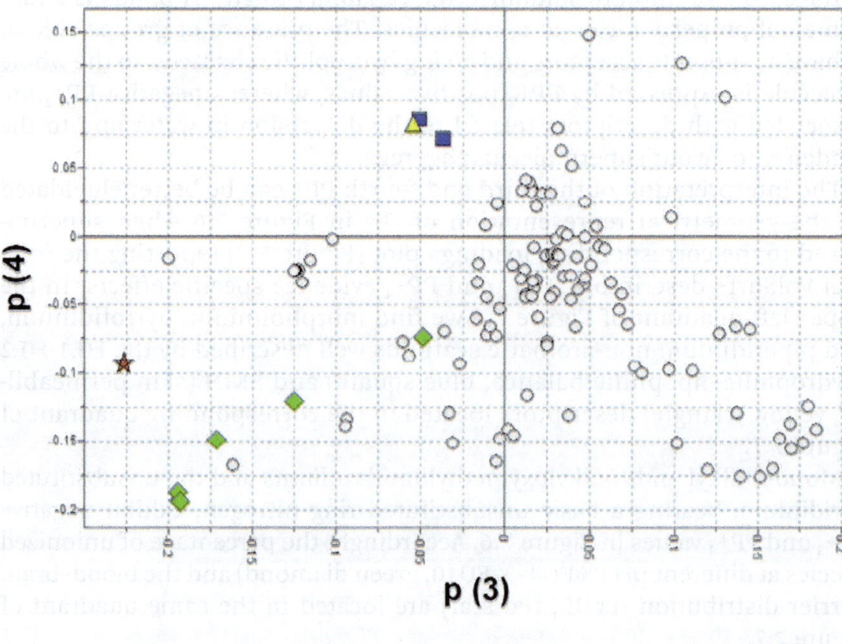

Figure 7.7 p(3) *vs* p(4) loadings plot reporting cations model descriptors ■ HL1, HL2; ◆ %FU4–%FU10; ▲ SKIN; ★ LgBB; ○ *others*. See Table 7.1 for descriptor code explanation. (Data from ref. 36.)

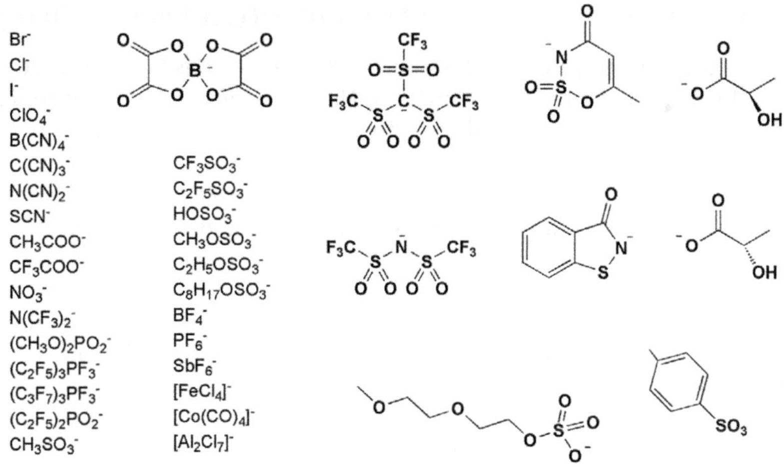

Scheme 7.3 Anions for which PP– values are available in ref. 36.

The increase of hydrophobic interactions capability in imidazolium and pyridinium cations on increasing substituent alkyl chain length, explained by hydrophobic descriptors, has been also discussed.[36]

The PCA of a matrix including 38 common ILs anions (objects) and 48 VolSurf+ non-zero descriptors (for a detailed explanation see ref. 36) gave a four significant PC model explaining 74% of the variance. The resulting PP– numerical values for 38 anions (Scheme 7.3) are reported in ref. 36.

$B(CN)_4^-$ and $C(CN)_3^-$ anions have very negative $PP{-}_1$ and very positive $PP{-}_2$ values, whereas fluorophosphates and anions bearing carboxylic groups exhibit high positive $PP{-}_1$. The highest negative $PP{-}_1$ and $PP{-}_2$ values are exhibited for nitrate and halides, respectively, whereas both the above principal properties are negative for $[Al_2Cl_7]^-$ and $[FeCl_4]^-$.[36]

Negative $PP{-}_3$ values account for the hydrophilic anions character, in particular indicating the tendency to give H-bonds in the interaction with water, while positive values, related to the dimension, size and shape of anionic structures, are found for tris[(trifluoromethyl)sulfonyl]methide, long alkyl chains sulfates and trifluorotris(pentafluoroethyl) and trifluorotris(heptafluoropropyl)phosphates.[36]

Positive $PP{-}_4$ values indicate the presence of both hydrophobic and hydrophilic domains within the same anionic structures, defining the amphiphilic character of anions such as octyl sulfate and tosylate, while high negative $PP{-}_4$ values, related to the extension of the hydrophilic region with respect the whole surface, are exhibited by PF_6^-, SbF_6^- and BF_4.[36]

The availability of five *in silico* structural descriptors (PP+) for 218 heterocyclic cations and four descriptors (PP–) for 38 inorganic or organic anions, covering respectively a five-dimensional cations chemical space and a four-dimensional anions chemical space, provides in principle descriptors for as many as 8284 ILs.

7.3.4 Balancing Simplicity and Complexity: A Real Challenge

The importance of simplicity in disseminating science has been pointed out by several scientists from Isaac Newton ("Nature is pleased with simplicity. And nature is no dummy") to Albert Einstein ("If you can't explain to a six year old, you don't understand").

Nowadays the multiplicity of intercorrelated complex phenomena cannot always be tackled by a simple approach, therefore aiming at simplifying the complex is presently a real challenge for scientists. In this context, the adopted PLS approach appears to comply with Maeda's topical statement "The simplest way to achieve simplicity is through thoughtful reduction".[37] Following these leads the PLS models using VolSurf+ descriptors will be reported below in order of increasing complexity. Recourse to more complex modelling will be dictated by failure of simpler models, according to the modelling strategy depicted in Figure 7.8.

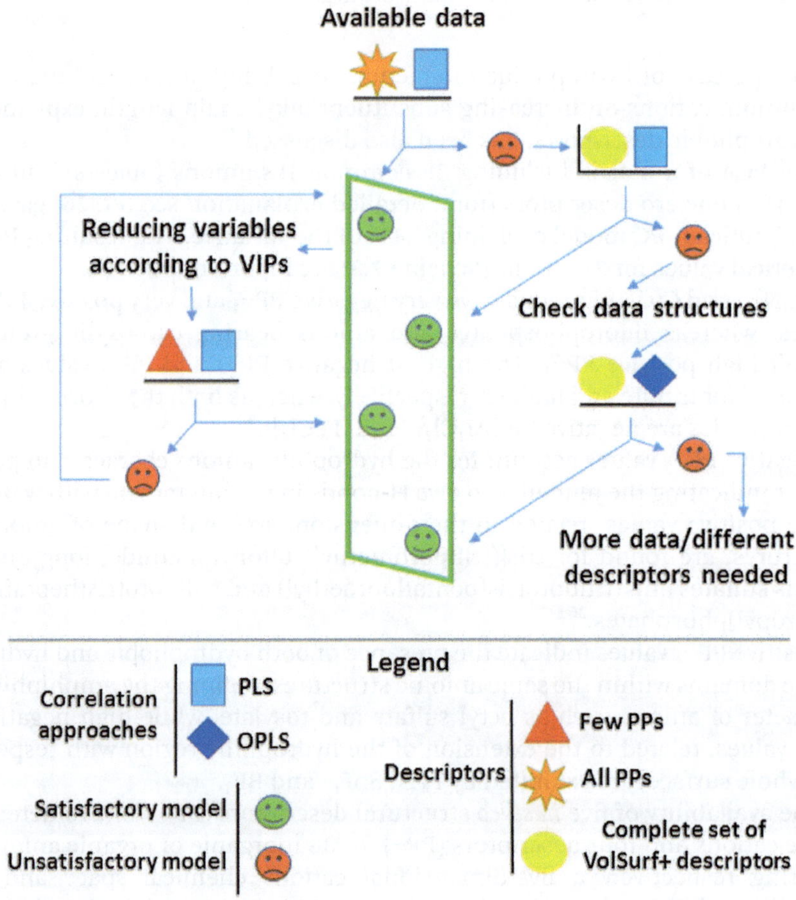

Figure 7.8 ILs data-driven modelling strategy using VolSurf+ descriptors.

7.3.4.1 PLS Modelling with Few VolSurf+ PPs

The adopted modelling strategy aims at achieving, whenever possible, the simplest linear model including only a few independent descriptors. In the treatment of available data a significant reduction in the number of descriptors can be achieved by adopting nine ILs PPs and PLS (orange star and blue square, modelling legend in Figure 7.8). When a satisfactory model is obtained, a further reduction in the number of descriptors can be achieved if elimination of those exhibiting low VIP values does not affect significantly the model performances. A very fortunate example is provided by *Vibrio fischeri* toxicities, for which a three descriptors PLS model (red triangle, blue square in Figure 7.8) exhibited statistical parameters comparable with those of the nine descriptors model and a satisfactory correlation plot (Figure 7.9) spanning over 5 log units.[38]

In this case toxicity values for as many as 520 ILs, including five aromatic and non-aromatic heterocyclic cationic scaffolds, different side chains and different anions ranging from halogens to oxygen, sulfur and nitrogen containing ones, could be easily predicted by the following three parameter equation:

$$\mathrm{Log}[(\mathrm{EC}_{50})_{\mathrm{pred.}}] = 0.264(\mathrm{PP}+_1) - 0.091(\mathrm{PP}+_2) - 0.240(\mathrm{PP}-_3) + 2.219 \quad (7.7)$$

As previously pointed out, the validity of the predictions is limited within the experimental space determined by the cations and the anions present in the ILs learning set and deviations from the correlation could be rationalized on the basis of structural considerations.[38]

Similar simple models with four descriptors were obtained for viscosity and density,[39] physico-chemical properties of paramount importance for

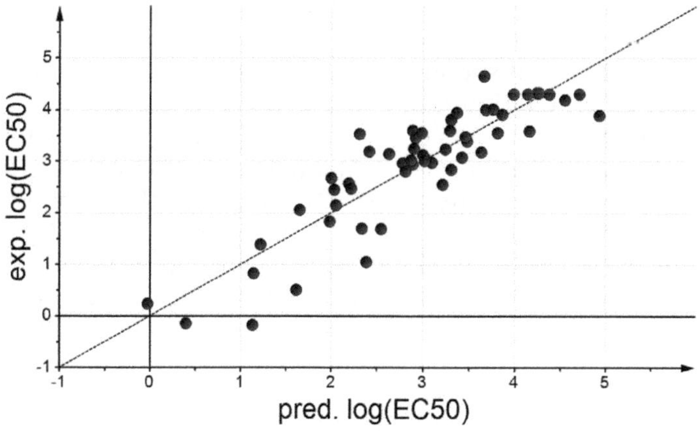

Figure 7.9 *Vibrio fischeri* correlation plot where the experimental values are compared to the corresponding predicted ones. Reproduced from ref. 38 with permission from the Royal Society of Chemistry.

many ILs applications. A low viscosity is generally requested to use an IL as a solvent, to minimize pumping costs and increase mass transfer rates, while higher viscosities may be required by other applications such as lubrication or use in membranes. Density is a crucial property required in the industrial process design of equipment such as condensers, reboilers, liquid/liquid two-phase mixer–settler units, in liquid metering calculations and in material and energy balances involving liquids, vapour–liquid and liquid–liquid separation processes.

A great advantage of data-driven approaches dealing with few descriptors is to provide experimentalists the possibility to evaluate the descriptors relevance and make for themselves an appropriate selection for design purposes.

It is recognized that anions have a high effect on ILs viscosity, and PLS modelling in principal properties, indicating $PP-_2$ as the most important descriptor, provides a quantitative estimate of this effect – the higher the $PP-_2$ value, the lower the viscosity.[39] Accordingly, a high viscosity is observed for ILs with chloride anion (*e.g.* 2.53 cP at 50 °C for 1-ethyl-3-methylimidazolium chloride having $PP-_2 = -5.73$),[36,39] while higher $PP-_2$ values correspond to lower viscosity ILs (*e.g.* 1-ethyl-3-methylimidazolium tetrafluoroborate and nitrate, 1.20 and 1.42 cP at 50 °C, respectively, having $PP-_2 = -0.06$ and -0.46).[36,39] Tetracyanoborate and tricyanomethanide exhibit very high $PP-_2$ values (7.57 and 6.89, and viscosity values of 19.8 cP at 20 °C and 18.0 cP at 22 °C respectively),[36,40,41] both outside the anions experimental space explored by the PLS model, and are expected to have a very low viscosity, even lower than that of all ILs in the learning set. Notably, both these hydrophobic anions have been reported to be used to generate hydrophobic ILs[42] and a low viscosity seems to be associated with their use in dye-sensitized solar cells.[40,41] Interestingly the ability of the above anions to determine very low ILs viscosity, highly relevant for important task-specific applications, could be easily unravelled on the basis of this soft model.[39]

In the case of viscosity, a difference with theory-driven approaches is that models, and consequently predictions, are made at a given temperature. However, different soft models allow predictions at each temperature and provide interpretation hints on the effect of temperature variation. In particular, the coefficients of the four parameter equations for the soft models pointed out that the effect of anions is related to their hydrophilic/hydrophobic character, to the size/shape ratio and to the capability to form H-bonds as donor or acceptor.[39] The effect of cations solubility, size, flexibility, molecular weight could also be evidenced, directional polar interactions, such as H-bond interactions, being less efficient on increasing temperature.[39]

7.3.4.2 PLS Modelling with Nine VolSurf+ PPs

Reduction of the number of PPs according to VIP values does not always provide satisfactory models. In these cases, according to the strategy depicted in Figure 7.8, available data can be modelled using all nine VolSurf+ PPs (orange star, blue square in Figure 7.8). Examples of such a situation is the modelling of above 200 experimental cyto- and enzyme-toxicities, expressed as IPC-81 rat cell line growth and acetylcholinesterase inhibition respectively, providing predictions for a set of 520 ILs.[36] This result indicates priorities in planning expensive ILs experimental toxicity tests to comply with the hazard assessment required by the REACH regulation.[10]

For ILs industrial uses a low toxicity has to be accompanied by a high efficiency in task-specific applications requiring the modelling of ILs physico-chemical properties such as heat capacity (C_p) required for heat exchanger and solar energy storage devices. In this context, PLS modelling using all nine VolSurf+ PPs provided an excellent correlation plot (Figure 7.10) for the prediction of C_p values of ILs.[43]

In silico predictions, validated by experimental C_p measurements for an external set of ILs including literature[44] values, provided a set of structurally different ILs with similar C_p values. Among them, an experimentalist has the opportunity to select a task-specific optimal IL according not only to its efficiency, but also to its environmental and economic sustainability.

7.3.4.3 PLS Modelling with the Full Set of VolSurf+ Descriptors

In cases when PLS models with PPs (orange star, blue square in Figure 7.8) do not provide satisfactory correlations or a more detailed interpretation of the structural descriptors is needed, PLS modelling with the full set of VolSurf+ descriptors (yellow circle, blue square in Figure 7.8) can be adopted. This approach turned out to be suitable to establish a QSPR correlation with aquatic toxicity scores spanning eleven log units for over one hundred ILs.[13]

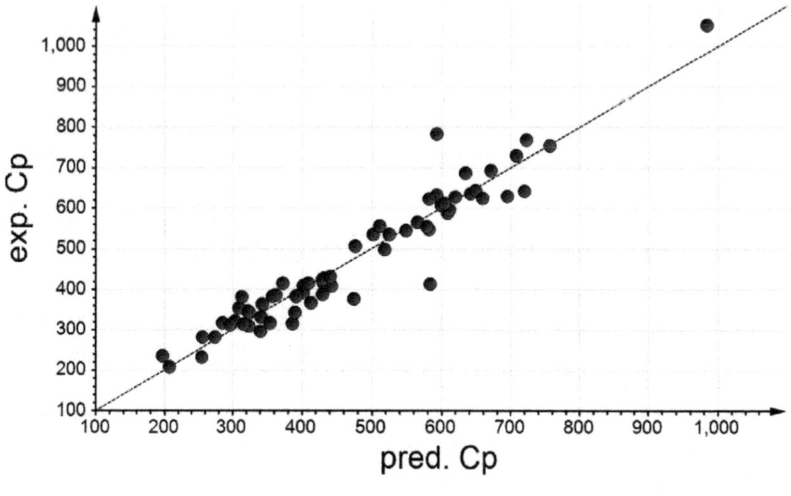

Figure 7.10 Heat capacity (C_p) correlation plot where the experimental values are compared to the corresponding predicted ones. Reproduced from ref. 43 with permission from the Royal Society of Chemistry.

In addition to predictions, the PLS model provided an insight into the ILs structural features (and descriptors) related to ILs aquatic toxicities allowing a useful insight into these relationships.

Non-aromatic cationic scaffolds such as piperidinium, pyrrolidinium and morpholinium exhibited lower toxicities than pyridiniums and imidazoliums, for which also side chain effects were evidenced.[13] Cations toxicity is mainly related to size-describing descriptors as well as to molecular hydrophobic volume and surface, Log*P* and Log*D*, molecular flexibility and protein binding, with all variables increasing aquatic toxicity. On the other hand, molecular diffusivity, water thermodynamic solubility, and molecular bio-metabolic stability decrease toxicity.

As expected, the anions played a minor role, although the approach evidenced a non-secondary role of the anions in determining aquatic toxicity, an effect previously reported to be negligible.[45,46] Big anions with wide hydrophilic and hydrophobic areas distant from the molecular barycentre influence positively the aquatic toxicity scores, increasing ILs environmental sustainability.

VolSurf+ descriptors, accounting for both cationic and anionic structural features of ILs, were used to develop a PLS model able to establish a QSPR correlation with solvatochromic dye Nile Red polarity (E_{NR}) providing an *in silico* ILs polarity database for over 100 ILs.[47] The model predictions were validated by experimental E_{NR} values for 18 ILs with different cationic and anionic structural features including piperidinium, pyrrolidinium and imidazolium cationic scaffolds with different alkyl side chain length and eight anions ranging from tetrafluoroborate to oxygenated anions.

7.3.4.4 OPLS Modelling with the Full Set of VolSurf+ Descriptors

In some cases the application of the PLS approach is not able to provide satisfactory models, due to a particular data structure where the orthogonal information contained in the **X**-block is strong. In such cases a different chemometric tool called orthogonal PLS (OPLS)[48–50] is more suitable to handle the data matrices by discriminating between the predictive and orthogonal **X**-variation.

Two examples of OPLS modelling are reported below, regarding the prediction of ILs decomposition temperature and conductivity.

The decomposition temperature of ILs is an essential property to evaluate their thermal stability and therefore the industrial applicable temperature range. PLS modelling of an **X**-matrix including 35 objects (ILs) and 224 variables (128 cation descriptors, 48 anion descriptors and 48 cross terms) and the decomposition temperature as the response provided a non-significant first PLS component, a second significant PLS one and a third non-significant PLS component.[39] Such a data structure could be better modelled by adopting the OPLS approach resulting in a better Q^2 value (0.795). The OPLS model was supported by external validation and provided, in addition to the predictive OPLS component, seven statistically significant orthogonal PLS components and a very satisfactory correlation plot (R^2 = 0.96) spanning a temperature range of 300 K.[39]

ILs conductivity is another important property for ILs application as electrolytes in electrochemical devices. For example, in electrolysis solutions for batteries larger ionic conductivities are required. Similarly to the previous case the adoption of an externally validated OPLS approach achieved a higher prediction ability than PLS (Q^2 = 0.833) for 39 ILs and a very satisfactory correlation plot (R^2 = 0.98) spanning 2.5 log conductivity units (S m^{-1}).[39]

7.4 Further Extension of the Approach for Smart ILs and Materials Design

The multivariate approach has been applied to ILs by our group in the past few years. The first multivariate insight on ILs focused on toxicity[51] aiming at compacting toxicity tests into aquatic and bacterial/fungi toxicities. We then developed new *in silico* descriptors[13,36] which turned out to be useful in developing QSPR multivariate models (taking into account the effect of both cations and anions simultaneously) by means of PLS. Using the same descriptors and the same correlation tool we then moved to QSPR soft models for the prediction of ILs physico-chemical properties such as polarity[47] and heat capacity.[43] We then extended the models to other relevant ILs physico-chemical properties such as viscosity, density, decomposition temperature and conductivity.[39] The importance of the above properties for ILs smart design in relation to specific applications has been pointed out in previous sections.

However, the PCA/PLS approach is very general as demonstrated by other applications ranging from drug identification[52–55] to food chemistry[56] and cultural heritage,[57] from NMR shifts[58] to the concept of aromaticity,[59] from genome based cancer research[60–66] to leukaemia.[66,67] All these applications used different multivariate approaches in relation to the specific problem under examination. In this context, it is conceivable that a multivariate approach could be successfully adopted for smart material design, provided that suitable process descriptors are identified.

7.5 Conclusions and Outlook

The results discussed in this chapter point out that smart design of ILs can be achieved *in silico* by means of appropriate approaches using user-friendly computer programs identifying the most promising problem solutions among millions of possibilities. In our opinion, human supervision is essential for the success of the design, which provides experimenters with a dramatic reduction in the number of experiments leading to successful targets. Although there is no guarantee that all designed compounds will work, a rational design allows finding of needles in haystacks but their performance has to undergo experimental validation. A similar situation is encountered in the field of synthesis, where chemists are now able to plan multistep syntheses with the aid of *in silico* programs. A fundamental contribution in the field was provided by Grzybowski[68,69] who developed computer programs aiming at identifying retrosynthetic routes based on factors such as cost, reagent availability, number of steps, and was able to find in a few hours a pathway for the synthesis of epicolactone which took chemists three years. His approach was recently commented on as a "Revolutionary computer program could change chemistry forever"[70] concluding that "If someone's designing a new drug or material, you can work out what molecule you really want ... and then a chemist can make it in a few weeks". Although this statement might be overoptimistic, there is no doubt that in the future chemistry cannot do without a smart design of properties and syntheses providing shortcuts to problem solutions.

References

1. J. P. Hellett and T. Welton, *Chem. Rev.*, 2011, **111**, 3508.
2. K. R. Seddon, A. Stark and M.-J. Torres, *Pure Appl. Chem.*, 2000, **72**, 2275.
3. J. M. Juran, *Managerial Breakthrough*, McGraw-Hill Book Company, London, 1st edn, 1964.
4. G. E. P. Box, W. G. Hunter and J. S. Hunter, *Statistics for Experimenters*, John Wiley & Sons, New York, 1978.
5. N. R. Draper and H. Smith, *Applied Regression Analysis*, John Wiley & Sons, Inc., New York, 3rd edn, 1998.
6. L. Eriksson, E. Johansson, N. Kettaneh-Wold, J. Trygg, C. Wikstrom and S. Wold, *Multi- and Megavariate Data Analysis*, Umetrics AB, Umea, 2nd edn, 2006.
7. S. Wold, K. Esbensen and P. Geladi, *Chemom. Intell. Lab. Syst.*, 1987, **2**, 37.
8. S. Wold, M. Sjöström and L. Eriksson, *The Encyclopedia of Computational Chemistry*, ed. P. V. R. Schleyer, John Wiley & Sons Ltd, Chichester, UK, 1998, pp. 2006–2020.
9. A. Leo, C. Hansch and D. Elkins, *Chem. Rev.*, 1971, **71**, 525.
10. REACH regulations, available at: http://echa.europa.eu/regulations/reach/.
11. J. C. Dearden, *IJQSPR*, 2016, vol. 1, p. 1.
12. G. Cruciani, P. Benedetti, G. Caltabiano, D. F. Condorelli, C. G. Fortuna and G. Musumarra, *Eur. J. Med. Chem.*, 2004, **39**, 281.
13. A. Paternò, G. Bocci, L. Goracci, G. Musumarra and S. Scirè, *SAR QSAR Environ. Res.*, 2016, **27**, 1.
14. M. Arshadi and P. Kebarle, *J. Phys. Chem.*, 1970, **74**, 1483.
15. *SYBYL-X 1.3*, Tripos, available at: http://www.tripos.com.

16. R. A. Fisher and W. A. MacKenzie, *J. Agric. Sci.*, 1923, **13**, 311.
17. R. Carlson and J. E. Carlson, *Design and Optimisation in Organic Synthesis, Volume 24*, Elsevier, Amsterdam, Second revised and enlarged edn, 2005, vol. 16, pp. 351–401.
18. F. P. Ballistreri, C. G. Fortuna, G. Musumarra, D. Pavone and S. Scirè, *Arkivoc*, 2002, (pt 11), 54.
19. R. Carlson, M. P. Prochazka and T. Lundstedt, *Acta Chem. Scand., Ser. B*, 1988, **42**, 145.
20. R. Carlson, M. P. Prochazka and T. Lundstedt, *Acta Chem. Scand., Ser. B*, 1988, **42**, 157.
21. R. Carlson, T. Lundstedt, A. Nordahl and M. P. Prochazka, *Acta Chem. Scand., Ser. B*, 1986, **40**, 522.
22. C. G. Fortuna, G. Musumarra, M. Nardi, A. Procopio, G. Sindona and S. Scire, *J. Chemom.*, 2006, **20**, 418.
23. M. Skagerberg, D. Bonelli, S. Clementi, G. Cruciani and C. Ebert, *Quant. Struct.-Act. Relat.*, 1989, **8**, 32.
24. S. Hellberg, M. Sjöström and S. Wold, *Acta Chem. Scand., Ser. B*, 1986, **40**, 135.
25. S. Hellberg, M. Sjöström and S. Wold, *J. Med. Chem.*, 1987, **30**, 1126.
26. M. Skagerberg, M. Sjöström and S. Wold, *J. Chemom.*, 1990, **4**, 241.
27. G. Cruciani, M. Baroni, E. Carosati, M. Clementi, R. Valigi and S. Clementi, *J. Chemom.*, 2004, **18**, 146.
28. L. Caruso, G. Musumarra and A. R. Katritzky, *Quant. Struct.-Act. Relat.*, 1993, **12**, 146.
29. S. Clementi, G. Cruciani, P. Fifi, D. Riganelli, R. Valigi and G. Musumarra, *Quant. Struct.-Act. Relat.*, 1996, **15**, 108.
30. E. Fischer and V. F. Hess, *Ber. Dtsch. Chem. Ges.*, 1884, **17**, 559.
31. B. Robinson, *The Fischer Indole Synthesis*, John Wiley & Sons Ltd, Chichester, 1982, p. 489.
32. M. P. Prochazka and R. Carlson, *Acta Chem. Scand.*, 1989, **43**, 651.
33. M. P. Prochazka and R. Carlson, *Acta Chem. Scand.*, 1990, **44**, 614.
34. M. Sandberg, L. Eriksson, J. Jonsson, M. Sjostrom and S. Wold, *J. Med. Chem.*, 1998, **41**, 2481.
35. A. Duran, I. Zamora and M. Pastor, *J. Chem. Inf. Model.*, 2009, **49**, 2129.
36. A. Paternò, G. Bocci, G. Cruciani, L. Goracci, S. Sciré and G. Musumarra, *SAR QSAR Environ. Res.*, 2016, **27**, 221.
37. J. Maeda, *The Laws of Simplicity*, MIT Press, 2006, available at: https://mitpress.mit.edu/books/laws-simplicity.
38. A. Paternò, S. Scire and G. Musumarra, *Toxicol. Res.*, 2016, **5**, 1090.
39. A. Paternò, L. Goracci, S. Scire and G. Musumarra, *ChemistryOpen*, 2017, **6**, 90.
40. D. Kuang, P. Wang, S. Ito, S. M. Zakeeruddin and M. Gratzel, *J. Am. Chem. Soc.*, 2006, **128**, 7732.
41. Y. Yoshida, K. Muroi, A. Otsuka, G. Saito, M. Takahashi and T. Yoko, *Inorg. Chem.*, 2004, **43**, 1458.

42. M. Marszalek, Z. Fei, D.-R. Zhu, R. Scopelliti, P. J. Dyson, S. M. Zakeeruddin and M. Gratzel, *Inorg. Chem.*, 2011, **50**, 11561.
43. A. Paternò, R. Fiorenza, S. Marullo, G. Musumarra and S. Sciré, *RSC Adv.*, 2016, **6**, 36085.
44. Y. U. Paulechka, *J. Phys. Chem. Ref. Data*, 2010, **39**, 033108.
45. S. Stolte, J. Arning, U. Bottin-Weber, A. Muller, W.-R. Pitner, U. Welz-Biermann, B. Jastorff and J. Ranke, *Green Chem.*, 2007, **9**, 760.
46. S. Bruzzone, C. Chiappe, S. E. Focardi, C. Pretti and M. Renzi, *Chem. Eng. J.*, 2011, **175**, 17.
47. A. Paternò, F. D'Anna, C. G. Fortuna and G. Musumarra, *Tetrahedron*, 2016, **72**, 3282.
48. J. Trygg and S. Wold, *J. Chemom.*, 2002, **16**, 119.
49. J. Trygg, *J. Chemom.*, 2002, **16**, 283.
50. J. Trygg and S. Wold, *J. Chemom.*, 2003, **17**, 53–64.
51. A. Paternò, F. D'Anna, G. Musumarra, R. Noto and S. Sciré, *RSC Adv.*, 2014, **4**, 23985.
52. G. Musumarra, G. Scarlata, G. Cirma, G. Romano, S. Palazzo, S. Clementi and G. Giulietti, *J. Chromatogr.*, 1984, **295**, 31.
53. G. Musumarra, G. Scarlata, G. Romano, S. Clementi and S. Wold, *J. Chromatogr. Sci.*, 1984, **22**, 538.
54. G. Musumarra, G. Scarlata, G. Cirma, G. Romano, S. Palazzo, S. Clementi and G. Giulietti, *J. Chromatogr.*, 1985, **350**, 151.
55. G. Musumarra, G. Scarlata, G. Romano, G. Cappello, S. Clementi and G. Giulietti, *J. Anal. Toxicol.*, 1987, **11**, 154.
56. G. Alberghina, L. Caruso, S. Fisichella and G. Musumarra, *J. Sci. Food Agric.*, 1991, **56**, 445.
57. G. Musumarra and M. Fichera, *Chemom. Intell. Lab. Syst.*, 1998, **1998**(44), 367.
58. G. Musumarra, S. Wold and S. Gronowitz, *Org. Magn. Reson.*, 1981, **17**, 118.
59. A. R. Katritzky, P. Barckzynski, G. Musumarra, D. Pisano and M. Szafran, *J. Am. Chem. Soc.*, 1989, **111**, 7.
60. G. Musumarra, D. F. Condorelli, A. S. Costa and M. Fichera, *J. Comput.-Aided Mol. Des.*, 2001, **15**, 219.
61. G. Musumarra, D. F. Condorelli, S. Scirè and A. S. Costa, *Biochem. Pharmacol.*, 2001, **62**, 547.
62. G. Musumarra, V. Barresi, D. F. Condorelli, C. G. Fortuna and S. Scirè, *Comput. Biol. Chem.*, 2005, **29**, 183.
63. V. Barresi, C. G. Fortuna, R. Garozzo, G. Musumarra, S. Scirè and D. F. Condorelli, *Mol. BioSyst.*, 2006, **2**, 231.
64. J. Hofmann, J. Easmon, G. Puerstinger, G. Heinisch, M. Jenny, A. A. Shtil, M. Hermann, D. F. Condorelli, S. Scirè and G. Musumarra, *Invest. New Drugs*, 2009, **27**, 189.
65. C. G. Fortuna, V. Barresi, C. Bonaccorso, G. Consiglio, S. Failla, A. Trovato-Salinaro and G. Musumarra, *Eur. J. Med. Chem.*, 2012, **47**, 221.

66. C. G. Molteni, G. Cazzaniga, D. F. Condorelli, C. G. Fortuna, A. Biondi and G. Musumarra, *QSAR Comb. Sci.*, 2009, **28**, 822.
67. G. Musumarra, D. F. Condorelli and C. G. Fortuna, *Comb. Chem. High Throughput Screening*, 2011, **14**, 36.
68. M. Kowalik, C. M. Gothard, A. M. Drews, N. A. Gothard, A. Weckiewicz, P. E. Fuller, B. A. Grzybowski and K. J. M. Bishop, *Angew. Chem., Int. Ed.*, 2012, **51**, 7928.
69. S. Szymkuć, E. P. Gajewska, T. Klucznik, K. Molga, P. Dittwald, M. Startek, M. Bajczyk and B. A. Grzybowski, *Angew. Chem., Int. Ed.*, 2016, **55**, 5904.
70. M. Gunther, *Chem. World*, 2016, **13**(9), 10.

CHAPTER 8

Applications of Ionic Liquids in Organic Electronic Devices

SATORU OHISA* AND JUNJI KIDO*

Department of Organic Materials Science, Graduate School of Organic Materials Science, Yamagata University, Yonezawa, Yamagata, 992-8510, Japan
*E-mail: s.ohisa@yz.yamagata-u.ac.jp, kid@yz.yamagata-u.ac.jp

8.1 Introduction

Ionic liquids (ILs) are salts that exist in the liquid state at temperatures of below 100 °C and have relatively weak coulombic interactions between cations and anions, giving them distinctive physical properties.[1–6] The melting points of ILs are extremely low compared to conventional inorganic ionic compounds such as NaCl (m.p. = 801 °C). Such high melting points of inorganic ionic compounds originate from large energetic stabilization due to coulombic interaction among the ions. In organic salts, the coulombic interaction is also dominant among all intermolecular interactions; therefore, lowering the energetic stabilization due to the coulombic interaction is indispensable for lowering the melting points of ionic compounds to near room temperature. Effective means of weakening the attractive interaction include adoption of asymmetric molecular skeletons inhibiting the molecular packing accompanying the large energetic stabilization, and delocalization of ionic charge to wider molecular orbital space. Representative cations are alkyl substituted imidazolium, pyridinium, pyrrolidinium,

Smart Materials No. 28
Ionic Liquid Devices
Edited by Ali Eftekhari

Published by the Royal Society of Chemistry, www.rsc.org

piperidinium, ammonium, and phosphonium. For example, 1-butyl-3-methylimidazolium (BMIM) is one of the most utilized cations in ILs. The asymmetric cation skeleton with the two different length alkyl chains inhibits molecular packing and the positive ionic charge is delocalized over the imidazolium ring, making BMIM a suitable cation for ILs. Representative anions are halogen anions (*e.g.* Cl^-, Br^-, and I^-), fluorinated inorganic anions (*e.g.* BF_4^-, and PF_6^-), non-fluorinated inorganic anions (*e.g.* NO_3^-, and NO_2^-), fluorinated organic anions (*e.g.* $CF_3SO_3^-$, $(CF_3SO_2)_2N^-$ (TFSI), $(CF_3SO_2)_3C^-$, and $(C_2F_5SO_2)_2N^-$), and non-fluorinated organic anions (*e.g.* $(CN)_2N^-$, CH_3COO^-). For example, in TFSI, the CF_3SO_2 electron-withdrawing substituent attached to N^- delocalizes the negative ionic charge to the CF_3SO_2 substituent, resulting in weak coulombic interaction with cations. From these effects, [BMIM] [TFSI] has the low melting point of −2 °C and exists in the liquid state at room temperature. Unlike conventional molecular liquids, these ILs composed of cations and anions show marked features: high fire retardancy; very low volatility; high ionic conductivity; wide electrochemical window (depending on IL); and high adjustability of these physical properties and functionality by modification of ions. Furthermore, polymerized ILs (PILs) possessing backbones with pendant substituents including IL units have been focused on recently.[7–10] The main merit of the polymerization is solidification suppressing fluidity. In the case that fluidity of ILs causes some problems, solidification of the ILs is desired. Solidification also gives mechanical stability to ILs. This solidification seems to be opposite to the tide of lowering the melting points of ionic compounds. Among several properties, ionic conductivity is one of those most influenced by solidification. In fact, generally, PILs show smaller ionic conductivities than the corresponding monomer ILs. However, an IL attached to the sidechains gives a certain degree of the IL features to the corresponding PIL, giving a PIL with among the highest ionic conductivities. PILs as solid electrolytes with high ionic conductivity have been studied in recent years. These features are advantages in utilizing the ILs in electronic device applications.[4,8,11] High fire retardancy lowers ignition capability during device driving with heat generation, making the ILs suitable components requiring high reliability. In a lithium ion battery, mostly, carbonate solvents have been used as the electrolyte solutions. However, carbonate solvents have a high ignition capability with the risk of fire accident. If ILs are adopted as the electrolyte solutions, the problem can be avoided. Very low volatility can also lower the risk of accidents in the lithium ion battery. In overheated cells, volatile organic solvents vaporize and increase the inner cell pressure, causing cell rupture. Outside of a partial exception, ILs cannot vaporize even in vacuum. Torimoto and Kuwabata *et al.* observed ILs by scanning electron microscope (SEM) requiring high vacuum condition.[5] Insulators such as ILs normally require metal deposition on them to prevent static charge, generating artifacts on images. Surprisingly, it was possible to observe the ILs clearly without metal deposition. This surprising result suggests a new application of

ILs as an alternative material to metal in preventing static charge in SEM observations. The same authors also found that metal nanoparticles are synthesized in ILs by metal deposition on ILs in vacuum, suggesting that nanoparticles composed of metals that are reactive with ambient air can be synthesized by this method.[5] High ionic conductivity is an important property in electronic devices such as lithium ion batteries and fuel cells requiring efficient ion transportation. The electrochemical window is the electric potential range over which the material is not oxidized or reduced. A wide electrochemical window is indispensable for devices requiring driving under high electric potential. Furthermore, there are infinite types of ions constituting ILs. A slight structural change of ions can impart a large property change to the ILs. ILs with desired properties can be obtained by designing the ions. Therefore, ILs are called designer solvents. If one IL does not solve a technical problem, a slight modification of the ions in the IL may possibly do so.

In recent years, applications of ILs in organic electronic devices have been focused on. Organic electronic devices are devices showing some kind of response to electric stimulations, such as current conduction, light emission, color change, and mechanical transformation, or devices showing some kind of electric response to stimulations such as light irradiation or mechanical transformation. Organic compounds are soft materials, unlike inorganic compounds, enabling fabrication of devices with flexibility, and stretchability. Recent intense research and developments have promoted practical applications of the devices. A representative example is organic light emitting device (OLED) displays for TVs and mobile phones. In other examples, organic field effect transistor (OFET) switching the amount of conduction current, solar cells generating electromotive force, electrochromic devices showing color changes, and soft actuators showing mechanical transformation have been studied. Here, ILs are organic compounds, and have high compatibility with organic electronic materials. Applications of ILs in organic electronic devices have been expected; in fact, in recent years, several applications of ILs have been reported in organic electronic devices, OLEDs,[12–26] light-emitting electrochemical cells (LECs),[27–34] solar cells,[23,35–57] OFETs,[58–78] electrochromic devices,[9,10,79–87] and soft actuators.[87–96] ILs with outstanding features are useful in the realization of novel functions. In this chapter, we review these applications of ILs in organic electronic devices.

8.2 Applications of ILs in Organic Electronic Devices

8.2.1 Light-emitting Devices

Light-emitting devices emit light when an electric field is applied to them. Two types of light-emitting devices are well known. One is OLEDs devised by Tang *et al.*[97] Another is LECs devised by Pei and Heeger *et al.*[98] In this section, we review applications of ILs in OLEDs and LECs.

8.2.1.1 OLEDs

OLEDs consist of an anode, a cathode, and organic functional layers sandwiched by the two electrodes (Figure 8.1a). When an electric field is applied, holes are injected from the anode to the organic layer, whereas electrons are injected from the cathode to the organic layer. The injected carriers, holes and electrons, move toward the corresponding opposite electrode, and meet to form excitons that relax *via* a radiation process in the light emitting layer (EML). To achieve highly efficient OLEDs, high luminous efficiency and low driving voltage are required. The external quantum efficiency (EQE) is one of the most important luminous efficiency characteristics. It is obtained from the ratio of the number of extracted photons into air to the number of injected holes and electrons. It is expressed as follows:

$$\mathrm{EQE}(\%) = \eta_o \times \gamma \times \zeta_s \times q$$

where, η_o, γ, ζ_s, and q are light out-coupling efficiency from the inside of the device to air, charge carrier balance, light-emitting exciton generation probability, and radiation quantum yield, respectively; η_o is generally 20–30% without any out-coupling technology; ζ_s depends on light-emitter type, and is 25% for a conventional fluorescent emitter[99–101] or 100% for phosphorescent[102–105] or thermally delayed fluorescent (TADF) emitter.[106–110] Both γ and q depend on the materials utilized and device design, and can be up to 100%. Another important characteristic is the driving voltage. Low driving voltage is achieved by efficient carrier injection and carrier transportation. To obtain highly efficient OLEDs, the adoption of a multilayer structure is highly useful.[104,111–113] In multilayered OLEDs, carrier injection, transportation and light-emitting functions are separated into multiple organic functional layers such as hole injection layer (HIL), hole-transporting layer (HTL), EML, electron-transporting layer (ETL), and electron injection layer (EIL). Figure 8.1(b) shows the

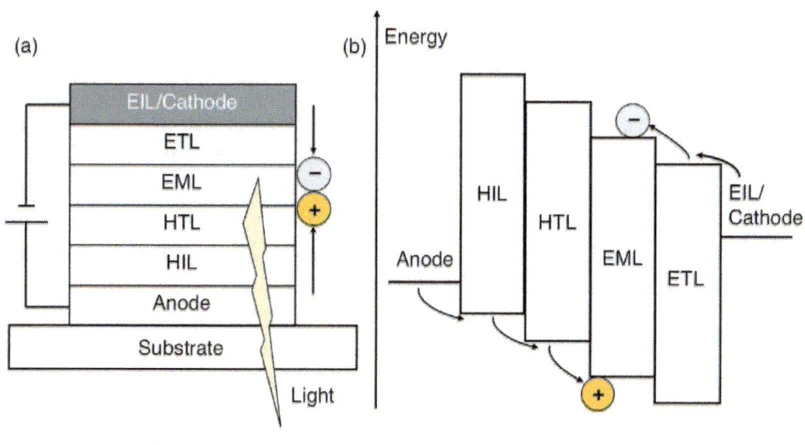

Figure 8.1 (a) OLED structure. (b) Energy diagram of a multilayer OLED.

energy diagrams of multilayer OLEDs. Insertion of HIL, HTL, EML, and ETL improves the carrier delivery between the layers. Generally, the multilayer structure is fabricated by a vacuum evaporation process. Recent materials and device designs have pushed up the OLED performance to a practicable level for displays and lighting. In a research step, white OLEDs attained high power efficiency of 139 lm W^{-1},[112] which is above that of a fluorescent tube. Presently, one of the largest problems that has to be overcome is lowering fabrication cost. As one means of solving this problem, solution-processing enabling large area coating has been focused on.[99,102,104,105,114,115] Unfortunately, for a long time, the solution-processed OLEDs have largely lagged behind the evaporation-processed OLEDs in terms of performance due to several reasons such as difficulty of multilayer stack formation. However, recent intense research and developments are improving the solution-processed OLED performances.

ILs are mainly employed as the electron injection layer in solution-processed OLEDs.[12–17,19–26,116] Generally, large carrier injection barrier exists between the electrode and the organic functional layer. Insertion of the carrier injection layer reduces the carrier injection barrier and facilitates carrier injection into the organic functional layer. One type of carrier injection layers leads to dipole polarization at the interface between the electrode and the organic functional layer, causing the work function (WF) shift and thereby reducing the injection barrier. Ishii *et al.* have explained these effects.[117] Larger dipole polarization formation further reduces the injection barrier. ILs have the potential for large dipole polarization formation because of their ionic nature. Materials including ILs are classified into three groups, (i) small ionic molecules,[12,16,20–24] (ii) conjugated polyelectrolytes (CPEs),[13–15,17,25,26] which include π-conjugated moieties such as fluorenes in the main chain and ionic moieties such as ammonium halides in the side chain, and (iii) non-conjugated polyelectrolytes (NCPEs).[19]

Small ionic molecules were mainly employed as surface modificator on n-type semiconductor ZnO layer, which works like an electrode, to improve electron injection efficiency in inverted stack structure OLEDs. Kim and Song *et al.* fabricated devices with a structure of [fluorine tin oxide (FTO)/ZnO/IL/a light emitting polymer (SY)/MoO_3/Au] (Figure 8.2(a), (b)).[23] Here, FTO, IL, SY, MoO_3, and Au are a cathode, an EIL, an EML, an HIL, and an anode, respectively. The ZnO layer was formed by spray pyrolysis from zinc acetate dehydrate/methanol precursor solutions at 400 °C onto FTO substrate. Then, an IL, 1-phenyl-3-methyl-imidazolium cation with chloride anion, was dissolved in water and spin-coated onto the ZnO layer. The SY was dissolved in chlorobenzene and spin-coated onto the ZnO/IL layer. MoO_3 and Au were thermally evaporated onto the SY layer. A large electron injection barrier exists between the ZnO conduction band level (4.0 eV) and SY LUMO level (2.7 eV), making direct electron injection from ZnO to SY difficult. The IL modified the ZnO surface and formed dipole polarization between ZnO and SY, facilitating electron injection from ZnO to SY (Figure 8.2(c)). The device with IL showed a low turn-on voltage of 3.2 V, whereas the device without IL showed a higher turn-on voltage of 8.0 V. This lowering of the driving voltage resulted

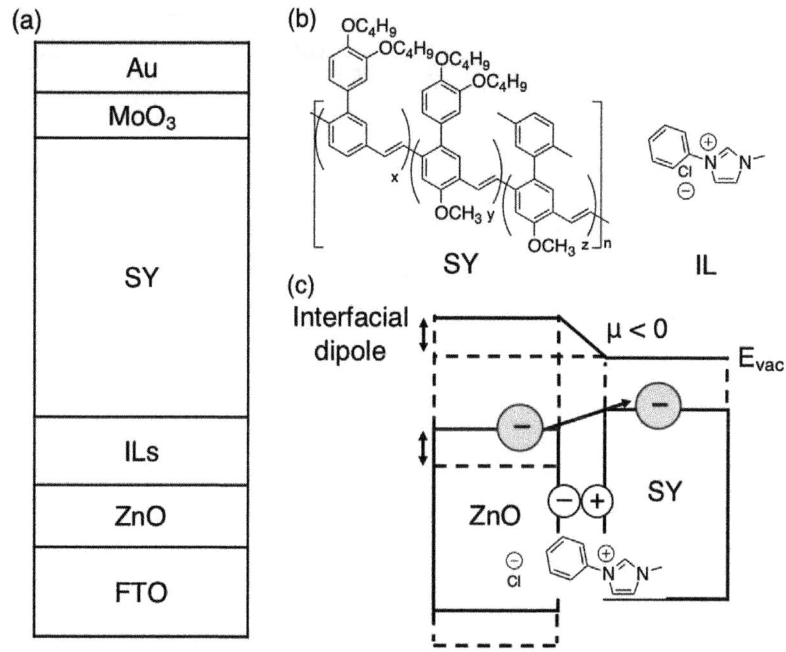

Figure 8.2 (a) OLED structure. (b) Molecular structures of a light-emitting polymer, SY, and 1-phenyl-3-methyl-imidazolium chloride IL. (c) Energy diagram and workfunction shift of the IL-modified ZnO/SY layer. Reproduced from ref. 23 with permission from the Royal Society of Chemistry.

from the improved electron injection efficiency. The modified device with ILs showed higher device characteristics with a luminance of 7500 cd m^{-2} at 14.0 V, and luminous efficiency of 2.7 cd A^{-1} at 14.0 V, which corresponds to approximately a forty-fold increase compared to that of the unmodified device. Bolink *et al.* fabricated devices with the above structure, except for the ZnO/IL layer fabrication method.[16] Instead, a ZnO precursor solution (1 wt% of zinc acetate dehydrate in ethanol : water (3 : 1)) was prepared. ILs were dissolved in 2-ethoxyethanol at a concentration of 50% and mixed with the ZnO precursor solution. The ILs used were [tetrabutylphosphonium][BF_4] (IL-A), [tetrabutylammonium][PF_6] (IL-B), and [1-butyl-3-methylimidazolium][Br] (IL-C). The mixed solutions were spin-coated on ITO substrate and annealed at 200 °C on a hotplate for 30 min, giving the ZnO:IL composite layers. Figure 8.3(a)–(c) shows the current density–voltage–luminance–current efficiency characteristics. The device without ILs showed a maximum luminance of only 150 cd m^{-2} at 10 V because of the large electron injection barrier between ZnO and SY. The device with IL-A (50% content) showed a maximum luminance of 13 500 cd m^{-2} at 10 V. The devices with IL-B and IL-C showed maximum luminances of 4300 and 2100 cd m^{-2}, respectively, at 8 V. These results indicate that the addition of ILs greatly improved the electron injection efficiency.

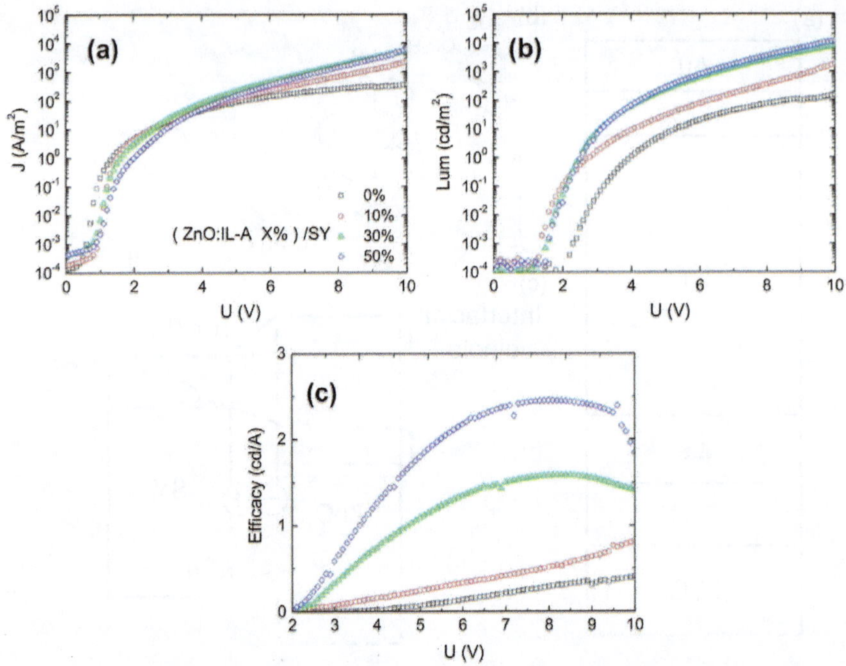

Figure 8.3 (a) Current density–applied voltage, (b) luminance–applied voltage, and (c) luminous efficiency–applied voltage characteristics with ZnO:tetrabutylphosphonium BF_4^- IL blend films using different IL content. Reprinted from *Organic Electronics*, **14** (1), H. Brine, J. F. Sánchez-Royo, H. J. Bolink, Ionic liquid modified zinc oxide injection layer for inverted organic light-emitting diodes, 164–168, Copyright (2013) with permission from Elsevier.[16]

CPE and NCPE are employed as ETL or EIL in regular stack structure OLEDs. These materials can be dissolved in polar solvents. Generally, polymer EML molecules can be dissolved in aromatic solvents and cannot be dissolved in polar solvents. Therefore, CPE and NCPE can be stacked onto EML by solution-process without dissolution of EML. In addition, polar solvents such as water and alcohol are green solvents, meeting criteria for a sustainable society. CPEs were reported by many groups, for example Cao *et al.*[13] and Bazan and Heeger *et al.*[26] The cationic conjugated alternating copolymer, poly{[9,9′-bis(6′-(*N*,*N*,*N*-trimethylammonium)hexyl)fluorene-2,7-diyl]-*alt*-[2,5-bis(*p*-phenylene)-1,3,4-oxadiazole]} ($PFON^+(CH_3)_3I^-$–PBD), consisting of alternating fluorene and phenylene-oxazole-phenylene in the main chain, was reported.[26] The synthesized CPE has π-conjugated moieties and charge transporting ability. This polymer includes ion pairs of quaternary ammonium cations and I^- anions. The device with a structure of [indium tin oxide (ITO)/poly(3,4-ethylenedioxythiophene)-poly(styrene sulfonate) (PEDOT:PSS)/light emitting polymer/$PFON^+(CH_3)_3I^-$–PBD as an ETL/Ba/Al] was fabricated. $PFON^+(CH_3)_3I^-$–PBD was spin-coated onto light emissive polymer, blue-light-emitting poly(9,9-dioctylfluorenyl-2,7-diyl) (PFO), from methanol solution to give a thickness of 30 nm. The device with the CPE ETL showed a driving voltage of 3 V, whereas the device without the CPE ETL showed a driving voltage of 5 V at 0.1 cd m^{-2}. At 6 V, the device with the CPE ETL showed a luminance of 3450 cd m^{-2}, whereas the device without the CPE ETL showed only 30 cd m^{-2}. These results indicated that insertion of the CPE ETL is very effective in improving device performance. Similar improvements were observed in a device with the light emitting-polymer poly[2-methoxy-5-(2-ethylhexyloxy)-1,4-phenylenevinylene] (MEH-PPV). These improvements resulted from a reduction of electron injection barrier from metal to the ETL compared to that from metal to the light-emitting polymers.

Woo and Lee *et al.* synthesized a CPE series (P1–P5) (Figure 8.4(a)) with an identical poly(fluorene-*co*-phenylene) (PFP) π-conjugated backbone but different numbers of cationic charged quaternary ammonium functionalities, attached to the alkoxy side chains, with Br^- counter-ions.[17] They changed the ionic density (ID) of polymers from ID = 2 for P1 to ID = 6 for P5 to investigate

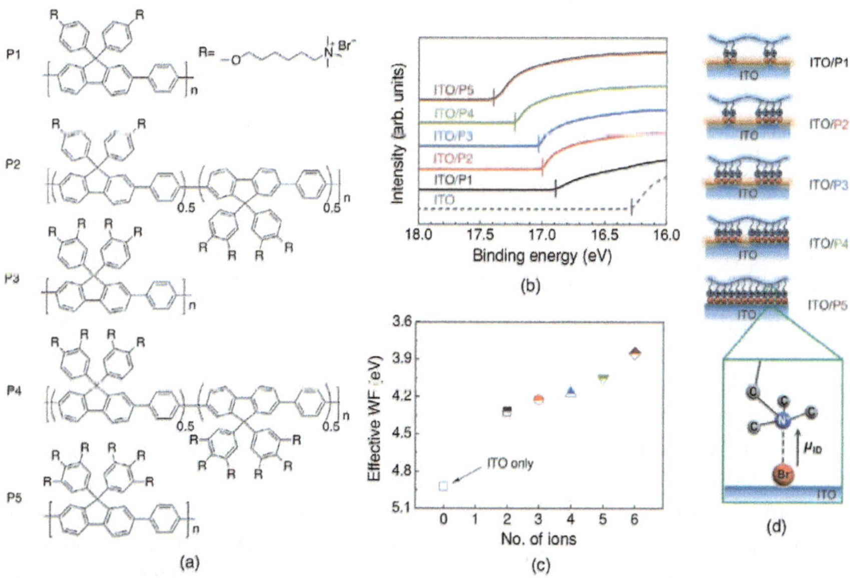

Figure 8.4 (a) Series CPEs polymers structures (P1–P5). (b) Ultraviolet photoelectron spectra of ITO electrodes covered with series CPEs within the range 16–18 eV (electron cut-off region). (c) Changes in the effective work-function (WF) of the CPE-covered ITO electrodes as a function of ionic density (ID). (d) Schematic illustrations of the proposed configuration of ionic functionalities on the ITO surface with increasing ID (P1 → P5) and an enlarged view of the ITO/CPE interface containing the ion-induced dipole μ_{ID}. Orange lines indicate the uncovered region of the ITO surface. Reprinted with permission from John Wiley and Sons (ref. 17). Copyright 2014 Wiley-VCH Verlag GmbH & Co. KGaA, Weinheim.

the influence of the ID on the WF shift in the CPE-covered ITO electrode. Here, the difference in ID did not influence the π-electronic structure of the main backbone. The authors used ultraviolet photoelectron spectroscopy (UPS) to investigate the WF shift. The CPEs (2 nm thick) were cast from methanol/isopropanol (2 : 1 by volume) solution onto pre-cleaned ITO/glass substrates. Figure 8.4(b), (c) shows the WF evaluation results for P1–P5. The CPE-covering of ITO reduced the WF of ITO substrates by dipole formation. In solution-process coating, the Br^- anion first attached to the ITO surface because of the ability to move in the solution compared with in the cationic polymer (Figure 8.4(d)). Next, the cationic polymer tethered to the Br^- on the ITO surface by the large coulombic interaction and the dipole was formed on the ITO surface. By increasing the IDs, the ITO WF was increasingly lowered. For the P5/ITO electrode, the effective WF is approximately 3.9 eV, which is almost 1 eV lower than that of the bare ITO. OLEDs using the CPE EILs with a structure of [ITO/PEDOT:PSS/MEH-PPV/CPE/Al] were fabricated. Aluminium worked as a cathode. The device with CPE (P3) (32 nm)/Al showed superior performance compared to the device with the low WF metal Ca (WF = 2.9 eV)/Al cathode, indicating that the insertion of the CPE is effective in improving electron injection. The ID effects of CPE EILs were investigated. At a CPE film thickness of 32 nm, the device with P3 showed the lowest driving voltage among the five CPEs, indicating that the driving voltage was determined by not only the ID effects but also other factors. The device characteristics behavior also suggested that the simple ion migration model could not explain the operation mechanism (Figure 8.5). In the ion migration model, the device with larger ID should show the lowest driving voltage. The authors explained the experimental results by assuming molecular reorientation at interfaces (Figure 8.5). When an electric field is applied to the devices, Br^- moves towards the MEH-PPV/CPE interface accompanied by ammonium cation attached to the side chain, whereas the ammonium cations move towards the CPE/Al cathode interface accompanied by Br^-. These side chain reorientations with the ionic motion generate an electric double layer (EDL) at the interfaces, resulting in improved electron injection. They assumed that the side chains with larger ID than P3 were difficult to reorient, resulting in less effective EDL than that of P3.

In applications of NCPE, we applied poly(*N*-alkyl-4-vinyl-pyridinium iodide)s PILs for EILs (Figure 8.6(a)).[19] The alkyl chains were propyl, butyl, and hexyl chains, and the corresponding PILs are referred to as P-C3, P-C4, and P-C6, where the monomer unit of P-C3 has a melting point of 81 °C, showing the IL feature. The PILs were synthesized *via* a simple one-step process from poly(4-vinylpyridine) and iodoalkanes. We fabricated OLEDs with a structure of [ITO/PEDOT:PSS/poly(9,9-dioctyl-fluorene-*co*-*N*-(4-butylphenyl)-diphenylamine) (TFB)/poly[(9,9-di-*n*-octylfluorenyl-2,7-diyl)-*alt*-(benzo[2,1,3]-thiadiazol-4,8-diyl)] (F8BT)/EILs/Al]. Here, PEDOT:PSS, TFB, and F8BT are HIL, HTL, and EML, respectively. Applied EILs were P-C3, P-C4, and P-C6 with a thickness of 5 nm, and 8-quinolinolato lithium (Liq), which is a conventional electron injection material showing good electron injection

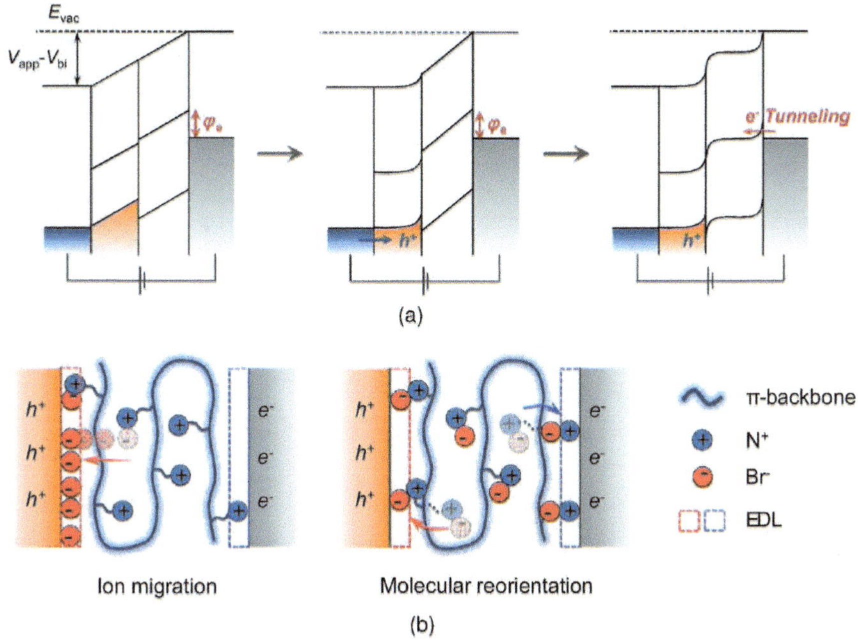

Figure 8.5 (a) Schematic energy level diagrams of an ITO/PEDOT:PSS (blue)/MEH-PPV (orange)/CPE/Al (grey) device under an applied bias (V_{app}) in the electron injection process. V_{bi} represents the built-in-potential of the OLED. (b) Schematic illustrations of a partial cross section of the MEH-PPV/CPE/Al layers in the OLED under the V_{app} using (left) the ion migration model and (right) molecular reorientation model. Reprinted with permission from John Wiley and Sons (ref. 17). Copyright 2014 Wiley-VCH Verlag GmbH & Co. KGaA, Weinheim.

characteristics, with a thickness of 1 nm. Figure 8.6(b)–(e) shows the OLED characteristics and the molecular distribution images in the PIL layer during device operation. The voltages at a luminance of 1 cd m^{-2} of the OLEDs were 2.5 V for Liq, 2.8 V for P-C6, 2.9 V for P-C4, and 3.4 V for P-C3, indicating that Liq is the most effective electron injection material at 1 cd m^{-2}. However, at 1000 cd m^{-2}, which is a practical luminance in application, the OLEDs showed driving voltages of 4.2 V for P-C6, 4.6 V for P-C4, 5.7 V for Liq, and 6.4 V for P-C3, indicating that P-C6 is the most effective electron injection material at 1000 cd m^{-2}. Among the PIL-based OLEDs, the devices with longer alkyl chains showed lower driving voltages. We assumed that the electric field induced polarization, which resulted from ionic motion, is related to the alkyl chain length dependence of the device characteristics. The interactions between cations and anions, and the thermal chain motions facilitating ionic motion, were investigated. Pyridinium iodide salts are known to show charge transfer (CT) absorption from I^- to pyridinium cation based on the CT interaction being closely related to ionic interactions. The HOMO of the PILs locates at I^-. To investigate the CT interaction, the ultraviolet-visible

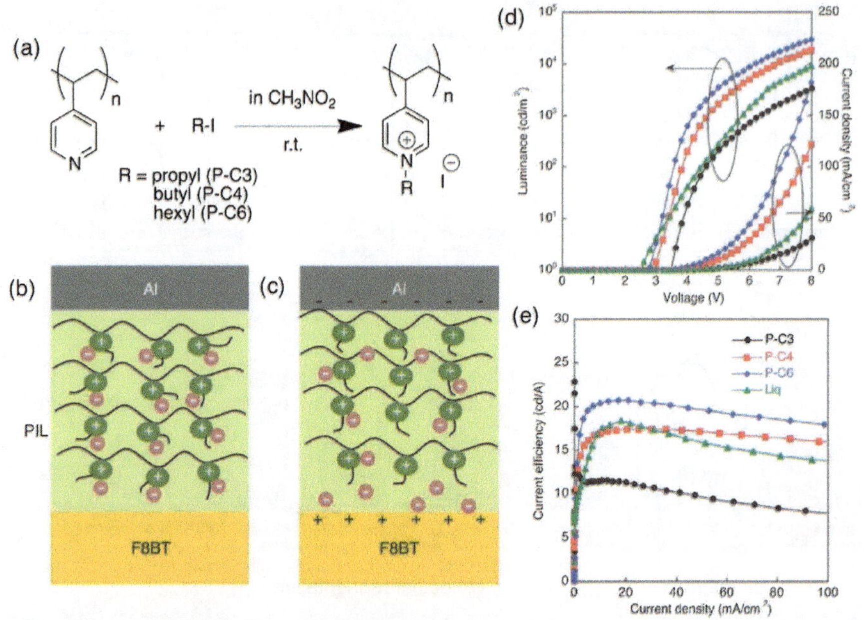

Figure 8.6 (a) Synthesis scheme and molecular structures of P-C3, P-C4 and P-C5 PILs. Electric field induced polarization images in the PIL layer (b) without an applied electric field and (c) with an applied electric field. (d) Current density–voltage–luminance characteristics. (e) Current density–current efficiency characteristics. Reprinted from ref. 19 with permission from the Royal Society of Chemistry.

(UV-vis) absorption spectra of the PIL films were measured. All PILs showed CT absorption bands located at the same wavelength region, suggesting that there are similar degrees of CT interactions in all PILs. The ionization potential (I_p) of the PILs was measured by photoelectron yield spectroscopy (PYS). All PILs showed the same I_p values of 6.0 eV, indicating that the alkyl chains do not affect the I_p values. The electron affinity (EA) value of the iodine atom in a vacuum is known to be 3.1 eV, corresponding to the I_p value of I^- in vacuum. The difference in the I_p values between the PILs and I^- in vacuum originates from the electrostatic stabilization energy for the I^- in the PIL films. Thus, the stabilization energy of the I^- in PILs is 2.9 eV, which is smaller than that of NaCl (7.9 eV), proving the weak ionic interaction in the PILs. Then, the thermal behavior of the PILs was investigated. The three PILs were spin-coated onto F8BT polymer films and dried in a vacuum oven at 30 °C. The PIL film morphology was investigated by atomic force microscopy (AFM). The surface roughness, Ra, values were 0.367 nm for P-C3, 0.406 nm for P-C4, and 0.392 nm for P-C6. Then, the films were annealed at 110 °C in a nitrogen atmosphere and the film morphology was investigated by AFM. The Ra values were 0.354 nm for P-C3, 0.431 nm for P-C4, and 0.495 nm for P-C6. In the P-C6 film with the longest alkyl chain, many aggregates were generated by thermal annealing, resulting in a larger Ra value. These results suggest that a PIL with a longer alkyl chain length has a larger thermal chain mobility on F8BT. From these results, we assumed that one of the mechanisms by which the PIL-based OLEDs with a longer alkyl chain length showed a lower driving voltage arises through the difference in thermal chain mobility, which is related to ionic motion facility. These considerations are consistent with the above reports by Woo and Lee *et al.* in that the chain motion facility influenced the device characteristics. In addition, another mechanism was proposed. The electrostatic repulsive force among the same kind of ions becomes stronger in the electric field induced polarized state than it is in the state under no electric field. A longer alkyl chain can work more efficiently as a spacer reducing the repulsive force. We concluded that the two mechanisms worked efficiently in lowering the driving voltages in the PIL-based OLEDs.

In applications of ILs in OLEDs, ILs were employed as EILs, greatly improving electron injection efficiency and charge carrier balance, leading to high luminous efficiency. The origin of electron injection efficiency improvement is dipole polarization formation and the molecular reorientation leading to the WF shifts. The degree of WF shift depends on the ILs. Developments of ILs showing larger dipole polarization are desired in future work.

8.2.1.2 LECs

LECs consist of an anode, a cathode, and an organic functional layer including mobile ions sandwiched by the two electrodes (Figure 8.7(a)).[32,118–127] Recent research and developments are summarized by Bolink *et al.*[32] LECs have a similar device architecture to OLEDs, and the characteristic evaluation methods of LECs are same as those of OLEDs. The important difference

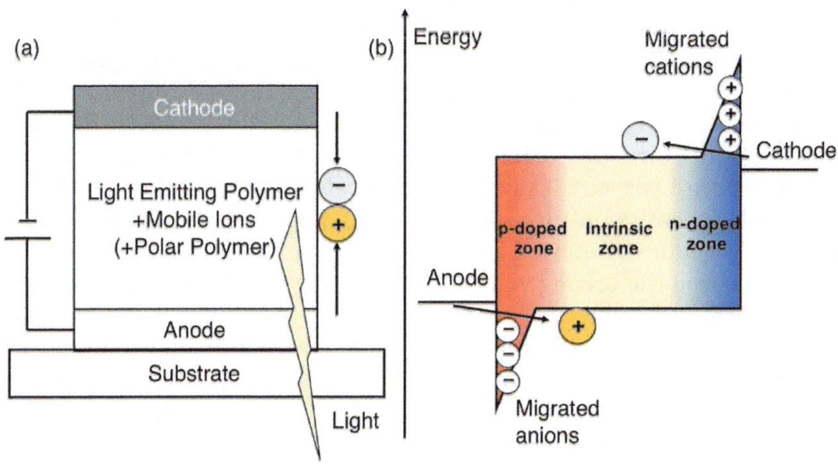

Figure 8.7 (a) LEC structure. (b) Energy diagram of a LEC with an applied voltage.

between OLEDs and LECs is whether the light-emitting layer includes mobile ions. For example, the organic functional layer includes a light emitting polymer, Li-salts, and a polar polymer such as poly(ethylene oxide) (PEO). These materials are dissolved in solution, and coated onto the electrode substrate to give the organic layer. One of the most prominent features of LECs is enabling low driving-voltage without multilayer stack structure. When an electric field is applied to a LEC, the applied electric field dissociates the Li-salts into Li^+ and anions, which migrate towards the cathode and the anode sides in the organic functional layer. The cations make the light-emitting polymer n-doped, whereas the anions make the light-emitting polymer p-doped. The doping facilitates carrier injection from the electrodes into the light-emitting polymer layer. Figure 8.7(b) shows the energy diagram of LECs. Under the application of the electric field, a quasi-multilayer stack structure, [p-doped-layer/intrinsic layer/n-doped-layer] is formed. The injected holes and electrons meet to form excitons emitting light. The doping enables light-emission at a voltage corresponding to near band-gap energy of the light-emitting polymer even when an air stable high WF cathode was utilized. The ionic conductivity of mobile ions is an important factor for fast response and low voltage driving. The light-emitting polymer is generally hydrophobic, whereas the salts are hydrophilic, causing phase separation in the film, making the ionic movements in the hydrophobic light-emitting polymer difficult. Hence, polar polymer PEO is added to suppress the phase separation and to improve the ionic conductivity. Merits of LECs compared to OLEDs are the extremely simple fabrication process and air stability of electrodes. On the other hand, LEC performances are not yet as well developed as OLEDs.

There are several reports on applications of ILs in LECs.[27–34,128–130] For example, Li *et al.* added only ILs as the mobile ion sources to the light-emitting polymer, MEH-PPV, instead of the combination of air sensitive Li-salts and polar polymers.[31] The added ILs were 1-methyl-3-alkylimidazolium salts with the alkyl substituents of butyl (BMIM), dodecyl (DMIM), tetradecyl (TMIM), or hexadecyl (HMIM) and the anions PF_6^- or BF_4^-. These ILs were hydrophobic and compatible with MEH-PPV without PEO, except for [BMIM][PF_6]. This is a good result because the inclusion of PEO in films accelerates device deterioration. LECs with a structure of [ITO/MEH-PPV:ILs/Al] were fabricated. For comparison, a device without ILs was also fabricated. In the LEC using [DMIM][PF_6], the light emission turn-on voltage was near 3 V. On the other hand, the device without ILs showed a turn-on voltage of 12 V. The addition of ILs largely improved the driving voltage. The melting points of ILs greatly influenced the device characteristics. The LEC using [DMIM][BF_4], which has lower a melting point than [DMIM][PF_6], showed better performance than the LEC using [DMIM][PF_6]. Device driving at high temperature (80 °C) also gave better performance compared to the driving at room temperature. These results showed that the ionic motion facility is an important factor in the LEC driving mechanism. As the applied voltage increases, the doped zone in the light-emitting polymer broadens, resulting in formation of a quasi-two-layer structure of [p-doped-layer/n-doped-layer]. Generally, the p- or n-doped light-emitting polymer does not emit light. On the other hand, insufficient doping gives a large driving voltage. Therefore, control of the doped zone breadth is required for stable and low voltage driving. To control the doped zone, it is desired that the dopant ILs are fastened at a definite position at room temperature (room temperature frozen-junction). For the fastening, the utilization of ILs having higher melting points than room temperature is preferred. LECs using [TMIM][PF_6] (m.p. = 73 °C) and [HMIM][PF_6] (m.p. = 83 °C) were fabricated. First, the devices were driven at high temperature for doping of MEH-PPV by ion migration, and then cooled to room temperature to fasten the ions to a definite position in MEH-PPV. The driving voltage of the cooled devices were sufficiently low; however, unfortunately, even at room temperature, the doped ions relaxed and could not be fastened at the definite position, resulting in unstable device driving. Some groups reported polymerizable ILs for fixed junction LECs.[27,32–34] Leger *et al.* developed ILs with the ability of ion-pair monomers to form covalent bonds to achieve ion immobilization.[34] The three ILs developed had a common anion, allyl sulfonate, whereas the cation consists of trialkyl allyl (or vinyl)-substituted quaternary nitrogen (Figure 8.8(a)). The alkyl chain lengths attached to the quaternary nitrogen were varied as follows: ATOAAS, C8; ATBAAS, C4: TMVAAS, C1. To form a covalent bond, both the cations and anions contained allylic or vinylic functionality. ATOAAS and ATBAAS have melting points of 15–20 °C, whereas TMVAAS has a melting

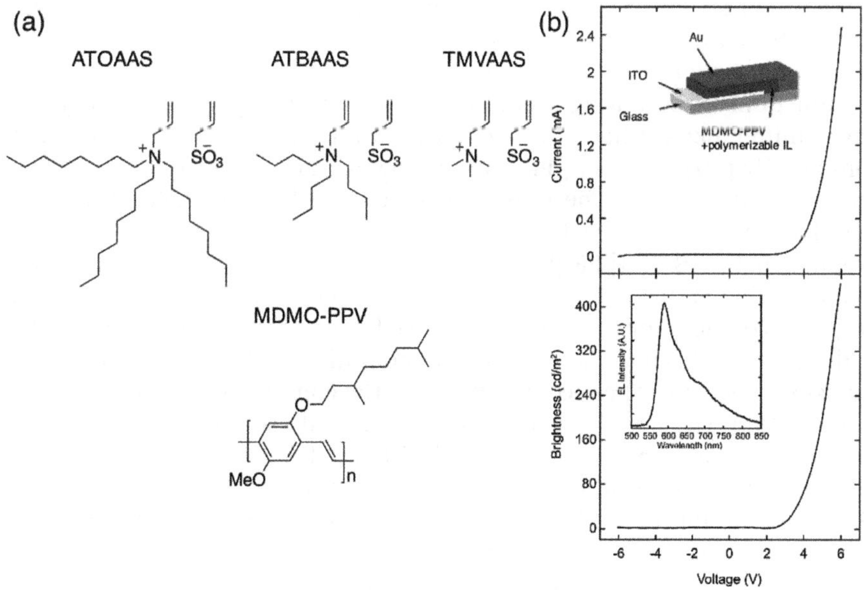

Figure 8.8 (a) Polymerizable ionic liquids and light-emitting polymer, MDMO-PPV, moleccular structures. (b) LEC structure, light emission spectrum and current–voltage and brightness–voltage characteristics. Reprinted with permission from ref. 34. Copyright (2010) American Chemical Society.

point of 65–68 °C. LECs with a structure [ITO/poly[2-methoxy-5-(3′,7′-dimethyloctyloxy)-*p*-phenylenevinylene (MDMO-PPV) as the red-orange light emitting polymer doped with ILs/Au] were fabricated and evaluated (Figure 8.8(b)). MDMO-PPV and the ILs were dissolved in chlorobenzene and spin-cast onto patterned ITO glass substrates. Then, an Au contact was evaporated onto the organic layer. Phase separation between MDMO-PPV and the ILs was minimal because of good compatibility among them. The devices were charged at 4–6 V at room temperature to dope the MDMO-PPV by ion migration. After 3–5 min, a steady current was achieved in the LEC operation, and then the voltage was swept from positive bias to negative bias. In the positive bias region, the devices showed a low light emission turn-on voltage of around 2.5 V, indicating that MDMO-PPV was doped by the ILs during the charging. The brightness of the LEC with ATOAAS was around 400 cd m^{-2} at 6 V. In contrast, in the negative bias region, current flow was negligible and the LECs showed diode-like behavior suggesting that the dopant ions were fastened to a definite position in MDMO-PPV. Probably, covalent-chemical bonds were formed between the ionic dopant containing allylic or vinylic functionality and MDMO-PPV. The authors assumed that the chemical bond formation was initiated by the formation of radicals during electrochemical doping.

To improve light emission in LECs, utilization of a phosphorescent ionic transition-metal complex (iTMC) is very effective.[28–30,32] An example iTMC is a salt of a cationic iridium complex emitting phosphorescence and an anion such as PF_6^-. An iTMC also dissociates into the corresponding cation and anion under an electric field. The dissociated ions move towards the opposite electrode, resulting in formation of EDL and p-type and n-type zones, facilitating carrier injection at the electrode interfaces. An iTMC combines the functions of both a light emitter and dissociative salts in the polymer light emitter LECs. Choe *et al.* added ILs to a layer of a cationic iridium complex $[Ir(ppz)_2(dmphen)]PF_6$ containing 5,6-dimethyl-1,10-phenanthroline ancillary ligand.[28] The utilized ILs were 1-ethyl-3-methylimidazolium hexafluorophosphate ($[EMIM][PF_6]$), 1-butyl-3-methylimidazolium hexafluorophosphate ($[BMIM][PF_6]$), and 1-hexyl-3-methylimidazolium hexafluorophosphate ($[HMIM][PF_6]$). The influence of addition of ILs on the optical properties of the complex were investigated by UV-vis spectra measurements. Addition of the ILs into the complex did not influence the shape of the light emission spectrum. Then, compatibility of the complex and ILs was investigated by AFM. The complex film on PEDOT:PSS had a surface with root-mean-square (RMS) roughness of 4.83 nm. The blend film of ILs to the complex at 1 : 1 molar ratio on PEDOT:PSS had a rms roughness of 5.50 nm for $[BMIM][PF_6]$, 5.53 nm for $[EMIM][PF_6]$, and 8.38 nm for $[HMIM][PF_6]$. The addition of ILs to the complex did not cause significant aggregation because of the hydrophobic nature of the ILs. LECs with a structure of [ITO/PEDOT:PSS/$[Ir(ppz)_2(dmphen)]PF_6$: ILs/Al] were fabricated. $[Ir(ppz)_2(dmphen)]PF_6$] was dissolved in acetonitrile solution and three different ILs were added at a 1 : 1 molar ratio of $[Ir(ppz)_2(dmphen)]PF_6]$: [IL]. The solutions were spin-coated onto the PEDOT:PSS layer and baked at 80 °C for 1 h. An Al cathode was evaporated onto the organic layer. In the device without ILs, the luminance and current density increased very slowly with an increase in the applied voltage up to 6.0 V. The device achieved a maximum luminance of 4052 cd m^{-2} at 9 V. In the devices with ILs, the luminances increased expeditiously at lower voltage but sacrificed their stability at above 6.5 V. Among the ILs, the device with [BMIM][PF_6] showed the maximum luminance of 3850 cd m^{-2} at 6.5 V, which is lower than that of the device without ILs.

Although it is not included in LEC devices, Moon, Lodge and Frisbie fabricated an electrochemiluminescent (ECL) device with a simple single organic layer sandwiched by two electrodes (Figure 8.9(a)–(c)).[131] They employed the organic layer of a $Ru(bpy)_3Cl_2$ emitter dispersed in ion gels consisting of a polystyrene-*block*-poly(methyl methacrylate)-*block*-polystyrene triblock copolymer denoted as SMS(15-81-15) and [EMIM][TFSI] IL. A $Ru(bpy)_3Cl_2$,

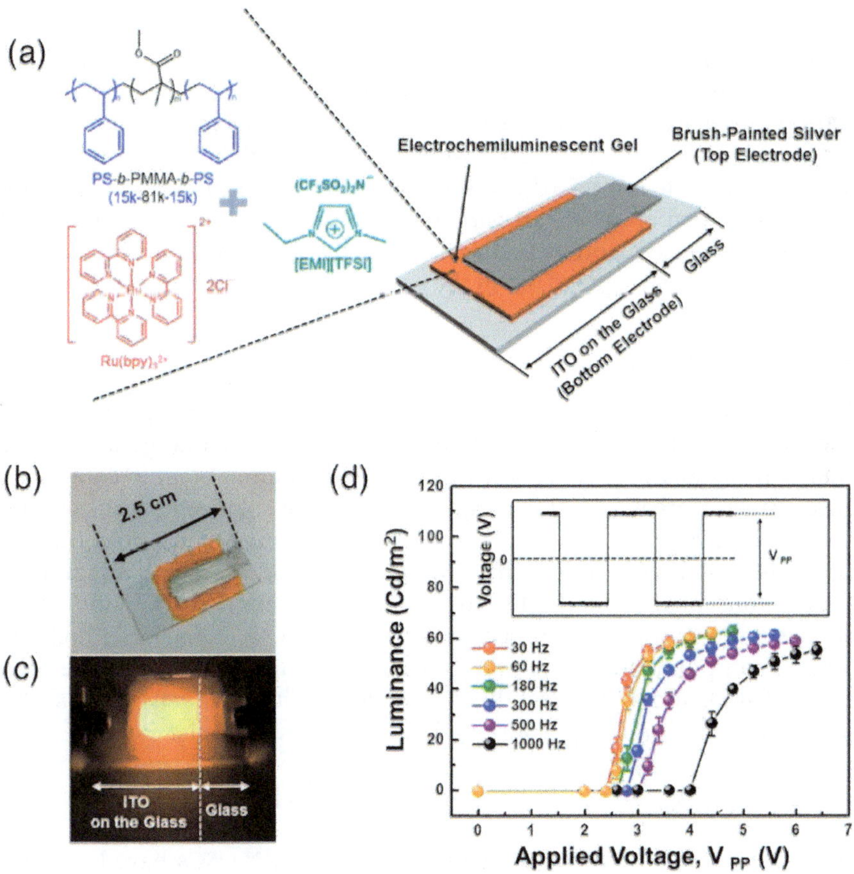

Figure 8.9 (a) Schematic of an ECL device with an emissive gel layer composed of $Ru(bpy)_3Cl_2$/PS-b-PMMA-b-PS/[EMI][TFSI] and a brush-painted Ag electrode. (b, c) Photographs of the ECL device. (d) Dependence of the peak luminance on the applied voltage at AC input frequencies from 30 to 1000 Hz. Reprinted with permission from ref. 131. Copyright (2014) American Chemical Society.

SMS(15-81-15), and [EMIM][TFSI] mixture was dissolved in ethyl acetate at a weight ratio of 1 : 4 : 16, and spin-casted onto an ITO electrode to give a film about 30 μm thick. The cast film was dried at 30 °C for 24 h under reduced pressure to remove the residual solvent. An Ag electrode was deposited onto the ion gel layer by brush-painting of Ag paint. When 60 Hz AC voltage was applied to the device, red-orange light emission from the Ru complex was observed at the very low turn-on voltage of 2.6 V (Figure 8.9(d)). The origin of the light emission is not carrier recombination but electrochemiluminescence. First, when a relative negative bias is applied to the ITO electrode, the Ru complex cation ($Ru(bpy)_3^{2+}$) near the ITO electrode is reduced to $Ru(bpy)_3^{+}$ by the ITO electrode. Then, a relatively positive bias is applied to the ITO electrode, and the $Ru(bpy)_3^{2+}$ diffuses toward to the ITO electrode and is oxidized to $Ru(bpy)_3^{3+}$. The $Ru(bpy)_3^{+}$ collides with $Ru(bpy)_3^{3+}$ and reacted to give the excited Ru complex ($Ru(bpy)_3^{2+*}$), which relaxes *via* a light emission process. ILs played a role as Ru-complex ion conductive medium.

In applications of ILs for LECs, ILs have been employed as p- or n-dopants to light-emitting material to improve the carrier injection characteristics. The hydrophobic nature removed the necessity for a polar polymer such as PEO. The most important problem of LECs with ILs is their stability. Some approaches have been made to improve the stability. For example, one is fixation of the ions after ionic migration. There are various reports on fixation methods other than the one mentioned report by Leger *et al.*[34] Given that stability improvements are being made, the widespread use in the near future of LECs is expected.[30,32,132]

8.2.2 Solar Cells

Solar cells are an energy conversion device, from photon energy into electric energy. Si-based solar cells are already used worldwide and generate green energy from solar light. Recently, solar cells including organic compounds have attracted great attention as a next-generation solar cell because of their potential for high energy conversion efficiency, light weight, flexibility, and large area manufacturing facility by solution process, which suggest new applications of the solar cells, for example, attachment to round-shaped houses and car roofs. There are major three types of solar cells including organic compounds: organic thin film solar cells (OSCs); dye sensitized solar cells (DSSCs); and organometallic halide perovskite solar cells (PeSCs). This section reviews some applications of ILs in these solar cells.

8.2.2.1 OSCs

Since the developments of OSCs by Tang in 1986,[133] OSCs have been studied intensely. At present, high performance OSCs consist of, for example, polymer donor such as thiophene-based polymers and fullerene derivatives acceptor materials for the light absorption layer (Figure 8.10).[134–140] Light is absorbed by the materials and excitons are generated. The generated excitons dissociate into hole and electron pairs at the donor/acceptor interface. In organic materials, the dissociative force is smaller than that in inorganic

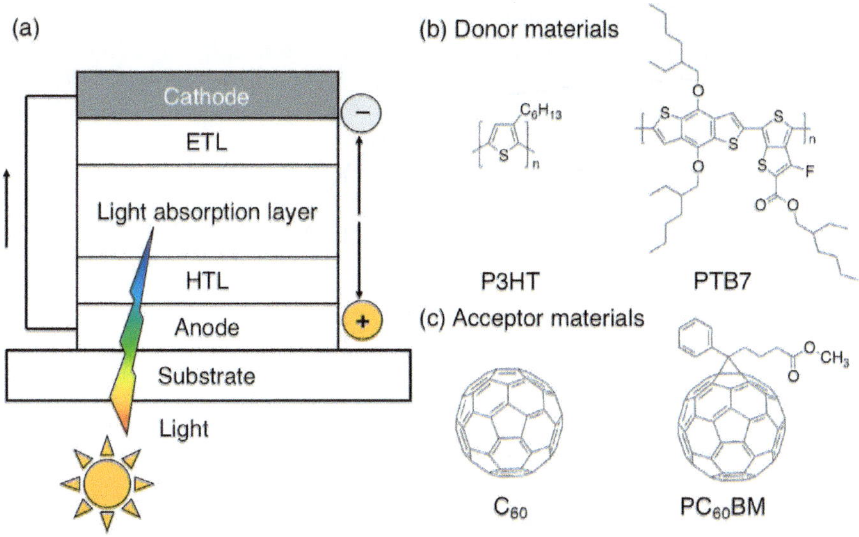

Figure 8.10 (a) OSC structure. (b) Representative thiophene-based donor materials. (c) Representative fullerene-based acceptor mateials.

materials because of the small dielectric constant, which relatively destabilizes the dissociated charge carriers. Therefore, large-area donor/acceptor interface formation was desired to increase the opportunity for exciton dissociation. The requirements led to bulk heterojunction devices with a large area interface. However, a large area interface also increases the opportunity for carrier recombination, thereby deteriorating the carrier extraction to electrodes. Therefore, optimization of the donor/acceptor interface is required. To extract the generated carriers efficiently, high carrier mobility layers are also required. Insertion of HTL and ETL between the light absorption layer and electrodes is favorable. To obtain energy from sun, the light absorber should have strong light absorption coefficients and a wide absorption wavelength region that overlaps with the emission spectrum of the sun. Current density (J)–voltage (V) characteristics during light irradiation to devices define the solar cell characteristics. Figure 8.11 shows the J–V characteristics. Current density at non-bias is called short circuit current density (J_{SC}). The voltage found with a non-conducting-current is called the open circuit voltage (V_{OC}). Output power is expressed by multiplication of voltage by current. Current density and voltage at the maximum power output are denoted as J_{max} and V_{max}, respectively. Fill factor (FF) and power conversion efficiency (PCE) are expressed as follows:

$$\mathrm{FF} = (J_{max} \times V_{max})/(J_{SC} \times V_{OC})$$

$$\mathrm{PCE}\ (\%) = \{(J_{max} \times V_{max})/P_{inc}\} \times 100$$

where, P_{inc} is the irradiation light power. The PCE is the measure representing the device efficiency.

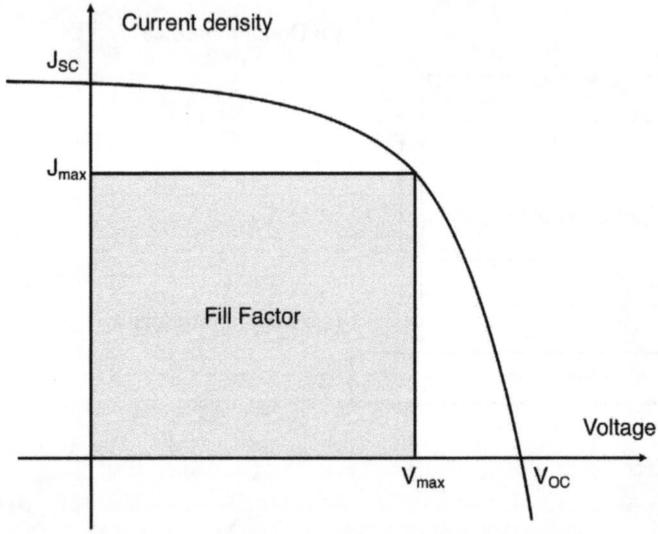

Figure 8.11 Current density–voltage characteristics in a solar cell.

There are several application reports on the usage of ILs in OSCs.[23,37–40,51] To extract the generated carriers efficiently, the interfacial engineering at the electrode/carrier-transporting layer is also important. Zhang and Li *et al.* reported IL application as interfacial modifier in OSCs.[38] They fabricated a device with the structure [ITO/ZnO/[BMIM][BF_4]/PTB7:PC_{71}BM or PTB7-Th: PC_{71}BM/MoO_3/Ag]. The ZnO layer was formed from the precursor consisting of zinc acetate dehydrate and ethanolamine in 2-methoxyethanol. The precursor solution was spin-coated onto the ITO substrate and annealed at 200 °C. [BMIM][BF_4] dissolved in methanol was spin-coated onto the ZnO film and annealed at 100 °C for 10 min. The PTB7:PC_{71}BM or PTB7-Th:PC_{71}BM dissolved solution was spin-coated onto the modified ZnO. MoO_3 and Ag electrodes were added by evaporation. The PTB7:PC_{71}BM-based devices showed stabilized PCEs of 8.65% for the IL modified device, and 7.51% for the unmodified device. PTB7-Th:PC_{71}BM-based devices showed stabilized PCEs of 9.56% for the IL modified device, and 8.94% for the unmodified device. The best PCE of the PTB7-Th:PC_{71}BM-based devices was 10.15%, which is among the highest reported PCEs for a single-junction device through a solution process. The PCE improvements by IL modification resulted mainly from the J_{SC} and FF improvements. The J_{SC} and FF improvement mechanisms were investigated. By impedance analysis, in the modified device, a lower series resistance (R_s) of 3.46 Ω cm^2 and a higher shunt resistance (R_{sh}) of 16.20 kΩ cm^2 were obtained compared to those of the unmodified device (R_s of 10.28 Ω cm^2 and R_{sh} of 4.53 kΩ cm^2). These changes implied that the leakage current was improved by the IL modification, enhancing charge extraction. Electron-only devices with a structure of [ITO/ZnO or (ZnO/[BMIM][BF_4])/PTB7:PC_{71}BM/Ca/Al] were used to examine the electron mobility.

The obtained electron mobilities were 2.47×10^{-4} cm^2 V^{-1} s^{-1} for the ZnO device, and 4.91×10^{-4} cm^2 V^{-1} s^{-1} for ZnO/[BMIM][BF_4] device. The improvement suggested the more efficient electron transfer occurs at the ZnO/[BMIM][BF_4]/PTB7:$PC_{71}BM$ interface. Kelvin probe force microscopy (KFM) was used to evaluate WF values. The measurements gave WF values of −4.44 eV for the ZnO and −4.35 eV for the ZnO/[BMIM][BF_4] surfaces. The EA value of electron-transporting $PC_{71}BM$ was −4.3 eV, which is closer to the ZnO/[BMIM][BF_4] level than ZnO, suggesting that more efficient electron transfer occurs at the (ZnO/[BMIM][BF_4])/PTB7:$PC_{71}BM$ interface. The WF shift originates from the spontaneous polarization of [BMIM][BF_4] on the ZnO layer. These interfacial property changes are the reasons why the modification of ZnO by [BMIM][BF_4] improved the OSCs efficiencies.

8.2.2.2 *DSSCs and PeSCs*

Present DSSCs are called Grätzel cells, after they were invented by Grätzel in 1991.[141–144] The PCEs of DSSCs are high, exceeding 10%. The Grätzel cell usually consists of a transparent mesoporous TiO_2 on conductive glass as a photoelectrode, a Ru-complex light sensitizing dye, an electrolyte solution generally including I^-/I_3^-, and an opposed electrode. Light is absorbed to the Ru-complex and then the generated exciton dissociates to give a hole and electron pair at the Ru-complex/TiO_2 interface. The generated electron travels in the TiO_2 semiconductor, whereas the oxidized Ru-complex is reduced *via* oxide-reduction reaction with the electrolyte. The oxidized electrolyte moves towards the opposed electrode in the electrolyte solution and receives electrons from the opposed electrode. ILs including I^- have been utilized as the electrolyte or electrolyte solution in DSSCs, and have achieved a degree of success in improving the device performance.[35,41,45,48–50,54–56,145–151] In 2009, Miyasaka *et al.* discovered efficient organometallic halide perovskite compounds, $CH_3NH_3PbBr_3$ and $CH_3NH_3PbI_3$, for DSSC dyes and achieved a PCE of 3.8% (Figure 8.12).[152] The perovskite compounds were obtained by

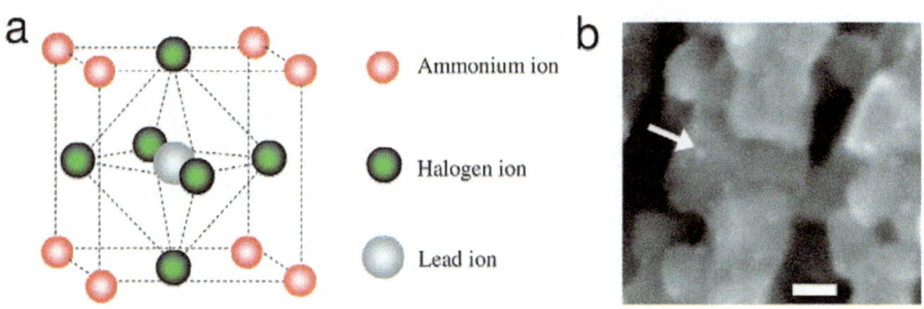

Figure 8.12 (a) Crystal structures of perovskite compounds. (b) SEM image of particles of nanocrystalline $CH_3NH_3PbBr_3$ deposited on the TiO_2 surface. The arrow indicates a partice, and the scale bar shows 10 nm. Reprinted with permission from ref. 152. Copyright (2009) American Chemical Society.

thermal reaction from the precursor $PbBr_2$ or PbI_2 and methylammonium halides (MABr or MAI). In 2012, Miyasaka and Snaith *et al.* developed PeSCs using the $CH_3NH_3PbI_3$ perovskite and achieved a PCE of 10.9% based on the device structure of [FTO/compact TiO_2/mesoporous TiO_2 with $CH_3NH_3PbI_3$/2,2′,7,7′-tetrakis-(*N*,*N*-di-*p*-methoxyphenylamine)-9,9′-spirobifluorene (Spiro-OMeTAD)/Ag].[153] Here, TiO_2 and Spiro-OMeTAD work as electron- and hole-transporting materials, respectively. Since the discovery of PeSCs, an enormous number of researchers have energetically studied and improved the performances of PeSCs.[43,52,57,154–159] At present, the best performance of a PeSCs is above 20%, which is almost comparable to that of the Si-based solar cells. Such a high PCE results from the strong light absorption strength, the wide light absorption wavelength region, long exciton, hole, and electron diffusion lengths (1 μm in certain cases), and high V_{OC} (up to *ca* 1.1 V) implying a small voltage loss. One of the difficulties in fabricating high performance PeSCs is control of perovskite film quality, which greatly influences the exciton and carrier transportation. There are several reports on IL applications in PeSCs.[36,42–44,46,47,52,53,57] Generally, inappropriate thermal annealing causes incomplete reaction or the decomposition of the perovskite films. Wiesner and Estroff *et al.* applied methylammonium formate (MAF) IL as a solvent of the perovskite film precursor solution.[36] They dissolved PbI_2 and MAI at the molar ratio of 1 : 1 at 30 wt% in MAF and spin- or drag-coated the solution to give a film 500–700 nm thick. The deposited film was thermally annealed to obtain the reacted perovskite film under nitrogen atmosphere. The perovskite films crystal growth was evaluated by *in situ* synchrotron wide-angle X-ray scattering. Direct conversion from the precursor into the perovskite crystal was observed in the measurement time scale. Five thermal annealing conditions (50 °C 10 min., 75 °C 10 min, 75 °C 100 min, 75 °C 1000 min, and 135 °C 10 min) gave only the perovskite crystal phase. The precursors or decomposed compounds were not observed. This high robustness of crystal growth probably originates from the IL physical properties. MAF IL has extremely low volatility, and physical properties such as viscosity are almost identical in value over a wide thermal temperature region, which has a good effect on the robustness over a wide range of thermal annealing. Another proposed reason is that the methylammonium cation in MAF has a good effect on crystal growth. Grätzel and Abate *et al.* reported IL control of crystal growth to enhance planar PeSCs efficiency.[44] They fabricated a device with a structure of [FTO/compact SnO_2/$CH_3NH_3PbI_3$/HTM including Spiro-OMeTAD/Ag]. They added a small amount of MAF IL (few mol%) to the precursor solution for the perovskite formation consisting of MAI, PbI_2, and anhydrous dimethyl sulfoxide (DMSO). The solution was spin-coated onto the compact SnO_2 layer and thermally annealed at 100 °C for 1 h in a nitrogen-filled glovebox. The perovskite conversion reaction with MAF was slower than that without MAF (Figure 8.13(a)). The device with MAF showed a larger stabilized PCE of 18.8% than that without MAF (18.0%). The device improvements were attributed to the better quality of the film with MAF. SEM images revealed that the film with MAF had larger crystals, over 150 nm, whereas the film

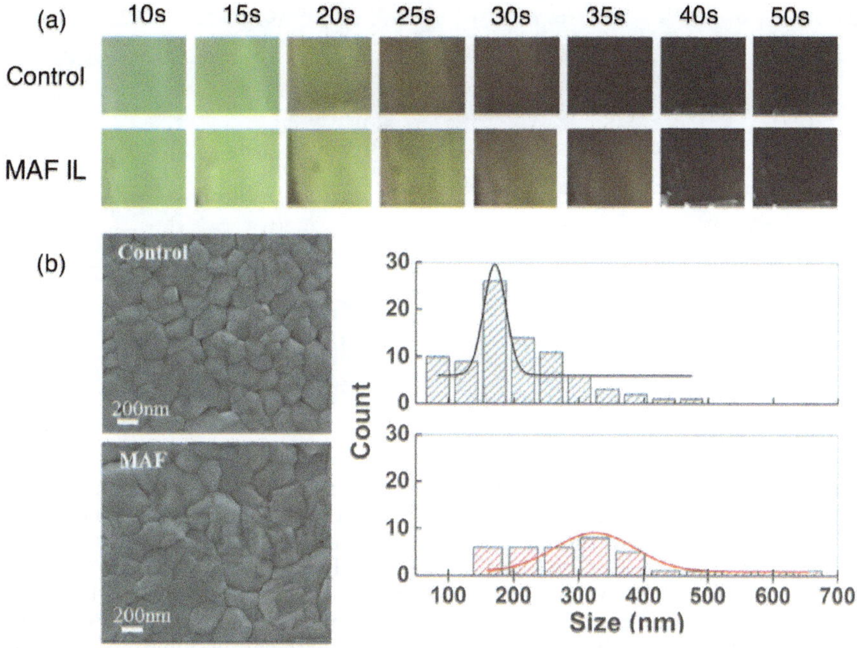

Figure 8.13 Morphological analysis of the perovskite films with and without MAF IL. (a) Pictures of perovskite films collected at subsequent delay times after placing the substrates on a hotplate at 100 °C. (b) (left) Top-view SEM images of perovskite films after 1 h annealing at 100 °C. (right) Grain sizes distribution as estimated from the SEM images using Nano measurer 1.2 software. Reprinted with permission from John Wiley and Sons (ref. 44). Copyright 2016 Wiley-VCH Verlag GmbH & Co. KGaA, Weinheim.

without MAF showed a wider crystal size distribution including 50–150 nm (Figure 8.13(b)). The AC voltage frequency response of the films was investigated. The film with MAF showed stronger response at high frequency 200 kHz and weaker responses in the lower frequency region than that without MAF. The faster response implies higher carrier mobility. Hence, the authors assumed that the carrier mobility was improved in the larger crystal size perovskite film, resulting in the improved device performance.

To improve the device efficiency, interfacial engineering for better carrier delivery is also important in the PeSCs. Li and Liu *et al.* reported that ILs are a very effective interfacial modifier at the TiO_2/perovskite interface in the planar type PeSCs (Figure 8.14).[57] They fabricated a device with the structure of [compact TiO_2/[BMIM][BF_4]/$CH_3NH_3PbI_3$/poly[bis(4-phenyl)(2,4,6-trimethylphenyl)amine] (PTAA)/Au] (Figure 8.14(b)). [BMIM][BF_4] was spin-coated onto the compact TiO_2 layer and annealed at 80 °C for 10 min. The device with IL showed a higher PCE of 19.59% than that of the device without IL (14.90%) at forward bias scan (Figure 8.14(c), (e)). This PCE value was higher than the reported best record of the planar type PeSCs. Figure 8.14(d) shows

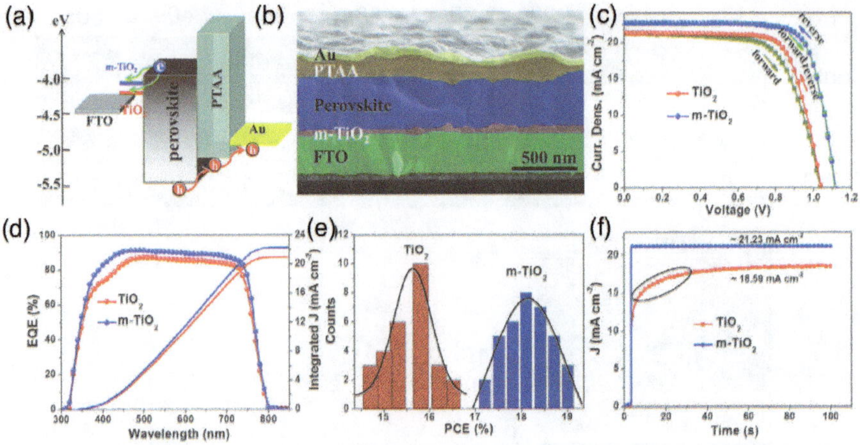

Figure 8.14 (a) Illustration of energy level diagram of the PSC. (b) Representative cross-sectional SEM view of the PSC device. (c) *J–V* characteristics under both reverse and forward scan directions. (d) Corresponding EQE curves of the best devices based on different ETLs. (e) The PCE distribution histogram of the PSCs using TiO_2 and m-TiO_2 ETLs. (f) Photocurrent density measured as a function of time for the TiO_2 and m-TiO_2 devices. Reprinted from ref. 57 with permission from the Royal Society of Chemistry.

the external quantum efficiency (EQE) spectra and the integrated product of the EQE curves. Here, the EQE represents the conversion ratio of the irradiated photons into the extracted carrier. The J_{sc} data calculated from the EQE data agreed with the *J–V* characteristics derived J_{sc} values. One of the major challenges with the planar type PeSCs is suppression of large hysteresis between the forward and backward voltage scans. Surprisingly, the insertion of IL eliminated the hysteresis, whereas the device without IL showed large hysteresis (Figure 8.14(c)). Three factors cause large hysteresis: (i) ion migration; (ii) high-trap state density for charge carriers at the perovskite interface; (iii) unbalanced carrier transport. They investigated the effects of the IL modification on these factors. The modified TiO_2 (m-TiO_2) showed a smoother surface (RMS = 9.05 nm) than the bare TiO_2 (RMS = 11.57 nm), improving the TiO_2/perovskite contact. A certain amount of IL on the TiO_2 was confirmed by X-ray photoelectron spectroscopy (XPS). The WF shift was investigated by KFM. The bare TiO_2 showed a WF of 4.26 eV, whereas the m-TiO_2 showed a WF of 4.01 eV resulting from the spontaneous polarization of [BMIM][BF_4] (Figure 8.14(a)). BF_4^- coordinated to TiO_2, and reduced the WF of TiO_2. The permanent dipoles formed point outwards in free space from the TiO_2 surface. The reduced WF value of the m-TiO_2 is better matched to the perovskite layer than that of the bare TiO_2. Electron mobility of the ETLs was obtained by the space charge limited current (SCLS) method using an electron-only device structure. The obtained electron mobilities were 3.83×10^{-4} cm^2 V^{-1} s^{-1} for the bare TiO_2, and 1.36×10^{-3} cm^2 V^{-1} s^{-1} for the m-TiO_2. The hole mobility of PTAA is on the order of 10^{-3}. The IL modification improved the charge balance. These changes probably largely improved the device efficiency and the hysteresis characteristics. Figure 8.14(f) shows photocurrent density measured as a function of time for the TiO_2 and m-TiO_2 devices. The m-TiO_2 device showed an instant photocurrent rise and constant values up to 100 s, whereas the TiO_2 device showed a slow photocurrent rise to the constant photocurrent value. The difference results from the hysteresis behaviors among the devices.

In applications of ILs in solar cells, ILs are utilized as interfacial modifier or perovskite crystal growth controlling-agent. The high PCE performance was achieved by using the ILs in the OSCs due to the better interface formation. The best PCE performance was achieved by using the ILs in the planar type due to the better interface formation and the high-quality perovskite film. ILs are very effective for achieving high performance solar cells. The emergence of devices with further improved efficiency is confidently expected.

8.2.3 OFETs

OFETs can switch on and off conduction current states in electric circuits and have attracted considerable attention because of their potential for applications in large-area, flexible, low-cost electronics.[160–164] OFETs consist of gate, source, and drain electrodes, a gate dielectric, and a hole-conducting (p-type) or an electron conducting (n-type) organic semiconductor placed on a substrate. There are two types of OFETs. One is a bottom-gate type OFET, where the gate electrode is underneath the organic semiconductor. The other is a top-gate type OFET, where the gate electrode is above the organic semiconductor (Figure 8.15). For simplicity, here, an explanation is limited to the top-gate type OFET. Figure 8.15 shows the top-gate type OFET structure. The gate dielectric is sandwiched by the gate electrode and the organic semiconductor. The gate electrode is above the gate dielectric and the organic semiconductor is underneath the gate dielectric. The drain and grounded electrodes contact the organic semiconductor. The gate and drain electrodes are connected to the source electrode *via* respective power sources. In OFET devices, the output is currents between the source and drain electrodes (I_D).

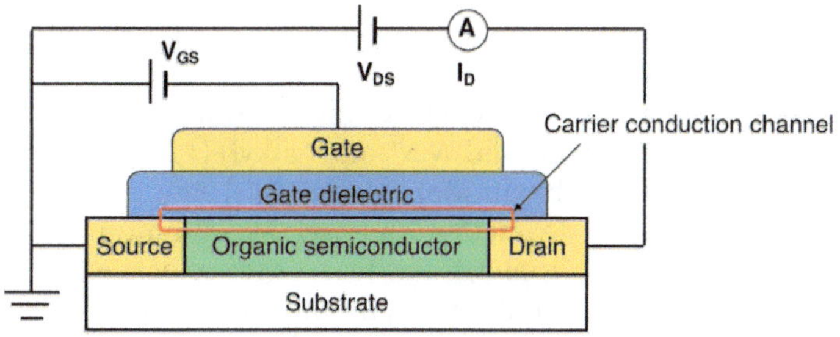

Figure 8.15 Top-gate type OFET device structure.

The I_D is not controlled by bias of the drain electrode to the source electrode (V_{DS}), but also bias of the gate electrode to the source electrode (V_{GS}). The I_D is regulated by changing the V_{GS}. When V_{GS} is applied to the device, the gate dielectric is electrically polarized. The electric polarized gate dielectric induces electric potential changes at the interface between the organic semiconductor and the gate dielectric. A positive V_{GS} stabilizes electrons at the interface in the organic semiconductor, causing electron accumulation and facilitating electron conduction along the interface, whereas a negative V_{GS} stabilizes holes at the interface in the organic semiconductor, causing hole accumulation and facilitating hole conduction along the interface. Therefore, the main path of the conduction current I_D is along the gate dielectric/organic semiconductor interface because of the carrier conduction facility. The electric polarization degree is an important parameter determining the device characteristics. The electric polarization degree depends on the capacitance of the gate dielectric layer (C). In the region where V_{DS} is much less than V_{GS}, the I_D is expressed as follows:

$$I_D = (W/L)\mu C(V_{GS} - V_T)V_{DS}$$

where W is the channel width, L is the channel length, μ is the carrier mobility, and V_T is the threshold voltage. To obtain low voltage driving OFETs, large μ and C values are necessary. There are various reports on IL applications in OFETs.[58–62,64–69,71,73–77,165,166] ILs have potentials for realizing a gate dielectric with a large capacitance. In ILs-based gate dielectrics, electric double layers (EDLs) are formed at the gate electrode/gate dielectric and the gate dielectric/organic semiconductor interfaces because of the high ionic conductivity and the electric potential mainly drops at the interfaces. In fact, ILs or PILs have been mainly employed as liquid or solid gate electrolyte dielectrics in OFETs. Recent research has been well reviewed by Frisbie and Lodge *et al.*[64] The FETs using electrolytes such as ILs are called, for example, electrolyte-gated transistors (EGTs). IL-based gate dielectrics have a huge capacitance, on the order of 1–10 μF cm^{-2}, which exceeds the conventional 'high-k' dielectrics such as Ta_2O_5 by a factor of 10, realizing the low voltage driving OFETs. One reason why the IL-based gate dielectric shows such huge capacitance values is the thinness of the formed EDL, on the order of 1 nm, because the capacitance value is almost inversely proportional to the EDL thickness. In addition, in using the IL-based gate dielectric, there are far more merits than just the huge capacitance. In this section, some examples of ILs based-gate dielectrics OFETs are reviewed.

Ono and Morpurgo *et al.* reported high-performance n-type OFETs with IL-based gate dielectric and *N*,*N*″-bis(*n*-alkyl)-(1,7 and 1,6)-dicyanoperylene-3,4 : 9,10-bis(dicarboximide)s (PDIF-CN_2) single-crystals (Figure 8.16).[68] They employed (*N*-methyl-*N*-propyl-pyrrolidinium [TFSI] ([P13][TFSI])) as the IL. In OFET applications, fast response of the gate dielectrics to applied AC voltage is required. Hence, the authors first investigated the frequency response of the capacitance values by impedance analyzers from 0.1 Hz to 10 MHz with a 5 mV AC voltage (Figure 8.16(c)). The capacitance value

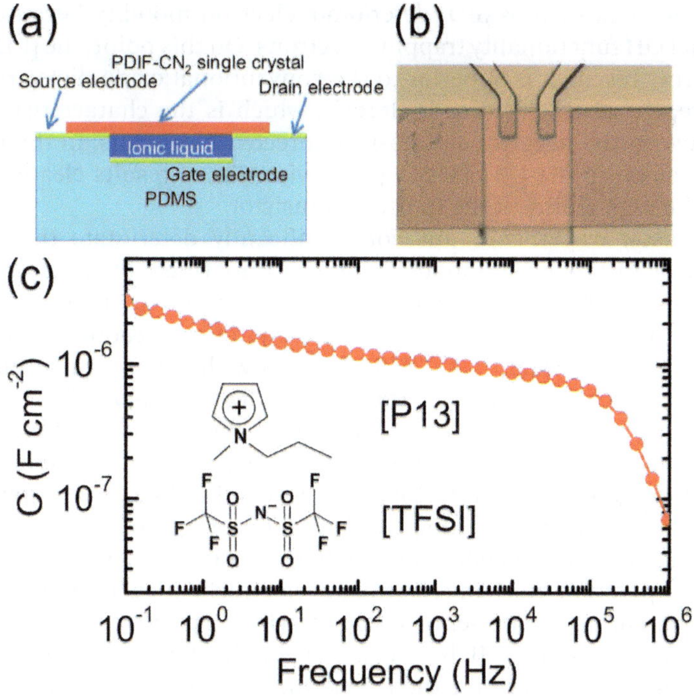

Figure 8.16 (a) Schematic structure of the IL gated devices. (b) Optical microscope image of a PDIF-CN2 crystal/[P13][TFSI] transistor. (c) Frequency dependence of the capacitance of [P13][TFSI] electric double layer, measured by the AC impedance analysis. (Inset) Chemical structure of [P13][TFSI]. Reprinted from ref. 68, with permission of AIP Publishing.

did not significantly decrease with increasing frequency and it remained close to 1 μF cm^{-2} even at 0.1 MHz. This value is sufficiently high and implies that the [P13][TFSI]-based gate dielectric is practical enough in terms of frequency response. They then fabricated a bottom-gate type OFET with or without [P13][TFSI] gate dielectric (Figure 8.16(a), (b)). In the device without the IL, air works as the gate dielectric. The gate, source, and drain electrodes consist of Ti/Au films. The gap between the gate electrode and the PDIF-CN_2 single-crystal was kept at 25 μm to easily insert [P13][TFSI] IL into the gap by capillary forces. The air gate device was characterized in advance. The device showed the high V_T of 20 V, and a mobility value with air as gate dielectric in the range of 1.5–5.5 $cm^2\ V^{-1}\ s^{-1}$, which is among the best electron mobilities reported for PDIF-CN_2 single-crystal transistors. Then, the [P13][TFSI] gate device was characterized. The device showed an extremely low V_T of 0.1 V and a sufficiently large current was obtained even at a V_{GS} of 0.5 V. The device also showed an electron mobility of 1.0–5.0 $cm^2\ V^{-1}\ s^{-1}$, indicating that the [P13][TFSI] gate dielectric did not deteriorate the interfacial state of PDIF-CN_2 in terms of electron conduction facility. It is well known that generally used oxide dielectrics such as SiO_2 deteriorate electron mobility because of the interfacial OH functionality trapping electrons. On this point, the [P13][TFSI] used in this research is superior to the conventional oxide dielectrics. Furthermore, the characteristics hysteresis, which is the characteristic difference between the negative and positive directed bias scans, in the V_{GS} scan was very small in the [P13][TFSI] gate device. These results clearly demonstrated the high ability of the IL-based transistor.

One reason why the IL did not significantly deteriorate the electron mobility of PDIF-CN_2 originates from the impermeable character of the single-crystal organic semiconductor. The large lattice energy and dense solid-state structure of the single crystal prevent the permeation of ILs into the PDIF-CN_2 single crystal.[68,72,165,166] Such EGTs with impermeable interfaces are called electric double layer transistors (EDLTs) (Figure 8.17).[64] Figure 8.17 also shows the EDLT energy diagrams in the forward bias-state (accumulation mode) and the backward bias-state (depletion mode). When an ion permeable organic semiconductor is employed, the ILs easily migrate to the organic semiconductor, causing the formation of a diffused interface with a smaller capacitance value. There are two ways to overcome this problem. One is an operation mechanism utilizing electrochemical doping by the diffused ions. For an n-type semiconductor, when a positive V_{GS} is applied to the device, cations in the IL-based gate dielectric migrate toward the n-type organic semiconductor, and then penetrate into the semiconductor. The penetrative diffusion makes the organic semiconductor n-doped. Electron conductivity increases by the n-type doping. This type of OFET is well known and is called organic electrochemical transistors (OECTs) (Figure 8.17). Figure 8.17 shows the OECTs energy diagrams in the accumulation mode and the depletion mode. One of the merits of OECTs is that larger currents can be obtained than the device utilizing EDL gating. In the EDL gating, the conductive channel is limited near the gate dielectric/organic semiconductor interface. In contrast, in the OECT mechanism, the conductive channel is the doped-zone and is wider than that of the EDL gating one. The wider conductive channel is advantageous in obtaining large currents. One of the demerits of OECTs is the large characteristics hysteresis. Chemical reactions and molecular structure changes of the organic semiconductor occur during doping and de-doping, causing the hysteresis. Furthermore, generally, both cations and anions diffuse into the organic semiconductor, causing unexpected operation behavior. Another method is suppression of the ion diffusion by utilizing PILs. In the case of n-type organic semiconductors, utilization of a PIL consisting of cations attached to a polymer backbone and small molecule anions is effective for diffusion suppression.[63,70,73,77] When a positive bias is applied to the device, the cations attached to polymers backbone almost cannot migrate into the semiconductor layer because of the large bulkiness. Frisbie and Lodge *et al.* synthesized a triblock copolymer (PS-PIL-PS) composed of styrene and 1-[(2-acryloyloxy)ethyl]-3-butylimidazolium [TFSI] PIL and applied it to n-type OFETs as the gate dielectric (Figure 8.18).[77] In this PIL copolymer, the imidazolium cations are covalently connected to the polymer

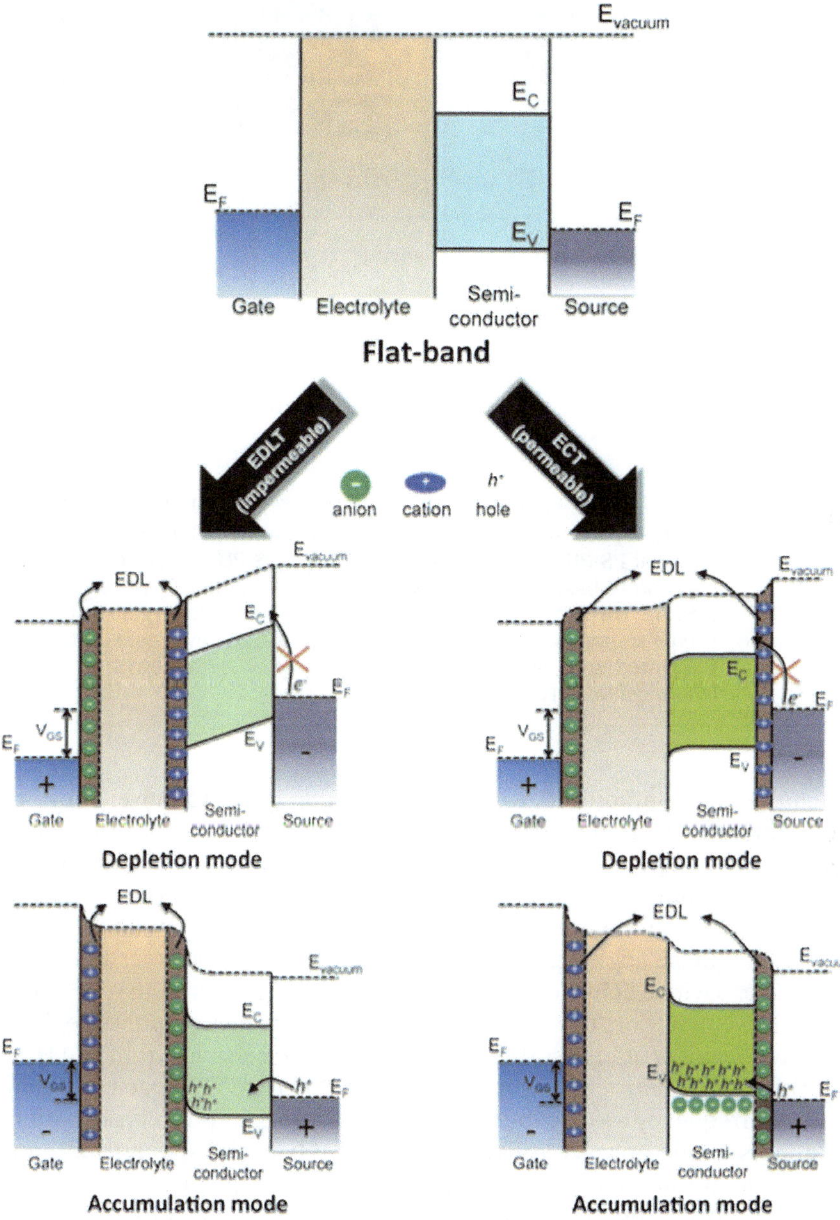

Figure 8.17 Energy band diagrams for the three principal states (flat-band, depletion, and accumulation) of gate/electrolyte/semiconductor interfaces in p-type EDLTs and ECTs. Reprinted with permission from John Wiley and Sons (ref. 64). Copyright 2013 Wiley-VCH Verlag GmbH & Co. KGaA, Weinheim.

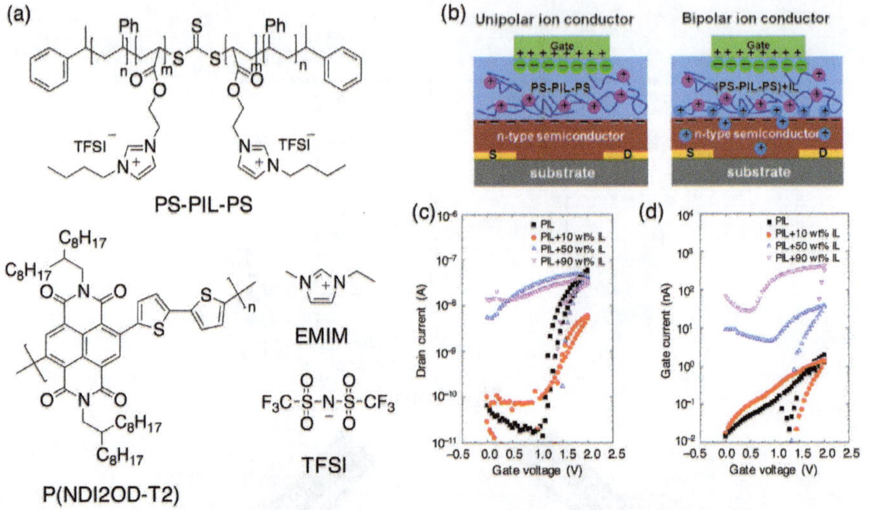

Figure 8.18 (a) Molecular structures of PS-PIL-PS, P(NDI2OD-T2) and [EMIM][TFSI]. (b) Schematic cross section showing the polarization in unipolar (neat PS-PIL-PS) and bipolar (IL-doped PS-PIL-PS) electrolyte gated transistors based on n-type semiconductors under positive gate voltage. (c) Transfer characteristics and (d) gate current–gate voltage of n-type transistors with neat PS-PIL-PS polyelectrolyte gate dielectrics. Reprinted with permission from ref. 77. Copyright (2015) American Chemical Society.

side chains and immobilized, while the TFSI anion is free to move. They first investigated the AC frequency response of the polymer with an impedance analyzer. They fabricated a measurement sample with a structure of [Au/PS-PIL-PS (500 nm)/Au]. The PS-PIL-PS was spin-coated onto the bottom Au electrode, and then the top Au electrode was evaporated onto the PS-PIL-PS layer. In the phase angle *vs.* frequency plots, phase angles lower than −45° represent capacitive behavior, whereas phase angles greater than −45° represent resistive behavior. The plots showed three frequency response regions, capacitive behavior at higher than 120 kHz, resistive behavior from 850 Hz to 120 kHz, and capacitive behavior at lower than 850 Hz. The capacitive behavior at high frequency was attributed to the dipolar relaxation between cations and anions of the polyelectrolyte. The resistive behavior was attributed to the anion migration. The capacitive behavior at low frequency was attributed to the EDL formation. The frequency dependence of the capacitance was evaluated. The capacitance values increased with decreasing frequency and are greater than 1 μF cm^{-2} at frequencies below 10 kHz. This value is high enough to apply PS-PIL-PS as the gate dielectric. They fabricated the n-type OFETs with the PS-PIL-PS gate dielectric. When a bias is applied to the device, it is anticipated that only TFSI anions migrate (unipolar ion conduction). Solution-processable poly{[*N*,*N*′-bis(2-octyldodecyl)-naphthalene-1,4,5,8-bis (dicarboximide)-2,6-diyl]-*alt*-5,5′-(2,2′-bithiophene)} (P(NDI2OD-T2)) was selected as the n-type semiconductor polymer (Figure 8.18(a)). Source and drain electrodes (5.0 nm Cr/30.0 nm Au) were patterned on SiO_2/Si substrates using photolithography. P(NDI2OD-T2) was spin-coated onto the substrate and annealed at 105 °C for 10 min. The PS-PIL-PS gate dielectric was spin-coated onto the P(NDI2OD-T2) layer to give a film thickness of 500 nm. An Al gate electrode was deposited on the P(NDI2OD-T2) layer by an evaporation process. Figure 8.18(c) shows the gate-voltage–drain-current characteristics of the devices. From device characterization analysis, the device showed an ON/OFF current ratio of 2×10^3 and a threshold voltage V_T of −0.4 V. A certain amount of hysteresis was observed probably because of the relatively low ionic conductivity of the PIL. A saturation carrier mobility of 0.008 $cm^2\ V^{-1}\ s^{-1}$ was also obtained and the mobility value is comparable to previously reported values for n-type P(NDI2OD-T2) OFETs. These results clearly demonstrated that the unipolar ion conductive PIL gate dielectric is effective for the OFET driving. Then, they mixed small molecule ILs, [EMIM][TFSI], with the PIL gate dielectric to investigate the influence of free cations on the device characteristics. With increasing mixing ratio of [EMIM][TFSI] to the PIL, the OFF current increased and the ON/OFF ratio decreased to *ca* 2, indicating that the transistor feature disappeared. Figure 8.18(d) shows the gate-voltage–gate-current characteristics of the devices with the ILs. The addition of ILs clearly increased leakage currents. This implied the free EMIM migrated to the P(NDI2OD-T2) and deteriorated the transistor characteristics. This result clearly suggests that the unipolar ion conducting PIL gate dielectric is necessary.

In applications of ILs in OFETs, ILs were employed as the gate dielectrics. ILs are compatible with hydrophobic organic semiconductors and migrate to the organic semiconductor layer during device operation. There are two ways to prevent the IL migration. One is utilization of an impermeable organic semiconductor such as a single crystal. Another way is to use PILs as the gate dielectric. In the mentioned case, immobile cations attached to the polymer backbone did not diffuse to the organic semiconductor layer because of the bulkiness. The former way is limited to specific types of organic semiconductor materials, whereas the latter way is applicable for various polymers. The research and development of the PILs as the gate dielectrics are desired.

8.3 Outlook for the Future

In this chapter we have covered several IL applications in organic electronic devices. As reviewed, ILs are very effective for obtaining high performance devices. First, ILs have been used as surface modifier to control the WF value of the electrode or inorganic semiconductor material in OLEDs or solar cells. Second, the high ionic mobility was effective in LECs or OECLs to realize a low driving voltage. Third, the high capacitance originating from the narrow EDL formation was very useful for the gate dielectric in OFETs. Last, ILs were used as perovskite crystal growth controlling agent due to the extremely stable thermal physical properties. Although not introduced in this chapter, ILs can be a very good dispersion solvent for carbon nanotubes or graphene, which are expected to be widely used materials for next-generation devices.[167,168] ILs have infinite possibility to control physical properties because of the infinite combination of cations and anions. Further advance of IL application is confidently expected.

References

1. J. S. Wilkes and M. J. Zaworotko, *J. Chem. Soc., Chem. Commun.*, 1992, **13**, 965–967.
2. J. C. Ma and D. A. Dougherty, *Chem. Rev.*, 1997, **97**, 1303–1324.
3. R. D. Rogers and K. R. Seddon, *Science*, 2003, **302**, 792–793.
4. M. Armand, F. Endres, D. R. MacFarlane, H. Ohno and B. Scrosati, *Nat. Mater.*, 2009, **8**, 621–629.
5. T. Torimoto, T. Tsuda, K. Okazaki and S. Kuwabata, *Adv. Mater.*, 2010, **22**, 1196–1221.
6. H. Jin, B. O'Hare, J. Dong, S. Arzhantsev, G. A. Baker, J. F. Wishart, A. J. Benesi and M. Maroncelli, *J. Phys. Chem. B*, 2008, **112**, 81–92.
7. J. Yuan, D. Mecerreyes and M. Antonietti, *Prog. Polym. Sci.*, 2013, **38**, 1009–1036.
8. D. Mecerreyes, *Prog. Polym. Sci.*, 2011, **36**, 1629–1648.
9. A. S. Shaplov, R. Marcilla and D. Mecerreyes, *Electrochim. Acta*, 2015, **175**, 18–34.
10. R. Marcilla, F. Alcaide, H. Sardon, J. A. Pomposo, C. Pozo-Gonzalo and D. Mecerreyes, *Electrochem. Commun.*, 2006, **8**, 482–488.
11. D. R. MacFarlane, N. Tachikawa, M. Forsyth, J. M. Pringle, P. C. Howlett, G. D. Elliott, J. H. Davis, M. Watanabe, P. Simon and C. A. Angell, *Energy Environ. Sci.*, 2014, **7**, 232–250.
12. D. G. Georgiadou, M. Vasilopoulou, L. C. Palilis, I. D. Petsalakis, G. Theodorakopoulos, V. Constantoudis, S. Kennou, A. Karantonis, D. Dimotikali and P. Argitis, *ACS Appl. Mater. Interfaces*, 2013, **5**, 12346–12354.
13. H. Wu, F. Huang, Y. Mo, W. Yang, D. Wang, J. Peng and Y. Cao, *Adv. Mater.*, 2004, **16**, 1826–1830.
14. J. H. Seo, R. Yang, J. Z. Brzezinski, B. Walker, G. C. Bazan and T.-Q. Nguyen, *Adv. Mater.*, 2009, **21**, 1006–1011.
15. F. Huang, L. T. Hou, H. B. Wu, X. H. Wang, H. L. Shen, W. Cao, W. Yang and Y. Cao, *J. Am. Chem. Soc.*, 2004, **126**, 9845–9853.
16. H. Brine, J. F. Sánchez-Royo and H. J. Bolink, *Org. Electron.*, 2013, **14**, 164–168.
17. B. H. Lee, I. H. Jung, H. Y. Woo, H.-K. Shim, G. Kim and K. Lee, *Adv. Funct. Mater.*, 2014, **24**, 1100–1108.
18. T. Kim, M. Suh, S. J. Kwon, T. H. Lee, J. E. Kim, Y. J. Lee, J. H. Kim, M. Hong and K. S. Suh, *Macromol. Rapid Commun.*, 2009, **30**, 1477–1482.
19. S. Ohisa, Y.-J. Pu and J. Kido, *J. Mater. Chem. C*, 2016, **4**, 6713–6719.

20. K.-W. Tsai, T.-F. Guo, A. K. Y. Jen and T.-C. Wen, *J. Mater. Chem. C*, 2014, **2**, 272–276.
21. S.-N. Hsieh, S.-W. Hsiao, T.-Y. Chen, C.-Y. Li, C.-H. Lee, T.-F. Guo, Y.-J. Hsu, T.-L. Lin, Y. Wei and T.-C. Wen, *J. Mater. Chem.*, 2011, **21**, 8715.
22. C.-H. Wu, C.-Y. Chin, T.-Y. Chen, T.-F. Guo, C.-H. Lee, T.-L. Lin, A. K. Y. Jen and T.-C. Wen, *J. Mater. Chem. C*, 2014, **2**, 4805–4811.
23. B. R. Lee, H. Choi, J. SunPark, H. J. Lee, S. O. Kim, J. Y. Kim and M. H. Song, *J. Mater. Chem.*, 2011, **21**, 2051.
24. K.-W. Tsai, C.-H. Wu, J.-Y. Jan, Y.-J. Hsu, T.-F. Guo and T.-C. Wen, *J. Mater. Chem. C*, 2016, **4**, 8559–8564.
25. S.-H. Oh, D. Vak, S.-I. Na, T.-W. Lee and D.-Y. Kim, *Adv. Mater.*, 2008, **20**, 1624–1629.
26. W. Ma, P. K. Iyer, X. Gong, B. Liu, D. Moses, G. C. Bazan and A. J. Heeger, *Adv. Mater.*, 2005, **17**, 274–277.
27. S. Tang, K. Irgum and L. Edman, *Org. Electron.*, 2010, **11**, 1079–1087.
28. C. D. Sunesh, S. Ok, G. Mathai and Y. Choe, *Thin Solid Films*, 2013, **531**, 530–534.
29. S. T. Parker, J. D. Slinker, M. S. Lowry, M. P. Cox, S. Bernhard and G. G. Malliaras, *Chem. Mater.*, 2005, **17**, 3187–3190.
30. R. n. D. Costa, A. Pertegás, E. Ortí and H. J. Bolink, *Chem. Mater.*, 2010, **22**, 1288–1290.
31. C. H. Yang, Q. J. Sun, J. Qiao and Y. F. Li, *J. Phys. Chem. B*, 2003, **107**, 12981–12988.
32. S. B. Meier, D. Tordera, A. Pertegás, C. Roldán-Carmona, E. Ortí and H. J. Bolink, *Mater. Today*, 2014, **17**, 217–223.
33. S. Tang and L. Edman, *Electrochim. Acta*, 2011, **56**, 10473–10478.
34. I. V. Kosilkin, M. S. Martens, M. P. Murphy and J. M. Leger, *Chem. Mater.*, 2010, **22**, 4838–4840.
35. S. Ito, S. M. Zakeeruddin, P. Comte, P. Liska, D. B. Kuang and M. Grätzel, *Nat. Photonics*, 2008, **2**, 693–698.
36. D. T. Moore, K. W. Tan, H. Sai, K. P. Barteau, U. Wiesner and L. A. Estroff, *Chem. Mater.*, 2015, **27**, 3197–3199.
37. G. H. Kim, H. B. Kim, B. Walker, H. Choi, C. Yang, J. Park and J. Y. Kim, *ACS Appl. Mater. Interfaces*, 2013, **5**, 1757–1760.
38. W. Yu, L. Huang, D. Yang, P. Fu, L. Zhou, J. Zhang and C. Li, *J. Mater. Chem. A*, 2015, **3**, 10660–10665.
39. P. Fu, L. Huang, W. Yu, D. Yang, G. Liu, L. Zhou, J. Zhang and C. Li, *Nano Energy*, 2015, **13**, 275–282.
40. C. H. Wu, C. Y. Chin, T. Y. Chen, S. N. Hsieh, C. H. Lee, T. F. Guo, A. K. Y. Jen and T. C. Wen, *J. Mater. Chem. A*, 2013, **1**, 2582–2587.
41. P. Wang, S. M. Zakeeruddin, P. Comte, I. Exnar and M. Grätzel, *J. Am. Chem. Soc.*, 2003, **125**, 1166–1167.
42. H. S. Kim, J. W. Lee, N. Yantara, P. P. Boix, S. A. Kulkarni, S. Mhaisalkar, M. Grätzel and N. G. Park, *Nano Lett.*, 2013, **13**, 2412–2417.
43. D. Yang, R. Yang, X. Ren, X. Zhu, Z. Yang, C. Li and S. F. Liu, *Adv. Mater.*, 2016, **28**, 5206–5213.

44. J.-Y. Seo, T. Matsui, J. Luo, J.-P. Correa-Baena, F. Giordano, M. Saliba, K. Schenk, A. Ummadisingu, K. Domanski, M. Hadadian, A. Hagfeldt, S. M. Zakeeruddin, U. Steiner, M. Grätzel and A. Abate, *Adv. Energy Mater.*, 2016, **6**, 1600767.
45. S. Sun, J. Song, R. Feng and Z. Shan, *Electrochim. Acta*, 2012, **69**, 51–55.
46. X. Huang, H. Guo, K. Wang and X. Liu, *Org. Electron.*, 2017, **41**, 42–48.
47. M. Shahiduzzaman, K. Yamamoto, Y. Furumoto, T. Kuwabara, K. Takahashi and T. Taima, *RSC Adv.*, 2015, **5**, 77495–77500.
48. M. Marszalek, F. D. Arendse, J.-D. Decoppet, S. S. Babkair, A. A. Ansari, S. S. Habib, M. Wang, S. M. Zakeeruddin and M. Grätzel, *Adv. Energy Mater.*, 2014, **4**, 1301235.
49. P. Wang, S. M. Zakeeruddin, J.-E. Moser and M. Grätzel, *J. Phys. Chem. B*, 2003, **107**, 13280–13285.
50. D. Kuang, S. Uchida, R. Humphry-Baker, S. M. Zakeeruddin and M. Grätzel, *Angew. Chem., Int. Ed. Engl.*, 2008, **47**, 1923–1927.
51. Y. Li, X. Li, X. Liu, L. Zhu, W. Zhang and J. Fang, *J. Phys. Chem. C*, 2016, **120**, 26244–26248.
52. G. D. Niu, X. D. Guo and L. D. Wang, *J. Mater. Chem. A*, 2015, **3**, 8970–8980.
53. Q. Wu, W. Zhou, Q. Liu, P. Zhou, T. Chen, Y. Lu, Q. Qiao and S. Yang, *ACS Appl. Mater. Interfaces*, 2016, **8**, 34464–34473.
54. S. M. Zakeeruddin and M. Grätzel, *Adv. Funct. Mater.*, 2009, **19**, 2187–2202.
55. D. B. Kuang, P. Wang, S. Ito, S. M. Zakeeruddin and M. Grätzel, *J. Am. Chem. Soc.*, 2006, **128**, 7732–7733.
56. W. Zhang, C. Yuan, J. Guo, L. Qiu and F. Yan, *ACS Appl. Mater. Interfaces*, 2014, **6**, 8723–8728.
57. D. Yang, X. Zhou, R. Yang, Z. Yang, W. Yu, X. Wang, C. Li, S. Liu and R. P. H. Chang, *Energy Environ. Sci.*, 2016, **9**, 3071–3078.
58. Y. Kaji, K. Ogawa, R. Eguchi, H. Goto, Y. Sugawara, T. Kambe, K. Akaike, S. Gohda, A. Fujiwara and Y. Kubozono, *Org. Electron.*, 2011, **12**, 2076–2083.
59. E. Schmidt, S. Shi, P. P. Ruden and C. D. Frisbie, *ACS Appl. Mater. Interfaces*, 2016, **8**, 14879–14884.
60. T. L. Atallah, M. V. Gustafsson, E. Schmidt, C. D. Frisbie and X. Y. Zhu, *J. Phys. Chem. Lett.*, 2015, **6**, 4840–4844.
61. K. H. Lee, M. S. Kang, S. Zhang, Y. Gu, T. P. Lodge and C. D. Frisbie, *Adv. Mater.*, 2012, **24**, 4457–4462.
62. T. Fujimoto, M. M. Matsushita and K. Awaga, *Appl. Phys. Lett.*, 2010, **97**, 123303.
63. S. Fabiano, S. Braun, M. Fahlman, X. Crispin and M. Berggren, *Adv. Funct. Mater.*, 2014, **24**, 695–700.
64. S. H. Kim, K. Hong, W. Xie, K. H. Lee, S. Zhang, T. P. Lodge and C. D. Frisbie, *Adv. Mater.*, 2013, **25**, 1822–1846.
65. M. J. Panzer and C. D. Frisbie, *Adv. Mater.*, 2008, **20**, 3177–3180.
66. J. H. Cho, J. Lee, Y. He, B. S. Kim, T. P. Lodge and C. D. Frisbie, *Adv. Mater.*, 2008, **20**, 686–690.

67. S. Ono, S. Seki, R. Hirahara, Y. Tominari and J. Takeya, *Appl. Phys. Lett.*, 2008, **92**, 103313.
68. S. Ono, N. Minder, Z. Chen, A. Facchetti and A. F. Morpurgo, *Appl. Phys. Lett.*, 2010, **97**, 143307.
69. J. Lee, L. G. Kaake, J. H. Cho, X. Y. Zhu, T. P. Lodge and C. D. Frisbie, *J. Phys. Chem. C*, 2009, **113**, 8972–8981.
70. Y. He, P. G. Boswell, P. Buhlmann and T. P. Lodge, *J. Phys. Chem. B*, 2007, **111**, 4645–4652.
71. S. A. Algarni, T. M. Althagafi, P. J. Smith and M. Grell, *Appl. Phys. Lett.*, 2014, **104**, 182107.
72. Y. Takeyama, S. Ono and Y. Matsumoto, *Appl. Phys. Lett.*, 2012, **101**, 083303.
73. L. Herlogsson, X. Crispin, S. Tierney and M. Berggren, *Adv. Mater.*, 2011, **23**, 4684–4689.
74. J. H. Cho, J. Lee, Y. Xia, B. Kim, Y. He, M. J. Renn, T. P. Lodge and C. Daniel Frisbie, *Nat. Mater.*, 2008, **7**, 900–906.
75. J. Ko, S. J. Lee, K. Kim, E. Lee, K.-H. Lim, J.-M. Myoung, J. Yoo and Y. S. Kim, *J. Mater. Chem. C*, 2015, **3**, 4239–4243.
76. S. Bubel, M. S. Menyo, T. E. Mates, J. H. Waite and M. L. Chabinyc, *Adv. Mater.*, 2015, **27**, 3331–3335.
77. J. H. Choi, W. Xie, Y. Gu, C. D. Frisbie and T. P. Lodge, *ACS Appl. Mater. Interfaces*, 2015, **7**, 7294–7302.
78. W. Shi, J. Ye, J. G. Checkelsky, C. Terakura and Y. Iwasa, *Adv. Funct. Mater.*, 2014, **24**, 2005–2012.
79. C. Pozo-Gonzalo, D. Mecerreyes, J. A. Pomposo, M. Salsamendi, R. Marcilla, H. Grande, R. Vergaz, D. Barrios and J. M. Sánchez-Pena, *Sol. Energy Mater. Sol. Cells*, 2008, **92**, 101–106.
80. Y. Pang, H. Xu, X. Li, H. Ding, Y. Cheng, G. Shi and L. Jin, *Electrochem. Commun.*, 2006, **8**, 1757–1763.
81. M. Deepa, A. Awadhia and S. Bhandari, *Phys. Chem. Chem. Phys.*, 2009, **11**, 5674–5685.
82. Y. Pang, X. Li, H. Ding, G. Shi and L. Jin, *Electrochim. Acta*, 2007, **52**, 6172–6177.
83. L. Ma, Y. Li, X. Yu, Q. Yang and C.-H. Noh, *Sol. Energy Mater. Sol. Cells*, 2009, **93**, 564–570.
84. A. S. Shaplov, D. O. Ponkratov, P. H. Aubert, E. I. Lozinskaya, C. Plesse, F. Vidal and Y. S. Vygodskii, *Chem. Commun.*, 2014, **50**, 3191–3193.
85. A. Šurca Vuk, V. Jovanovski, A. Pollet-Villard, I. Jerman and B. Orel, *Sol. Energy Mater. Sol. Cells*, 2008, **92**, 126–135.
86. D. Zhou, R. Zhou, C. Chen, W. A. Yee, J. Kong, G. Ding and X. Lu, *J. Phys. Chem. B*, 2013, **117**, 7783–7789.
87. W. Lu, A. G. Fadeev, B. H. Qi, E. Smela, B. R. Mattes, J. Ding, G. M. Spinks, J. Mazurkiewicz, D. Z. Zhou, G. G. Wallace, D. R. MacFarlane, S. A. Forsyth and M. Forsyth, *Science*, 2002, **297**, 983–987.
88. T. Sugino, K. Kiyohara, I. Takeuchi, K. Mukai and K. Asaka, *Sens. Actuators, B*, 2009, **141**, 179–186.

89. L. Lu and W. Chen, *Adv. Mater.*, 2010, **22**, 3745–3748.
90. I. Takeuchi, K. Asaka, K. Kiyohara, T. Sugino, N. Terasawa, K. Mukai and S. Shiraishi, *Carbon*, 2009, **47**, 1373–1380.
91. T. Fukushima, K. Asaka, A. Kosaka and T. Aida, *Angew. Chem., Int. Ed. Engl.*, 2005, **44**, 2410–2413.
92. M. Cho, H. Seo, J. Nam, H. Choi, J. Koo and Y. Lee, *Sens. Actuators, B*, 2007, **128**, 70–74.
93. K. Mukai, K. Asaka, K. Kiyohara, T. Sugino, I. Takeuchi, T. Fukushima and T. Aida, *Electrochim. Acta*, 2008, **53**, 5555–5562.
94. N. Terasawa, I. Takeuchi, H. Matsumoto, K. Mukai and K. Asaka, *Sens. Actuators, B*, 2011, **156**, 539–545.
95. K.-S. Kwon and T. N. Ng, *Org. Electron.*, 2014, **15**, 294–298.
96. F. Vidal, C. Plesse, D. Teyssié and C. Chevrot, *Synth. Met.*, 2004, **142**, 287–291.
97. C. W. Tang and S. A. Vanslyke, *Appl. Phys. Lett.*, 1987, **51**, 913–915.
98. Q. Pei, G. Yu, C. Zhang, Y. Yang and A. J. Heeger, *Science*, 1995, **270**, 719.
99. Y.-J. Pu, T. Chiba, K. Ideta, S. Takahashi, N. Aizawa, T. Hikichi and J. Kido, *Adv. Mater.*, 2015, **27**, 1327–1332.
100. B. R. Lee, E. D. Jung, J. S. Park, Y. S. Nam, S. H. Min, B. S. Kim, K. M. Lee, J. R. Jeong, R. H. Friend, J. S. Kim, S. O. Kim and M. H. Song, *Nat. Commun.*, 2014, **5**, 4840.
101. Y.-J. Pu, G. Nakata, F. Satoh, H. Sasabe, D. Yokoyama and J. Kido, *Adv. Mater.*, 2012, **24**, 1765–1770.
102. S. Ohisa, Y.-J. Pu, N. L. Yamada, G. Matsuba and J. Kido, *Nanoscale*, 2017, **9**, 25–30.
103. K. Udagawa, H. Sasabe, C. Cai and J. Kido, *Adv. Mater.*, 2014, **26**, 5062–5066.
104. S. Ohisa, S. Kagami, Y.-J. Pu, T. Chiba and J. Kido, *ACS Appl. Mater. Interfaces*, 2016, **8**, 20946–20954.
105. T. Chiba, Y.-J. Pu and J. Kido, *Adv. Mater.*, 2015, **27**, 4681–4687.
106. Q. S. Zhang, B. Li, S. P. Huang, H. Nomura, H. Tanaka and C. Adachi, *Nat. Photonics*, 2014, **8**, 326–332.
107. Y. Seino, S. Inomata, H. Sasabe, Y.-J. Pu and J. Kido, *Adv. Mater.*, 2016, **28**, 2638–2643.
108. H. Uoyama, K. Goushi, K. Shizu, H. Nomura and C. Adachi, *Nature*, 2012, **492**, 234–238.
109. K. Goushi, K. Yoshida, K. Sato and C. Adachi, *Nat. Photonics*, 2012, **6**, 253–258.
110. T. A. Lin, T. Chatterjee, W. L. Tsai, W. K. Lee, M. J. Wu, M. Jiao, K. C. Pan, C. L. Yi, C. L. Chung, K. T. Wong and C. C. Wu, *Adv. Mater.*, 2016, **28**, 6976–6983.
111. H. Sasabe and J. Kido, *J. Mater. Chem. C*, 2013, **1**, 1699–1707.
112. K. Kato, T. Iwasaki and T. Tsujimura, *J. Photopolym. Sci. Technol.*, 2015, **28**, 335–340.
113. T. Chiba, Y.-J. Pu, R. Miyazaki, K. Nakayama, H. Sasabe and J. Kido, *Org. Electron.*, 2011, **12**, 710–715.

114. S. Ohisa, G. Matsuba, N. L. Yamada, Y.-J. Pu, H. Sasabe and J. Kido, *Adv. Mater. Interfaces*, 2014, **1**, 1400097.
115. N. Aizawa, Y.-J. Pu, M. Watanabe, T. Chiba, K. Ideta, N. Toyota, M. Igarashi, Y. Suzuri, H. Sasabe and J. Kido, *Nat. Commun.*, 2014, **5**, 5756.
116. T. Chiba, Y.-J. Pu and J. Kido, *J. Mater. Chem. C*, 2015, **3**, 11567–11576.
117. H. Ishii, K. Sugiyama, E. Ito and K. Seki, *Adv. Mater.*, 1999, **11**, 972.
118. N. Siraj, F. Hasan, S. Das, L. W. Kiruri, K. E. Steege Gall, G. A. Baker and I. M. Warner, *J. Phys. Chem. C*, 2014, **118**, 2312–2320.
119. S. van Reenen, R. A. J. Janssen and M. Kemerink, *Adv. Funct. Mater.*, 2012, **22**, 4547–4556.
120. J. Mindemark, S. Tang, J. Wang, N. Kaihovirta, D. Brandell and L. Edman, *Chem. Mater.*, 2016, **28**, 2618–2623.
121. S. Tang, J. Mindemark, C. M. G. Araujo, D. Brandell and L. Edman, *Chem. Mater.*, 2014, **26**, 5083–5088.
122. Y. Shao, G. C. Bazan and A. J. Heeger, *Adv. Mater.*, 2007, **19**, 365–370.
123. R. D. Costa, E. Orti, H. J. Bolink, F. Monti, G. Accorsi and N. Armaroli, *Angew. Chem., Int. Ed. Engl.*, 2012, **51**, 8178–8211.
124. K. J. Chee, V. Kumar, C. V. Nguyen, J. Wang and P. S. Lee, *J. Phys. Chem. C*, 2016, **120**, 11324–11330.
125. C. D. Ertl, L. Gil-Escrig, J. Cerda, A. Pertegas, H. J. Bolink, J. M. Junquera-Hernandez, A. Prescimone, M. Neuburger, E. C. Constable, E. Orti and C. E. Housecroft, *Dalton Trans.*, 2016, **45**, 11668–11681.
126. T. Hu, L. He, L. Duan and Y. Qiu, *J. Mater. Chem.*, 2012, **22**, 4206–4215.
127. A. K. Pal, D. B. Cordes, A. M. Slawin, C. Momblona, E. Orti, I. D. Samuel, H. J. Bolink and E. Zysman-Colman, *Inorg. Chem.*, 2016, **55**, 10361–10376.
128. H. J. Lee, B. R. Lee, J. S. Park, S. O. Kim, J. Y. Kim and M. H. Song, *Appl. Phys. Lett.*, 2011, **98**, 253309.
129. R. Marcilla, D. Mecerreyes, G. Winroth, S. Brovelli, M. D. R. Yebra and F. Cacialli, *Appl. Phys. Lett.*, 2010, **96**, 043308.
130. J. H. Shin, S. Xiao, A. Fransson and L. Edman, *Appl. Phys. Lett.*, 2005, **87**, 043506.
131. H. C. Moon, T. P. Lodge and C. D. Frisbie, *J. Am. Chem. Soc.*, 2014, **136**, 3705–3712.
132. D. Tordera, S. Meier, M. Lenes, R. D. Costa, E. Orti, W. Sarfert and H. J. Bolink, *Adv. Mater.*, 2012, **24**, 897–900.
133. C. W. Tang, *Appl. Phys. Lett.*, 1986, **48**, 183–185.
134. K. Cnops, B. P. Rand, D. Cheyns, B. Verreet, M. A. Empl and P. Heremans, *Nat. Commun.*, 2014, **5**, 3406.
135. C. Gu, Y. Chen, Z. Zhang, S. Xue, S. Sun, C. Zhong, H. Zhang, Y. Lv, F. Li, F. Huang and Y. Ma, *Adv. Energy Mater.*, 2014, **4**, 1301771.
136. M. C. Scharber and N. S. Sariciftci, *Prog. Polym. Sci.*, 2013, **38**, 1929–1940.
137. V. Vohra, K. Kawashima, T. Kakara, T. Koganezawa, I. Osaka, K. Takimiya and H. Murata, *Nat. Photonics*, 2015, **9**, 403–408.
138. Y. Zang, C. Z. Li, C. C. Chueh, S. T. Williams, W. Jiang, Z. H. Wang, J. S. Yu and A. K. Jen, *Adv. Mater.*, 2014, **26**, 5708–5714.

139. L. Huo, T. Liu, X. Sun, Y. Cai, A. J. Heeger and Y. Sun, *Adv. Mater.*, 2015, **27**, 2938–2944.
140. B. Kan, Q. Zhang, M. Li, X. Wan, W. Ni, G. Long, Y. Wang, X. Yang, H. Feng and Y. Chen, *J. Am. Chem. Soc.*, 2014, **136**, 15529–15532.
141. B. O'Regan and M. Grätzel, *Nature*, 1991, **353**, 737–740.
142. P. Wang, S. M. Zakeeruddin, I. Exnar and M. Grätzel, *Chem. Commun.*, 2002, 2972–2973.
143. M. K. Wang, C. Gratzel, S. M. Zakeeruddin and M. Grätzel, *Energy Environ. Sci.*, 2012, **5**, 9394–9405.
144. J. Gong, J. Liang and K. Sumathy, *Renewable Sustainable Energy Rev.*, 2012, **16**, 5848–5860.
145. P. Wang, S. M. Zakeeruddin, R. Humphry-Baker and M. Grätzel, *Chem. Mater.*, 2004, **16**, 2694–2696.
146. X. Chen, J. Zhao, J. Zhang, L. Qiu, D. Xu, H. Zhang, X. Han, B. Sun, G. Fu, Y. Zhang and F. Yan, *J. Mater. Chem.*, 2012, **22**, 18018–18024.
147. M. Takahashi, K. Sato, S. Sakurai and K. Kobayashi, *Electrochim. Acta*, 2016, **219**, 766–772.
148. V. Saavedra Becerril, D. Franchi and M. Abrahamsson, *J. Phys. Chem. C*, 2016, **120**, 20016–20023.
149. Y. F. Lin, C. T. Li, C. P. Lee, Y. A. Leu, Y. Ezhumalai, R. Vittal, M. C. Chen, J. J. Lin and K. C. Ho, *ACS Appl. Mater. Interfaces*, 2016, **8**, 15267–15278.
150. W. Kubo, S. Kambe, S. Nakade, T. Kitamura, K. Hanabusa, Y. Wada and S. Yanagida, *J. Phys. Chem. B*, 2003, **107**, 4374–4381.
151. T. He, Y. F. Wang and J. H. Zeng, *ACS Appl. Mater. Interfaces*, 2015, **7**, 21381–21390.
152. A. Kojima, K. Teshima, Y. Shirai and T. Miyasaka, *J. Am. Chem. Soc.*, 2009, **131**, 6050–6051.
153. M. M. Lee, J. Teuscher, T. Miyasaka, T. N. Murakami and H. J. Snaith, *Science*, 2012, **338**, 643–647.
154. D. Q. Bi, W. Tress, M. I. Dar, P. Gao, J. S. Luo, C. Renevier, K. Schenk, A. Abate, F. Giordano, J. P. C. Baena, J. D. Decoppet, S. M. Zakeeruddin, M. K. Nazeeruddin, M. Grätzel and A. Hagfeldt, *Sci. Adv.*, 2016, **2**, e1501170.
155. W. S. Yang, J. H. Noh, N. J. Jeon, Y. C. Kim, S. Ryu, J. Seo and S. I. Seok, *Science*, 2015, **348**, 1234–1237.
156. N. Ahn, D. Y. Son, I. H. Jang, S. M. Kang, M. Choi and N. G. Park, *J. Am. Chem. Soc.*, 2015, **137**, 8696–8699.
157. M. Saliba, T. Matsui, K. Domanski, J. Y. Seo, A. Ummadisingu, S. M. Zakeeruddin, J. P. Correa-Baena, W. R. Tress, A. Abate, A. Hagfeldt and M. Grätzel, *Science*, 2016, **354**, 206–209.
158. J. W. Jung, C. C. Chueh and A. K. Jen, *Adv. Mater.*, 2015, **27**, 7874–7880.
159. X. Li, D. Q. Bi, C. Y. Yi, J. D. Decoppet, J. S. Luo, S. M. Zakeeruddin, A. Hagfeldt and M. Grätzel, *Science*, 2016, **353**, 58–62.
160. M. Muccini, *Nat. Mater.*, 2006, **5**, 605–613.
161. H. Sirringhaus, *Adv. Mater.*, 2005, **17**, 2411–2425.
162. D. Braga and G. Horowitz, *Adv. Mater.*, 2009, **21**, 1473–1486.

163. C. R. Newman, C. D. Frisbie, D. A. da Silva, J. L. Bredas, P. C. Ewbank and K. R. Mann, *Chem. Mater.*, 2004, **16**, 4436–4451.
164. G. Horowitz, *Adv. Mater.*, 1998, **10**, 365–377.
165. Y. Xia, J. H. Cho, J. Lee, P. P. Ruden and C. D. Frisbie, *Adv. Mater.*, 2009, **21**, 2174–2179.
166. T. Uemura, M. Yamagishi, S. Ono and J. Takeya, *Appl. Phys. Lett.*, 2009, **95**, 103301.
167. T. Fukushima, A. Kosaka, Y. Ishimura, T. Yamamoto, T. Takigawa, N. Ishii and T. Aida, *Science*, 2003, **300**, 2072–2074.
168. M. Matsumoto, Y. Saito, C. Park, T. Fukushima and T. Aida, *Nat. Chem.*, 2015, 7, 730–736.

CHAPTER 9

Applications of Ionic Liquid Materials in Microfluidic Devices

TUGÇE AKYAZI[†a], JANIRE SAEZ[a], ALEXANDRU TUDOR[†b], COLM DELANEY[b], WAYNE FRANCIS[b], DERMOT DIAMOND[b], LOURDES BASABE-DESMONTS[c,d], LARISA FLOREA*[b] AND FERNANDO BENITO-LOPEZ*[a]

[a]Analytical Microsystems & Materials for Lab-on-a-Chip Group (AMMa-LOAC), Microfluidics Cluster UPV/EHU, Analytical Chemistry Department, University of the Basque Country UPV/EHU, Vitoria-Gasteiz, Spain; [b]INSIGHT Centre for Data Analytics, National Centre for Sensor Research, Dublin City University, Dublin 9, Ireland; [c]BIOMICs Microfluidics, Microfluidics Cluster UPV/EHU, Lascaray Ikergunea Research Center, University of the Basque Country UPV/EHU, Vitoria-Gasteiz, Spain; [d]Ikerbasque, Basque Foundation for Science, 48011 Bilbao, Spain
*E-mail: larisa.florea@dcu.ie, fernando.benito@ehu.eus

9.1 Introduction

Ionic liquids (ILs) are drawing increasing interest both in academia and in industry as confirmed by the growing number of publications and patents in the area.[1,2] Ionic liquids are salts, completely constituted of ions with melting

[†]Both authors contributed equally to this book chapter.

Smart Materials No. 28
Ionic Liquid Devices
Edited by Ali Eftekhari

Published by the Royal Society of Chemistry, www.rsc.org

temperatures below 100 °C, which is a result of their low-charge density and low symmetry ions.[1,3–5] ILs are categorised as "green" solvents since they are, potentially, green alternatives to volatile organic compounds due mainly to their two outstanding properties: negligible volatility and conventional non-flammability.[4–7] Their unique properties are not limited to non-volatility and non-flammability; they have an excellent solvation ability for organic, inorganic and organometallic compounds with improved selectivity, high thermal stability (decomposition temperatures around 300–500 °C), high chemical stabilities (extremely redox robust), and lastly, high ionic conductivity, all of which highly extends the variety of their applications.[1–3,8–15] Nevertheless, other properties, such as biodegradability and toxicity have not yet been successfully undertaken, and they should be considered in IL applications.[7]

One of the main advantages concerning the applicability of ILs is the ability to tailor their physical and chemical properties (such as their polarities and affinities, their thermo-physical properties, biodegradation ability or toxicological features, as well as their hydrophobicity and solution behaviours) by appropriate manipulation of the cation/anion chemical structure.[16–21] This feature gives them the name of "designer solvent" and favours their use, particularly in the extraction, separation and analysis of value-added compounds from biomass.[1–3,21,22] Moreover, these tuneable properties are enabling rapid advances in devices and processes for the production, storage and efficient use of energy.[1,23]

In many applications, the immobilisation of ILs in a solid or semisolid substrate, while keeping their specific properties, is a main requirement in order to generate useful devices. This is possible by the introduction of a new class of hybrid materials, ionogels. Ionogels preserve the important properties of the ILs (liquid-like dynamics and ion mobility) in a solid or a gel like structure, enabling easy shaping, manipulation and integration, increasing remarkably the potential application of ILs in fundamental areas such as energy, environment and analysis.[24,25]

In recent years, researchers decided to make use of the exceptional features of ILs and ionogels in the microfluidics area and many papers that include the incorporation of ILs materials in microfluidic systems have been published. "Lab-on-a-chip" (LOC)/"micro-total analysis systems" (μTAS) or microfluidic (continuous flow, microarray and droplet-based) analysis systems enable the manipulation of fluids at small scales (from a few micrometres up to a millimetre) and small volumes (nL to μL). Microfluidic devices run a series of fluidic unit operations on a platform that is designed with a well-defined fabrication technology and provides a consistent approach for miniaturisation, integration, automation and parallelisation of bio-chemical processes.[26,27] They have the greatest capability of integration of multiple functional elements into a small structure to produce absolute sample-in/answer-out systems.[26] The incorporation of functionalities such as liquid handling, temperature control and detection components for sensing allows fast analysis and improves selectivity compared to conventional devices.[26,27] Nevertheless, the development of fully integrated microfluidic devices is still facing some compelling obstacles, including fluidic control, miniaturisation and high costs.[26,28] Considering the critical requirement of fluid control and fluid transport processes and high detection performance within these platforms, microfluidic devices have been adopted with a wide range of passively or actively controlled, high performance components, such as microfluidic separators, actuators (valves and pumps), reactors, sample/biomolecule storage, sensing elements, which although they are sophisticated and achieve their tasks efficiently are also very costly, greatly increasing the price and so decreasing the final market possibilities of microfluidic devices.

Ionic liquids have excellent properties that could be used to improve microfluidic devices, since they are low-cost, easily obtainable materials. Their features make them very good and multifunctional candidates for improving the capabilities of microfluidic devices by building miniaturised/passive microfluidic elements for fluid control, sensing, sample storage, microfluidic separation, microreactors (nanoparticle synthesis), power generators, temperature controllers, electrowetting of surface, *etc.*, while decreasing the manufacturing costs compared with conventional devices and widening the market possibilities. In this chapter we will review the fundamental applications of ionic liquids within microfluidic devices. We will summarise ILs integration in microfluidic devices (Table 9.1) as well as their reported applications in lab-on-a chip devices and systems.

Table 9.1 Ionic liquid names and abbreviations used in this chapter.

Ionic liquid name	Abbreviation
Triisobutyl(methyl) phosphonium tosylate	$[P_{1,4,4,4}]$[Tos]
Tetrabutylphosphonium dicyanamide	$[P_{4,4,4,4}]$[DCA]
Trihexyltetradecyl-phosphonium dicyanamide	$[P_{6,6,6,14}]$[DCA]
Trihexyltetradecyl-phosphonium bis(trifluoromethanesulfonyl) imide	$[P_{6,6,6,14}][Ntf_2]$
Trihexyltetradecyl-phosphonium dodecylbenzenesulfonate	$[P_{6,6,6,14}]$[DBSA]
Trihexyltetradecyl-phosphonium chloride	$[P_{6,6,6,14}]$[Cl]
1-Methyl-3-octylimidazolium tetrafluoroborate	$[OMIM][BF_4]$
1-Ethyl-3-methyl imidazolium ethyl sulfate	$[EMIM][EtSO_4]$
1-Ethyl-3-methyl imidazolium tetrafluoroborate	$[EMIM][BF_4]$
1-Ethyl-3-methylimidazolium dicyanamide	[EMIM][DCA]
1-Butyl-3-methylimidazolium tetrafluoroborate	$[BMIM][BF_4]$
1-Ethyl-3-methylimidazolium methyl sulfate	$[EMIM][MeSO_4]$
1-Butyl-3-methylimidazolium hexafluorophosphate	$[BMIM][PF_6]$
1-Butyl-3-methylimidazolium bis(trifluoromethanesulfonyl) imide	$[BMIM][NTf_2]$
1-Butyl-3-methylimidazolium hydrogen sulfate	$[BMIM][HSO_4]$
1-Hexyl-3-methylimidazolium bis(trifluoromethanesulfonyl) imide	$[HMIM][NTf_2]$
1-Buty1-butyl-4-methylpyridinum tetrafluoroborate	$[BMPy][BF_4]$
1-Butyl-3-methylimidazolium dodecane-sulfonate	[BMIM][DoS]

9.2 Ionic Liquids Materials as Actuators

9.2.1 Microvalves

In 2007 Sugiura *et al.*[29] proposed photo-responsive microfluidic hydrogel valves by copolymerising *N*-isopropylacrylamide (NiPAAm) with a photochromic acrylic benzospiropyran ester (BSP) moiety. Prior to photopolymerisation, the cocktail mixture of the monomer and the photochromic unit was dissolved in 1-butanol, together with a UV initiator and crosslinker. The photoresponse mechanism of these hydrogels comes as a result of the benzospiropyran moiety present in the copolymer matrix. When the copolymer is kept in the dark and exposed to an aqueous solution of HCl, the benzospiropyran moiety protonates, changing its conformation to the protonated merocyanine form. When in this conformation, the presence of charges on its backbone contributes to it being more hydrophilic than the closed benzospiropyran conformation. Thus, it absorbs more water and, by irradiating it with white light, it can be reversed to the more hydrophobic benzospiropyran form, together with the release of water. The authors have shown that valves obtained from this material can perform, if preconditioned overnight in acidic conditions, to open and stop the flow inside a poly(dimethyl siloxane) (PDMS) microfluidic device.

Building on this work, Benito-Lopez *et al.*[30] synthesised photo-responsive ionogel valves in microfluidic channels by copolymerising NiPAAm and BSP using a mixture of 1-butanol and phosphonium ionic liquids as the solvent. Four different ILs were used, namely triisobutyl(methyl)phosphonium tosylate ($[P_{1,4,4,4}][Tos]$), trihexyltetradecyl-phosphonium dicyanamide ($[P_{6,6,6,14}][DCA]$), trihexyltetradecyl-phosphonium bis(trifluoromethanesulfonyl)imide ($[P_{6,6,6,14}][Ntf_2]$), and trihexyltetradecyl-phosphonium dodecylbenzenesulfonate ($[P_{6,6,6,14}][DBSA]$). The resulting crosslinked polymers were left to swell in 1 mM HCl solution for 2 h to reach their maximum swelling capabilities. Following this, they were irradiated with a white light LED until they shrank to their minimum size. In all cases, the time needed for this operation was 100 s. The hydrogel that exhibited maximum shrinking was the control hydrogel, followed by the $[P_{6,6,6,14}][DCA]$, $[P_{1,4,4,4}][Tos]$, $[P_{6,6,6,14}][DBSA]$, and $[P_{6,6,6,14}][Ntf_2]$, respectively. Following this, the monomer mixtures were polymerised in microfluidic devices that featured inlet ports connected to five microchannels which converged into a single microchannel with an outlet port. In each of the five microchannels there was a circular reservoir with a diameter of 500 μm and a height of 175 μm where the monomer mixtures were photopolymerised. The microchannels were 500 μm wide and 50 μm high. After photopolymerisation, the microfluidic device was flushed with 1 mM HCl to swell the hydrogels. After the hydrogels were swollen and were blocking the channels, a vacuum was applied to the outlet to facilitate the flow and water containing different coloured dyes was placed at each of the inlets. The valves were actuated using a white light LED with a power output of 1 mW cm^{-2}. Actuation times were 2 s for the control hydrogel, 4 s for the $[P_{6,6,6,14}][DCA]$, 18 s for the $[P_{1,4,4,4}][Tos]$, 44 s for the $[P_{6,6,6,14}][DBSA]$, and 49 s for the $[P_{6,6,6,14}][Ntf_2]$ ionogel (Figure 9.1). This example shows that through

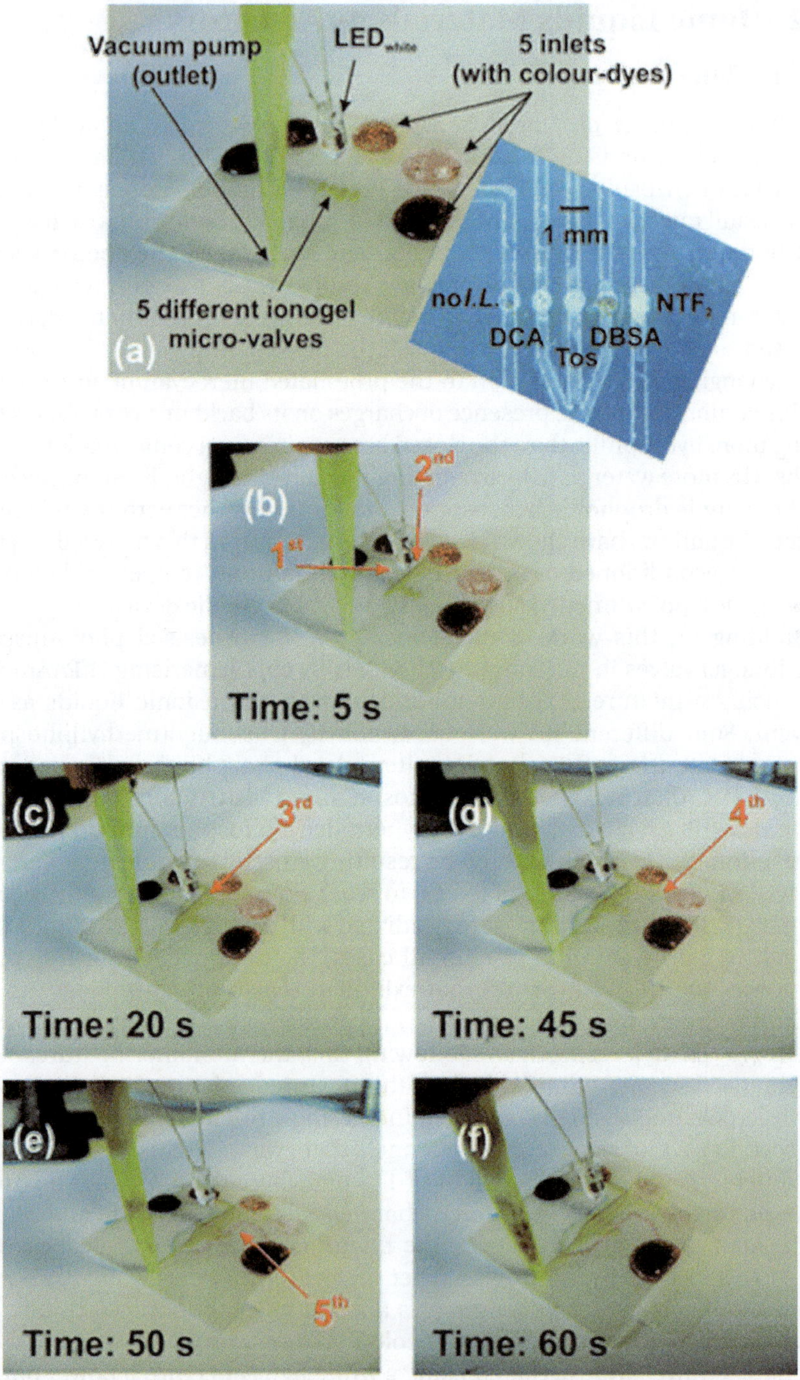

Figure 9.1 Image of the microfluidic manifold showing the performance of the ionogel microfluidic valves: (a) microvalves closed under the applied vacuum. White light is applied for the time specified in each picture.

the incorporation of different ionic liquids of varying hydrophilic/hydrophobic character, inside ionogel matrices, actuation times can be modulated on demand. The time for the hydrogels to revert back to their original size is approximately 30 min, which led the researchers to conclude that this type of valve would be best suited for single-use devices.

A more comprehensive study of such photo-responsive ionogel materials was later performed by Czugala *et al.*[31] who studied the photo-responsive behaviour of pNiPAAm-*co*-BSP ionogels made using phosphonium ILs as solvents. Three different phosphonium ILs were chosen for this study: trihexyltetradecyl-phosphonium chloride $[P_{6,6,6,14}][Cl]$, $[P_{6,6,6,14}][DCA]$ and $[P_{6,6,6,14}][Ntf_2]$, and ionogel actuators were photopolymerised in four different shapes of varying surface area to volume ratios (SA/V): rings, 250 μm discs, 500 μm discs, and lines, respectively. Their change in size was determined by the change in height, as measured using digital microscopy. By measuring the amount of hydration for each ionogel shape, it was determined that the $[P_{6,6,6,14}][Ntf_2]$ swelled the most, at a value of 109–180% of the initial gel height (immediately after photopolymerisation), followed by $[P_{6,6,6,14}][DCA]$ ionogels (40–58%), and lastly the $[P_{6,6,6,14}]Cl$ ionogels (20–27%). Following this, the shrinking behaviour of the ionogels was analysed and it was determined that after 30 min of white light irradiation the $[P_{6,6,6,14}][Ntf_2]$ ionogels shrunk by 108%, 15% of their initial height, the $[P_{6,6,6,14}][DCA]$ ionogels shrunk by 42%, reaching also 15% of their initial height, while the $[P_{6,6,6,14}]Cl$ ionogels shrunk only by 16%, reaching 4% of their initial height (Figure 9.2). The kinetics of the swelling and shrinking behaviours were also determined and in each case the $[P_{6,6,6,14}][Ntf_2]$ ionogels exhibited the highest rates of swelling and shrinking at $5.3 \pm 0.1 \times 10^{-2}$ s and $29 \pm 4 \times 10^{-2}$ s, followed by $[P_{6,6,6,14}][DCA]$ at $4.5 \pm 0.3 \times 10^{-2}$ s and $8.3 \pm 0.9 \times 10^{-2}$ s, and $[P_{6,6,6,14}]Cl$ at $3.9 \pm 0.2 \times 10^{-2}$ s and $9 \pm 2 \times 10^{-2}$ s. The high swelling and shrinking values attributed to the $[P_{6,6,6,14}][Ntf_2]$ are believed to stem from the fact that this ionic liquid possesses a highly delocalised charge on the S–N–S backbone of the Ntf_2^- anion, which makes it interact less with the charged moieties of the polymer. This leads to more freedom for the polymer backbone to interact with the hydration medium. In the case of the $[P_{6,6,6,14}][Cl]$, this effect is inhibited by the localised charge on the chloride ion, thus associating more strongly to the polymer backbone. The $[P_{6,6,6,14}][DCA]$ ionogels would have an intermediate behaviour between these two states. Based on these results, the $[P_{6,6,6,14}][Ntf_2]$ ionogels were incorporated in a glass-poly(dimethylsiloxane) microfluidic device as light actuated valves. Using a fibre optic to irradiate with white light, the ionogel valve opened after 180 s, allowing liquid to pass through the microfluidic channel.[31]

(b) Hydrogel (no IL present) valve is first to actuate followed by ionogels incorporating $[DCA]^-$ (c), $[Tos]^-$ (d), $[DBSA]^-$ (e), $[NTf2]^-$ (f). Numbers and arrows indicate when the channel is filled with the dye due to microvalve actuation. Reproduced from ref. 30 with permission from the Royal Society of Chemistry.

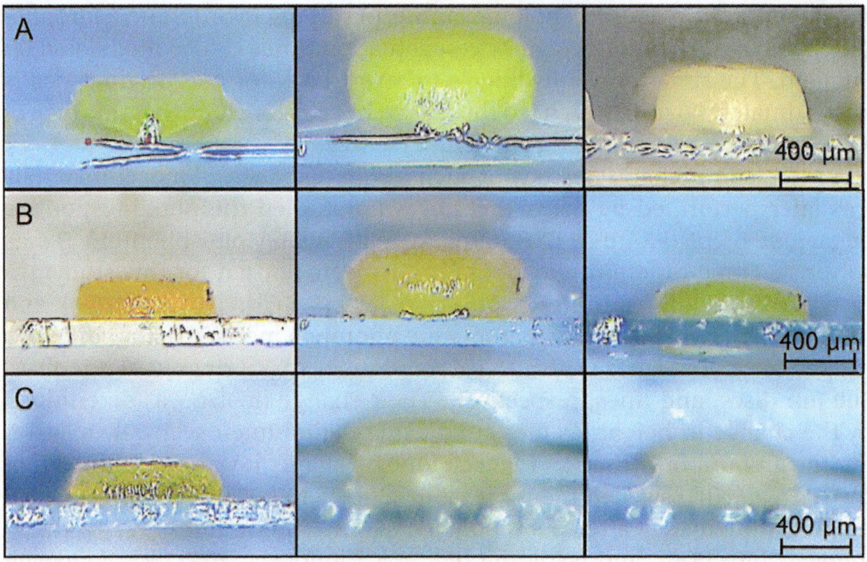

Figure 9.2 Microscope images of ionogel discs made of: (a) [$P_{6,6,6,14}$][NTf_2], (b) [$P_{6,6,6,14}$][DCA] and (c) [$P_{6,6,6,14}$][Cl] after photopolymerisation (left); swelling in 1 mM HCl solution for 2 h (middle) and shrinking upon white light irradiation (right). Reprinted from *Sensors and Actuators B: Chemical*, **194**, M. Czugala, C. O'Connell, C. Blin, P. Fischer, K. J. Fraser, F. Benito-Lopez, D. Diamond, Swelling and Shrinking Behaviour of Photoresponsive Phosphonium Based Ionogel Microstructures, 105–113, Copyright (2014), with permission from Elsevier.

An application of these materials was demonstrated by Czugala *et al.*[32] by using the [$P_{6,6,6,14}$][DCA] ionogels as photo-responsive valves in a microfluidic analysis platform for the detection of nitrite anions in water. The nitrite assay was performed using the Griess reagent and the change in colour was determined using a paired emitter detector diode (PEDD) arrangement integrated in the microfluidic holder. The emitter diode has a wavelength of 540 nm, while the detector diode has its maximum absorption at 660 nm. Part of the light produced by the emitter diode is absorbed by the Griess-nitrite complex, which has its maximum absorption at 547 nm, while the rest reaches the detector diode and is transformed into a photocurrent. The amount of photocurrent generated is proportional to the concentration of nitrite within the sample. By connecting the PEDD setup to a microcontroller fitted with a wireless radio antenna, the data was sent to a PC, where it was stored and analysed. The monomer mixture was photopolymerised using UV light in a circular reservoir with a radius of 500 µm and a height of 225 µm. The reservoir sat at the junction of a Y-shaped microchannel that separated the sample from the Griess reagent, Figure 9.3(a). To operate the valve, the microchannel was filled with 1 mM HCl and left to swell for 2 h in the dark. The opening of

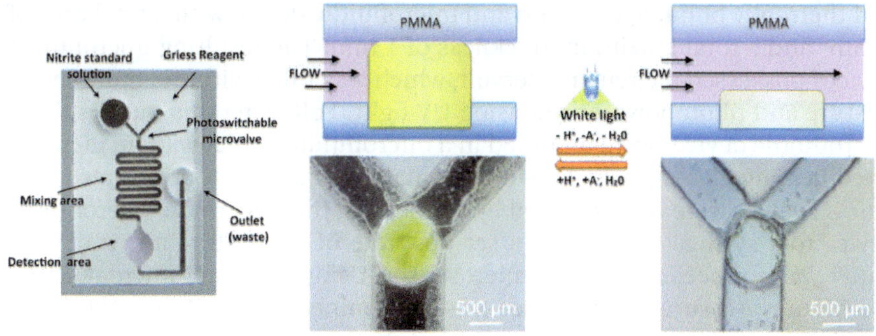

Figure 9.3 Picture of the microfluidic device fabricated in PMMA : PSA polymer by CO_2 laser ablation (a). (b) Schematic (top) and images (bottom) of the photoresponsive microvalve in closed (left) and opened (right) state. Reprinted from *Talanta*, **116**, M. Czugala, C. Fay, N. E. O'Connor, B. Corcoran, F. Benito-Lopez, D. Diamond, Portable Integrated Microfluidic Analytical Platform for the Monitoring and Detection of Nitrite, 997–1004, Copyright (2013), with permission from Elsevier.

the valve was performed with irradiation from a white light LED that was also mounted on the microfluidic device holder. By irradiating with a power of 1 mW cm^{-2}, the ionogel valve opened after 30 ± 5 s (number of measurements, n = 3) (Figure 9.3(b)) and was operated at a pressure of 25 mbar; pressures higher than 31 ± 4 mbar (number of measurements, n = 3) deformed the materials and the valves failed. After the ionogel shrunk, the reagent mixture was pumped through the microchannel where it mixed and moved to the detection area for analysis. The calibration curve was made using concentrations of nitrite from 0.2 to 1.2 mg L^{-1}, in 0.2 mg L^{-1} steps and performed in triplicate. This yielded a R^2 value of 0.98, a level of detection (LOD) of 34.0 ± 0.1 μg L^{-1} and a level of quantification (LOQ) of 115 ± 3 μg L^{-1}, compared to a R^2 value of 0.99, LOD value of 1.50 ± 0.02 μg L^{-1}, and a LOQ of 14.8 ± 0.2 μg L^{-1} for a UV-Vis spectrophotometer. The results obtained for the microfluidic platform are lower than the detection limits set by the World Health Organisation. Following this, freshwater samples from the Tolka River in Dublin, Ireland were analysed with both the portable platform and the UV-Vis spectrometer. The close nature of both sets of results proved the suitability of such a microfluidic device and detector for the accurate monitoring of nitrates in real-life samples.

Thermo-responsive actuation was also proposed for microvalves fabricated using poly(N-isopropylacrylamide) polymer (pNiPAAm) gels.[33] To synthesise these materials, all the components for the monomer mixture were dissolved in 1-ethyl-3-methylimidazolium ethyl sulfate [EMIM][$EtSO_4$]. The addition of this IL, as demonstrated previously by Gallagher *et al.*,[34] improves the swelling and shrinking capabilities of the pNiPAAm materials, together with lowering the LCST. The microfluidic devices in this study were fabricated using a cutting plotter to cut 100 μm thick layers of cyclic olefin polymer (COP) and thermally bonded, to obtain and microfluidic device with a total area of 1 mm^2 and a total maximum thickness of 1 mm. The resulting microfluidic devices possessed a circular reservoir, which was filled with the monomeric mixture and photopolymerised using UV light. Following this, the resulting microfluidic devices were mounted in a microfluidic holder with an incorporated heating element which could thermally actuate the ionogel valve. The ionogel valves were characterised by connecting the inlet of the microfluidic device to a syringe pump with a set flow rate of 1000 nL min^{-1}, while the outlet was connected to a flow microsensor. Using this experimental setup, the failure pressure of the valves was determined to be 1100 ± 100 mbar (n = 5) and they successfully operated at 200 mbar after being exposed to pressures higher than 1100 mbar. Setting the heating element of the microfluidic holder at a temperature higher than 50 °C, the valves opened after 4 ± 1 s (n = 5), after the temperature of the ionogel passed its LCST. Recovery was achieved in 32 ± 2 s (n = 5), after the temperature dropped below the LCST of the ionogel valves (Figure 9.4). Furthermore, after ten repetitions, there was no discernible drop in valve performance, showing the potential of these materials as cost-effective reversible valves.

Electro-actuation of IL based microvalves has also been studied. Ghamsari *et al.*[35] demonstrated the use of Bucky gels based on the 1-ethyl-3-methylimidazolium tetrafluoroborate ionic liquid ([EMIM][BF_4]), poly(vinylidene fluoride) (PVDF), and single-walled carbon nanotubes (SWCNTs), respectively, as low-voltage microvalve actuators. Bucky gels are gel-like mixtures of carbon nanotubes and ILs which benefit from both the high electrical conductivity

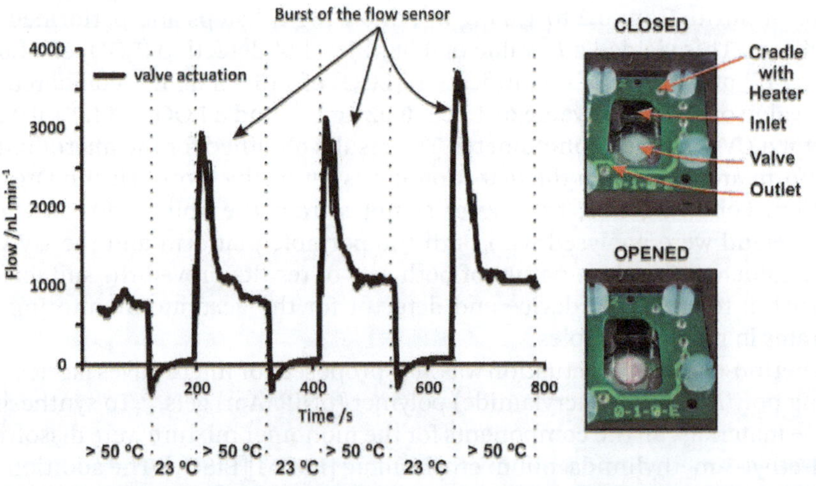

Figure 9.4 Flow profile during three full actuation cycles (left) and photo of the chip containing the thermo-actuated valve and the chip holder with integrated heaters at the bottom (right). High flow spikes are due to the stabilisation of the microflow sensor after opening of the valve. Reproduced from ref. 33 with permission from the Royal Society of Chemistry.

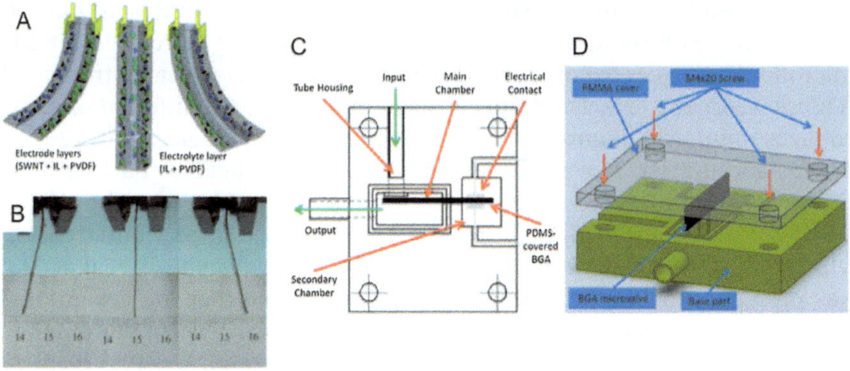

Figure 9.5 (a) BGA bending motion as a result of ion transfer between layers; direction of bending can be reversed by changing the polarity of the applied potential; (b) BGA strip bended (10 V, 0.1 Hz). (c) Base part and (d) 3D view of the flow regulator assembly. Reprinted with permission from Application of Bucky Gel in Fabrication of a Low-Voltage Rapid Microvalve for Flow Regulation, A. K. Ghamsari, E. Zegeye, Y. Jin, E. Woldesenbet, *ACS Appl. Mater. Interfaces*, 2013, 5, 5408–5412. Copyright (2013) American Chemical Society.

associated with carbon nanotubes and the properties of ILs, such as high temperature and electrochemical stability. Bucky gel actuators (BGAs) are composite materials that feature a polymer electrolyte core, in this case an ionogel of [EMIM][BF_4] and PVDF, inserted between two layers of electrodes made out of Bucky gels. Applying voltage to this composite will make it bend in the direction of the applied voltage (Figure 9.5(a) and (b)). The electrode components composite was made by mixing all the aforementioned constituents with dimethylacetamide (DMAC) in a ball mill until a black gel mass was obtained. The resulting gel was cast in PDMS moulds and dried until all the DMAC had evaporated. The ionogel layer was fabricated in the same way, without the addition of SWCNTs to the constituent mixture. The resulting layers were then hot-pressed together and covered in a layer of PDMS to increase the adhesion of the resulting BGA to the walls of the microfluidic device, which would ensure better sealing during operation. The actuation properties of the BGA are dependent on the total thickness of the device and on the ratio between the electrolyte layers. Three devices were tested to determine which generates the maximum amount of force by application of a voltage sweep between 4 and 10 V. The devices had thicknesses of 281.9 µm (BGA1), 322.6 µm (BGA2), and 393.7 µm (BGA3), and thickness ratios of 0.73, 1.20, and 0.87, respectively. The results indicated that BGA3 generated a force of 80 mN, which was the highest generated force of the three BGAs. BGA1 and BGA2 generated forces of 38 and 41 mN, respectively. By enclosing the BGAs in a PDMS layer, the forces generated when a voltage is applied are increased between 14% and 23%. Taking into account that the BGAs will be used in an aqueous medium, a voltage sweep between 2 and 10 V confirmed that there are no bubbles formed due to the hydrolysis. All three BGAs were fitted to microfluidic devices to cover an inlet channel that was fabricated from a tube (Figure 9.5(c) and (d)). The tests consisted of using three different operating voltages, namely 5, 8 and 10 V, and six different frequencies: 250, 125, 100, 50, 25, and 0 mHz. For all the BGAs, the results indicated that the higher the voltages and the lower the frequency, the better they are suited for use as microfluidic valves. The best results were obtained at 10 V and 0 mHz, at which the flow rate was reduced by 93%. In all experiments, a leakage flow was present, which led the researchers to determine that the design of the device can be improved to minimise the reoccurrence of this phenomenon.

9.2.2 Passive Pumps

Another application of ILs and ionogels in microfluidic devices is as passive pumps, several examples of which have been proposed in recent years. In the field of paper-based microfluidic devices, Akyazi *et al.*[36] demonstrated the use of pNiPAAm based ionogels as delaying pumps. Paper is becoming a material of choice in microfluidics due to its ubiquity and low cost, together with attractive physical properties, such as good flexibility, low thickness, low weight, and ability to wick fluids. Its fibrous nature also presents additional challenges with isotropic wicking. This study discusses the use of ionogels integrated in paper microfluidic devices to manipulate the wicking properties of these devices. These ionogels were made by dissolving NiPAAm and a crosslinker unit in two different ILs, [EMIM][$EtSO_4$] and [$P_{6,6,6,14}$][DCA], respectively. The paper microfluidic devices (μPADs) were fabricated by ink stamping in triplicate the channels on Whatman Filter paper. After the designs were stamped and dried, the ink provided hydrophobic barriers, which kept the flow of liquid between its limits. The ionogels were photopolymerised at the ends of the microfluidic devices immediately after application, so that most of the ionogel was formed at the surface of the μPAD. This ensures that the swelling is uniform and the hydrogel will not deteriorate during swelling, while also still being attached to the μPAD. The swelling results indicate that the ionogel synthesised with [$P_{6,6,6,14}$][DCA] absorbed the most water compared to the [EMIM][$EtSO_4$] ionogel and the μPAD with no ionogel. Following this, investigations were made to explore the effect of adding an ionogel to one of the channels of a Y-shaped μPAD. The solutions used for this test consisted of a pH 13 NaOH solution and a pH 2 H_2SO_4 solution which had methyl red dissolved in it. The pH 2 solution was added to the right branch of the μPAD, with no ionogel, while the pH 13 solution was added to the left branch, with the [$P_{6,6,6,14}$][DCA] ionogel (Figure 9.6). Another Y-shaped μpad without any hydrogels was used as the control. In the case of the control μPAD, the meeting point of the solutions was determined to be in the centre branch of the μPAD, where the colour of the yellow pH 2 solution changed to purple when it interfaced with the pH 13 solution. In the case of the μPAD that had the ionogel, the interface between the two solutions formed in the left branch of the μPAD, close to where the [$P_{6,6,6,14}$][DCA] ionogel resided.

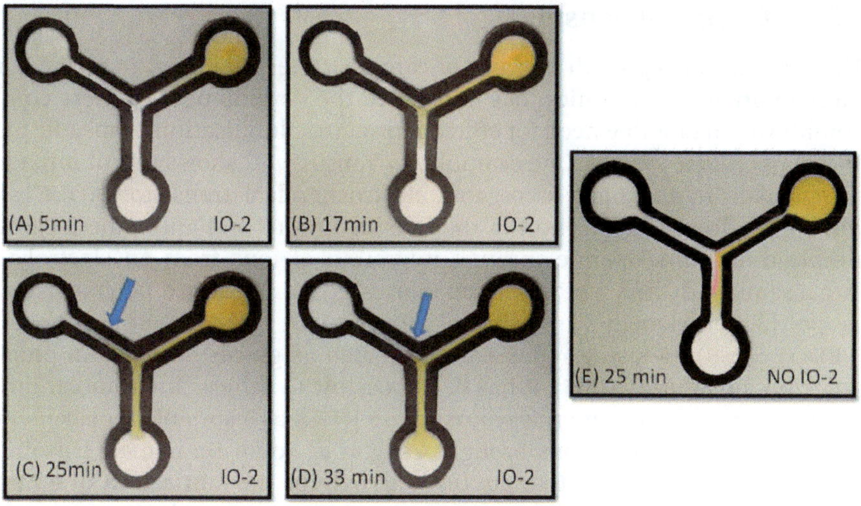

Figure 9.6 A Y-shaped μPAD with photopolymerised $[P_{6,6,6,14}]$[DCA] ionogel in the left side inlet after (a) 5, (b) 17, (c) 25, and (d) 33 min, while injecting NaOH solution (pH 13) into the left inlet (purple) and phenol red pH indicator and H_2SO_4 solution (pH 2) (yellow) into the right inlet (see text for details). Phenol red gives a pink colour change in a basic environment. (e) A Y-shaped μPAD with no ionogel. Reprinted from *Sensors and Actuators B: Chemical*, **233**, T. Akyazi, J. Saez, J. Elizalde, F. Benito-Lopez, Fluidic Flow Delay by Ionogel Passive Pumps in Microfluidic Paper-based Analytical Devices, 402–408, Copyright (2016), with permission from Elsevier.

After the initial formation of the interface between the two solutions, the interface moved towards the centre branch of the μPAD, due to the constant flow of solution coming from the $[P_{6,6,6,14}]$[DCA] ionogel, after it reached its swelling equilibrium. The use of high and low pH solutions emphasises the fact that the ionogels maintain their function even under harsh conditions. The addition of these materials to paper microfluidic devices increases the usefulness of these devices by decreasing the complexity associated with fluid manipulation, while also being cheap enough to keep costs low.

9.3 Ionic Liquids for Sensing

The use of ionic liquids for sensing in microfluidic devices continues to gain traction, owing primarily to the ability of these materials to offer a matrix which is capable of responding to chemical and physical stimuli. The wide electrochemical windows, high conductivity, propensity to stabilise enzymes and liquid state at room temperature have carved a particular niche for these exciting new materials in the field of sensing chemical and physical changes.[37] Their ability to immobilise molecules for use in pH analysis, catalysis and electrochemistry has also brought additional application as biomolecular sensors.[38]

9.3.1 Chemical Sensing

The serendipitous growth of point-of-care (POC) technologies, in particular through organic electronics, has buttressed the development of these compounds with a tangible need for efficient protein solubilisation in specific pH and temperature ranges. One example, by Yang *et al.*[39] shows recent inroads being made in the field of organic electrochemical transistors (OECTs), which have found application in the sensing of ions and antibodies. These simple transistors operate through migration of ions from an electrolyte into a semiconductor, which is often fabricated from a doped polymer, such as poly(3,4-ethylenedioxythiophene) doped with poly(styrene sulfonate) (PEDOT : PSS). By using redox enzymes, such as glucose oxidase in phosphate buffer solution (PBS), it has been possible to achieve micromolar limits of detection.[39] Recent endeavours to use RTILs as a suitable replacement for aqueous electrolytes has brought Yang *et al.* to an IL, namely triisobutyl(methyl)phosphonium tosylate $[P_{1,4,4,4}][Tos]$, for the fabrication of a new generation of OECTs (Figure 9.7). The hydrophilic nature of the IL, attributed to the tosylate anion, ensures that when patterned over the active area of the OECT the material subsequently acts as a reservoir for the enzyme and

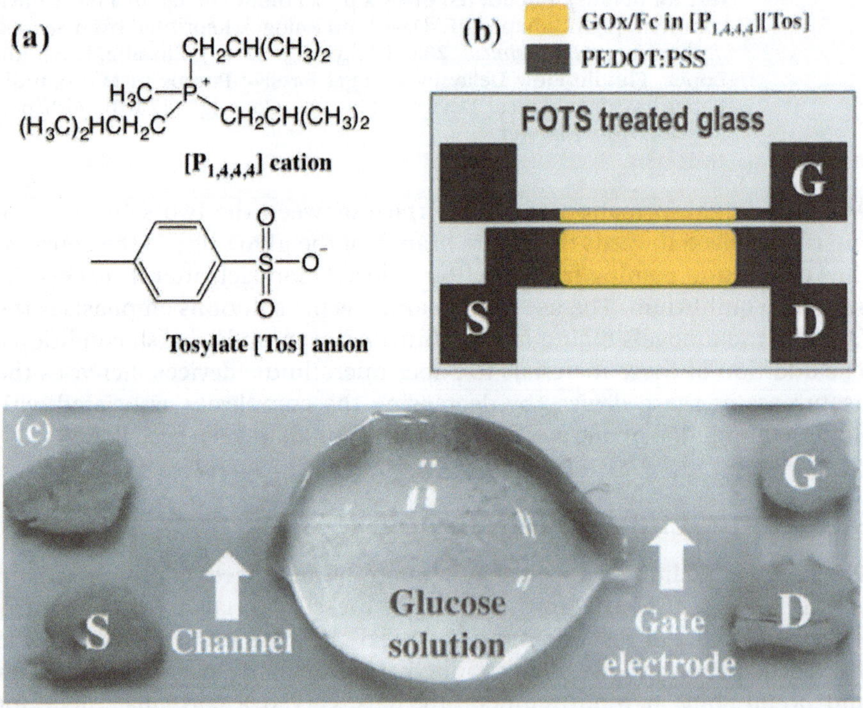

Figure 9.7 (a) Chemical structure of $[P_{1,4,4,4}][Tos]$. (b) Schematic of the OECT. (c) Image of the OECT with a drop of glucose solution added. Reproduced from ref. 39 with permission from the Royal Society of Chemistry.

mediator. Upon mixing, the mediator (ferrocene in this case) dissolves, while the enzyme remains dispersed. The presence of this dispersion can have a positive impact on the lifetime of the device by inhibiting a change in the secondary enzyme structure.[40] Figure 9.7(c) shows that the analyte, in this case a glucose solution, forms directly over the area predefined by trideca-fluoro-1,1,2,2-tetra-hydrooctyl trichlorosilane (FOTS) template. The device, which operates in the 10^{-7}–10^{-2} M range, well within clinical ranges found in the blood (2–30 mM) and saliva (0.008–0.21 mM),[41] shows great potential for use as a low cost, disposable, POC device.

This work has also been developed to sense for other clinically relevant analytes, such as lactate, through incorporation of lactate oxidase (LOx) into an OECT device.[42] In this instance a flexible ionogel based NIPAAm, *N*,*N*-methylene-bis(acrylamide) (MBAAm) and [EMIM][$EtSO_4$] incorporating the LOx enzyme was polymerised on the OECT device. As in the case for the glucose sensor, introduction of the specific analyte (in this case lactate) results in an increase in the drain current, which can be directly correlated to lactate concentration, as seen in Figure 9.8(a). Figure 9.8(b) shows a prototype fabricated from Parylene worn on the forearm.[43] Such a flexible prototype, coupled with detection levels suitable for use in the clinical range, show extremely exciting potential in the fields of sport science and patient care. The conformability of an ionogel incorporated into the OECT device offers flexibility not previously possible with more rigid conjugated polymers.

A true understanding of the effect that ILs have on enzymes has recently been developed by Curto *et al.*,[44] in particular for choline-based ILs containing LOx enzyme. They conclude that hydrated ILs can provide the necessary hydrogen bonding for necessary stabilisation of proteins, in addition to controlling the proton buffering within the medium. Interestingly, when stored in choline chloride, over a 140 day period at 5 °C, 80% of the initial activity of LOx was retained.

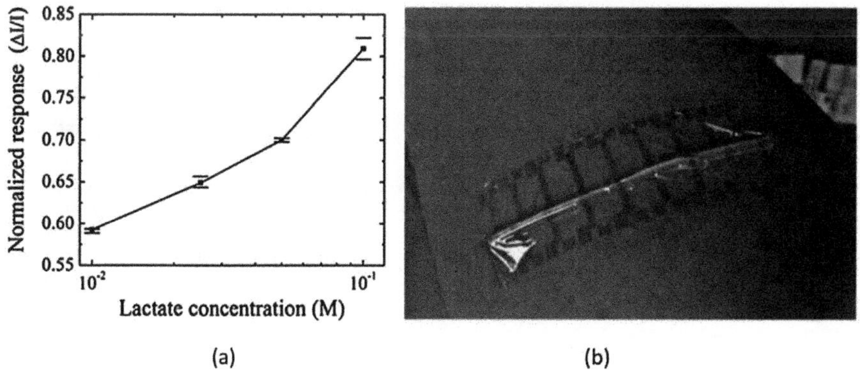

Figure 9.8 (a) Normalized response *vs* lactate concentration for the OECT and (b) flexible OECT on the forearm. Reproduced from ref. 43 with permission from the Royal Society of Chemistry.

In a similar fashion to the immobilisation of enzymes, ionic liquids can also be used to stabilise other sensing molecules such as dyes. Through ion-pair interactions, a charged dye molecule can be held in the IL or ionogel matrix without leaching. Czugala *et al.* have developed a direct application of such a system, through the use of a centrifugal disc with functionalised ionogel sensing areas.[45] The ionogel, based on poly(*N*-isopropyl-acrylamide) and *N*,*N*′-methylene-bis(acrylamide), Figure 9.9(a), is used to entrap the ionic liquid, tetrabutylphosphonium dicyanamide [$P_{4,4,4,4}$][DCA], and the dye molecule, Bromocresol purple (BCP), shown in Figure 9.9(b) and (c), respectively. Photopolymerisation yielded an ionogel, which responds to a variation in pH with the colour change shown in Figure 9.9.

By development of a light emitting diode (LED) based detector it is possible to generate a colorimetric assay. Optimisation of the concentration of BCP concluded with a 6×10^{-3} M concentration which was used to generate calibration curves across a range of pH values (Figure 9.10). Moreover, through

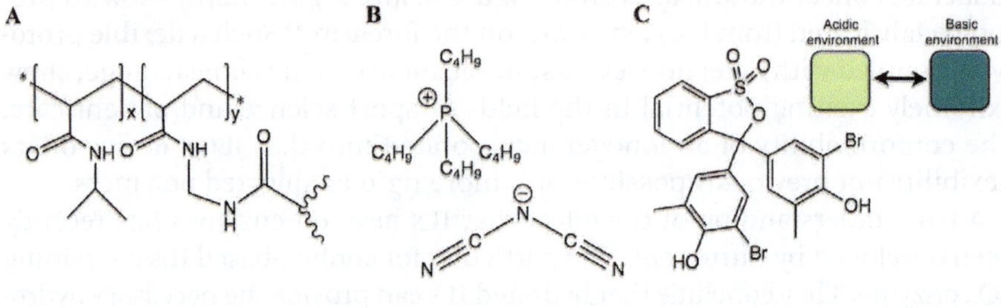

Figure 9.9 Chemical structures of (a) *N*-isopropyl-acrylamide and *N*,*N*-methylene-bis(acrylamide) crosslinked polymer; (b) ionic liquid tetrabutylphosphonium dicyanamide [$P_{4,4,4,4}$][DCA] and (c) Bromocresol Purple, showing colour changes in acidic and basic environments. Reproduced from ref. 45 with permission from the Royal Society of Chemistry.

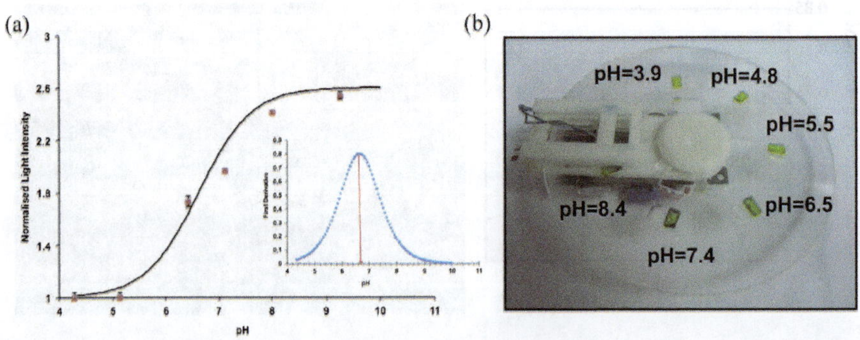

Figure 9.10 (a) Calibration curve of the sensing area of the microfluidic device using pH buffer solutions. (b) Image of the CD platform with the sensing area. Reproduced from ref. 45 with permission from the Royal Society of Chemistry.

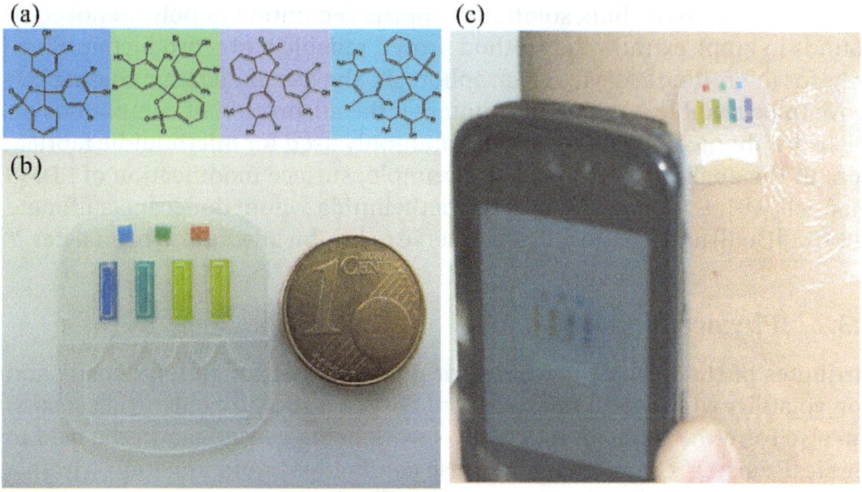

Figure 9.11 (a) Chemical structures of the pH sensitive dyes: Bromophenyl Blue (BPB), Bromocresol Green (BCG), Bromocresol Purple (BCP) and Bromothymol Blue (BTB). (b) Fabricated micro-fluidic device. (c) Smartphone application imaging on-body device.

incorporation of these ionogel materials into a centrifugal disc device, it was possible to include a full colorimetric assay in a CD platform.

In a similar fashion, Curto *et al.*[46] have extended the use of IL encapsulated pH responsive dyes to yield a simple barcode device which is capable of measuring sweat pH in real time, using colorimetric imaging through a mobile phone application (Figure 9.11). In a bid to further extend the lifetime of the microfluidic device, immobilisation of the ionogel on a poly(methyl methacrylate) (PMMA) substrate was achieved through the use of water plasma treatment, followed by silanisation. The ionogel containing the various dye molecules could then be directly bonded to the functionalised surface, through covalent attachment. By applying an algorithm which mapped to the hue saturation value (HSV) colour space it was then possible to use the mobile phone application to generate calibration curves with $R^2 > 0.995$.

Not only can ionic liquids be used as vehicles for entrapping, immobilising and stabilising sensing materials, such as enzymes, dyes and stimuli responsive materials, they can also be used for their direct interactions with target molecules, most notably in the fields of capillary and microfluidic device electrophoresis. Using ILs as the supporting electrolyte or as additives to the running buffer can dramatically increase the speed and efficiency of operation, in addition to broadening the range of compounds which can be separated using electrophoresis. When used as background electrolytes, in a similar fashion to alkylammonium salts, ILs based on the 1-alkyl-3-methylimidazolium cation were seen to behave as electroosmotic flow modifiers through interaction of the cation either by coating the capillary wall or by migrating into the bulk solution.[47] For the separation of polyphenol compounds in grape extracts the method proved reliable and reproducible. Similarly, for microfluidic device electrophoresis, the use of dynamic coating can prove to be a viable solution to counteract the adsorption of compounds on the hydrophobic polymer materials commonly used for microfluidic fabrication. In the analysis of proteins, for example, surface modification of PDMS channels with ILs, such as 1-butyl-3-methylimidazolium dodecane-sulfonate ([BMIM][DoS]) and [EMIM][BF_4], can serve to inhibit adsorption of analytes.[48]

9.3.2 Physical Sensing

Attributes of ILs, such as good electrical conductivity, high ion density and non-volatility which form the basis for new generations of chemical sensors, can also form the basis for much novel work being carried out in the field of physical sensing. Confinement of these conductive liquid materials within a microfluidic channel can generate cheap, pliable and accurate strain-sensors for monitoring motion. These sensors, previously fabricated by casting conductive materials such as silver nanowire, graphene films and carbon nanotubes or through the use of liquid metals within microfluidic channels have generated much recent attention, with the advent of the ubiquitous wearable device. Previous iterations, however, have suffered from complicated fabrication techniques which can result in large hysteresis, resulting in the inability to undergo multiple deformations.

Fabrication of a flexible PDMS channel, filled with a single IL or binary mixture thereof, can be used to fabricate thin wearable strain sensors. Matching the refractive indices of the binary mixture can generate a transparent strain sensor, used to measure stretching, bending and other motions. When pressure is applied to the flexible membrane a geometric change within the channel results in an increase of pressure within the channel, thereby leading to a measurable variation in the electrical properties of the ionic liquid (Figure 9.12). These devices can therefore be considered as constant electrofluidic resistors.

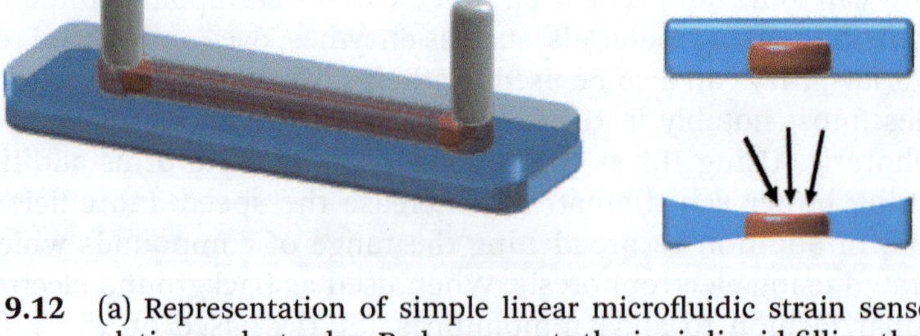

Figure 9.12 (a) Representation of simple linear microfluidic strain sensor with platinum electrodes. Red represents the ionic liquid filling the microfluidic channel. (b) Two-dimensional representation of the geometric deformation of the microfluidic channel when pressure is applied.

Yoon *et al.*[49] described a simple linear channel system of 400 μm width and 70 μm height, filled with a binary mixture (51 : 49 molar ratio) of 1-butyl-3-methylimidazolium bis(trifluoromethanesulfonyl)imide ([BMIM][Ntf$_2$]) and 1-butyl-3-methylimidazolium acetate ([BMIM][Ac]). Using platinum electrodes at a constant voltage of 1 V (DC) it was possible to monitor current under a range of tensile strain from 10% to 25% for multiple cycles. At low strain speeds minimal hysteresis was observed, attributed to the dynamic nature of the IL, thereby eliminating any possible disconnection in the circuit.

This concept has been extended upon in publications such as that by Wu *et al.*[50] to fabricate novel pressure sensors using ILs in similar flexible PDMS channels. Through the use of a bilayer microfluidic device it is possible to use one microfluidic channel as a pressure sensor at a specific location in the second. This change in pressure in the channel filled with 1-ethyl-3-methylimidazolium dicyanamide [EMIM][DCA] can be directly related to electrical resistance within the circuit. This change in resistance can be accurately measured through the design of a Wheatstone bridge (Figure 9.13). Using pressurised gas and liquid in the microfluidic channel it was possible to characterise the sensor, which exhibited impressive thermal stability over extended periods of time. These results displayed much promise for the application of such a device in LOC devices.

The ability to develop these systems to measure stretch and shear within microfluidic devices has extended their application to devices for microfluidic cell culture, such as organ-on-a-chip. These devices can provide a greater understanding of analogous biological processes by monitoring cells *in vitro*, through real time flow monitoring. Liu *et al.*[51] have developed a microfluidic device for endothelial cell culturing with an embedded pressure sensor (Figure 9.14). Again, by measuring the variation in electrical properties generated

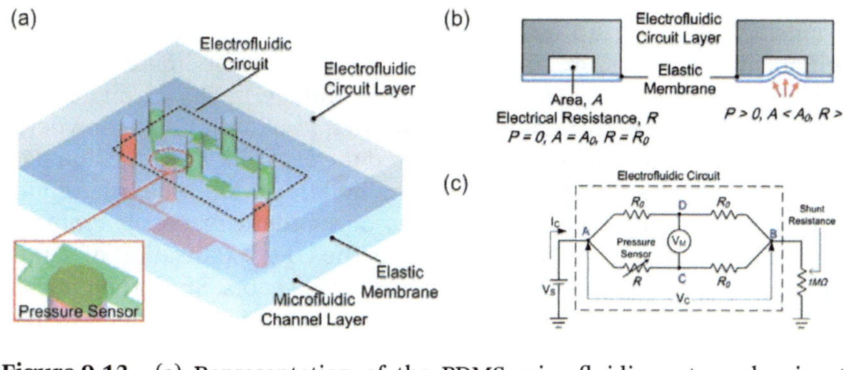

Figure 9.13 (a) Representation of the PDMS microfluidic system showing the microfluidic channel and electrofluidic circuit layers. (b) Conversion of mechanical pressure into electrical resistance. (c) Wheatstone bridge electrofluidic circuit. Reproduced from ref. 50 with permission from the Royal Society of Chemistry.

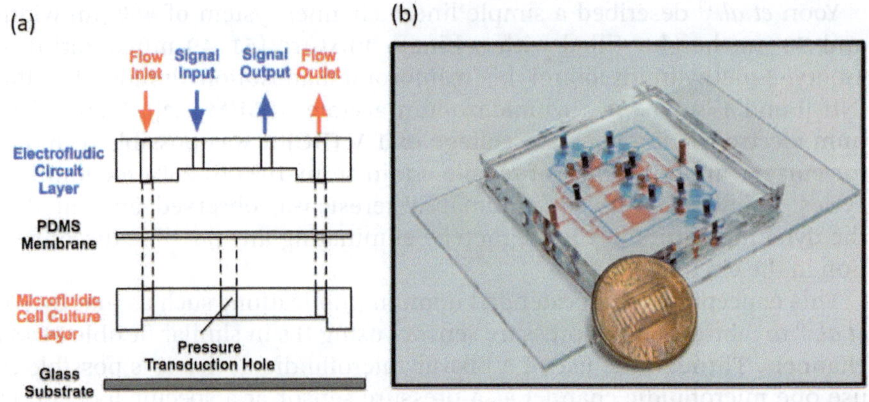

Figure 9.14 (a) Representation of the microfluidic cell culture showing electrofluidic circuit layer and a microfluidic cell culture layer and glass substrate. (b) Image of the fabricated device showing the electrofluidic circuit channels (blue) and microfluidic cell culture channels (red). Reproduced from ref. 51 with permission from the Royal Society of Chemistry.

from a change in the geometry of the soft-polymer channel filled with [EMIM][DCA], it is possible to determine the pressure within the microfluidic device (Figure 9.14). Such a measurement is critical for the understanding of the surrounding environment of endothelial cells which are commonly exposed to stress, stretch and hydrostatic pressure.

9.4 Ionic Liquids for Reagent Storage

Due to their versatility, promising solvation properties and interactions with solute species, ionic liquids have been long proposed as alternative solvents.[52–54] Building up on this and adding their high thermal stability, negligible vapour pressure and enzymatic stability, ionic liquids and ionic liquid gels have been recently proposed as storage media in microfluidic devices. Weidmann *et al.*[55] proposed in 2012 a microfluidic deposition device using ionic liquid matrices which displayed the capability of ionic liquids as matrices for the analysis of biomolecules by MALDI-MS. The microfluidic spotting device addressed several issues of standard protocols employed for MALDI-MS sample preparation such as the co-crystallisation of sample and matrix, clogging and heterogeneity of sample spots. Since the ionic liquids did not solidify during the measurement, the sample spots remained homogeneous.

Using an ink-jet printing approach, UV-cured ionogel-based microarrays were fabricated and used for long-term reagent storage of biotin-647 in a LOC device.[56] The ionogel cocktail was based on crosslinked *N*-isopropylacrylamide (NIPAM) and [EMIM][EtSO$_4$] ionic liquid. Biotin-647 was added

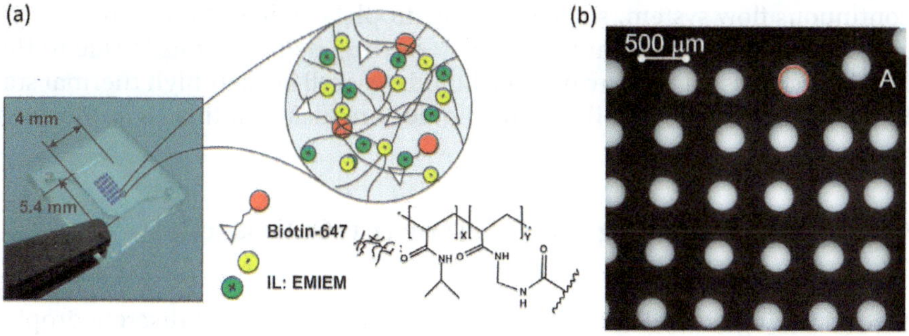

Figure 9.15 (a) Inkjet-printed ionogel microarray in the microfluidic device and representation of the composition of the ionogel spots. (b) Fluorescence images of biotin-647 inside an ionogel-based microarray. Reprinted from Biomolecule Storage on Non-modified Thermoplastic Microfluidic Chip by Ink-jet Printing of Ionogels, M. Tijero, R. Díez-Ahedo, F. Benito-Lopez, L. Basabe-Desmonts, V. Castro-López, A. Valero, *Biomicrofluidics*, 2015, 9, 044124, with the permission of AIP Publishing.

to this cocktail and the mixture was printed on 6 × 6 array of 300 µm circles (Figure 9.15a) on a variety of untreated substrates such as cyclic olefin copolymer (COC), cyclic olefin polymer (COP), and polypropylene (PP). Following this, the samples were exposed to UV light at 365 nm for 15 s to photopolymerise the printed cocktail and create the final ionogel matrix (Figure 9.15(b)).

The viability of this ionogel microarray was demonstrated for biotin-647 storage for over 1 month at room temperature while also proving to effectively keep the activity of biotin-647 under these conditions.

9.5 Ionic Liquids in Segmented Flow Microfluidics

Segmented flow systems offer many exciting and advantageous opportunities in the microfluidic field. Unlike continuous flow microfluidics (which use one single continuous flow), segmented flow devices typically rely on two distinct methods in order to move reagents throughout the microfluidic devices. In the first method, two immiscible flows are forced together at a T-junction, which results in one of the flows becoming the carrier while the second forms droplets.[57] The second method involves the direct manipulation of discrete individual droplets across different types of interfaces (solid–liquid, liquid–air).[58,59] All segmented flow systems are characterised by a high surface area to volume ratio which results in high heat and mass transfer rates; however, the biggest advantage of these types of flows is the compartmentalisation of reagents into droplets, which allows for control over both the internal and external environment which the reagents are exposed to. Dissociation of reagents from the external environment allows for numerous applications which are not otherwise able to be performed using a continuous flow system, such as having droplets acting as micro-reactors,[60] cargo transporters,[61] dynamic sensors[62] or drug delivery units.[63] Due to the highly customisable nature of ionic liquids, as well as their high thermal stability and non-combustibility, ionic liquids are ideal candidates for droplet/segmented flow regimes.

9.5.1 Electrowetting on Dielectric (EWOD) Based Microfluidics

Electrowetting on dielectric is the phenomenon by which a discrete droplet can be electrically actuated across a series of electrodes. EWOD is one of the main forms of actuation associated with digital microfluidics and the process involves altering the wettability of a surface *via* an externally applied electrical field. When the electric field is applied to the device, charges accumulate at the interface between the droplet and solid surface, decreasing the interfacial tension and thus lowering the contact angle between the droplet and the surface.[64] By addressing sequential electrode pairs in the system, an interfacial tension gradient can be created between neighbouring electrodes, which the droplet will follow to a predetermined destination. Traditionally, aqueous based droplets have been used in EWOD systems; however, due to their high thermal stability, low combustibility and minimal tendencies to corrode the metal parts of the devices, many groups have looked to ILs to replace them.[65–67]

Li *et al.*[68] used high speed video microscopy to investigate the dynamic electrowetting and dewetting capabilities of ILs. In this study they employed five popular imidazolium-based ILs as probe liquids, namely 1-butyl-3-methylimidazolium tetrafluoroborate ([BMIM][BF_4]), 1-butyl-3-methylimidazolium hexafluorophosphate ([BMIM][PF_6]), [BMIM][NTf_2], 1-hexyl-3-methylimidazolium bis(trifluoromethanesulfonyl)imide ([HMIM][NTf_2]) and 1-methyl-3-octylimidazolium tetrafluoroborate ([OMIM][BF_4]). The focus of this study was to observe the electrowetting at a fixed potential, followed by the dewetting once the potential had been removed. The electrowetting experiments were performed on a fluoropolymer surface (low wettability) under a fixed DC potential of 120 V. Once the electric field was introduced, the base area of the ILs increased exponentially until a maximum area was reached; once the electric field was removed the base area decreased exponentially until they assumed their initial position (Figure 9.16).

The IL droplets electrowetting occurred roughly twice as fast as the subsequent retraction, due to the bulky ions within the ILs, which makes it easier to arrange the ions than to destroy their arrangement. The results showed good reversibility which was related to the low contact angle hysteresis of the ionic liquids used. This study showed that ILs can undergo reversible electrowetting behaviour, which added to the fact that ILs show good chemical stability, a large viscosity range and have varying surface tensions make them ideal for use as electrowetting agents in a variety of applications such as variable focus lenses,[69] RC filters[70] and microreactors.[60]

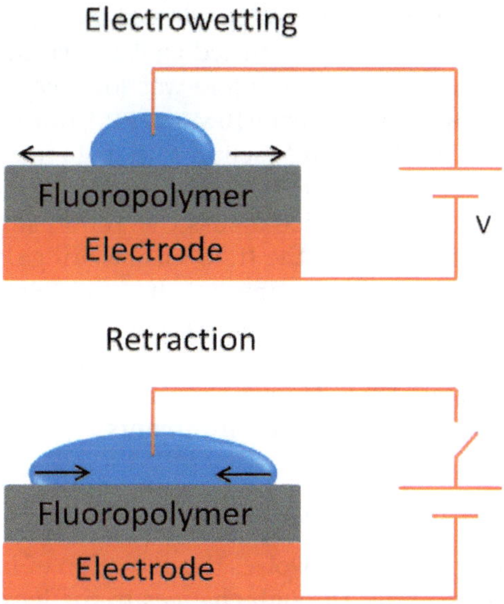

Figure 9.16 Illustration of ILs electrowetting and retraction behaviour in the presence (top) and absence (bottom) of a DC potential of 120V. Reprinted with permission from Dynamic Electrowetting and Dewetting of Ionic Liquids at a Hydrophobic Solid–Liquid Interface, H. Li, M. Paneru, R. Sedev and J. Ralston, *Langmuir*, 2013, 29, 2631–2639. Copyright (2013) American Chemical Society.

Amin *et al.*[71] reported an interesting application for ILs which uses their electrowetting capabilities. In this study the group employs ILs as soft micro-grippers under conditions of high temperatures and high vacuum, *via* electrowetting actuation. The group proposed the use of ILs instead of aqueous droplets as water based systems have a limited functional temperature and vacuum levels while also showing a tendency to erode electrowetting devices. ILs, on the other hand, have negligible vapour pressure and high chemical and thermal stability which make them suitable candidates for use in high temperature and vacuum environments. Micro-grippers are typically used in micro-assembly tasks, where a micro-gripper should be able to pick up, hold and release an item. Classically, capillary forces are used; this is achieved by creating a liquid bridge between the object and the gripper; the capillary forces generated have enough force to lift objects weighing a few milligrams. However, because the capillary force remains constant once the bridge is formed it is difficult to then release the item. The capillary force has to be strong enough to pick up and hold the object for it to be positioned correctly, then the force has to be reduced sufficiently to where it can be released. Amin *et al.*[72] demonstrated that electrowetting could be used to achieve this process, in a procedure similar to the example by Li *et al.*[68] described above. Once the voltage was applied, the surface became hydrophilic, lowering the contact angle of the liquid, which resulted in the formation of a bridge with the object. The capillary forces generated allowed for the object to be picked up and held. When the voltage was turned OFF, the surface became hydrophobic and the contact angle of the liquid was lowered, thus releasing the object. The group showed that by using [BMIM][PF_6], this process could work in temperatures of up to 110 °C and vacuums up to 24 inch-Hg.

The device functions as follows: firstly the gripper is moved towards the object and a high voltage is applied across the Teflon surface. This makes the surface hydrophilic, allowing the IL to form a bridge, thus creating the maximum force and picking up the object. When the voltage is reduced, the surface becomes hydrophobic, increasing the contact angle of the IL with the surface, reducing the applied capillary forces and releasing the object. The experiments indicated that ILs were ideal for use as micro-grippers in high temperature and high vacuum environments.

9.5.2 Chemotactic Ionic Liquids

Another segmented flow regime involves the actuation of individual droplets across the liquid–air interface through changes in the local surface tension, which is brought upon by the triggered release of a surfactant from the droplet itself. Surfactants are long chained amphiphiles which have hydrophilic "heads" and hydrophobic "tails". When a surfactant is released into an aqueous solution (below the critical micelle concentration) it will interact with polar molecules present at the surface of the solution and will interrupt the attractive forces felt by surface molecules, distributing the local surface tension and creating flows within the bulk solution. Fluid flows from areas of low surface tension to high surface; this is known as the Marangoni effect. By designing systems in which surfactants are asymmetrically released from the droplet into the aqueous phase *via* external stimulation, droplets have been produced that can solve complex maze designs[73] or can be actuated through white light irradiation.[74] To date, surface tension driven segmented flow systems have mostly focused on aqueous or organic solvent based droplets, which are subject to evaporation and combustion. ILs have a broad range of surface tensions and many have showed surfactant properties,[75] while also being highly customisable and overcoming evaporation due to their negligible vapour pressure, making them promising alternatives to aqueous/organic based droplets.

Francis *et al.*[76] were the first to report on the chemotactic movement of ionic liquid droplets composed of $[P_{6,6,6,14}]Cl$. The IL droplets used in this study moved across the liquid–air interface *via* the triggered release of $[P_{6,6,6,14}]^+$, a very effective cationic surfactant and cationic constituent of the IL (Figure 9.17).

When the $[P_{6,6,6,14}]^+$ ion diffuses from the droplet into the aqueous solution it interacts with the molecules at the surface, resulting in a decrease in the local surface tension. The rate of release of the surfactant depends on the solubility of the Cl^- counter-ion, as any loss of Cl^- ion must also result in an equivalent transfer of $[P_{6,6,6,14}]^+$ in order to maintain overall charge neutrality within the droplet. Therefore the rate of release of the

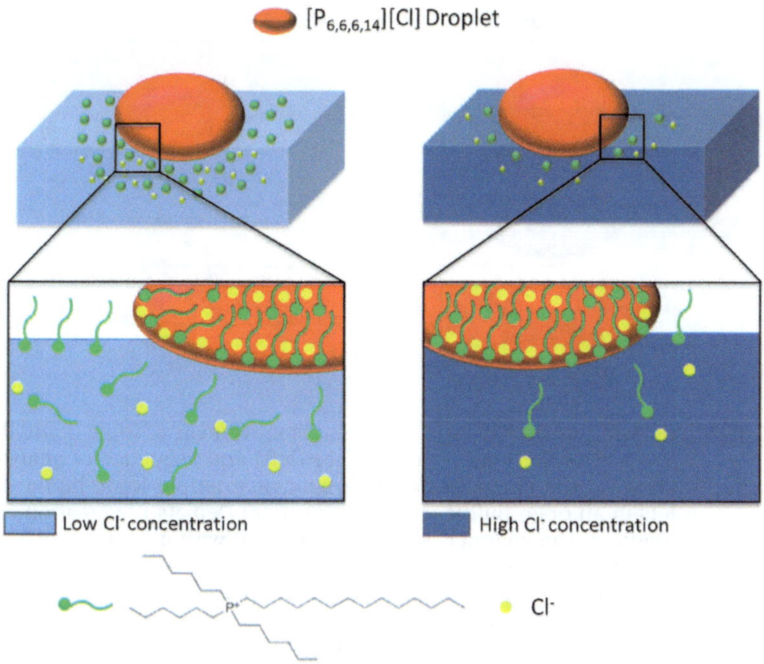

Figure 9.17 Cartoon illustrating the composition and relative solubility of $[P_{6,6,6,14}]$ [Cl] in solutions of 10^{-2} M NaOH and 10^{-2} M HCl. Reproduced from ref. 76 with permission from the Royal Society of Chemistry.

counter-ion was dependent on the local aqueous Cl^- concentration. Once a droplet of the IL was placed onto an aqueous solution with an imposed Cl^- concentration gradient there was an asymmetrical release of $[P_{6,6,6,14}]^+$ into the aqueous solution, which resulted in the formation of a surface tension gradient around the droplet. This created Marangoni like flows which caused the droplet to move towards areas of highest surface tension (Figure 9.18).

Francis *et al.*[77] have also demonstrated that ionic liquids can respond to electrical stimuli and show electrotactic movement. Droplet composition and movement remains the same as for the chemotactic ionic liquid droplets described above. In this case, 3D printed electrodes, which were embedded within 3D fluidic channels, were used to generate the required ionic gradients. Francis *et al.* electrotactically actuated droplets of $[P_{6,6,6,14}]$[Cl] across the liquid–air interface of 10^{-3} M NaCl electrolyte solutions. This was achieved by imposing an external electric field across the NaCl solution which resulted in the migration of the mobile ions towards their respective counter electrodes. This phenomenon resulted in the formation of a Cl^- gradient across the channel, between opposite electrodes. Once the potential was switched ON, an IL droplet placed at the cathode would asymmetrically release surfactant and move towards the anode. By simply reversing the polarity of the electrodes,

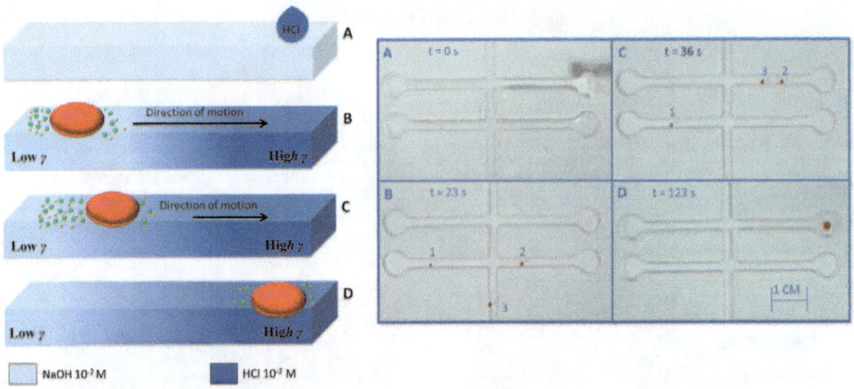

Figure 9.18 Illustration showing the chemotactic movement of ionic liquid droplets across the liquid–air interface (left) and sequence of snapshots showing the chemotactic movement of multiple ionic liquid droplets in an open fluidic channel (right). (a) Depicts the creation of Cl^- gradient; the channels were initially filled with a solution of 10^{-2} M NaOH and at the desired destination a few drops of a 10^{-2} M solution of HCl were placed. (b) Shows the initial placement of the ionic liquid droplet(s). (c) The droplet(s) are propelled towards the highest area of surface tension. (d) The droplet arrives at the desired destination. Reproduced from ref. 76 with permission from the Royal Society of Chemistry.

the gradient and ultimately the droplet movement could be reversed. Moreover, by selectively polarising appropriate electrode pairs the direction of the droplets could be changed and they could be steered into side channels at junctions (Figure 9.19). The major advantages of electrotactic movement over chemotactic movement of droplets is the ability for concentration gradients to be established and varied dynamically, and maintained for longer periods of time. This resulted in flexible control over the speed and direction of IL droplet movement.

The use of ILs allows the droplets to remain stable on the surface of the aqueous solutions for much longer than traditional droplets (based on organic solvents) due to their negligible vapour pressure. When coupled with the large (and ever growing) library of ILs available and the fact that many show surfactant properties, it makes them particularly promising for applications involving autonomous droplet systems such as cargo transporters, sensing units, self-repair agents and drug delivery mechanisms.

9.5.3 Ionic Liquids as Microreactors

As mentioned previously, one of the main advantages of segmented flows is the compartmentalisation of the reagents into individual discrete droplets. This has a number of unique properties. Firstly, it separates the reagents

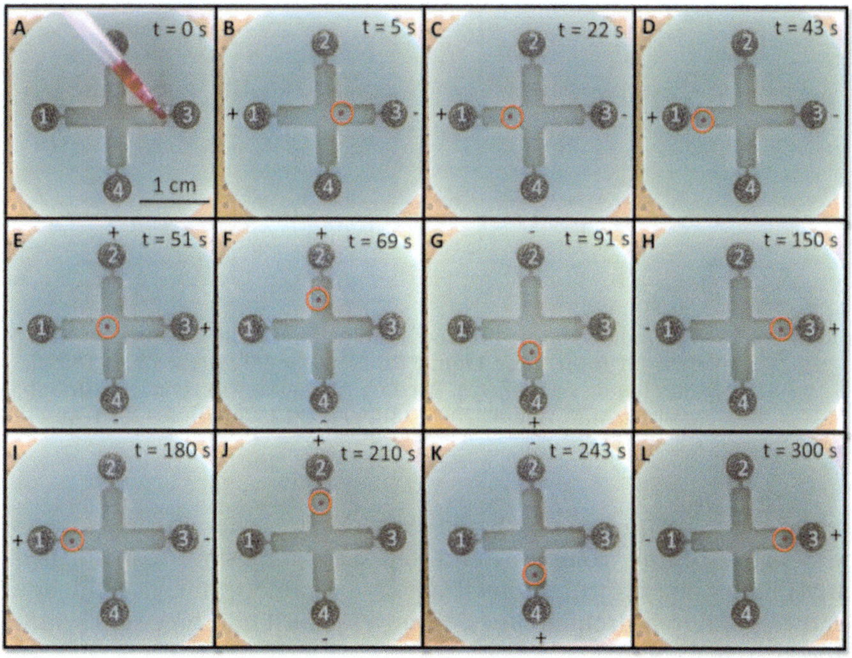

Figure 9.19 Series of snapshots which demonstrate the electrotactic movement of IL droplets. Initially the open fluidic channels are filled with a 10^{-3} M solution of NaCl and a potential difference of 9 V is applied across selected electrodes. (a) Introduction of the IL droplet. (b–d) Droplet moves away from cathode (3) to anode (1). (e) The polarity of electrodes (3) and (1) is reversed and the droplet begins to move to electrode (3). (f) As the droplet approaches the junction, the potential difference is applied between electrodes (2) and (4) and removed from electrodes (3) and (1). The droplet then begins to migrate toward anode (2). (g) Upon arriving at anode (2) the polarity of electrode (2) and (4) is reversed and the droplet moves towards the new anode (4). (h) Upon arriving at anode (4) the potential is again reversed and using a similar method to sequence (f) the droplet is returned to the starting position. *(i-l)* It shows the process repeated in the same run. Reprinted from *Sensors and Actuators B: Chemical*, **239**, W. Francis, K. Wagner, S. Beirne, D. L. Officer, G. G. Wallace, L. Florea, D. Diamond, Electrostatic Ionic Liquid Droplets, 1069–1075, Copyright (2017), with permission from Elsevier.

from the outside environment, and secondly it allows for the reagents to exist independently of each other within the same device, allowing the user to decide which reagents to mix, when and where. ILs particularly suit this type of application as they have been shown to be excellent solvents[78] and can be tailor-made to meet the requirements for many reactions.

Dubois *et al.*[60] reported on the use of IL droplets as microreactors. The ILs were actuated in EWOD devices, in a process similar to the examples described above in which a large number of IL droplets can be efficiently

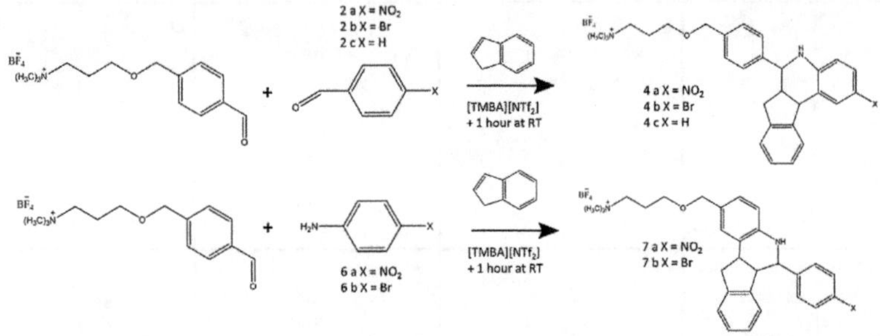

Scheme 9.1 Scheme for the synthesis of tetrahydroquinoline. Reprinted with permission from Ionic Liquid Droplet as e-Microreactor, P. Dubois, G. Marchand, Y. Fouillet, J. Berthier, T. Douki, F. Hassine, S. Gmouh, M. Vaultier, Anal. Chem., 2006, 78, 4909–4917. Copyright (2013) American Chemical Society.

moved and mixed on a Teflon based EWOD device. They also tested the viability of performing complex multi-step reactions using ILs (trimethyl-*N*-butylammonium bis(trifluoromethylsulfonyl)amide ([tmba][NTf_2])) by the synthesis of tetrahydroquinoline derivatives, which is a three-step synthetic process (Scheme 9.1).

As the reaction had two possible routes both were tested using the IL droplets. To perform the reaction on the EWOD device, a 0.2 μL IL droplet of [tmba][NTf_2] which contained 0.2 M of either of the task-specific onium salts (**1** or **5**, Scheme 9.1) and **10** equivalents of trifluoroacetic acid (TFA) was placed onto the device. It was then actuated towards a second IL droplet (which contained 2.5 equivalents of either the aniline derivative (**2**) or the benzaldehyde derivative (**6**)) by applying 55 V to the corresponding electrodes. Upon merging, the combined droplet was moved to, and merged with, a final IL droplet that contained an excess of indene (**3**). The droplet was then allowed to incubate for one hour at room temperature. It was then removed and tested *via* HPLC. The results indicated a near 100% conversion. Moreover, the group was able to perform the analysis on-device by actuating the final reaction droplet to an analytical area where an electrochemical analysis was performed using two gold wires. The signal reflected the concentration of the final product. This paper describes a powerful and flexible tool which takes advantage of the chemical stability, customisation and non-combustible properties of ILs to perform complex, small-scale organic synthesis on a EWOD device.

Marchand *et al.*[78] demonstrated the use of ILs as soft wall-free microreactors for organic synthesis *via* a click reaction. This work describes the synthesis of fluorescent 1,2,3-triazole by copper-catalysed 1,3 dipolar cycloaddition reaction of azides with terminal acetylenes. This reaction (Scheme 9.2) is a well-known click chemistry reaction and results in no by-products and excellent yields.

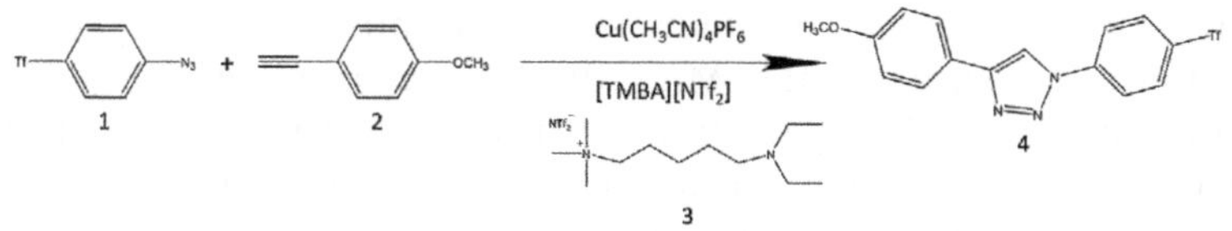

Scheme 9.2 Reaction scheme for the synthesis of fluorescent 1,2,3-triazole. Reprinted with permission from Organic Synthesis in Soft Wall-Free Microreactors: Real-Time Monitoring of Fluorogenic Reactions, G. Marchand, P. Dubois, C. Delattre, F. Vinet, M. Blanchard-Desce, M. Vaultier, Anal. Chem., 2008, 80, 6051–6055. Copyright (2013) American Chemical Society.

Actuation of the ILs was performed using an EWOD device and various mixing methods. To perform the reaction (without mixing) on the device, 0.2 μL of the IL ([tmba][NTf_2]) which contained 0.2 mol L^{-1} of the azide (**1**) and 0.01 mol L^{-1} of the catalyst ($Cu(CH_3CN)_4PF_6$) was actuated *via* electrowetting (55 V) towards, and mixed with, a second droplet of the same IL which contained 1 mol L^{-1} of the alkyne (**2**) and 0.03 mol L^{-1} of the support amine (**3**). The reaction was monitored *via* fluorescence spectroscopy and a 100% conversion rate was noted after 40 min, which was confirmed *via* HPLC. This reaction had a long reaction time compared to standard magnetic stirring methods, due to the viscosity of the ILs. To overcome this problem, the group tested a number of mixing methods; firstly, the stationary droplet was heated to 40 °C *via* a thermoresistor which was placed under the device. This created Marangoni flows within the droplet, due to the substrate being at a higher temperature than the air, causing the droplet to experience a temperature gradient, which in turn generates a flow circulation within the droplet. This was proven by monitoring added fluorescent markers to the droplet. The second method involved repeatedly moving the droplet backwards and forwards between a series of electrodes. The final mixing procedure involved positioning the droplet on a module that generated surface acoustic waves (SAW); these waves generate a random and chaotic flow within the droplet which accelerates the mixing of reagents. In all cases the required reaction time was reduced, with the shortest time of 21 min being achieved when mixing *via* the SAW module was employed. Increasing the temperature decreased the reaction time even further to the point where it was comparable with the orthodox stirring method. This study is important as it shows that ILs can be used for small-scale organic synthesis (*via* electrowetting) but it also describes methods to overcome issues with viscosity which can limit the use of ILs in synthesis.

9.6 Ionic Liquids for Separation in Microfluidics

ILs have attracted considerable interest as continuous phase materials for droplet generation,[79–86] as electrolyte modifiers,[87] as biocatalysts[10,11] and in recent years acting as the continuous phase in two-phase microfluidics devices.[90–94]

Droplet generation is the main technique applied for separation in microfluidics when using ILs. Microfluidic devices allow the generation of a large number of monodispersed droplets of controlled size. Water immiscible liquids, such as oils and organic solvents, have been used for the formation of micro-droplets in microchannels.[81] However, to overcome the limitations of using conventional non-polar solvents (polarity, biomolecule toxicity and device deformation), novel water-immiscible solvents, such as ionic liquids, are being introduced in droplet microfluidics. As an example, Khan *et al.*[80] have presented a general analytical method using imidazolium ionic liquids for "on-drop" reaction, separation and dynamic reporting.

Droplet generation can also be controlled by using magnetic fields. One such example has been reported by Luo *et al.*[82] who demonstrated the droplet formation of a paramagnetic ionic liquid within a magnetic field in a coaxial microfluidic device. Through the application of an external magnetic field they outlined that droplet generation, size and velocity could be controlled. Similarly, Loewe *et al.*[83] have reported the synthesis of monodisperse double emulsions based on organic components and paramagnetic ionic liquids. The external application of magnetic fields allowed for a selective and controlled manipulation of emulsion droplets.

Recently, ionic liquids have also been used as liquid electrodes modifiers.[87] Four ionic liquids, 1-butyl-4-methylpyridinum tetrafluoroborate ([BMPy][BF_4]), [BMIM][BF_4], 1-butyl-3-methylimidazolium hydrogen sulfate ([BMIM][HSO_4]), and 1-ethyl-3-methylimidazolium methyl sulfate ([EMIM][$MeSO_4$]) were applied as dynamic modifiers in glass and SU-8 microdevice electrophoresis. This modification allowed for the successful separation of catecholamine derivatives by exploiting the molecular interactions between ionic liquids and catecholamines. He *et al.*[88] developed a dielectrophoresis (DEP) device by using ionic liquids as liquid electrodes, which can be used to continuously separate polystyrene microbeads and PC-3 human prostate cancer cells based on positive DEP. The use of ionic liquids as liquid electrodes has the advantage of low cost and easy fabrication.

The stabilisation of enzymes within ILs has been discussed previously. This forms the basis for the use of ionic liquids as reaction media within enzymatic microreactors. They can facilitate the scale-up of microreactors to the industrial scale providing good adjustability and control over the process. This new field of research in microfluidics has found application in the racemic separation of ibuprofen from biocatalytic reactors.[89]

Ionic liquids have also been used as novel solvents to replace common volatile organic solvents in ionic liquid based aqueous two-phase systems (IL-ATPS).[90] The hydrophobicity and immiscibility of ionic liquids in water has been exploited for their use in the extraction of hydrophobic analytes. As an example, Su *et al.*[91] developed an IL-ATPS microfluidic device for the selective extraction of Bisphenol A (BPA) in aqueous samples using

N,N,N-trioctyl-ammonium propionate (TOAP) as the continuous phase. The device showed a 95% extraction of BPA in 2 s. Moreover, Znidarsic and Plazl *et al.*[92] combined the advantages of the IL-ATPS technique with microfluidics, for the continuous separation of proteins. The hydrophilic IL ([BMIM][BF_4]) and D-fructose were employed in an IL-ATPS system with bovine serum albumin (BSA) used as the protein model. Wang *et al.*[93] also worked on protein separation with the use of ionic liquids as extractants. In this case, a water soluble ionic liquid, 1-butyl-3-methylimidazolium dodecane-sulfonate ([BMIM][DoS]), was designed and applied to microdevice micellar electrokinetic chromatography (MEKC), possessing both surfactant and IL properties. [BMIM][DoS], which was used as the supporting electrolyte in a microfluidic PDMS device, was found to have a pH of 7.4, making it especially suitable for biological purposes, in addition to preventing adsorption of analytes onto the PDMS surface. Finally, Park *et al.* created an ATPS system for separation in microfluidics in which an IL-ATPS device was demonstrated for the effective purification of bacteriorhodopsin (BR) from *Halobacterium salinarium.*[94] The separation of gases has also been achieved using RTILs on polyethersulfone hollow fibers.[95] The generated membranes were successfully employed for the separation of CO_2 and N_2 using two different ionic liquids, [BIMIM][BF_4] and [BMIM][PF_6].

Despite the outlined applications, several issues still exist with the use of ILs in microfluidic device separations. These drawbacks, associated with high viscosities, poor fluid and mass transfer of ILs, can be circumvented through the use of microencapsulation. Luo *et al.*[96] prepared polysulfone microcapsules filled with [BMIM][PF_6] in a microfluidic device. [BMIM][PF_6] was successfully encapsulated by polysulfone to yield small microcapsules with excellent monodispersity and spherical morphology. The authors subsequently immobilised the same ionic liquids in microcapsules using a mixture of PDMS and *n*-butyl acetate as the continuous phase and polyacrylonitrile and ILs dissolved in *N,N*-dimethylformamide as the dispersed phase.[97] The use of ILs in microfluidic separation allows for the purification of target molecules with rapid purification times and increased yields, in addition to improving the performance of PDMS microfluidic devices by reducing absorption of molecules to the PDMS channel walls.[98]

9.7 Other Applications in Microfluidics

The broad nature of applications using ILs in microfluidic devices makes it difficult to efficiently categorise them independently. The following section presents a compendium of publications, which have not been covered hitherto, in which ILs play an important role for the development and application of microfluidic devices.

9.7.1 Ionic Liquids for Nanoparticle Synthesis in Microfluidics

A particular use for the incorporation of ionic liquids in microfluidic devices is the synthesis of metallic nanoparticles and inorganic nanomaterials. As nanoparticles are an emerging tool in many biological applications there is an increasing demand for the low cost, rapid and reproducible nanomanufacture methods for these materials.[99,100] Microfluidic devices are considered an advantageous option for manufacturing nanoparticles, through continuous flow synthesis. In addition, improved heat and mass transport, fast reagent mixing, continuous throughput, greater reaction control, and minimal solvent waste serve as extremely beneficial attributes.[101,102] These features result in tailor-made nanoparticles of controlled size distribution.[103–107] Recently, ionic liquids, such as those based on dialkylimidazolium cations, have demonstrated promise as both excellent solvents and stabilising ligands for metal nanoparticles while also exhibiting compatibility with PDMS microfluidic devices.[108–111]

Their high ionic charge, high dielectric constant, low interfacial tension and ability to form supramolecular hydrogen-bonded networks in the condensed phase make them ideal solvents for the synthesis and stabilization of metal nanoparticles.[108–112]

Malmstadt *et al.*[113] presented the synthesis of small monodisperse AuNPs using imidazolium-based ILs in a simple PDMS device. It was reported that the nanoparticles owed their high quality to the particle stabilisation effects of the IL solvent working cooperatively with the controlled and fast mixing from the microfluidic device. This technology was used for the fabrication of other high quality nanoparticles through control of reaction parameters, mixing rate and anion of the imidazolium-based IL. The same group fabricated gold and silver nanoparticles in an ionic liquid solvent using a simple droplet-based microfluidic device.[114] They presented a microfluidic platform that brings together a controlled microfluidic mixing with ionic solvents as in their previous work, but this time involving a simplified two-phase droplet flow for the continuous synthesis of high quality, small, monodisperse gold and silver nanoparticles (AuNPs and AgNPs, respectively). This method is reported as an inexpensive, rapid, and reproducible approach, which has minimal impact on the environment and opens the possibility of utilising such platforms for large-scale nano-manufacturing of nanoparticles.

After the synthesis of metal nanoparticles, the synergistic combination of ionic liquid and droplet microfluidic process was once again adopted for the nano-manufacturing of various inorganic materials by Kim and his co-workers.[115] Inorganic nanomaterials with unique properties and functions received much recent interest due to their wide variety of applications.[116–118] Despite the many efforts to achieve efficient, fast and continuous synthesis of nanostructured inorganic materials with uniform structure, shape and narrow size distribution, the synthetic process has always been a big challenge. This can be attributed to unmanageable mass and heat transfer and the harsh experimental conditions of high temperature and pressure.[115,119,120] Kim and his team presented a novel droplet–ionic liquid cooperated microfluidic (DIM) synthetic method, which reaps the benefits of both ionic liquids and droplet-assisted micro-reaction systems, enabling ultrafast, delicate and continuous synthesis of inorganic nanomaterials.[115] This method significantly reduced the reaction time from days to tens of minutes and produced a narrow size distribution and excellent crystalline qualities. As a proof of concept, three nanomaterials, ZSM-5, γ-AlOOH, and β-FeOOH, were synthesised with the DIM system with excellent quality.

9.7.2 Ionic Liquids for Building Microfluidics-based Power Generators

Kong *et al.*[121] introduced another outstanding application of ionic liquids in microfluidics by building an electrets-based microfluidic power generator (MPG) for harvesting vibrational energy using ionic liquids. It was demonstrated that the MPG with different ILs can generate varying amounts of power, depending on the different variation magnitude of the top contact area. An MPG using IL over water can operate for longer times in air over a wider range of operating temperatures. This power generator demonstrated the significant potential applications in harvesting low-frequency mechanical vibrations.

9.7.3 Ionic Liquids for Precise Temperature Control in Microfluidics

A novel application, to generate precise and accurate temperature control in microfluidic devices using Joule heating of ionic liquids, was presented by de Mello *et al.*[122] The relationship between the temperature and conductivity of ionic liquids allows for an internal temperature measurement, with the possibility for application within a thermocouple system. Microfluidic platforms with such fluidic heaters can be easily manufactured, using small amounts of ionic liquid and require no manipulation or replenishment of the heating medium once introduced into the heating microchannel. They also offer extremely long operational lifetimes (estimated order of hundreds of hours). This concept could be used in thermocycling applications, where integration of multiple heating channels along a co-running reaction microchannel should favour efficient and localised temperature control. Ongoing studies are focusing on the use of these heaters in microfluidic systems for small molecule synthesis within a temperature-controlled environment.

9.8 Conclusions

Lab-on-a-chip and micro-total analysis systems show great potential for the integration of multiple functional elements, to produce absolute sample-in/answer-out systems. However, due to the critical need for fluid control, fluid transport, separation and sensing, the need for high performance microfluidic components can provide significant obstacles for the development of low-cost, miniaturised microfluidic devices. Ionic liquid materials offer a solution for the fabrication of low cost and high performance microfluidic elements, which can improve the potential of microfluidic devices. Moreover, the promising results obtained from the use of ILs and microfluidics in separation science and chemical synthesis (nanoparticle generation) provide endless opportunity for this emerging area of research.

Acknowledgements

A. T., L. F., and D. D. are grateful for financial support from the Marie Curie Initial Training Network funded by the European Community's FP7 People Programme OrgBIO (Marie Curie ITN, GA607896) and Science Foundation Ireland (SFI) under the Insight Centre for Data Analytics initiative, Grant Number SFI/12/RC/2289. F.B.L. and J.S. acknowledge the Ramón y Cajal Programme (Ministerio de Economía y Competitividad) and also Marian M. De Pancorbo for letting them use her laboratory facilities at UPV/EHU. J. S., C. D., F. B. L and D. D. also acknowledge the European Union's Seventh Framework Programme for research, technological development, and demonstration; through the NAPES project grant agreement no. 604241.

References

1. M. Petkovic, K. R. Seddon, L. P. N. Rebelo and C. S. Pereira, *Chem. Soc. Rev.*, 2011, **40**, 1383–1403.
2. S. Zhang, J. Wang, X. Lu and Q. Zhou, *Structures and Bonding*, Springer-Verlag, Berlin-Heidelberg, 2014.
3. H. Passos, M. G. Freire and J. A. P. Coutinho, *Green Chem.*, 2014, **16**, 4786–4815.
4. A. Stark and K. R. Seddon, *Kirk-Othmer Encyclopaedia of Chemical Technology*, John Wiley & Sons, Inc., Hoboken, New Jersey, 2007.
5. K. R. Seddon, *J. Chem. Technol. Biotechnol.*, 1997, **68**, 351–356.
6. J. Dupont, P. A. Suarez and A. P. Umpierre, *Catal. Lett.*, 2000, **73**, 11–213.
7. B. Soares, H. Passos, C. S. R. Freire, J. A. P. Coutinho, A. J. D. Silvestre and M. G. Freire, *Green Chem.*, 2016, **18**, 4582–4604.
8. N. V. Plechkova and K. R. Seddon, *Chem. Soc. Rev.*, 2008, **37**, 23–150.
9. C. C. Weber, A. F. Masters and T. Maschmeyer, *Green Chem.*, 2013, **15**, 655–2679.

10. M. J. Earle, J. M. S. S. Esperanca, M. A. Gilea, J. N Canongia- Lopes, L. P. N. Rebelo, J. W. Magee, K. R. Seddon and J. A. Widegren, *Nature*, 2006, **439**, 831–834.
11. J. S. Wilkes, *Green Chem.*, 2002, **4**, 73–80.
12. J. M. S. S. Esperanca, J. N. Canongia Lopes, M. Tariq, L. M. N. B. F. Santos, J. W. Magee and L. P. N. Rebelo, *J. Chem. Eng. Data*, 2010, **55**, 3–12.
13. R. D. Rogers and K. R. Seddon, *Science*, 2003, **302**, 792–793.
14. F. Guo, S. Zhang, J. Wang, B. Teng, T. Zhang and M. Fan, *Curr. Org. Chem.*, 2015, **19**, 455–468.
15. J. Zhang and A. Bond, *Analyst*, 2005, **130**, 1132–1147.
16. L. P. N. Rebelo, J. N. C. Lopes, J. M. S. S. Esperanca, H. J. R. Guedes, J. Lachwa, V. Najdanovic-Visak and Z. P. Visak, *Acc. Chem. Res.*, 2007, **40**, 1114–1121.
17. T. Welton, *Chem. Rev.*, 1999, **99**, 2071–2083.
18. R. D. Rogers and K. R. Seddon, *Ionic Liquids 11 IB: Fundamentals, Progress, Challenges, and Opportunities*, American Chemical Society Symposium Series, Washington D.C, 2005.
19. J. Ranke, S. Stolte, R. Stoermann, J. Arning and B. Jastorff, *Chem. Rev.*, 2008, **107**, 2183–2206.
20. M. J. Earle and K. R. Seddon, *Pure Appl. Chem.*, 2000, **72**, 1391–1398.
21. M. G. Freire, A. F. M. Cláudio, J. M. M. Araújo, J. A. P. Coutinho, I. M. Marrucho, J. N. C. Lopes and L. P. N. Rebelo, *Chem. Soc. Rev.*, 2012, **41**, 4966–4995.
22. M. G. Freire, A. R. R. Teles, M. A. A. Rocha, B. Schröder, C. M. S. S. Neves, P. J. Carvalho, D. V. Evtuguin, L. M. N. B. F. Santos and J. A. P. Coutinho, *J. Chem. Eng. Data*, 2011, **56**, 4813–4822.
23. J. F. Wishart, *Energy Environ. Sci.*, 2009, **2**, 956–961.
24. J. Le Bideau, L. Viau and A. Vioux, *Chem. Soc. Rev.*, 2011, **40**, 907–925.
25. P. C. Marr and A. C. Marr, *Green Chem.*, 2016, **18**, 105–128.
26. C. T. Culbertson, T. G. Mickleburgh, S. A. Stewart-James, K. A. Sellens and M. Pressnall, *Anal. Chem.*, 2014, **86**, 95 118.
27. E. Livak-Dahl, I. Sinn and M. Burns, *Annu. Rev. Chem. Biomol. Eng.*, 2011, **2**, 325–353.
28. R. Byrne, F. Benito-Lopez and D. Diamond, *Mater. Today*, 2010, **13**, 16–23.
29. S. Sugiura, K. Sumaru, K. Ohi, K. Hiroki, T. Takagi and T. Kanamori, *Sens. Actuators, A*, 2007, **140**, 176–184.
30. F. Benito-Lopez, R. Byrne, A. Răduţă, N. Vrana, G. McGuinness and D. Diamond, *Lab Chip*, 2009, **10**, 195–201.
31. M. Czugala, C. O'Connell, C. Blin, P. Fischer, K. J. Fraser, F. Benito-Lopez and D. Diamond, *Sens. Actuators, B*, 2014, **194**, 105–113.
32. M. Czugala, C. Fay, N. E. O'Connor, B. Corcoran, F. Benito-Lopez and D. Diamond, *Talanta*, 2013, **116**, 997–1004.
33. F. Benito-Lopez, M. Antoñana-Díez, V. F. Curto, D. Diamond and V. Castro-López, *Lab Chip*, 2014, **14**, 3530–3538.

34. S. Gallagher, A. Kavanagh, B. Ziółkowski, L. Florea, D. R. MacFarlane, K. Fraser and D. Diamond, *Phys. Chem. Chem. Phys.*, 2014, **16**, 3610–3616.
35. A. K. Ghamsari, E. Zegeye, Y. Jin and E. Woldesenbet, *ACS Appl. Mater. Interfaces*, 2013, **5**, 5408–5412.
36. T. Akyazi, J. Saez, J. Elizalde and F. Benito-Lopez, *Sens. Actuators, B*, 2016, **233**, 402–408.
37. A. Kavanagh, R. Byrne, D. Diamond and K. J. Fraser, *Membranes*, 2012, **2**, 16.
38. K. Behera, S. Pandey, A. Kadyan and S. Pandey, *Sensors*, 2015, **15**, 29813.
39. S. Y. Yang, F. Cicoira, R. Byrne, F. Benito-Lopez, D. Diamond, R. M. Owens and G. G. Malliaras, *Chem. Commun.*, 2010, **46**, 7972–7974.
40. R. Madeira Lau, M. J. Sorgedrager, G. Carrea, F. van Rantwijk, F. Secundo and R. A. Sheldon, *Green Chem.*, 2004, **6**, 483–487.
41. M. Yamaguchi, M. Mitsumori and Y. Kano, *IEEE Eng. Med. Biol. Mag.*, 1998, **17**, 59–63.
42. D. Khodagholy, V. F. Curto, K. J. Fraser, M. Gurfinkel, R. Byrne, D. Diamond, G. G. Malliaras, F. Benito-Lopez and R. M. Owens, *J. Mater. Chem.*, 2012, **22**, 4440–4443.
43. D. Khodagholy, V. F. Curto, K. J. Fraser, M. Gurfinkel, R. Byrne, D. Diamond, G. G. Malliaras, F. Benito-Lopez and R. M. Owens, *J. Mater. Chem.*, 2012, **22**, 4440.
44. V. F. Curto, S. Scheuermann, R. M. Owens, V. Ranganathan, D. R. MacFarlane, F. Benito-Lopez and D. Diamond, *Phys. Chem. Chem. Phys.*, 2014, **16**, 1841–1849.
45. M. Czugala, R. Gorkin Iii, T. Phelan, J. Gaughran, V. F. Curto, J. Ducree, D. Diamond and F. Benito-Lopez, *Lab Chip*, 2012, **12**, 5069–5078.
46. V. F. Curto, C. Fay, S. Coyle, R. Byrne, D. Diamond and F. Benito-Lopez, *Wearable Micro-fluidic pH Sensing Device Based on Colorimetric Imaging Techniques*, Micrototal Analysis Systems, MicroTas 2011 Conference, 2–6 Oct. 2011, Seattle, Washington, USA, 2011, pp. 577–579.
47. E. G. Yanes, S. R. Gratz, M. J. Baldwin, S. E. Robison and A. M. Stalcup, *Anal. Chem.*, 2001, **73**, 3838–3844.
48. Y. Xu and E. Wang, *J. Chromatogr. A*, 2009, **1216**, 4817–4823.
49. S. G. Yoon, H. J. Koo and S. T. Chang, *ACS Appl. Mater. Interfaces*, 2015, **7**, 27562–27570.
50. C.-Y. Wu, W.-H. Liao and Y.-C. Tung, *Lab Chip*, 2011, **11**, 1740–1746.
51. M.-C. Liu, H.-C. Shih, J.-G. Wu, T.-W. Weng, C.-Y. Wu, J.-C. Lu and Y.-C. Tung, *Lab Chip*, 2013, **13**, 1743–1753.
52. T. Welton, *Chem. Rev.*, 1999, **99**, 2071–2084.
53. R. D. Rogers and K. R. Seddon, *Science*, 2003, **302**, 792–793.
54. W. Kunz and K. Häckl, *Chem. Phys. Lett.*, 2016, **661**, 6–12.
55. S. Weidmann, S. Kemmerling, S. Mdler, H. Stahlberg and R. Zenobi, *Eur. J. Mass Spectrom.*, 2012, **18**, 279.
56. M. Tijero, R. Díez-Ahedo, F. Benito-Lopez, L. Basabe-Desmonts, V. Castro-López and A. Valero, *Biomicrofluidics*, 2015, **9**, 044124.

57. V. van Steijn, M. T. Kreutzer and C. R. Kleijn, *Chem. Eng. Sci.*, 2007, **62**, 7505–7514.
58. J. Gong, *Lab Chip*, 2008, **8**, 898–906.
59. D. Baigl, *Lab Chip*, 2012, **12**, 3637–3653.
60. P. Dubois, G. Marchand, Y. Fouillet, J. Berthier, T. Douki, F. Hassine, S. Gmouh and M. Vaultier, *Anal. Chem.*, 2006, **78**, 4909–4917.
61. I. Moon and J. Kim, *Sens. Actuators, B*, 2006, **130**, 537–544.
62. V. Srinivasan, V. Pamula, M. Pollack and R. Fair, *A Digital Microfluidic Biosensor for Multianalyte Detection, Micro Electro Mechanical Systems, IEEE The Sixteenth Annual International Conference, Kyoto, Jan 19–23*, 2003, pp. 327–330.
63. Q. Xu, M. Hashimoto, T. T. Dang, T. Hoare, D. S. Kohane, G. M. Whitesides, R. Langer and D. G. Anderson, *Small*, 2009, **5**, 1575–1581.
64. K. Choi, A. Ng, R. Fobel and A. Wheeler, *Annu. Rev. Anal. Chem.*, 2012, **5**, 413–440.
65. Y. S. Nanayakkara, H. Moon, T. Payagala, A. B. Wijeratne, J. A. Crank, P. S. Sharma and D. W. Armstrong, *Anal. Chem.*, 2008, **80**, 7690–7698.
66. M. Paneru, C. Priest, R. Sedev and J. Ralston, *J. Am. Chem. Soc.*, 2010, **132**, 8301–8308.
67. S. Millefiorini, A. H. Tkaczyk, R. Sedev, J. Efthimiadis and J. Ralston, *J. Am. Chem. Soc.*, 2006, **128**, 3098–3101.
68. H. Li, M. Paneru, R. Sedev and J. Ralston, *Langmuir*, 2013, **29**, 2631–2639.
69. X. Hu, S. Zhang, C. Qu, Q. Zhang, L. Lu, X. Ma, X. Zhang and Y. Deng, *Soft Matter*, 2011, 7, 5941–5943.
70. Y. S. Nanayakkara, H. Moon and D. W. Armstrong, *ACS Appl. Mater. Interfaces*, 2010, **2**, 1785–1787.
71. A. Al Amin, A. Jagtiani, A. Vasudev, J. Hu and J. Zhe, *J. Micromech. Microeng.*, 2011, **21**, 125025.
72. A. Vasudev and J. Zhe, *Appl. Phys. Lett.*, 2008, **93**, 103503.
73. I. Lagzi, S. Soh, P. Wesson, K. Browne and B. Grzybowski, *J. Am. Chem. Soc.*, 2010, **132**, 1198–1199.
74. L. Florea, K. Wagner, P. Wagner, G. G. Wallace, F. Benito-Lopez, D. L. Officer and D. Diamond, *Adv. Mater.*, 2014, **26**, 7339–7345.
75. J. L. Anderson, V. Pino, E. C. Hagberg, V. V. Sheares and D. W. Armstrong, *Chem. Commun.*, 2003, **87**, 2444–2445.
76. W. Francis, C. Fay, L. Florea and D. Diamond, *Chem. Commun.*, 2015, **51**, 2342–2344.
77. W. Francis, K. Wagner, S. Beirne, D. L. Officer, G. G. Wallace, L. Florea and D. Diamond, *Sens. Actuators, B*, 2017, **239**, 1069–1075.
78. G. Marchand, P. Dubois, C. Delattre, F. Vinet, M. Blanchard-Desce and M. Vaultier, *Anal. Chem.*, 2008, **80**, 6051–6055.
79. X. Feng, Y. Yi, X. Yu, D.-W. Pang and Z.-L. Zhang, *Lab Chip*, 2009, **10**, 313–319.
80. Z. Barikbin, Md. T. Rahman, P. Parthiban, A. S. Rane, V. Jain, S. Duraiswamy, S. H. S. Lee and S. A. Khan, *Lab Chip*, 2010, **10**, 2458–2463.

81. S. Mashaghi, A. Abbaspourrad, D. A. Weitz and A. M. Van Oijen, *Trends Anal. Chem.*, 2016, **82**, 118–125.
82. J.-P. Huang, X.-H. Ge, J.-H. Xu and G.-S. Luo, *Chem. Eng. Sci.*, 2016, **152**, 293–300.
83. V. Misuk, A. Mai, K. Giannopoulos, F. Alobaid, B. Epple and H. Loewe, *Lab Chip*, 2013, **13**, 4542–4548.
84. L. Bai, Y. Fu, S. Zhao and Y. Cheng, *Chem. Eng. Sci.*, 2016, **145**, 141–148.
85. J. W. Hwang, J.-H. Choi, B. Choi, G. Lee, S. W. Lee, Y.-M. Koo and W.-J. Chang, *Korean J. Chem. Eng.*, 2016, **33**, 57–62.
86. D. Chatterjee, B. Hetayothin, A. R. Wheeler, D. J. King and R. L. Garrell, *Lab Chip*, 2005, **6**, 199–206.
87. I. Alvarez-Martos, F. J. Garcia-Alonso, A. Anillo, P. Arias-Abrodo, M. D. Gutierrez-Alvarez, A. Costa-Garcia and M. T. Fernandez-Abedul, *Sens. Actuators, B*, 2013, **188**, 837–846.
88. M. Sun, P. Agarwal, S. Zhao, Y. Zhao, X. Lu and X. He, *Anal. Chem.*, 2016, **88**(16), 8264–8271.
89. Y. S. Huh, Y.-S. Jun, Y. K. Hong and D. H. Kim, *J. Mol. Catal. B: Enzym.*, 2006, **43**, 96–101.
90. J.-P. Fan, J. Cao, X.-H. Zhang and X.-K. Ouyang, *Sep. Sci. Technol.*, 2012, **47**(12), 1740–1747.
91. L. Qi, Y. Wang, Y. Li, G. Zheng, C. Li and H. Su, *Anal. Bioanal. Chem.*, 2015, **407**, 3617–3625.
92. U. Novak, A. Pohar, I. Plazl and P. Znidarsic, *Sep. Purif. Technol.*, 2012, **97**, 172–178.
93. Y. Xu, J. Li and E. Wang, *J. Chromatogr. A*, 2008, **1207**, 175–180.
94. Y. S. Huh, C.-M. Jeong, H. N. Chang, S. Y. Lee, W. H. Hong and T. J. Park, *Biomicrofluidics*, 2010, **4**, 014103.
95. W. Lan, S. Li, J. Xu and G. Luo, *Ind. Eng. Chem. Res.*, 2013, **52**, 6770–6777.
96. W. W. Yang, Y. C. Lu, Z. Y. Xiang and G. S. Luo, *React. Funct. Polym.*, 2007, **67**, 81–86.
97. Z. Y. Xiang, Y. C. Lu, Y. Zou, X. C. Gong and G. S. Luo, *React. Funct. Polym.*, 2008, **68**, 1260–1265.
98. Y. Xu and E. Wang, *J. Chromatogr. A*, 2009, **1216**, 4817–4823.
99. C.-H. Chang, B. K. Paul, V. T. Remcho, S. Atre and J. E. Hutchison, *J. Nanopart. Res.*, 2008, **10**, 965.
100. Y. Song, J. Hormes and C. S. S. R. Kumar, *Small*, 2008, **4**, 698.
101. A. J. de Mello, *Nature*, 2006, **442**, 394.
102. H. Song, D. L. Chen and R. F. Ismagilov, *Angew. Chem., Int. Ed.*, 2006, **45**, 7336–7356.
103. Y. Song, H. Modrow, L. L. Henry, C. K. Saw, E. E. Doomes, V. Palshin, J. Hormes and C. S. S. R. Kumar, *Chem. Mater.*, 2006, **18**, 2817.
104. E. M. Chan, R. A. Mathies and A. P. Alivisatos, *Nano Lett.*, 2003, **3**, 199.
105. J. Wagner and J. M. Kohler, *Nano Lett.*, 2005, **5**, 685.
106. S. A. Khan, A. Gunther, M. A. Schmidt and K. F. Jensen, *Langmuir*, 2004, **20**, 8604.

107. S. Krishnadasan, R. J. C. Brown, A. J. deMello and J. C. deMello, *Lab Chip*, 2007, **7**, 1434.
108. J. Dupont and J. D. Scholten, *Chem. Soc. Rev.*, 2010, **39**, 1780.
109. M.-A. Neouze, *J. Mater. Chem.*, 2010, **20**, 9593–9607.
110. D. Marquardt, Z. Xie, A. Taubert, R. Thomann and C. Janiak, *Dalton Trans.*, 2011, **40**, 8290–8293.
111. J. N. Lee, C. Park and G. M. Whitesides, *Anal. Chem.*, 2003, **75**, 6544–6554.
112. Z. Ma, J. Yu and S. Dai, *Adv. Mater.*, 2010, **22**, 261–285.
113. L. L. Lazarus, A. S.-J. Yang, S. Chu, R. L. Brutchey and N. Malmstadt, *Lab Chip*, 2010, **10**, 3377–3379.
114. L. L. Lazarus, C. T. Riche, B. C. Marin, M. Gupta, N. Malmstadt and R. L. Brutchey, *ACS Appl. Mater. Interfaces*, 2012, **4**, 3077–3083.
115. P. H. Hoang, H. Park and D. Kim, *J. Am. Chem. Soc.*, 2011, **133**, 14765–14770.
116. K. J. Van Bommel, C. A. Friggeri and S. Shinkai, *Angew. Chem., Int. Ed.*, 2003, **42**, 980.
117. B. L. Cushing, V. L. Kolesnichenko and C. J. O'Connor, *Chem. Rev.*, 2004, **104**, 3893.
118. Y. Jun, J. Choi and J. Cheon, *Angew. Chem., Int. Ed.*, 2006, **45**, 3414.
119. A. Abou-Hassan, O. Sandre and V. Cabuil, *Angew. Chem., Int. Ed.*, 2010, **49**, 6268.
120. S. Duraiswamy and S. A. Khan, *Small*, 2009, **5**, 2828.
121. W. Kong, L. Cheng, X. He, Z. Xu, X. Ma, Y. He, L. Lu, X. Zhang and Y. Deng, *Microfluid. Nanofluid.*, 2015, **18**, 1299–1307.
122. A. J. de Mello, M. Habgood, N. L. Lancaster, T. Welton and R. C. R. Wootton, *Lab Chip*, 2004, **4**, 417–419.

CHAPTER 10

Recognition-based Smart Ionic Liquids

YI-PIN CHANG[a] AND YEN-HO CHU*[b]

[a]The Forsyth Institute, 245 First Street, Cambridge, MA 02142, USA; [b]Department of Chemistry and Biochemistry, National Chung Cheng University, 168 University Road, Minhsiung, Chiayi 62102, Taiwan, Republic of China
*E-mail: cheyhc@ccu.edu.tw

10.1 Introduction

Biomolecular recognition involves specific intermolecular forces between molecular units through non-covalent bonding such as the electrostatic effect, hydrogen bonding, hydrophobic interaction and metal coordination. While these direct forces mediate interactions between biomolecules in solution, solvents generally contribute an indirect effect on recognition events concomitantly.[1] In biomolecular recognition, the host and guest exhibit molecular complementarity which has been considered to be a critical phenomenon in the chemical origins of life.[2] Learning from nature, the application of molecular complementarity is ubiquitous in academia and industry, where solvents usually take part as simply a carrier or media and do not directly participate in chemical reactions or biomolecular recognition. Intriguingly, water acts as not only a solvent but also engages in numerous important biochemical reactions in living organisms, hence it is literally the essence of life. Is it possible to extract a specific biomolecule from the

Smart Materials No. 28
Ionic Liquid Devices
Edited by Ali Eftekhari

Published by the Royal Society of Chemistry, www.rsc.org

"primordial soup" into an organic phase and then recover it in the aqueous phase?

Extraction, separation and purification of biomacromolecules (*e.g.* proteins, nucleic acids, carbohydrates or lipids) have become essential daily routines in basic research and biopharmaceutical industry, in which the solvent plays a pivotal role. Among the purification methods, solvent extraction and affinity purification have been extensively used for decades. Affinity purification typically involves three steps for both column chromatography and batch method, *i.e.*, equilibrium of crude sample, washing of unbound components and elution of desired molecule. The biomolecule to be purified must have an inherent or engineered recognition site that binds to the affinity ligand on the solid phase such as the widely used purification of polyhistidine-tagged recombinant proteins by Ni(II)-chelated matrix.[3] The process can be considered an entrapment – the dissolved target biomolecule in aqueous phase is trapped on a solid phase, which relies on the specific and reversible interactions between the affinity tag and its binding counterpart. On the other hand, extraction is the pulling or drawing of certain compound(s) from one phase to another, where the distribution coefficient is the key factor for the selection of solvent(s). The most prominent and classical form of solid–liquid extraction is the brewing of tea or the percolation of coffee. In practice, liquid–liquid extraction or partitioning is commonly performed in two immiscible liquids, usually an organic solvent and water in a separatory funnel, as the "work-up" step to give the crude product after an organic synthesis. A rudimentary examination of the theoretical basis of both techniques suggests that the highly expandable surface area of the two-phase system and reversible biospecific interactions are the major advantages of solvent extraction and affinity purification, respectively. While each technique works well separately in general, hurdles and limitations can arise in some cases. Is it possible to develop a hybrid method based on smart solvents to have the best of both worlds?

Dissolution of compounds is of essential importance in all chemical processes, including purification methods (*e.g.* affinity purification, centrifugation, crystallization, electrophoresis and liquid–liquid extraction) and biosensing technology (*e.g.* electronic nose, quartz crystal microbalance and surface plasmon resonance). The dissolution of gases, liquids and solids into water, a liquid or other solvents is a phenomenon by which these initial states become solutes (dissolved constituents), *i.e.*, forming a solution. When a solution is formed, a substance is dissolved into a solvent, in which all of the constituents are uniformly distributed at the molecular level. A solvent is typically a liquid but can also be a solid or a gas. A metal alloy is an example of a solid solution, *e.g.* melchior is an alloy of copper and nickel, the high corrosion resistance and silvery color of which are preferred in the manufacture of household utensils. As for liquid and gas solutions, intermolecular non-covalent interactions are formed between the molecules and the solvent. The general rule of thumb for the selection of solvent is "like dissolves like" and the free energy of the overall process must be negative for dissolution to occur. Water is a polar solvent for polar molecules and a variety of biomolecules in living organisms dissolve in water. Organic solvents have various applications in chemical synthesis and purification methods. Understandably, the selection of solvents to meet criteria for a certain task is an art, including evaporation rate, solution viscosity, environmental concern and effectiveness of a solvent to dissolve adequately target molecule while leaving other compounds unaffected. Based on unique physical and chemical properties, ionic liquids or room temperature molten salts have been developed as a new choice of solvent in organic synthesis, biomass processing, solar energy and batteries, if not as a more environment-friendly substitute. In addition, this new class of "designer solvents" offers a number of very exciting applications in biomolecular recognition. Recent advances in ionic liquid and biosensing technology have allowed the development of an ultra-sensitive gas sensing system that can detect in real-time and discriminate volatile organic compounds (VOCs) chemoselectively.[4] These gas sensing ionic liquids serve as solvent, reactant and sensing material from physical, chemical and biosensing perspectives, respectively.[5–8] Notably, these reaction-based ionic liquids capture VOCs with irreversible binding interactions that contribute to the high sensitivity and selectivity. To develop an affinity purification or solvent extraction method, a reversible interaction system is required to recover the target molecule.

Liquid–liquid extraction is a method used to separate compounds based on their relative solubilities in two distinct immiscible solvents; thus, how to fine-tune non-covalent interactions between solute and solvent is the key question. Solubility is determined by the intermolecular interactions between the solute and the solvent, and the entropy change upon solvation. Temperature and pressure are two extrinsic factors that can change solubility. Depending on the physical property of the solute, solubility may increase or decrease with temperature, but the influence on solubility by pressure change is usually small and neglectable. In general, immiscible aqueous and organic solvents are used in solvent extraction based on the rule of thumb, *i.e.*, polar compounds are more soluble in aqueous solutions than they are in non-polar or organic solvents, and *vice versa*. As the saying goes, there is always an exception to the rule. Interestingly, liquid–liquid extraction is not impossible in non-aqueous systems despite the rarity. In a liquid battery system, which consists of two liquid metals in contact with a molten salt, one metal cation can be extracted from one phase to the other. This strong interaction between the liquid metals provides the thermodynamic driving force (cell voltage) for liquid metal batteries.[9] Liquid–liquid extraction from molten salts is also an important method in hydrometallurgy for the separation and purification of rare earth metals. In the extraction process, metal ions can be distributed between either an aqueous layer and an organic layer or two immiscible organic solvents. Recently, two mutually immiscible low temperature molten salts have been used as solvent extraction systems for separations of cobalt, nickel and mixtures of rare earth metals.[10,11] An intriguing example of liquid–liquid extraction for biomolecules is the purification of DNA from cell lysate by a polymer–salt separation system. The typical issue for plasmid purification is degradation of DNA by nucleases. It has been shown that nucleases can be partitioned into the heavy PEG phase and then be deactivated by the incorporated ligands known to bind and deactivate nucleases, while target DNA fragments can be harvested in the light sodium citrate phase.[12] Another elegant paradigm of solvent extraction for biomolecules is the biorefinery of lignocellulosic biomass by ionic liquids. The breakthrough of the dissolution of cellulose in ionic liquids provides new opportunities for the processing of this biopolymer, which may ultimately become a green and sustainable strategy.[13]

To date, ionic liquids have emerged as versatile solvents for a number of applications in protein chemistry, including protein catalysis, crystallization, extraction and stabilization.[14] Ionic liquids are organic molten salts with melting points below 100 °C, and the following major properties make ionic liquids distinct from other molecular liquids: neglectable vapor pressure, non-flammable, highly polar, high thermostability and usually immiscible with organic solvents. Interestingly, while a plethora of ionic liquids has been developed by chemists in laboratories over the last few decades, the "voila moments" of mixing ions and counterions have long existed inconspicuously in nature. Tawny crazy ants secret venom of their own to neutralize the other when under attack by the deadly venom of fire ants. A naturally occurring protic ionic liquid is formed upon the mixing of the alkaloid-based venom and the formic acid venom.[15] As ionic liquids are made up exclusively of ions, their solvent properties can be further altered and controlled intrinsically by the engineering of cations or anions. Recently, significant progress has been made in exploring functionalized ionic liquids through the incorporation of functional groups to equip artificial ionic liquids with tailored properties for bioanalytical separation and preconcentration.[16] Moreover, the development of protein purification methods through traditional or novel techniques has never faded away. Imidazolium-based ionic liquids have been used to extract bovine serum albumin, cytochrome *c*, γ-globulins and trypsin, and effects of pH values and temperature have been investigated.[17] As liquid–liquid extraction is very much like a tug of war between the selected solvents, the ability to direct the driving force for the unequal distribution of target molecule in solutions is the solution for an effective partition. It has been suggested that partition of proteins in aqueous two-phase systems is driven by a combination of electrostatic interactions, hydrophobic interaction and salting-out effect.[18] In the following sections, how extra non-covalent interactions are intrinsically incorporated into smart ionic liquids for the affinity extraction of selected proteins and peptides will be presented. The cartoon in Figure 10.1 vignettes the explicit and implicit intermolecular interactions that may be involved in protein extraction by affinity ionic liquid (AIL).

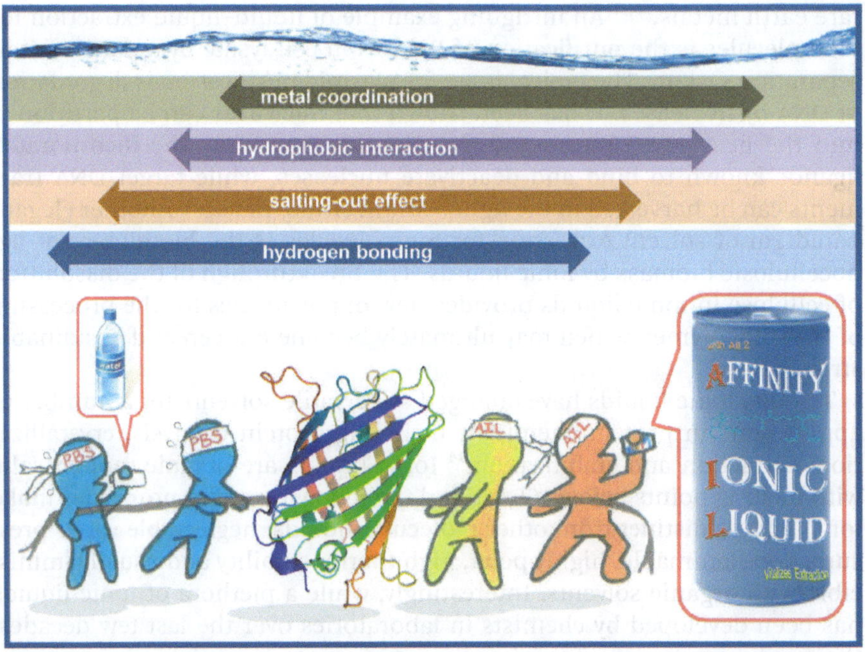

Figure 10.1 Illustrative representation of liquid–liquid extraction of protein by affinity ionic liquid and PBS buffer. The proposed intermolecular forces are not exhaustive but aim to superficially predict the driving forces that might participate in the tug of war of the two immiscible phases. Artwork by Jessica Lim.

10.2 Bicyclic Imidazolium Ionic Liquid for Affinity Purification

Affinity purification is a powerful and perhaps the most effective method for protein purification if the interaction of the target protein and the baiting molecule is highly specific. Each affinity method has its own set of ligand–target binding systems, which allows the purification of virtually any protein without knowing much of its biochemical properties, if the affinity tag can be captured by the ligand coupled on a matrix.[19] Although this technique was first introduced half a century ago, there remains room for improvement and new methods based on affinity tag, functionalized linker, solid support and solvent have been reported constantly.[20,21] The search for more robust and selective ligands has been a trend, such as the use of synthetic hapten for the non-chromatographic purification of a monoclonal antibody from biological fluids.[22] When weighing the pros and cons of affinity purification, the superior selectivity and high yield of columns and resins come at a high price. By contrast, liquid–liquid extraction does not require expensive solid supports, but the two immiscible solvents may not be specific enough and co-extraction is not uncommon. Ionic liquids, which possess remarkable solubility with a wide range of compounds, have shown their superior attributes toward extraction of metal ions and amino acids.[23] Moreover, ionic liquids are polar solvents and have also been demonstrated as excellent media for peptide synthesis.[24] An ionic liquid/aqueous two-phase system based on the hydrophilic ionic liquid 1-butyl-3-methylimidazolium chloride has been employed for the extraction of proteins from human body fluids.[25] In addition, ionic liquid solutions have also been used to break cell walls for the direct extraction of proteins from yeast, despite the harsh condition (100 °C) and thus, only the primary structures were retained.[26] Here, we present the first affinity ionic liquid (AIL) that is capable of extracting peptide and protein quantitatively from aqueous buffer by acting as a smart solvent and a functional ligand in liquid–liquid extraction and affinity purification, respectively.[27] In our initial attempt, a bicyclic imidazolium-based **AIL 1**, a fluorescein-labeled hexahistidine peptide FITC-(His)$_6$-NH$_2$ (**1**) and a His-tagged green fluorescent protein (**2**, GFP) were prepared for the first model system.

We previously reported a series of bicyclic imidazolium ionic liquids that were far more chemically stable than the common [bmim][PF$_6$] and [bdmim][PF$_6$] ionic liquids.[28] The chemical property of this class of ionic liquid is desirable for our first attempt and thus the bicyclic imidazolium core was incorporated for the design and synthesis of the AIL (Scheme 10.1).

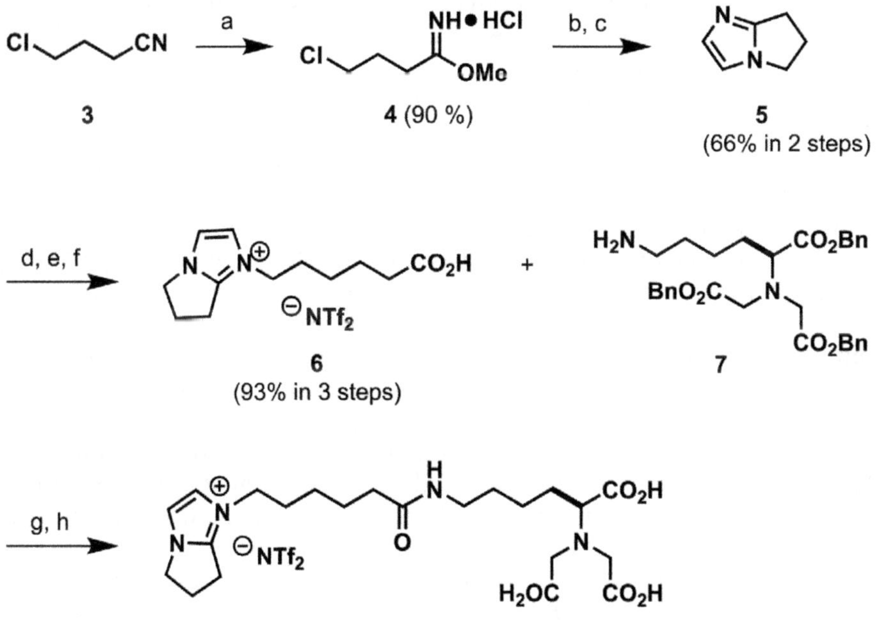

Scheme 10.1 Synthesis of affinity ionic liquid **AIL 1**. (a) HCl, MeOH, diethyl ether, −20 °C, 24 h; (b) H$_2$NCH$_2$CH(OMe)$_2$, Et$_3$N, CH$_2$Cl$_2$, reflux, 2 h; (c) HCO$_2$H, reflux, 20 h; (d) Br(CH$_2$)$_5$CO$_2$Bn, 80 °C, 2 h; (e) H$_2$, Pd(OH)$_2$/C, MeOH, room temperature (rt), 1 h; (f) LiNTf$_2$, H$_2$O, rt, 12 h; (g) EDC, DIEA, CH$_2$Cl$_2$, rt, 5 h; (h) H$_2$, Pd(OH)$_2$/C, MeOH, rt, 1 h.

Briefly, the synthesis of **AIL 1** commenced with commercially available 4-chlorobutyronitrile (**3**) to give the intermediate imidate (**4**) and then the key bicyclic imidazole (**5**) was prepared by the previously employed protocol (59% isolated yield in three steps).[28] Direct alkylation of the resulting imidazole with benzyl 6-bromohexanoate was performed under heated conditions. Subsequently, deprotection was achieved by catalytic hydrogenation and then an ion exchange in water readily gives the precursor ionic liquid acid (**6**) (93% isolated yield in three steps). Finally, EDC coupling of the precursor with a protected NTA amine (**7**) in dichloromethane completed the preparation of **AIL 1** (76% yield in two steps). The overall isolated yield of **AIL 1** was 42% with a total of eight steps. **AIL 1** was a viscous liquid at room temperature.

Affinity extraction of the fluorescent peptide **1** was performed in the ionic liquid solution (**AIL 1** dissolved in the ionic liquid [bdmim][NTf_2] (**8**)) and PBS buffer. Equal volumes of the organic layer and aqueous layer were mixed and then gently shaken to carry out the extraction at room temperature. As shown in Figure 10.2 (tube c), upon association with Ni(II), the **AIL 1** dissolved in ionic liquid **8** (bottom layer) recognized and captured the peptide **1** in PBS buffer (upper layer). The partition of peptide **1** was Ni(II)-chelated **AIL 1** dependent, as the upper layers of ionic liquid **8** only (tube a), ionic liquid **8** without Ni(II) (tube b) and ionic liquid **8** with Ni(II)-chelated **AIL 1** plus imidazole remained fluorescent. In other words, the affinity extraction of peptide **1** from the aqueous phase into the ionic liquid phase was solely due to affinity recognition by Ni(II)-chelated **AIL 1** and the binding was reversible. The quantitative extraction was easily distinguished with the naked eye through the distinct green fluorescence. Note that the fluorescence of peptide **1** in the ionic liquid phase was quenched completely, which is in agreement with a previous report that ionic liquid could act as fluorescence quencher.[29] Strikingly, not only was the peptide extracted into the organic layer, it was also soluble in the ionic liquid layer as no precipitation was observed at the aqueous–organic interface. From a receptor–ligand interaction perspective, the stoichiometry of the reversible binding is valuable to measure active receptor concentration and to probe ligand specificity. To determine the stoichiometry, affinity extractions of various concentrations of Fmoc-$(His)_6$-NH_2 peptide (**9**) in PBS buffer were performed with a fixed concentration of the ionic liquid solution. The Fmoc-$(Gly)_6$-NH_2 peptide (**10**) was used as the internal standard. Peptides **9** and **10** were used because they could be readily separated and quantified by HPLC. An expected 1 : 1 binding system was elucidated through a series of data obtained by C_{18}-HPLC, which demonstrated a stable peptide–ionic liquid complex was formed (Figure 10.3). This binding integrity based on non-covalent and reversible binding is the first step for the affinity purification of His-tagged proteins.

The direct one-step isolation and purification of His-tagged proteins from crude *Escherichia coli* lysate by AIL was the immediate goal of this biosensing system. GFP and its fusion variants have been used as fluorescent reporters

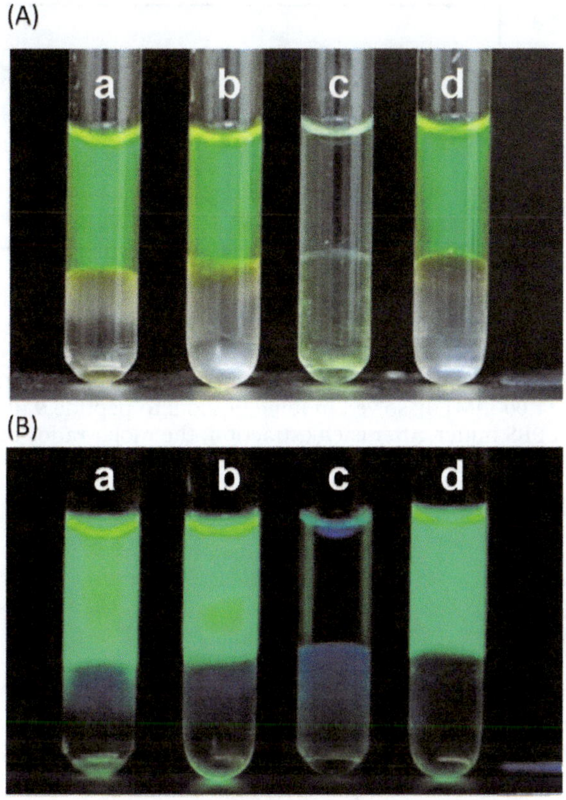

Figure 10.2 Affinity extraction of the fluorescent peptide **1** in PBS buffer (upper layer) by **AIL 1** solubilized in ionic liquid **8** (bottom layer). (A) and (B) Optical and fluorescence photographs, respectively. (a) peptide **1** (100 mM) in PBS buffer and ionic liquid **8** only; (b) peptide **1** (100 mM) in PBS buffer and ionic liquid **8** with **AIL 1** (20 mM) but without $NiCl_2$; (c) extraction of peptide **1** (100 mM) in PBS buffer into ionic liquid layer containing Ni(II)-chelated **AIL 1** (20 mM) in ionic liquid **8**; (d) competitive extraction of peptide **1** (100 mM) in ionic liquid layer into the PBS buffer using imidazole (500 mM). Note that the fluorescence of the peptide **1** was, in (c), quenched in ionic liquid (bottom layer) and, in (d), re-established in aqueous buffer (upper layer). Reproduced from ref. 27 with permission from The Royal Society of Chemistry.

in a variety of applications in gene expression, intracellular dynamics and protein localization, therefore His-tagged GFP was expressed in *E. coli* for the assessment of affinity purification.[30]

As shown in Figure 10.4, SDS-polyacrylamide gel electrophoresis demonstrated that His-tagged GFP was effectively extracted from the crude cell lysate into the ionic liquid layer and could be competitively back-extracted by imidazole to a newly prepared buffer solution. Notably, the extracted His-tagged GFP continues to emit fluorescence in the ionic liquid layer as shown

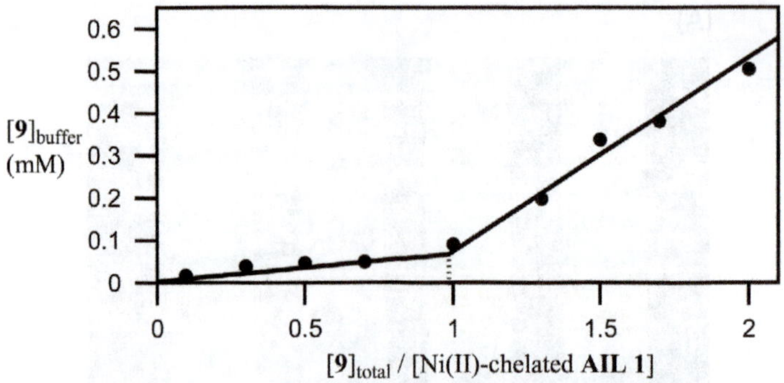

Figure 10.3 Determination of the binding stoichiometry of the Ni(II)-chelated **AIL 1** (1.00 mM) dissolved in ionic liquid **8** to peptide **9** (0.067–1.33 mM) in PBS buffer. After each extraction, the molar ratios of [**9**]/[**10**] in the aqueous layer were quantitatively determined at 280 nm by C_{18}-HPLC.

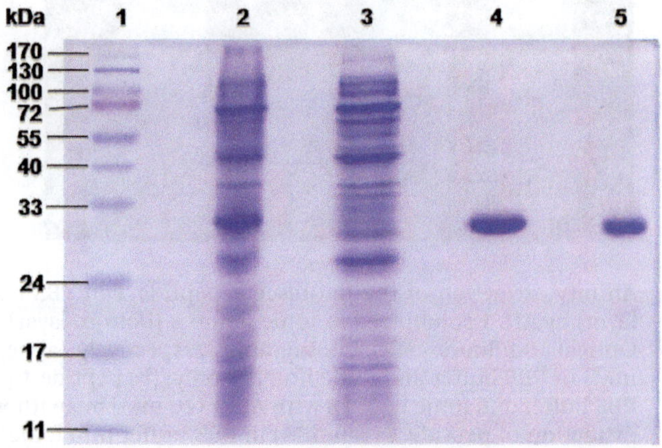

Figure 10.4 SDS-polyacrylamide gel electrophoresis of direct extraction of the His-tagged GFP in crude *Escherichia coli* cell lysate by the Ni(II)-chelated **AIL 1**. Lane 1, protein molecular weight markers (170, 130, 100, 72, 55, 40, 33, 24, 17 and 11 kDa); lane 2, total cell lysate; lane 3, the cell lysate after affinity extraction by Ni(II)-chelated **AIL 1**; lane 4, competitive extraction of His-tagged GFP in ionic liquid into aqueous buffer by imidazole; lane 5, purified GFP (2 g). Reproduced from ref. 27 with permission from The Royal Society of Chemistry.

in Figure 10.5. The unquenched fluorescence suggests that the conformation of the protein most likely remains intact, as only GFP with the correct tertiary structure can generate fluorescence.[31] Again, no precipitation of GFP at the water–ionic liquid interface was observed. Interestingly, upon affinity extraction, the protein-bound **AIL 1** uniformly dispersed in the ionic liquid

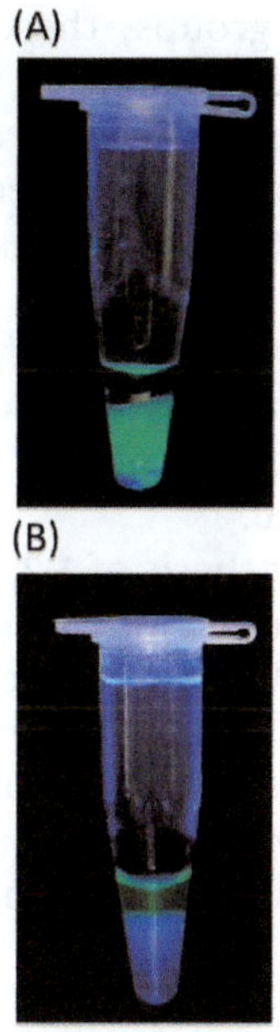

Figure 10.5 Direct affinity extraction of the His-tagged GFP from cell lysate: (A) before extraction and (B) after extraction. The upper layer was aqueous phase containing cell lysate (5 mL) in PBS (25 mL) and the bottom layer was Ni(II)-chelated **AIL 1** (89 mM) in ionic liquid **8** (30 mL). Reproduced from ref. 27 with permission from The Royal Society of Chemistry.

phase, but not in the aqueous layer. We reasoned that Ni(II)-chelated **AIL 1** has a strong affinity to His-tagged GFP and thus tightly coordinates with it to enable the complex formation, ultimately pulling the protein from the aqueous layer and resulting in quantitative transfer into the ionic liquid phase in suspension.

10.3 Crowned 1,2,3-Triazolium Ionic Liquid for Biomolecular Recognition

Crown ethers are macrocyclic compounds that bind certain cations tightly and form complexes. Along with the synthetic coronands, crown ethers have been employed in extraction systems that are selective for natural and synthetic molecules depending on the size of the crown rings.[32] The oxygen atoms are well positioned to coordinate with a cation in the interior of the ring, whereas the exterior of the ring is hydrophobic. Apart from its high affinity for cations, 18-crown-6 (18C6) is known to recognize the positively charged side chain of lysine, *i.e.*, protonated primary amine. The complex is stabilized through a hydrogen bond network between the three hydrogen atoms of protonated amine and three oxygen atoms of 18C6.[33]

By incorporating luminescent groups, these host molecules have been proven to be sensitive probes, as changes in fluorescence of the photoactive groups can be measured in the presence of very low concentrations of metal. On the basis of host–guest molecular recognition, a series of crowned 1,2,3-triazolium ionic liquids (**CIL 1–6**) was synthesized and found to be capable of quantitatively extracting peptides and proteins from the aqueous layer into the ionic liquid layer. This is the first example of biomolecular interaction of lysine- and arginine-containing peptides and proteins between CILs in pure ionic liquid phase. The chemical structures of **CIL 1–6** are shown in Figure 10.6.

We previously reported the synthesis of 1,2,3-triazolium ionic liquids and envisaged that the crown ether ring embedded in 1,2,3-triazolium ionic liquids could be readily assembled by a cycloaddition reaction.[34] Scheme 10.2 illustrates the synthesis of **CIL 1–6**; the crowned 1,2,3-triazoles **2a–f** could be assembled from *O*-(2-azidoethyl)-*O*′-propargyl-oligo(ethylene glycol)s **1a–f** *via* an intramolecular Huisgen [3+2] cycloaddition reaction in dilute toluene solution (12.5 mM) under reflux conditions. The six-step synthesis was

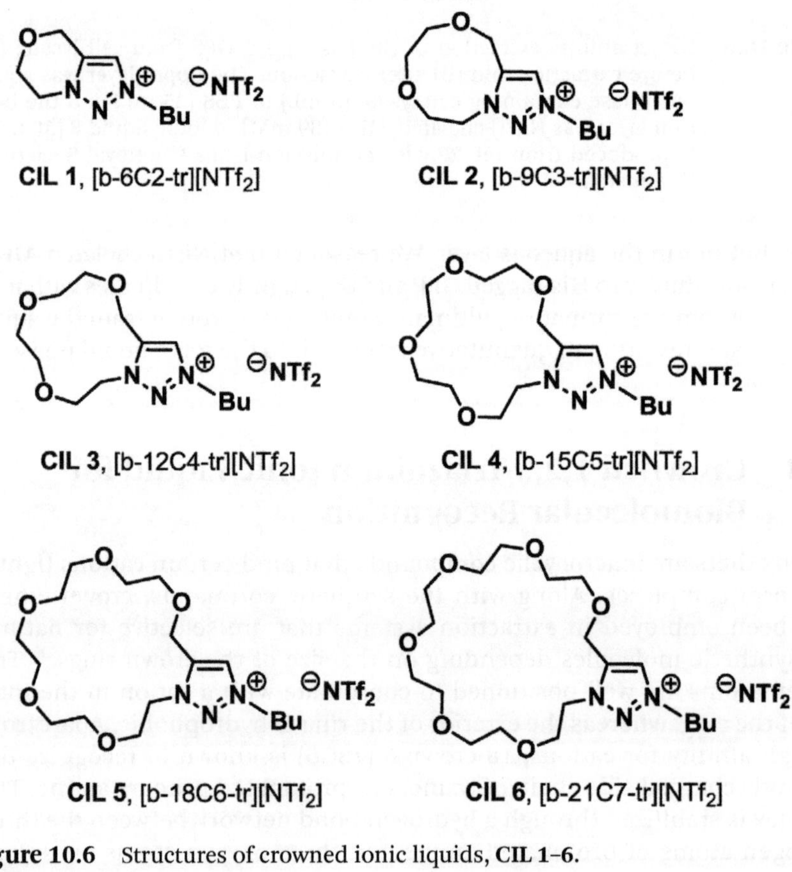

Figure 10.6 Structures of crowned ionic liquids, CIL 1–6.

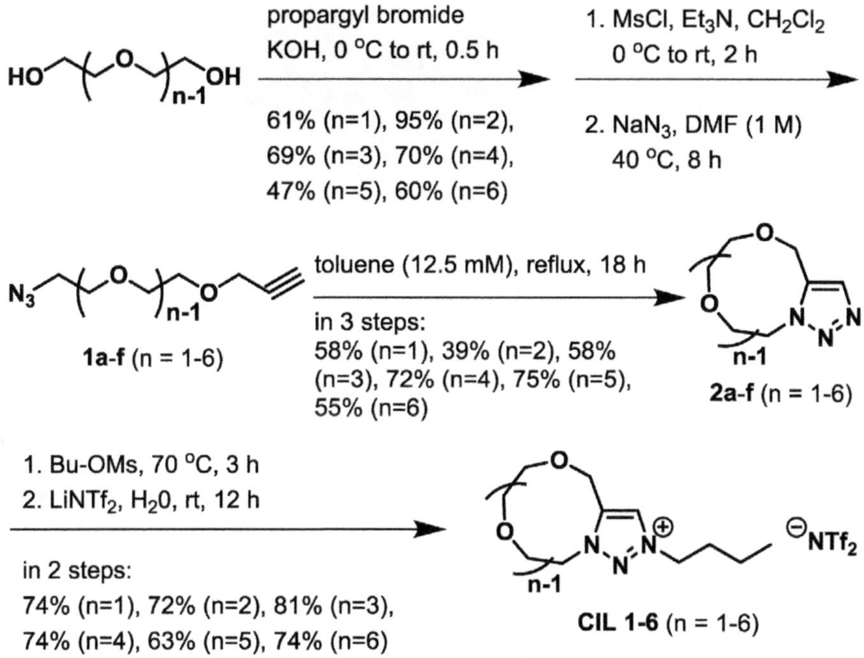

Scheme 10.2 Synthesis of crowned ionic liquids **CIL 1–6**. Please note that the numerical order of the structures is in accordance with that in ref. 35.

straightforward and the overall isolated yields for **CIL 1–6** were in the range of 22–37%. Notably, the small crowned ionic liquids **CIL 1–3** were purely in 1,5-isoform while the larger crowned ionic liquids **CIL 4–6** were composed of minor 1,4- and major 1,5-isomers. All **CIL 1–6** are pale yellow liquids at room temperature. The grafting of ethylene glycol moieties not only lowers the viscosity and melting temperature of these CIL, but also likely makes them potentially more biocompatible than alkyl substituted ionic liquids.[35]

Initially, the affinity extraction of the six N-dabsylated lysine/K-containing hexapeptide amides was performed in the six water-immiscible CILs and aqueous solution. Equal volumes of the CIL and water were first mixed, vigorously shaken and then centrifuged to carry out phase separation and extraction in room temperature. As shown in Figure 10.7, **CIL 4–6** chemoselectively recognized the ammonium group of all six K-containing peptides in water (upper layer) and effectively extracted them into the ionic liquid phase (bottom layer). Clearly, the efficiency of peptide extraction was highly ring size-dependent and regardless of the number of lysine in peptides. Using Dab-AKKKKK-NH_2 peptide as a representative example, the extraction of this peptide from the aqueous phase into the ionic liquid phase was a result of affinity recognition by **CIL 6** (Figure 10.8). In addition, the peptide in ionic liquid phase could be totally recovered into the aqueous phase by using 1 M KCl for the competitive extraction. Despite the seemingly unmatched size of ammonium cation of lysine and the crown rings of **CIL 4–6**, the effective affinity extraction is

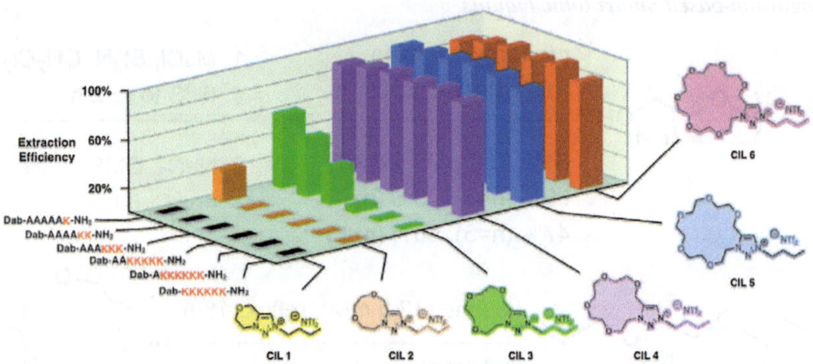

Figure 10.7 Screening of water-immiscible crowned ionic liquids (**CIL 1–6**, 10 μL) used to affinity extract N-dabsylated lysine/K-containing hexapeptide amides in aqueous solutions (1 mM in 10 mM Li_2SO_4 with 10% DMSO, 10 μL). The solution mixture was vigorously agitated and centrifuged to induce phase separation. The extraction efficiency was defined as $[1 - (A_{after}/A_{before})] \times 100\%$, where A_{before} and A_{after} were measured at 502 nm (NanoDrop 2000) before and after extraction by **CIL 1–6.** Estimated RSD, 10%. Reprinted with permission from ref. 35. Copyright (2016) American Chemical Society.

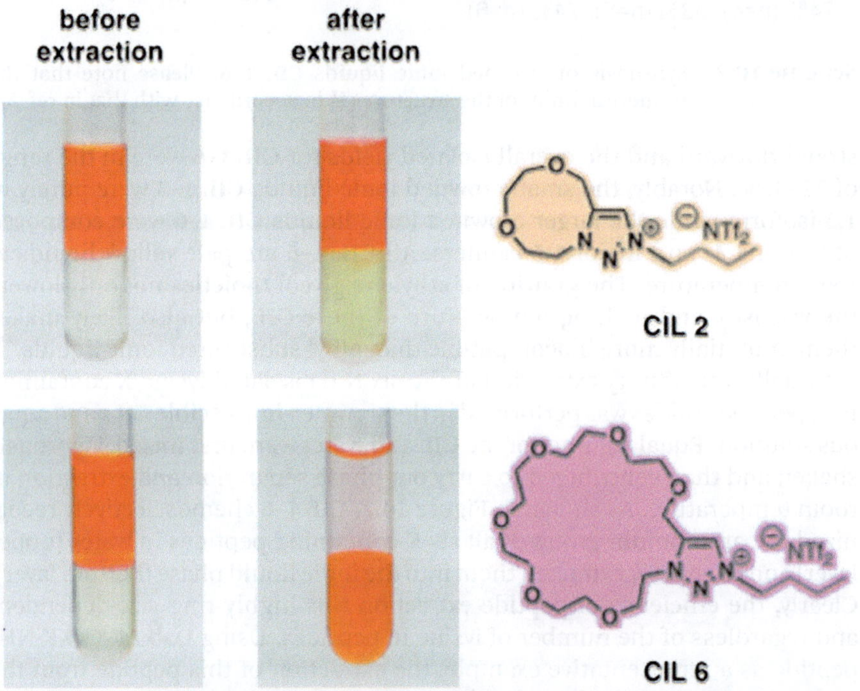

Figure 10.8 Optical photographs of affinity extraction of Dab-AKKKKK-NH_2 peptide (1 mM in 10 mM Li_2SO_4 with 10% DMSO, 100 μL) by **CIL 2** and by **CIL 6** (100 μL each). Competitive extraction of the peptide in **CIL 6** into the aqueous layer could be readily achieved by using KCl (1 M). Reprinted with permission from ref. 35. Copyright (2016) American Chemical Society.

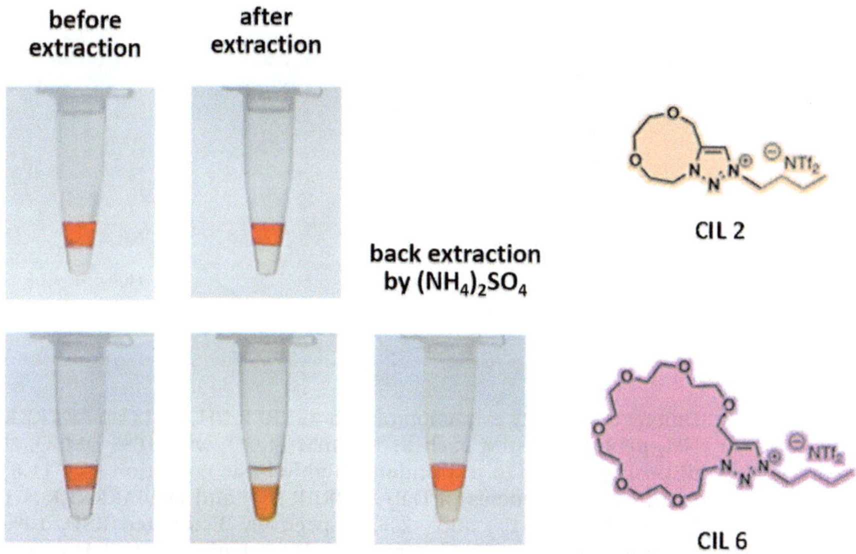

Figure 10.9 Optical photographs of affinity extraction of Dab-ARRRRR-NH_2 peptide (1 mM in 10 mM Li_2SO_4 with 10% DMSO, 20 μL) by **CIL 2** and by **CIL 6** (20 μL each). Competitive extraction of the peptide in ionic liquid phase back to the upper aqueous phase was achieved by using 1 M ammonium sulfate. Reprinted with permission from ref. 35. Copyright (2016) American Chemical Society.

likely due to the flexibility of the larger crown ethers, which can induce-fit and optimize its interactions, ultimately giving stable complexes. The affinity interaction is driven by both the thermodynamically favored ammonium ion complexation by ionic CILs (ion–dipole interactions and hydrogen bonds) and the solvation of the peptide ions by the ionic solvent. These dual properties of ionic liquids make them efficient solvents for the extraction of ionic species from aqueous solutions. The other positively charged amino acid was also included for the affinity extraction of **CIL 6**. As shown in Figure 10.9, the N-dabsylated ARRRRR-NH_2 hexapeptide amide was efficiently extracted into the ionic liquid layer from the aqueous layer. In addition, this peptide in **CIL 6** could be completely extracted back to the upper aqueous layer by using 1 M ammonium sulfate for the competitive extraction. Interestingly, the Dab-ARRRRRNH_2 peptide could be completely extracted into the ionic liquid layer by **CIL 4** but the Dab-AKKKKK-NH_2 peptide remained unaffected and was retained in the aqueous layer (Figure 10.10) due to the different pK_a values of lysine and arginine (10.5 and 12.5, respectively).

Next, myoglobin of equine skeletal muscle and cytochrome *c* from equine heart were selected for affinity extraction by CILs. Both proteins are small, single-chain heme proteins that are rich (>13%) in basic residues and their colors in solution are apparent to the naked eye. Based on the protein amount measured by SDS-polyacrylamide gel electrophoresis, myoglobin was readily

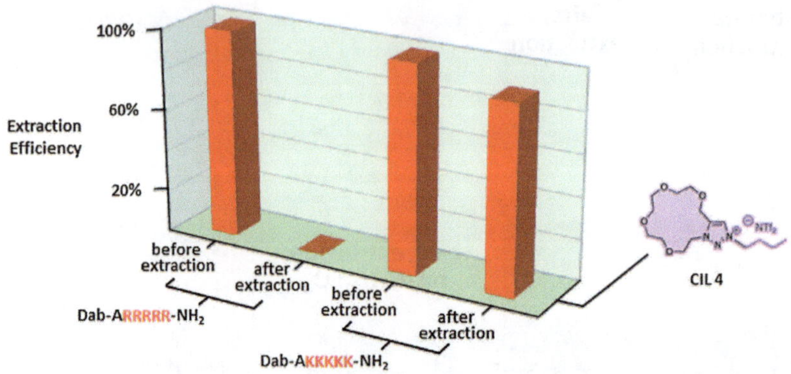

Figure 10.10 Differential affinity extraction of Dab-ARRRRR-NH$_2$ and Dab-AKKKKK-NH$_2$ peptides (1 mM each in 100 mM Li_2CO_3 with 10% DMSO, 20 μL) using **CIL 4** (20 μL) under a highly basic condition (pH 11.8). Extraction efficiencies of Dab-ARRRRR-NH$_2$ and Dab-AKKKKK-NH$_2$ by **CIL 4** were 91% and <2%, respectively. Estimated RSD, 10%. Reprinted with permission from ref. 35. Copyright (2016) American Chemical Society.

recognized and quantitatively extracted by **CIL 6** from the aqueous layer into the ionic liquid layer and could be competitively back-extracted by lysine (1 M) and fully recovered into the aqueous solution (Figure 10.11(A)). After centrifugal dialysis and protein refolding by ultrafiltration, the amount of myoglobin was fully reserved; however, its correct conformation was only partially restored based on the results of circular dichroism (Figure 10.11(B)). The results of affinity extraction, protein recovery and refolding for cytochrome *c* were similar to that of myoglobin. Despite this dissolution or solvation of protein at the molecular level being highly complex, both myoglobin and cytochrome *c* are soluble in the hydrophobic ionic liquid based on the results of circular dichroism. Strikingly, the strong affinity of **CIL 6** for K/R-rich proteins enables the complex formation and ultimately pulls the protein from the aqueous layer and quantitative transfer to the ionic liquid layer.

10.4 Bicyclic 1,2,3-Triazolium Ionic Liquid for Chemoselective Extraction

Chemoselective control, a skill that nature has practiced deftly for millions of years and scientists have been endeavoring to master, has significant implications in organic synthesis and molecular recognition.[36] Predicting chemoselectivity can be a formidable task due to the various factors involved such as the effect of solvent. From the point of view of organic synthesis, ionic liquids have emerged as superior alternatives to conventional molecular solvents for many organic reactions.[37,38] In the previous section, a series of CILs based on the 1,2,3-triazolium core were synthesized and used for the affinity extraction of K/R-containing peptides; here the preparation of a series

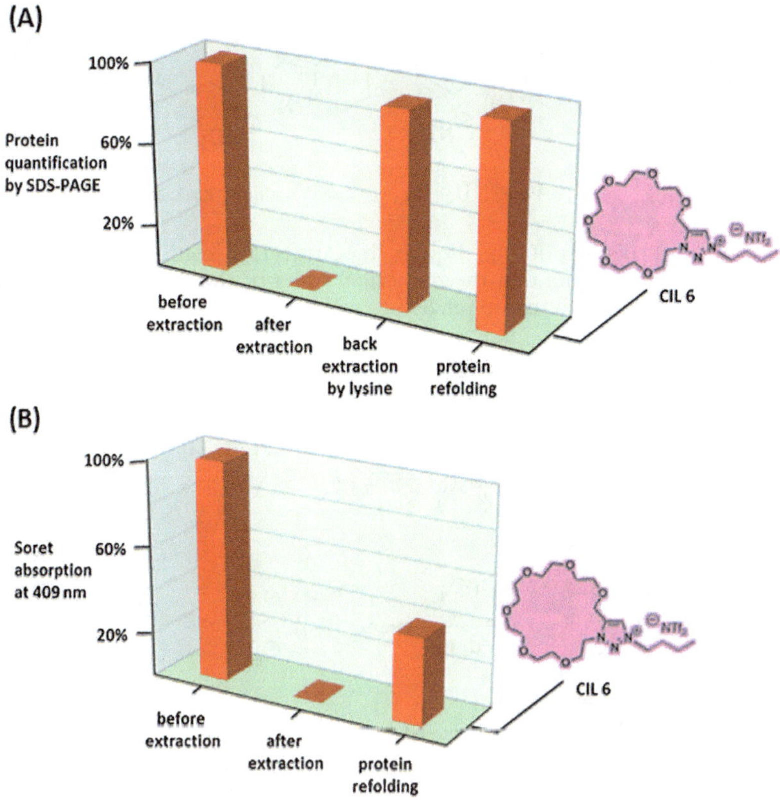

Figure 10.11 Extraction of myoglobin by **CIL 6**. (A) Protein measurements by SDS-PAGE of the upper aqueous solutions upon extracting myoglobin (0.2 mM, 10 μL) using **CIL 6** (10 μL): 100% (before extraction), 0% (after extraction), 97% (competitive extraction by 1 M lysine) and 94% (desalting and refolding after Amicon ultra-3K filtration). (B) Soret absorption measurements at 409 nm (NanoDrop 2000): 100% (before extraction), 0% (after extraction) and 38% (desalting and refolding after Amicon ultra-3K filtration). Estimated RSD, 20%. Reprinted with permission from ref. 35. Copyright (2016) American Chemical Society.

of bicyclic 1,2,3-triazolium ionic liquids and the chemoselective and affinity extraction of metal ions and the remaining positively charged histidine-containing peptide will be described.[39] The first two ionic liquids [b-4C-tr][NTf_2] and [e-4C-tr][NTf_2] required no chromatographic separation and were used as microwave absorbers and solvents for the synthesis of tryptanthrin (see ref. 39 for further information). The ionic liquid [e-4C-tr][NTf_2] was used to dissolved **AIL 3** for related affinity extractions. The chemical structures of these ionic liquids are shown in Figure 10.12.

Based on a common structural core of 4,5,6,7-tetrahydro[1,2,3]triazolo[1,5-*a*] pyridine, a bicyclic triazolium ionic liquid (**AIL 3**) was synthesized through an intramolecular Hüisgen [3+2] cycloaddition (a Cu-free click reaction).[40,41] The

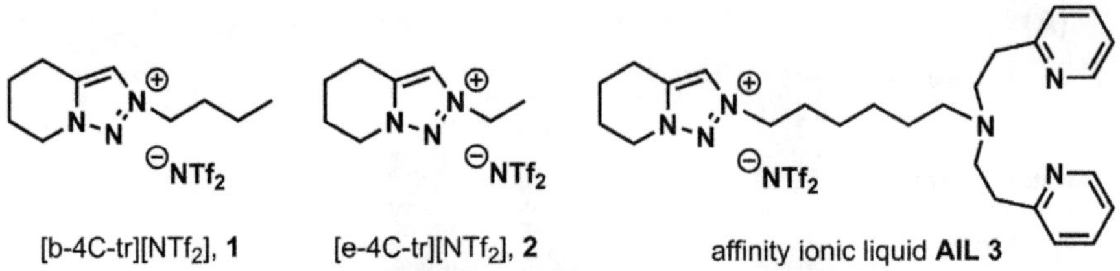

Figure 10.12 Structures of ionic liquids [b-4C-tr][NTf_2], [e-4C-tr][NTf_2] and **AIL 3**. Please note that the numerical order of the structures is in accordance with that in ref. 39.

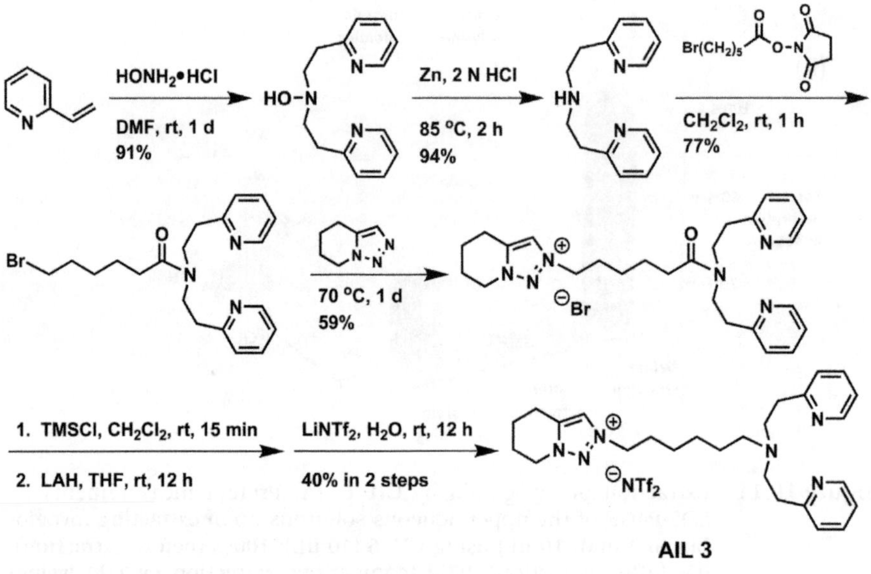

Scheme 10.3 Synthesis of affinity ionic liquid **AIL 3**. Please note that the numerical order of the structures is in accordance with that in ref. 39.

synthesis of **AIL 3** (Scheme 10.3) commenced with the Michael addition of hydroxylammonium chloride to 2-vinylpyridine in DMF at room temperature for 1 d (91% yield), followed by a zinc reduction under acidic and heating conditions (94% yield) to give the Cu(II)-recognition element, *N*,*N*-bis(2-(2-pyridyl) ethyl)amine (85% isolated yield in two steps). An amide bond coupling of the resulting bipyridylamine with *N*-succinimidyl 6-bromohexanoate was then carried out in dichloromethane at room temperature for 1 h (77% yield). Finally, direct alkylation with the bicyclo-triazole (59% yield), a LAH reduction and an ion exchange in water (40% yield in two steps) ultimately give **AIL 3**. The overall isolated yield was 16% over six steps starting from 2-vinylpyridine. **AIL 3** was obtained as a viscous, pale yellow liquid at room temperature.

Equal volumes of ionic liquid solution and water were first mixed, vigorously shaken and then centrifuged to carry out affinity extractions at room temperature. As shown in Figure 10.13, **AIL 3** chemoselectively recognized

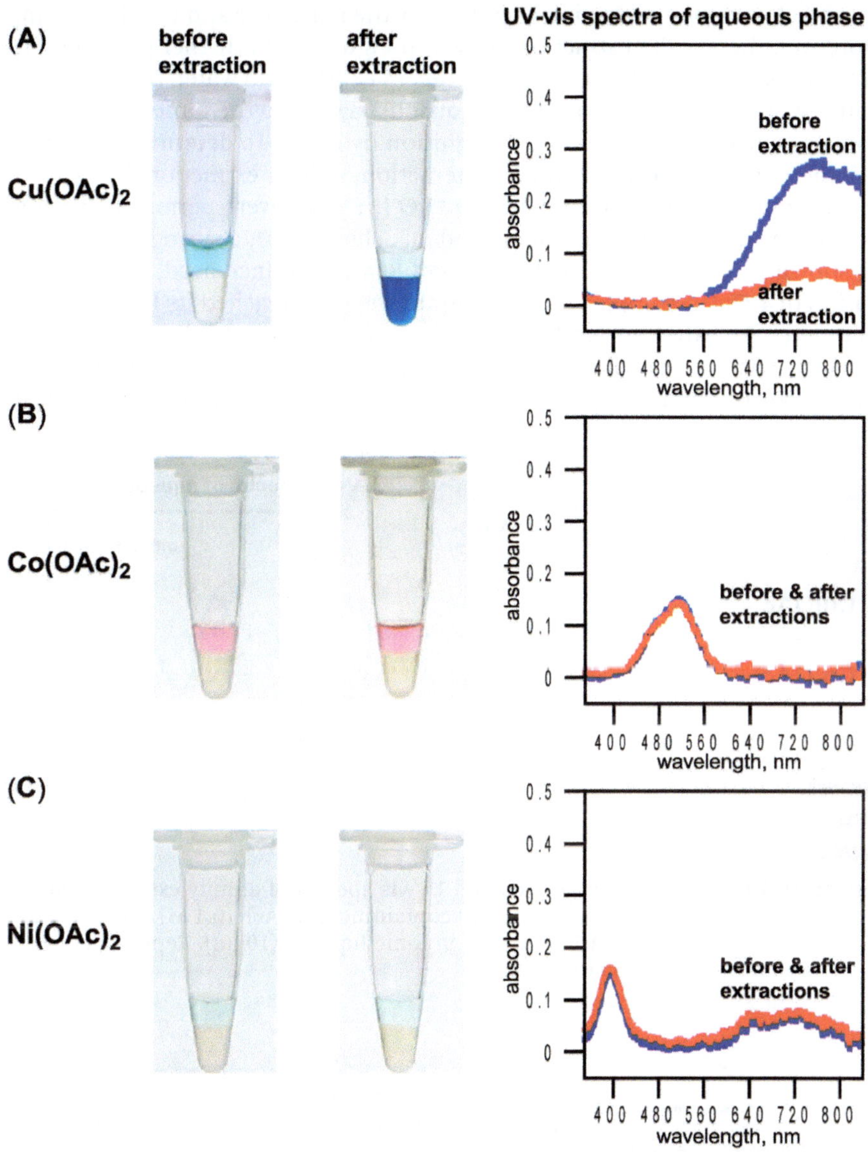

Figure 10.13 Optical photographs and UV-vis spectra of chemoselective extractions of (A) $Cu(OAc)_2$ (0.1 M), (B) $Co(OAc)_2$ (0.25 M), and (C) $Ni(OAc)_2$ (0.25 M) in H_2O (25 μL each) by **AIL 3** (0.25 M) in ionic liquid **2** (25 μL). UV-vis spectra (1 mm path length) of aqueous layers were obtained by a NanoDrop 2000 spectrophotometer. Reproduced with permission from ref. 39.

Cu(II) ion in the aqueous layer and effectively extracted the metal ion into the ionic liquid layer, whereas Co(II) and Ni(II) ions remained unaffected in the aqueous layer upon extraction as judged by the naked eye and UV-vis spectra. Moreover, the affinity extraction shown in Figure 10.14 further proved that, in the presence of both $Cu(OAc)_2$ and excess $Co(OAc)_2$ in water, the complete transfer of only Cu(II) ion from the aqueous layer into the ionic liquid solution was entirely due to affinity recognition by **AIL 3**. To determine the stoichiometry of **AIL-3**–Cu(II) binding interaction, affinity extractions of various concentrations of $Cu(OTf)_2$ in DMSO/H_2O (5 : 1, v/v) were performed with a fixed concentration of an affinity ligand (**4**, Scheme 10.4, a surrogate of **AIL 3**. As shown in Figure 10.15, as the ratio of [Cu(II)]/[**4**] increased, the concentration of **4**-Cu(II) complex increased until the ratio reached its binding stoichiometry. An expected 1 : 1 binding system was elucidated through a series of UV-vis spectra. It is likely that the **4**-Cu(II) complex adopts a square-pyramidal coordination geometry in which affinity ligand **4** binds at the basal

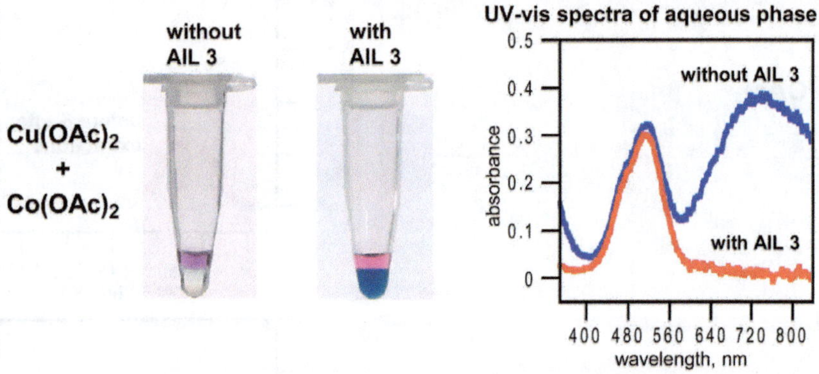

Figure 10.14 Optical photographs and UV-vis spectra of affinity extraction of an aqueous solution (10 μL) containing $Cu(OAc)_2$ (0.1 M) and $Co(OAc)_2$ (0.5 M) by **AIL 3** (1.0 M) in ionic liquid **2** (10 μL). Reproduced with permission from ref. 39.

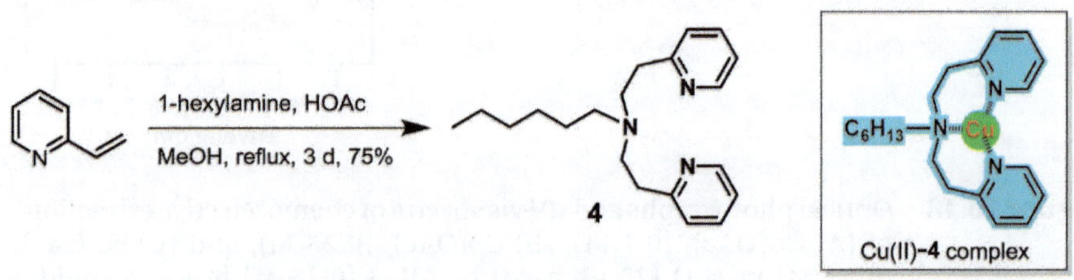

Scheme 10.4 Synthesis of affinity ligand **4** and the proposed structure of the Cu(II)-**4** complex. Please note that the numerical order of the structures is in accordance with that in ref. 39.

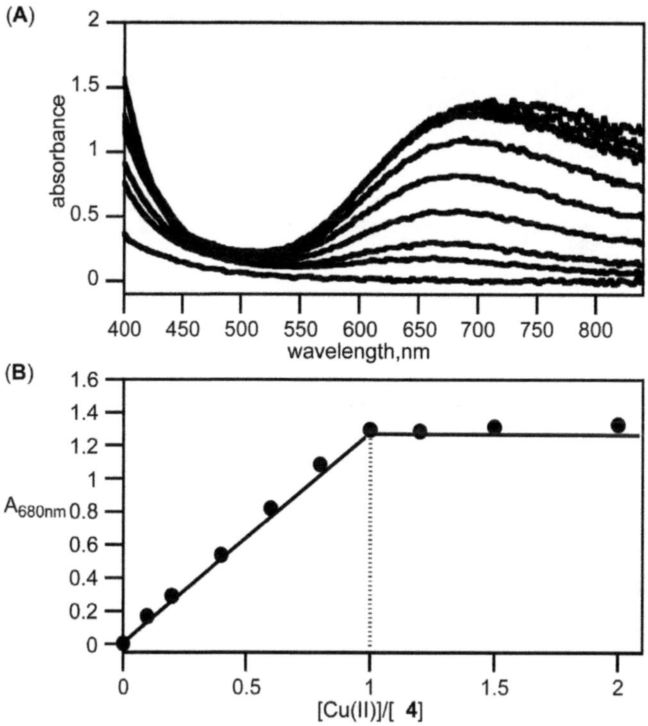

Figure 10.15 Measurement of the stoichiometry of affinity ligand **4** (100 mM) binding to $Cu(OTf)_2$ (0, 10, 20, 40, 60, 80, 100, 120, 150 and 200 mM) in DMSO/H_2O (5 : 1, v/v) solution. (A) UV-vis spectra (1 mm path length) of samples with various (Cu(II))/(**4**) molar ratios were scanned, collected and overlapped by a NanoDrop 2000 spectrophotometer. (B) Plot of max. absorbance at 680 nm *vs.* (Cu(II))/(**4**) was used to determine the binding stoichiometry of **4** with $Cu(OTf)_2$. Reproduced with permission from ref. 39.

plane and the two triflates occupy the remaining sites (Scheme 10.4). Cu(II)-iminodiacetic acid and related structural motifs are known to bind to histidine residues in peptides and proteins.[42] Therefore, we reasoned that the **AIL-3**–Cu(II) complex system should be an effective platform for the one-step affinity isolation and purification of H-containing peptide. As shown in Figure 10.16, a N-dabsylated RHHHH-NH_2 pentapeptide amide was quantitatively extracted by the **AIL-3**–Cu(II) complex in ionic liquid [b-4C-tr][NTf_2] from the aqueous phase into the ionic liquid phase. In addition, the affinity extraction of the H-containing peptide is Cu(II) ion dependent and the peptide could be easily recovered completely by back extraction with 1 M HCl (Figure 10.17).

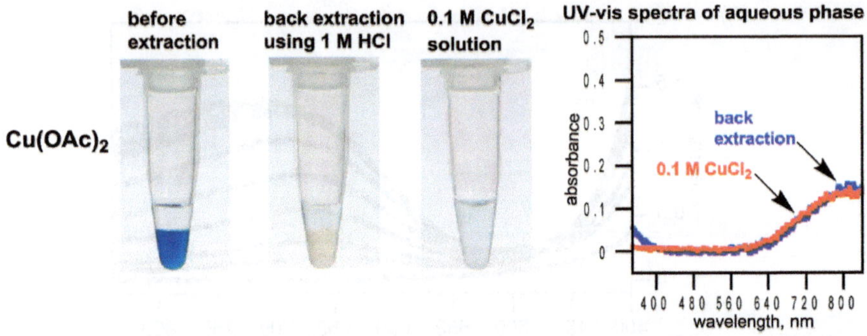

Figure 10.16 Optical photographs and UV-vis spectra of competitive extraction of $Cu(OAc)_2$ (0.1 M) and **AIL 3** (0.25 M) in ionic liquid **2** (25 μL) by 1 M HCl in H_2O (25 μL). A photograph of 0.1 M $CuCl_2$ aqueous solution (50 μL) is included for comparison. Reproduced with permission from ref. 39.

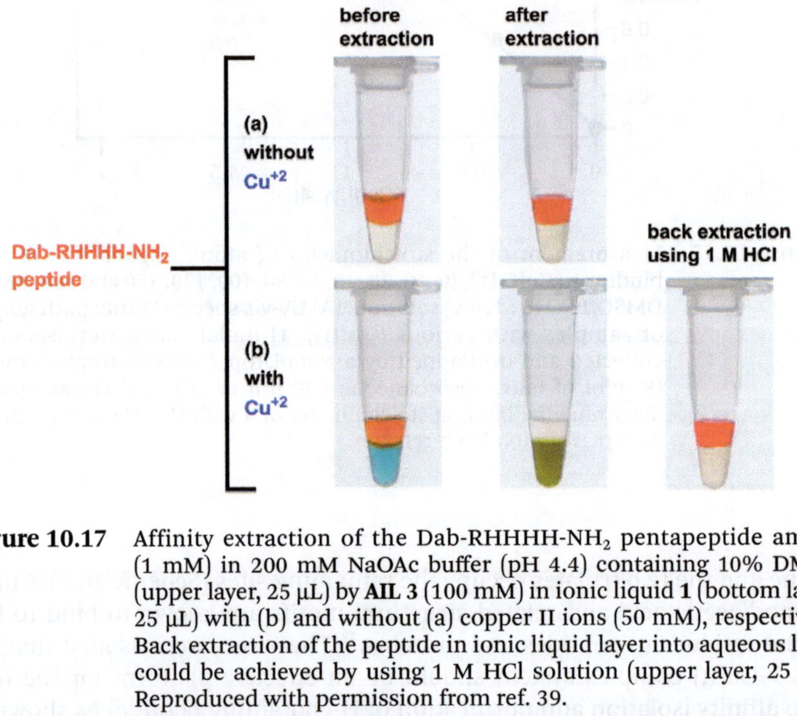

Figure 10.17 Affinity extraction of the Dab-RHHHH-NH_2 pentapeptide amide (1 mM) in 200 mM NaOAc buffer (pH 4.4) containing 10% DMSO (upper layer, 25 μL) by **AIL 3** (100 mM) in ionic liquid **1** (bottom layer, 25 μL) with (b) and without (a) copper II ions (50 mM), respectively. Back extraction of the peptide in ionic liquid layer into aqueous layer could be achieved by using 1 M HCl solution (upper layer, 25 μL). Reproduced with permission from ref. 39.

10.5 Conclusion

The ultimate goal for a biomolecular interaction purification system is ideally based on an "all or nothing" protocol, despite being thermodynamically unattainable when considering the law of mass action.[43] Nonetheless, affinity purification, which makes use of specific binding interactions between biomolecules, is one of the most effective techniques for protein purification. Each specific affinity system requires its own set of conditions and presents its own peculiar challenges for a given binding couple. Moreover, most purification methods involve some format of chromatography such that the target protein in solution (liquid phase) is separated based on differences in chemical or physical interaction such as ionic strength or pH value with the matrix (solid phase). The seemingly simple yet sophisticated setup of affinity chromatography could be problematic due to the poor choice of liquid phase, solid phase or even the spacer arm between affinity ligand and matrix. A new realm of potential purification method is hiding in plain sight. Ionic liquids, an exciting class of organic compound that could bypass potential problems of protein purification are being engineered for the development of recognition-based smart solvents. Affinity liquid–liquid extraction of peptides and proteins using a series of bicyclic imidazolium, crowned 1,2,3-triazolium and bicyclic 1,2,3-triazolium ionic liquids was demonstrated as an alternative for affinity purification, biomolecular recognition and chemoselective extraction, respectively.

With the ease of chemical synthesis of ionic liquids, we described the first AIL and demonstrated that Ni(II)-chelated **AIL 1** was capable of extracting and purifying His-tagged peptides and proteins from the aqueous phase into the ionic liquid phase. Strikingly, the His-tagged GFP directly purified from cell lysate remained fluorescent and therefore the Ni(II)-chelated **AIL 1** binding system holds great promise for a wide range of target biomolecules and potential applications in biotechnology. We reasoned that the hydrophobic ionic liquid has a lesser tendency to pull out the essential water from the protein surface and thus hydrogen bonds and other interactions within the protein could be maintained. Subsequently, Cu(II)-chelated AIL/IL and Zn(II)-chelated AIL/IL systems have also been used for the liquid–liquid extraction of His-tagged proteins.[44] In our crowned ionic liquid work, **CIL 4–6** of medium to large ring sizes were able to recognize myoglobin, cytochrome *c* and arginine- and lysine-containing peptides. The system utilizes water-immiscible CILs to extract target biomolecules from aqueous solution through the complexation of targeted ions with CILs. This was the first example of biomolecular recognition of proteins and K/R-containing peptides by CILs in pure ionic liquid phase. The CIL system may pave the way for the development of new biosensing technology, if proteins with higher refolding yields could be achieved in due course. Finally, the **AIL 3**-Cu(II) system based on a bicyclic 1,2,3-triazolium ionic liquid was used for the chemoselective extraction of Cu(II) and the first affinity extraction of His-containing peptides.

In conclusion, recognition-based smart ionic liquids can be used to isolate and enrich a target protein from a mixture into a solution, reduce the amount of the protein in a mixture, or distinguish what protein binds to a particular receptor, whereby the protein of interest is purified or identified by virtue of its tailored intermolecular binding interactions. In these processes, selective extraction of proteins in aqueous buffer is achieved by contacting a water-immiscible affinity ionic liquid. Therefore, the partition may be influenced by the nature and concentrations of the functional ionic liquid, pH value, temperature, molecular weight and the surface hydrophobicity of the biomolecules. It appears that the issue of protein refolding is on a case-by-case basis for now until a universal structural core can be incorporated as a solution. Overall, from a bioanalytical perspective, recognition-based ionic liquids have advantages over conventional affinity purification and liquid–liquid extraction techniques, which include better limits of detection, selectivity and sensitivity.

Acknowledgements

This work was supported by the Ministry of Science and Technology of Taiwan, ROC (MOST 103-2113-M-194-002-MY3). We thank Ms Jessica Lim at the Harvard Graduate School of Education for the artwork in Figure 10.1.

References

1. R. Baron and J. A. McCammon, *Annu. Rev. Phys. Chem.*, 2013, **64**, 151–175.
2. R. Root-Bernstein, *Acc. Chem. Res.*, 2012, **45**, 2169–2177.
3. P. Hengen, *Trends Biochem. Sci.*, 1995, **20**, 285–286.
4. Y. P. Chang, W. C. Liu, M. C. Tseng and Y. H. Chu, *Rev. Anal. Chem.*, 2015, **34**, 77–86.
5. H. Y. Li, T. H. Hsu, C. Y. Chen, M. C. Tseng and Y. H. Chu, *Analyst*, 2015, **140**, 6245–6249.
6. Y. L. Liu, M. C. Tseng and Y. H. Chu, *Chem. Commun. (Camb.)*, 2013, **49**, 2560–2562.
7. M. C. Tseng and Y. H. Chu, *Chem. Commun. (Camb)*, 2010, **46**, 2983–2985.
8. M. C. Tseng and Y. H. Chu, *Anal. Chem.*, 2014, **86**, 1949–1952.
9. H. Kim, D. A. Boysen, J. M. Newhouse, B. L. Spatocco, B. Chung, P. J. Burke, D. J. Bradwell, K. Jiang, A. A. Tomaszowska, K. Wang, W. Wei, L. A. Ortiz, S. A. Barriga, S. M. Poizeau and D. R. Sadoway, *Chem. Rev.*, 2013, **113**, 2075–2099.
10. S. Wellens, B. Thijs, C. Moller and K. Binnemans, *Phys. Chem. Chem. Phys.*, 2013, **15**, 9663–9669.
11. A. Rout, S. Wellens and K. Binnemans, *RSC Adv.*, 2014, **4**, 5753–5758.
12. F. Rahimpour, F. Feyzi, S. Maghsoudi and R. Hatti-Kaul, *Biotechnol. Bioeng.*, 2006, **95**, 627–637.
13. H. Wang, G. Gurau and R. D. Rogers, *Chem. Soc. Rev.*, 2012, **41**, 1519–1537.
14. R. Patel, M. Kumari and A. B. Khan, *Appl. Biochem. Biotechnol.*, 2014, **172**, 3701–3720.
15. L. Chen, G. E. Mullen, M. Le Roch, C. G. Cassity, N. Gouault, H. Y. Fadamiro, R. E. Barletta, R. A. O'Brien, R. E. Sykora, A. C. Stenson, K. N. West, H. E. Horne, J. M. Hendrich, K. R. Xiang and J. H. Davis Jr, *Angew. Chem., Int. Ed. Engl.*, 2014, **53**, 11762–11765.
16. L. B. Escudero, A. Castro Grijalba, E. M. Martinis and R. G. Wuilloud, *Anal. Bioanal. Chem.*, 2013, **405**, 7597–7613.
17. Y. Pei, J. Wang, K. Wu, X. Xuan and X. Lu, *Sep. Purif. Technol.*, 2009, **64**, 288–295.

18. S. Dreyer, P. Salim and U. Kragl, *Biochem. Eng. J.*, 2009, **46**, 176–185.
19. D. S. Waugh, *Trends Biotechnol.*, 2005, **23**, 316–320.
20. P. Cuatrecasas, M. Wilchek and C. B. Anfinsen, *Proc. Natl. Acad. Sci. U. S. A.*, 1968, **61**, 636–643.
21. C. L. Young, Z. T. Britton and A. S. Robinson, *Biotechnol. J.*, 2012, **7**, 620–634.
22. B. Bilgicer, S. W. Thomas 3rd, B. F. Shaw, G. K. Kaufman, V. M. Krishnamurthy, L. A. Estroff, J. Yang and G. M. Whitesides, *J. Am. Chem. Soc.*, 2009, **131**, 9361–9367.
23. C. F. Poole and S. K. Poole, *J. Chromatogr. A*, 2010, **1217**, 2268–2286.
24. C. Huo and T. H. Chan, *Chem. Soc. Rev.*, 2010, **39**, 2977–3006.
25. Z. Du, Y. L. Yu and J. H. Wang, *Chemistry*, 2007, **13**, 2130–2137.
26. L. Ge, X. T. Wang, S. N. Tan, H. H. Tsai, J. W. Yong and L. Hua, *Talanta*, 2010, **81**, 1861–1864.
27. M. C. Tseng, M. J. Tseng and Y. H. Chu, *Chem. Commun. (Camb.)*, 2009, 7503–7505.
28. H. C. Kan, M. C. Tseng and Y. H. Chu, *Tetrahedron*, 2007, **63**, 1644–1653.
29. M. Blesic, A. Lopes, E. Melo, Z. Petrovski, N. V. Plechkova, J. N. Canongia Lopes, K. R. Seddon and L. P. Rebelo, *J. Phys. Chem. B*, 2008, **112**, 8645–8650.
30. M. Zimmer, *Chem. Rev.*, 2002, **102**, 759–781.
31. S. Enoki, K. Saeki, K. Maki and K. Kuwajima, *Biochemistry*, 2004, **43**, 14238–14248.
32. G. W. Gokel, W. M. Leevy and M. E. Weber, *Chem. Rev.*, 2004, **104**, 2723–2750.
33. A. Spath and B. Konig, *Beilstein J. Org. Chem.*, 2010, **6**, 32.
34. M. C. Tseng, H. T. Cheng, M. J. Shen and Y. H. Chu, *Org. Lett.*, 2011, **13**, 4434–4437.
35. M. C. Tseng, T. C. Yuan, Z. Li and Y. H. Chu, *Anal. Chem.*, 2016, **88**, 10811–10815.
36. R. A. Shenvi, D. P. O'Mallcy and P. S. Baran, *Acc. Chem. Res.*, 2009, **42**, 530–541.
37. S. Sowmiah, V. Srinivasadesikan, M. C. Tseng and Y. H. Chu, *Molecules*, 2009, **14**, 3780–3813.
38. S. Sowmiah, C. I. Cheng and Y. H. Chu, *Curr. Org. Synth.*, 2012, **9**, 74–95.
39. H. Y. Li, C. Y. Chen, H. T. Cheng and Y. H. Chu, *Molecules*, 2016, **21**(10), 1355.
40. B. Schulze and U. S. Schubert, *Chem. Soc. Rev.*, 2014, **43**, 2522–2571.
41. P. Thirumurugan, D. Matosiuk and K. Jozwiak, *Chem. Rev.*, 2013, **113**, 4905–4979.
42. K. P. Carter, A. M. Young and A. E. Palmer, *Chem. Rev.*, 2014, **114**, 4564–4601.
43. E. O. Voit, H. A. Martens and S. W. Omholt, *PLoS Comput. Biol.*, 2015, **11**, e1004012.
44. W. Xu, H. Cao, G. Ren, H. Xie, J. Huang and S. Li, *Appl. Microbiol. Biotechnol.*, 2014, **98**, 5665–5675.

CHAPTER 11

Ionic Liquid-based Physical Sensors

YI-CHUNG TUNG*, HSIAO-MEI WU AND TSE-ANG LEE

Academia Sinica, Research Center for Applied Sciences, 128, Sec. 2, Academia Rd., Taipei 11529, Taiwan
*E-mail: tungy@gate.sinica.edu.tw

11.1 Introduction

Ionic liquids, also known as salts in liquid state, which are composed solely of ions, play an important role in various applications due to their unique material properties,[1] including electrical conductivity,[2,3] non-volatility,[4] thermal stability, and favorable solvating properties for a range of polar and non-polar compounds. The non-volatile property prevents ionic liquids from drying during operation. Most ionic liquids possess excellent thermal stability and non-flammability. They are stable until temperatures up to 400 °C.[5,6] Another important property of ionic liquids is their broad electrochemical window,[7] which is a measure of the oxidative and reductive stability. In addition, many kinds of air and water stable ionic liquids have been prepared and characterized[8] nowadays to reduce the moisture sensitivity in relation to the hygroscopic nature of salts. The above features make ionic liquids good solvents and electrolytes for various applications. However, most applications are limited to chemically related ones. With the recent development of wearable devices and flexible electronics, emerging conductive fluidic materials have been directly incorporated as elements for their mechanical flexibility

Smart Materials No. 28
Ionic Liquid Devices
Edited by Ali Eftekhari

Published by the Royal Society of Chemistry, www.rsc.org

and reversible deformability.[9] Among various conductive fluidic materials, ionic liquids have received increasing attention in recent decades for physical sensor applications because of several desired advantageous properties.

Physical sensors are transducers that convert physical stimulation or property variation into signals that can be easily observed or recorded (usually electrical signals). Physical sensors, such as thermometers, pressure sensors, accelerometers, and humidity sensors, have been widely used in various applications from daily life to specific industries.[10,11] Operation principles of the commonly used sensors rely on capacitive, resistive, inductive, piezoelectric, and electromagnetic sensing schemes. With the recent development of wearable devices, flexible electronics, and micro technologies, physical sensors with better sensitivity and capability that can be easily integrated into flexible substrates or microdevices are highly desired.[12,13] Because of the advantageous physical properties of ionic liquids they have been exploited in recent years to construct physical sensors. Current ionic liquid-based physical sensors mainly use capacitive and resistive sensing schemes. Below, this chapter will introduce the operation principles and applications of these physical sensors, and discuss their advantages.

11.2 Ionic Liquid Capacitive Sensors

Physical sensors converting physical stimulation into electrical signals by capacitance variation have been developed for various applications, such as proximity, position or displacement, humidity, fluid level, and acceleration. Ionic liquids have great mechanical flexibility, and high interfacial double layer capacitance at the ionic liquid–electrode interface due to their pure ionic composition. Therefore, ionic liquids have been integrated with deformable substrates to construct capacitive physical sensors with high sensitivity.

11.2.1 Operation Principle

A capacitor is a passive electrical component that stores electrical charge in an electrical field; it can be viewed as two electrodes separated by a dielectric region. For simplification, a capacitor that consists of two parallel electrodes with areas A separated by a distance d by a dielectric material with permittivity ε can be used for modeling as shown in Figure 11.1. The capacitance of the capacitor can be given as:

$$C = \varepsilon \frac{A}{d} \tag{11.1}$$

In a capacitive physical sensor, under a physical stimulation, the separation of the electrodes or permittivity of the dielectric material varies accordingly, resulting in a capacitance change of the capacitor. As a result, the physical stimulation can be quantified by characterizing the capacitance variation.

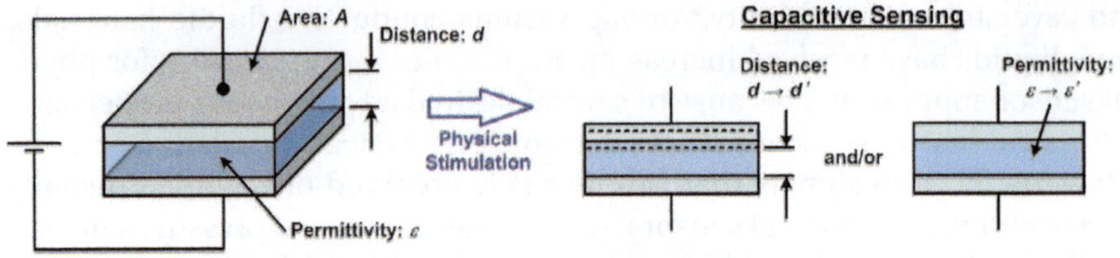

Figure 11.1 Simplified parallel-plate electrical capacitor model, and general operation principles of capacitive physical sensors.

Since ionic liquids have a wide electrochemical window, and mechanical flexibility, they have been used to construct new types of capacitive physical sensors. Unlike traditional capacitive pressure sensors using rigid materials, two soft constructs, ionic liquid and elastomeric materials, can be exploited to achieve ultra-high sensitivity. Two types of capacitive sensors have been developed in the literature. The first type of sensor is based on permittivity variation of the capacitor. In the sensor, an ionic liquid is used as the dielectric material in the capacitor. The capacitance is a result of interfacial double layer capacitance formed at the liquid–electrode interface. At the interface, ions in the ionic liquid collect in a thin layer in order to maintain electrostatic neutrality despite the energy difference between the two materials. The accumulated charge at the interface behaves like a capacitor with a very high density that is unachievable with traditional solid-state technology.[9] Under physical stimulation, ionic liquid moves between the electrodes to change the effective permittivity, resulting in a capacitance change. By measuring the capacitance variation, the physical stimulation magnitudes can be quantified. A second type of physical sensor relies on the deformation of electrodes and dielectric materials similar to the conventional capacitive sensors. By integrating ionic liquids with deformable substrates, the capacitance variation in an ionic-liquid based capacitor can be greater than that in conventional ones with solid electrodes.

11.2.2 Applications

11.2.2.1 Pressure Sensor

An ultra-high sensitivity sensor made of elastomeric materials using a capacitor with ionic liquid as the dielectric material has been constructed for pressure measurement (Figure 11.2).[14] In the sensor, an ionic liquid is filled into a sensing chamber connected with a microfluidic channel. When physical pressure is applied on the chamber, the ionic liquid is displaced outwardly into the microfluidic channel. A pair of coplanar electrodes is aligned at each side of the channel to sense displacement of the ionic liquid from the pressurized chamber by measuring the capacitance variation between the electrodes. The geometries of the device can be designed such that the fluidic displacement can be amplified to improve the device sensitivity. In contrast,

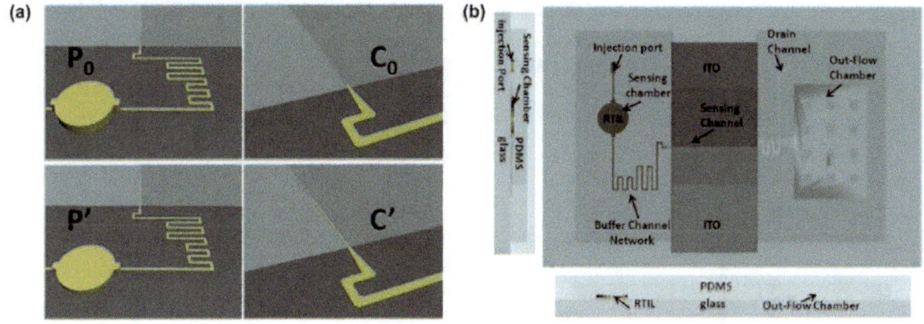

Figure 11.2 (a) Operation principle of ionic-liquid based capacitive pressure sensor, and (b) illustration of the pressure sensor device. © 2011 IEEE. Reprinted, with permission, from ref. 14.

when the applied pressure is released from the chamber, the ionic liquid is drawn back to the chamber and away from the electrodes by elastic recovery of the material.

The ionic liquid-based pressure-sensing device is made of an elastomeric material, polydimethylsiloxane (PDMS), and a glass substrate with indium tin oxide (ITO) coplanar electrodes. A PDMS layer with the sensing chamber and microfluidic channel patterns is fabricated using a well-developed soft lithography replica molding process.[15] In brief, a silicon wafer with multiple photoresist layers patterned using conventional photolithography processes is exploited as a master mold with positive relief patterns. The PDMS prepolymer is then poured into the mold, cured, and peeled from the mold. The ITO electrodes are deposited and patterned using photolithography and wet etching processes onto the glass substrate. The PDMS layer and the glass substrate are then irreversibly bonded using oxygen plasma surface treatment. Due to the selected materials, the entire device is optically transparent, which is highly desired for numerous applications. Figure 11.3 shows the measured capacitance at a frequency of 30 Hz under different applied pressure; the sensor can achieve a sensitivity of 522.5 pF mm Hg^{-1}. The ionic-liquid-based capacitive pressure sensor is ready to be integrated with other microfluidic and lab-on-a-chip systems for various applications.

11.2.2.2 Tactile Sensor

A flexible tactile sensor based on an iontronic microdroplet array (IMA) taking advantage of an ultra-large interfacial capacitance at the highly elastic ionic liquid droplet–electrode contact has been developed for physical sensing.[16] The constructed sensor is optically transparent, and composed of an array of nanoliter ionic liquid droplets sandwiched by two deformable polymer substrates with transparent ITO electrodes as shown in Figure 11.4. The interface between the droplet and the electrode forms an electrical double layer with relatively large capacitance. While the substrates deform under

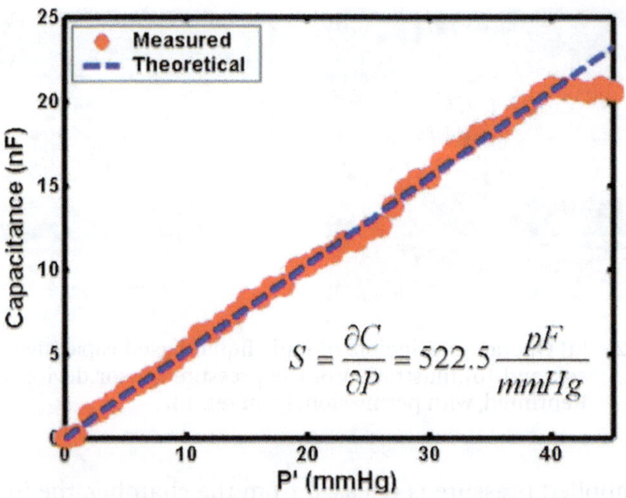

Figure 11.3 Measured capacitance *versus* pressure applied on the deformable chamber. © 2011 IEEE. Reprinted, with permission, from ref. 14.

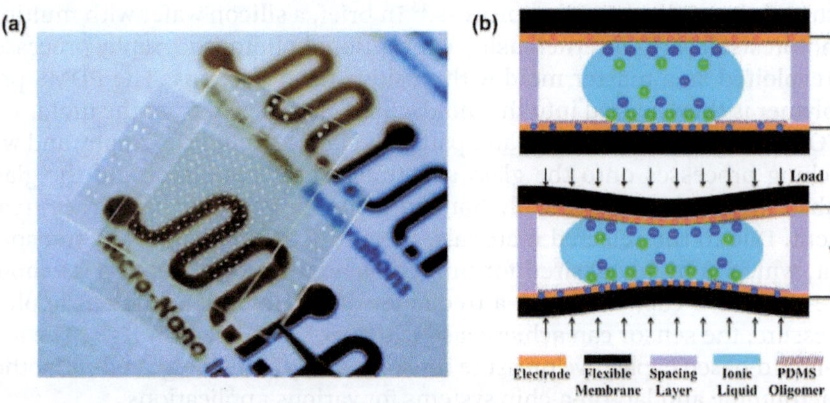

Figure 11.4 Photograph and operation principle of ionic liquid-based tactile sensor with an ionic microdroplet array (IMA). (a) The sensor is optically transparent and has 12 × 12 elements with a spatial resolution of 1 mm. (b) Cross-sectional sketch of the capacitive sensing scheme for the device. The ionic-liquid–electrode interfacial capacitance is established with an external voltage application. Mechanical deformation of the electrodes under physical stimulation results in ionic-liquid–electrode contact area expansion and increase of the capacitance. Reproduced from ref. 16 with permission from the Royal Society of Chemistry.

physical stimulation, the droplet–electrode contact expands along the circumference, resulting in dramatic capacitance variation. In addition, with a layer of hydrophobic coating on the electrodes, the hysteresis of the droplet deformation can be minimized to achieve fast response (on the order of milliseconds).

Fabrication of the tactile sensor starts with patterning conductive ITO electrodes on flexible polymer substrates (*e.g.* poly(ethylene terephthalate), PET, films) using conventional photolithography and wet etching processes. A spacing layer is then patterned on a UV-curable dry-film photoresist thermally laminated on top of the flexible substrate. To accurately position the ionic droplets on the substrate, a hydrophobic oligomer layer of PDMS is contact-printed on the substrate to create a high hydrophobicity contrast. Subsequently, a microfluidic impact printing technique is exploited to deposit an array of ionic liquid droplets on to the substrate.[17] To assemble the device, two substrates are aligned and bonded using oxygen plasma surface treatment.

The experimental characterization shows that the fabricated tactile sensor has great sensitivity (0.43 nF kPa^{-1}), low minimal detectable pressure (33 Pa), and fast response (a few milliseconds) due to the unique electrical and mechanical properties of ionic liquid. The sensor also has great long-term and temperature stability due to low vapor pressure and the stable material properties of the ionic liquid. For practical applications, the sensor has been demonstrated for wearable surface topology mapping and wrist pulse recording devices (Figure 11.5). In the experiments, the sensor successfully measured the pressure distribution on a

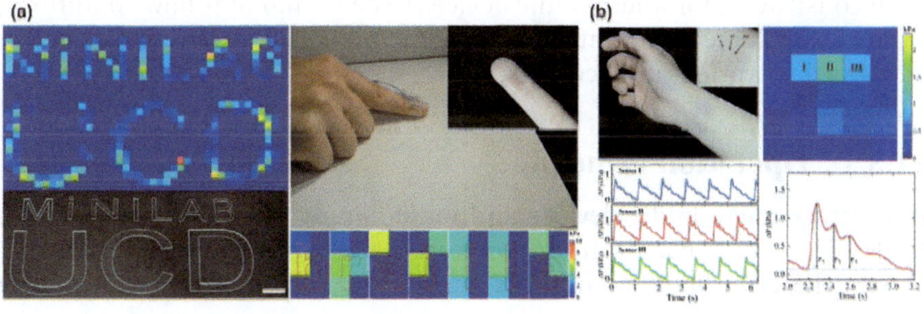

Figure 11.5 (a) Application of the ionic liquid-based tactile sensor to surface topology mapping. Topology of a plastic stamp (scale bar is 1 cm) can be successfully mapped using the developed device. The sensor can also be mounted on a fingertip to resolve raised dots of characters in a Braille book. (b) Real-time wrist pulse measurements using a flexible band with the ionic liquid-based tactile sensor for non-invasive pressure wave recording. The spatial distribution of the pulse pressure can be mapped in real-time with great temporal resolution to show three consecutive peaks of the recorder pressure wave in a cardiovascular cycle. Reproduced from ref. 16 with permission from the Royal Society of Chemistry.

surface, and scanned Braille characters in a fingertip configuration. In addition, the sensor has been applied to record non-invasive cardiovascular pressure with its high sensitivity and rapid response. With such great performance, the ionic liquid-based sensor can potentially enable a highly transformative platform of tactile sensing for applications such as robotics, medical prosthetics, and wearable devices.

11.3 Ionic Liquid Resistive Sensors

Instead of measuring capacitance variation, another commonly used physical sensor measures the change in resistance under physical stimulation. For simplification, the resistance (R) of a uniform conductor can be written as:

$$R = \rho \frac{l}{A} \tag{11.2}$$

where ρ is the resistivity, and l and A are the length and cross-sectional area of the conductor, respectively. Under physical stimulation, the geometry or resistivity of the conductor changes accordingly (Figure 11.6). Resistive sensing schemes have been utilized for a number of physical sensor applications, such as strain gauges, thermometers and photoresistors. In recent years, microfluidics research has attracted increasing attention for various applications, especially chemical and biomedical ones, due to its desired advantages, including low fabrication cost, low chemical consumption, small footprint, and great control in spatial and temporal domains.[18–20] Taking advantage the of unique material properties of ionic liquid, novel physical sensors have been constructed for microfluidic devices to better monitor flow conditions, and the new ionic liquid integrated microfluidic devices have been further developed for various applications.

11.3.1 Operation Principle

Integrating microfluidic devices and ionic liquids as a new concept, electrofluidic circuits, has been proposed in the construction of physical sensors.[21] The electrofluidic circuit is an electrical circuit constructed using

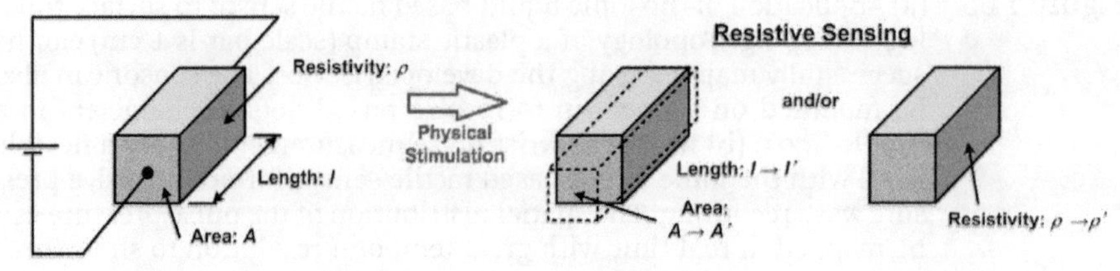

Figure 11.6 Simplified electrical resistor model, and general operation principles of resistive physical sensors.

ionic liquid-filled deformable microfluidic channels. By designing channel geometries, three basic electrofluidic components can be made to construct electrofluidic circuits: a constant electrofluidic resistor, a pressure regulated electrofluidic variable resistor, and a pressure controlled electrofluidic switch. In addition, electrofluidic circuits can be easily fabricated using a well-developed replica molding soft lithography process with an elastomeric material, polydimethylsiloxane (PDMS). The material provides the deformability and optical transparency that are advantageous for flexible devices and biomedical applications. Given below is the operation principle of each electrofluidic circuit component.

11.3.1.1 Constant Electrofluidic Resistor

An electric resistor with constant resistance can be constructed by a fluidic channel filled with ionic liquid due to electrical conductivity of ionic liquid. At low frequency range (in the order of hundred Hz), an ionic liquid-filled electrofluidic channel can be considered as a constant electrofluidic resistor as shown in Figure 11.7(a). The resistance value is proportional to the length of the conductive wire, and inversely proportional to the cross-sectional area of the wire. Through designing the geometries of the electrofluidic channel, the resistance value of the constant resistor can be varied.

11.3.1.2 Pressure Regulated Electrofluidic Variable Resistor

An electrofluidic channel with flexible walls can be regarded as a pressure regulated electrofluidic variable resistor. Figure 11.7(b) shows an example of the design – the ionic liquid filled channels (electrofluidic channels) are bonded with an elastic membrane at the bottom. Once the elastic membrane is deformed due to an applied pressure on the opposite side, the cross-sectional area of the electrofluidic channel is reduced. As a result, the electrical resistance of the electrofluidic channel is varied. While the area reduction, ΔA, is small compared to the initial area A, it causes the resistance to increase by:

$$\Delta R = \left(\frac{\rho l}{A^2}\right) \cdot \Delta A \tag{11.3}$$

For a small deformation of the elastic membrane, the reduction in cross-sectional area in the electrofluidic channel can be assumed to be proportional to the applied pressure according to the linear plate theory.[22] Consequently, the resistance change is linearly proportional to the area reduction and, further, linearly proportional to the applied pressure (P):

$$\Delta R \propto \Delta A \propto P \tag{11.4}$$

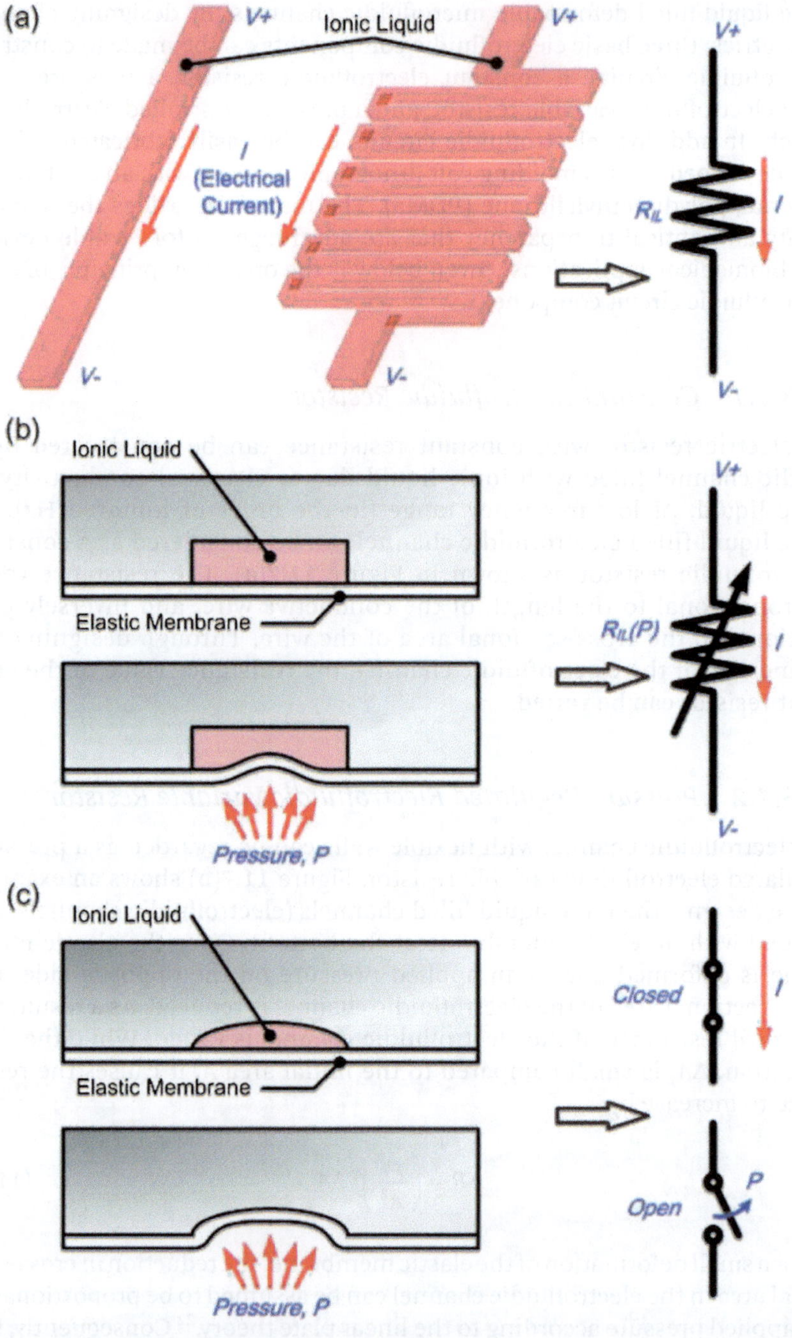

Figure 11.7 Design and symbols of three basic electrofluidic components: (a) constant electrofluidic resistor, (b) pressure regulated electrofluidic variable resistor, and (c) pressure controlled electrofluidic switch. Reproduced from ref. 38 with permission from the Royal Society of Chemistry.

11.3.1.3 Pressure Controlled Electrofluidic Switch

Furthermore, an electrofluidic channel with a rounded cross-section shape with flexible walls can be utilized to construct a pressure controlled electrofluidic switch (Figure 11.7(c)). Similar to a pressure regulated electrofluidic variable resistor, the microfluidic channel is bonded with an elastic membrane on the bottom side. Once the applied pressure value exceeds a threshold value, the membrane is deformed to fully seal the electrofluidic channel. Thus, the rounded cross-sectional shape is fully closed by the membrane,[23] which results in the electrical circuit being open (*i.e.* switch OFF).

11.3.2 Applications

11.3.2.1 Pressure Sensor

Ionic-liquid-based resistive pressure sensors have been designed using the concept of electrofluidic circuits. By designing the geometries of electrofluidic channels, the electrofluidic circuit can convert applied pressure into an electrical property change to achieve pressure-sensing capability. The first example is a pressure sensor that is embedded in a microfluidic device for flow monitoring. In fluidic mechanics, the ability to control and monitor the pressure inside microfluidic systems is essential for various applications. For instance, applying pressure enables the transfer of biological and chemical samples to specific locations at desired time points, manipulate flows, and separate materials.[24–34] By controlling the pressure, mixing and analysis can be performed more efficiently. Furthermore, pressure measurement has also been used to study the mechanical properties of cells flowing inside the microfluidic channel.[35–37] As a result, precise spatiotemporal pressure sensing is desired for understanding the fundamental phenomena in various applications, including cell culture and analysis, biological detection, and chemical synthesis.

The pressure sensor integrated microfluidic device consists of two PDMS layers with channel patterns: a microfluidic channel layer and an electrofluidic circuit layer (Figure 11.8). The two layers are separated by a PDMS membrane. The former channels are designed for the desired applications, whereas the latter channels serve to measure pressure. When the pressure within the fluidic channel changes in the microfluidic layer, the pressure will deform the PDMS membrane and further alter the cross sectional area of the electrofluidic channel. Finally, such an area change will cause the electrical resistance variation and affect the entire circuit's characteristics. As a result, through measuring the resistance change, the pressure within the microfluidic channel can be estimated. With the advantages of electrofluidic circuits, not only single components can be built, but also circuit functionalities can be embedded into the sample device. Here, pressure sensor designs with

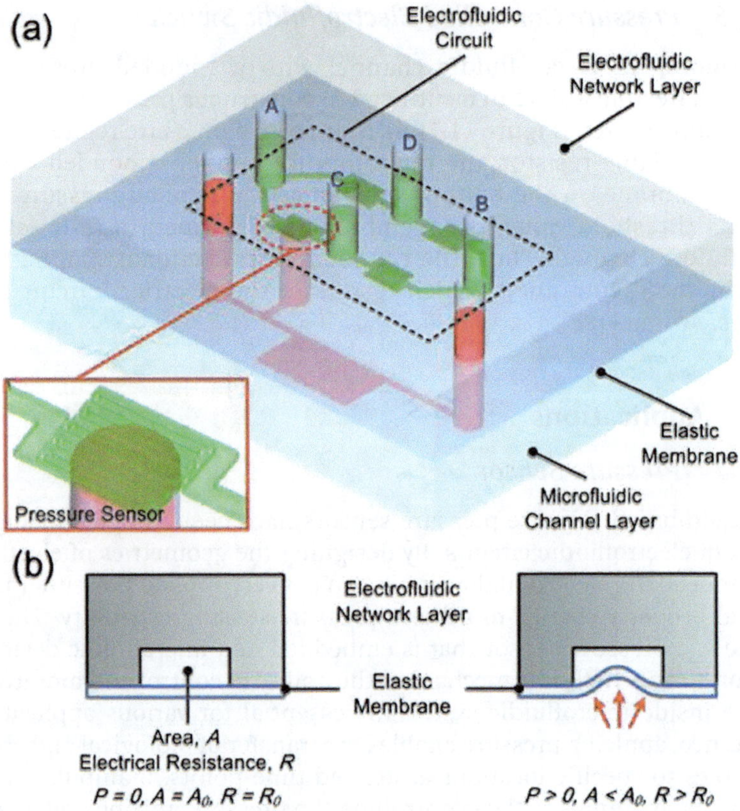

Figure 11.8 Illustration of the PDMS microfluidic system with an integrated electrofluidic circuit for pressure sensing. (a) Schematic of the system with a PDMS membrane sandwiched by two PDMS layers: microfluidic channel and electrofluidic circuit layers. (b) The working principle of the integrated pressure sensor converting mechanical pressure into electrical resistance variation. Reproduced from ref. 21 with permission from the Royal Society of Chemistry.

great temperature stability with analog and digital operation capabilities are introduced.

To accurately measure the resistance change, an electrofluidic Wheatstone bridge is adapted on the electrofluidic circuit layer. A Wheatstone bridge circuit is well known for its preciseness, and it can accurately measure a resistance change with less than only 1% of the nominal resistance error. In addition, the output voltage across the Wheatstone bridge circuit depends on the resistance variation of the resistors. Therefore, an electrofluidic circuit capable of measuring the analog operation of multiple pressures can be constructed. Ionic-liquid-based pressure sensors can be generally designed using one constant electrofluidic resistor and three pressure regulated electrofluidic variable resistors (Figure 11.9(a)).[38]

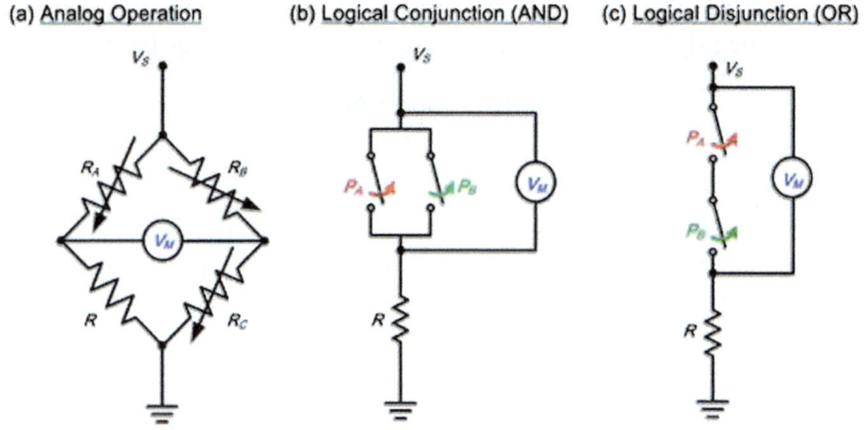

Figure 11.9 Schematics of the designed electrofluidic devices for (a) analog (addition and subtraction), (b) digital (AND), and (c) digital (OR) operations. Reproduced from ref. 38 with permission from the Royal Society of Chemistry.

According to Kirchhoff's circuit laws, while applying the supply voltage (V_S) across the entire Wheatstone bridge circuit, the gate voltage (V_M) across the two arms of the bridge can be derived as:

$$V_M = \left(\frac{R_C}{R_B + R_C} - \frac{R}{R_A + R}\right) \cdot V_S \tag{11.5}$$

where R_A, R_B, and R_C are the resistance of the pressure regulated electrofluidic variable resistors, and R is the original resistance of the constant electrofluidic resistor without pressure application. Since the geometries of four ionic-liquid resistors are designed to be identical in a Wheatstone bridge circuit, the measured gate voltage (V_M) approaches zero while no pressure is applied. On the other hand, when the resistors are pressurized their resistance values will vary according to the applied pressure. Hence, the resistance of the variable resistors can be written as:

$$\begin{cases} R_A = R + \Delta R_A \\ R_B = R + \Delta R_B \\ R_C = R + \Delta R_C \end{cases} \tag{11.6}$$

Substituting into eqn (11.5), the gate voltage across two arms of the Wheatstone bridge circuit can be derived as:

$$V_M = \left[\frac{R(\Delta R_A - \Delta R_B + \Delta R_C) + \Delta R_A \Delta R_C}{4R^2 + 2R(\Delta R_A + \Delta R_B + \Delta R_C) + (\Delta R_A \Delta R_B + \Delta R_A \Delta R_C)}\right] \cdot V_S \tag{11.7}$$

Because the resistance change, ΔR, is assumed to be relatively small compared to the original resistance, R, the measured gate voltage becomes:

$$V_M \simeq \left(\frac{\Delta R_A - \Delta R_B + \Delta R_C}{4R}\right) \cdot V_S \tag{11.8}$$

According to eqn (11.4) and (11.8), the resultant gate voltage is linearly proportional to the analog operation of the resultant resistance changes, linearly proportional to the area reduction, and further linearly proportional to the resultant applied pressures. As a result, the measured gate voltage of the designed Wheatstone bridge has the relationship:

$$V_M \propto (\Delta R_A - \Delta R_B + \Delta R_C) \propto (P_A - P_B + P_C) \tag{11.9}$$

For usual pressure sensing applications, the device is designed using one pressure regulated electrofluidic variable resistor and three constant electrofluidic resistors in a Wheatstone bridge circuit. The resistances of the resistors become R_A, and $R_B = R_C = R$, respectively. Similarly, the measured voltage across two arms of the Wheatstone bridge can be written as:

$$\Delta V_M = \left[\frac{1}{2} - \frac{R}{(R + \Delta R_A) + R}\right] \cdot V_S = \left(\frac{\Delta R_A}{4R + 2\Delta R_A}\right) \cdot V_S \tag{11.10}$$

while $\Delta R_A \ll R$, eqn (11.10) can be simplified as:

$$\Delta V_M \simeq \left(\frac{\Delta R_A}{4R}\right) \cdot V_S \tag{11.11}$$

By combining with eqn (11.3), the gate voltage variation originating from the pressure induced area reduction of the pressure-sensing resistor is:

$$\Delta V_M \simeq \left(\frac{V_S}{4R}\right) \cdot \left(\frac{\rho l}{A^2}\right) \cdot \Delta A \tag{11.12}$$

In summary, the theoretical derivation shows that the measured gate voltage of the designed electrofluidic circuit will be linearly proportional to the analog operation of the applied pressures, $(P_A - P_B + P_C)$ for the general case. The derivation has also proven that the gate voltage is also linearly proportional to the cross-sectional area reduction in the channel and further linearly proportional to the applied pressure according to the linear plate theory[22] while applying the device for other applications.

Electrofluidic circuits not only perform analog operations as described previously, but also accomplish digital operations between applied pressures and output an electrical signal. The digital operation electrofluidic circuits are composed of constant electrofluidic resistors and pressure controlled switches. Here, two types of electrofluidic circuits for basic binary digital operations are exploited, including logical conjunction (AND) and disjunction (OR). For these two operations, two pressure controlled electrofluidic switches and one constant electrofluidic resistor are arranged as shown in Figure 11.9(b) and (c). To measure the output voltage, V_M, an AC voltage, V_S, is applied. For AND operation, only when both switches are turned off (open circuit) is the output voltage equal to the applied voltage. Otherwise, the resultant voltage approaches zero. For OR operation, turning off either or both switches causes the output voltage to equal the applied voltage. Since without any pressure applied the switches are turned on, the output voltage is approximately zero due to short circuit. Based on a similar design concept, the exclusive disjunction (XOR) operation can also be achieved using electrofluidic circuits.[38]

Electrofluidic circuits composed of the three types of components can be fabricated on PDMS substrates with other microfluidic systems by a well-developed soft lithography replica molding process without further assembly and sophisticated cleanroom fabrication processes. The general fabrication process of an electrofluidic circuit-integrated microfluidic device can be divided into three major steps: electrofluidic layer fabrication, membrane bonding, and microfluidic channel layer assembly (Figure 11.10). For fabrication of electrofluidic and microfluidic layers, the designed features are first patterned on silicon wafers as molds using photoresists by conventional photolithography. For channels with rounded cross-sections, molds made by reflowing the patterned positive photoresist layers using elevated temperature can be used. PDMS prepolymer is poured into the molds and cured at elevated temperature (usually 60–150 °C). After curing, the interconnection holes for electrofluidic channels are punched using a biopsy punch. Then, the electrofluidic layer is permanently bonded with the elastic membrane by surface oxygen plasma treatment (Figure 11.10(b)). An elastic PDMS membrane with specific thickness can be prepared by spinning the PDMS prepolymer on top of a substrate.

To seamlessly integrate electrofluidic circuits into the PDMS microfluidic systems, two designs have been developed. One design has punch inlets and outlets through all layers that are bonded not only with each other but also with a glass substrate by surface oxygen plasma treatment (Figure 11.10(d)). The other design has punch inlets and outlets through only the electrofluidic circuit layer and the membrane layer first. Then, the two layers are bonded with a microfluidic channel layer by surface oxygen plasma treatment (Figure 11.10(e)). The former design provides better flexibility for designing the electrofluidic circuit, since it can eliminate the interference of the electrofluidic circuits from the underneath microfluidic channels. In the latter design, pressure sensing can be achieved with a faster response. However, the electrofluidic channels need to be designed so that they do not overlap with other microfluidic channel patterns in order to eliminate undesired interferences. After the device fabrication, ionic liquid is then filled into the electrofluidic channels.

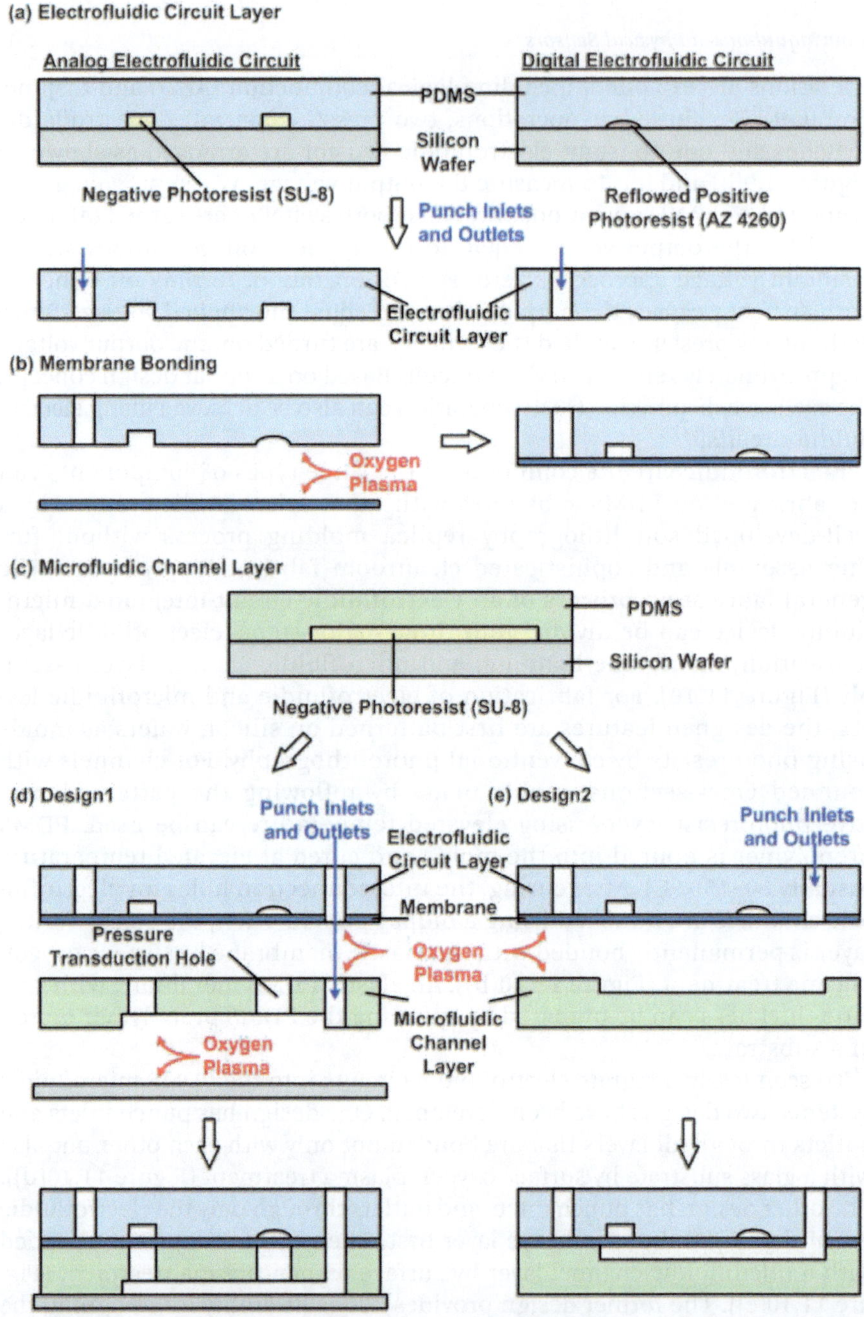

Figure 11.10 Fabrication process of an electrofluidic circuit-integrated microfluidic device. (a) Fabrication of electrofluidic circuit layers for analog and digital operations. (b) Bonding of elastic membranes. (c) Fabrication of a microfluidic layer, (d) Assembly process for design1 with pressure transduction holes for pressure sensing, and (e) assembly process for design2 with electrofluidic and microfluidic channels face-to-face bonded with an elastic membrane sandwiched in between. Reproduced from ref. 38 with permission from the Royal Society of Chemistry.

The seamless integration of ionic-liquid-based electrofluidic circuits into a PDMS microfluidic system provides a sensor capable of monitoring pressure with long-term and temperature stabilities, since the device possesses the advantages of the ionic liquid (*e.g.*, low vapor pressure and thermal stability). A device with three electrofluidic constant resistors and a pressure regulated electrofluidic resistor is utilized to verify the stabilities of the device. The results shown in Figure 11.11 suggest that the pressure sensor has great sensing linearity ($R^2 > 0.99$ for applied pressure up to 36 psi) and high temperature (25–100 °C) and long-term (up to 8 days) stabilities.[21]

Furthermore, a device capable of performing pressure sensing with analog and digital operation has also been demonstrated. In the experiments, one set of analog experiments (three pressure regulated electrofluidic variable

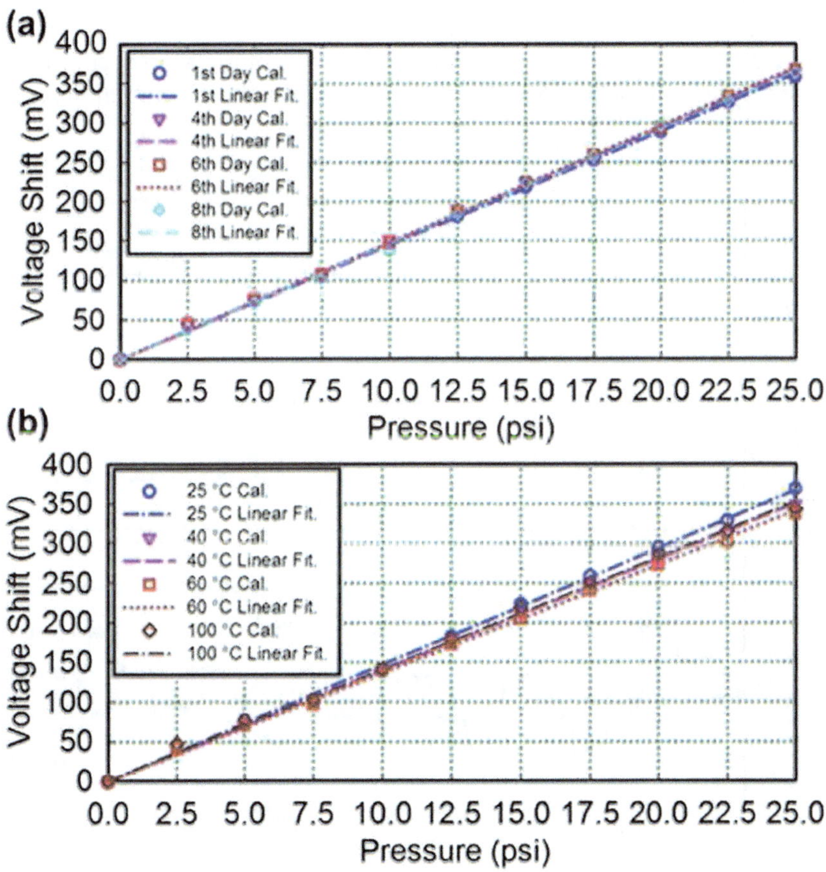

Figure 11.11 (a) Long-term stability measurement: the results and linear regressions for an eight day calibration. (b) Temperature stability measurement: the results and linear regressions for the calibration conducted under 25 (room temperature), 40, 60, and 100 °C. Reproduced from ref. 21 with permission from the Royal Society of Chemistry.

resistors R_A, R_B, and R_C applied with different pressures (P_A, P_B, and P_C)) and two sets of digital experiments performing AND and OR operations have been demonstrated. As derived in eqn (11.9), the electrofluidic circuit is designed to output an electrical signal representing an analog operation, ($P_A - P_B + P_C$), of the applied pressure. Figure 11.12 shows the three different pressures applied to the electrofluidic variable resistors. The pneumatic pressure was maintained constant for at least 5 min during each stage to assure complete deformation of the elastic membrane. Figure 11.12(b) shows the results of analog operation of various pressure combinations. Here, the output voltages of the horizontal dotted lines correspond to addition and subtraction. The output voltage is linearly related to ($P_A - P_B + P_C$). The results suggest that the designed electrofluidic circuit is capable of detecting the physically applied pressures, conducting analog addition and subtraction operations, and outputting consistent electrical signals. The correlation coefficient of a linear regression analysis, R^2, is estimated to be larger than 0.999, indicating the great linearity of the electrofluidic circuits for pressure sensing and analog operation. Moreover, with the benefits of the material properties of the ionic liquid, the experimental results

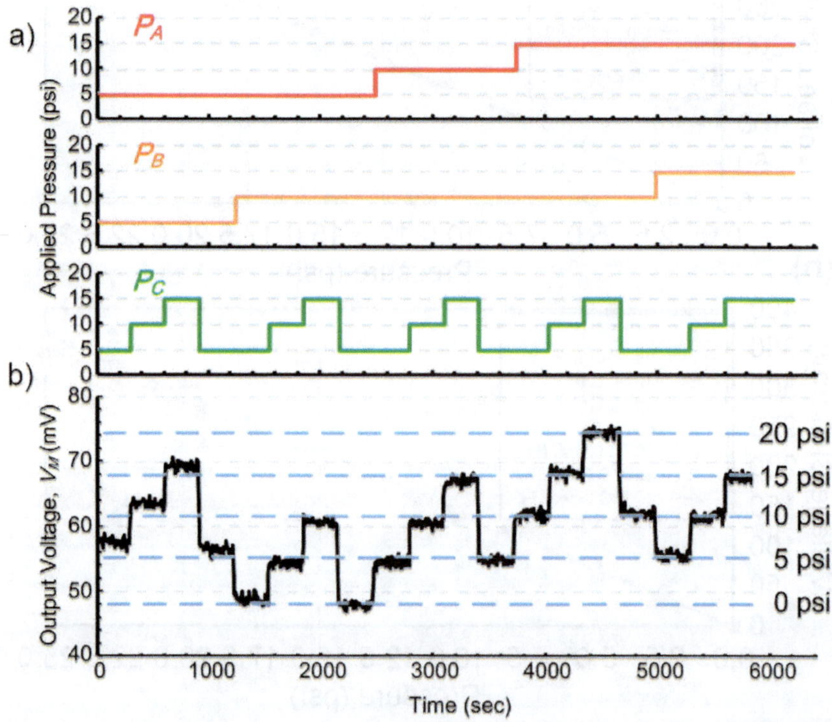

Figure 11.12 Experimental results of analog operation electrofluidic circuit characterization. (a) Time domain application scheme of each pressure (P_A, P_B, and P_C), and the calculated value of ($P_A - P_B + P_C$) for analog operation. (b) Time domain amplitude of the output electrical signal experimentally measured from the electrofluidic circuit. Reproduced from ref. 38 with permission from the Royal Society of Chemistry.

show a stable performance of the device for continuous two-hour operation. Compared to conventional devices that require multiple pressure sensors, analog circuits, and a number of electrical and fluidic interconnections, the electrofluidic circuit developed in this study is capable of seamlessly integrating pressure sensors and analog operation circuits into a microfluidic system with minimum interconnections. Therefore, the setup can easily be scaled up to accommodate various functions for microfluidics applications without sophisticated fabrication processes and tedious interconnections.

In addition, Figure 11.13 shows the results of pressure sensing with AND and OR digital operations. The output voltages show that the devices successfully work as designed in performing the AND and OR operations respectively. For AND operation, the output voltage remains zero as expected (Figure 11.13(a)) while there is only one input pressure (either A or B). When both electrofluidic switches were turned off by applying pressures greater than the threshold value, the amplitude of the output voltage was approximately equal to the applied voltage. For OR operation, the output voltage equaled the applied one, while either or both electrofluidic switches were turned off (Figure 11.13(b)).

The experimental characterization shows that the designed microfluidic device is not only capable of transferring physical pressure applications to electrical signals, but is also capable of performing digital operation to the applied pressures. The seamless integration of pressure transducers and electrical circuit digital functionalities into microfluidic systems using electrofluidic circuits provides additional desired functionalities and makes large-scale integration feasible. Therefore, the ionic-liquid-integrated microfluidic devices can be used for various advanced on chip applications such as cell culture and analysis and biological detection. For example, the developed pressure sensor integrated devices have been utilized for viscosity measurement and cell in-plane elasticity measurement (Figure 11.14). In the viscosity measurement, fluid builds up hydrostatic pressure at upstream when flowing thorough a fluidic channel. The pressure is related to the viscosity of the fluid; therefore, by measuring the pressure, the viscosity can be estimated.[21] For the cell in-plane elasticity measurement, the cell attachment varies the flexural rigidity of the flexible wall of the pressure regulated electrofluidic variable resistor in the pressure sensor, and further changes the sensitivity of the sensor. As a result, the in-plane elasticity of the cells can be calculated from the sensitivity variation of the sensor.[39]

11.3.2.2 Flexible Key Pad

Due to the mechanical flexibility of PDMS and ionic liquids, a flexible and optically transparent keypad device has been developed based on the electrofluidic circuit concept.[40] The device consists of three PDMS layers: top channel layer, bottom channel layer, and a membrane layer sandwiched in between (Figure 11.15). On the top and bottom channel layers, the electrofluidic circuits containing resistors (channel with squared channel cross-section) and switches (channel with rounded cross-section) are designed to detect the button at which column and row is pressed.

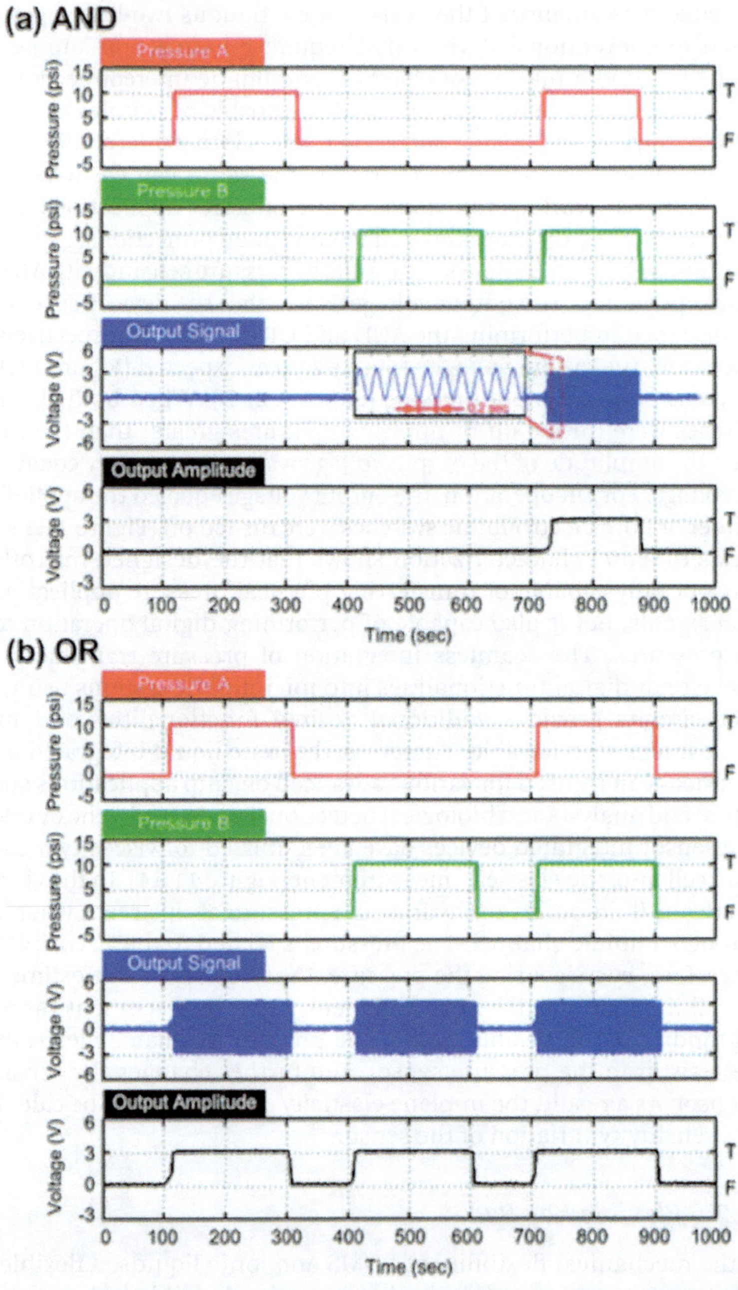

Figure 11.13 Experimental results of digital operation electrofluidic circuit characterization. Plots of applied pressures, output electrical signal, and amplitude of the output signal in time domain for (a) AND and (b) OR operations. Reproduced from ref. 38 with permission from the Royal Society of Chemistry.

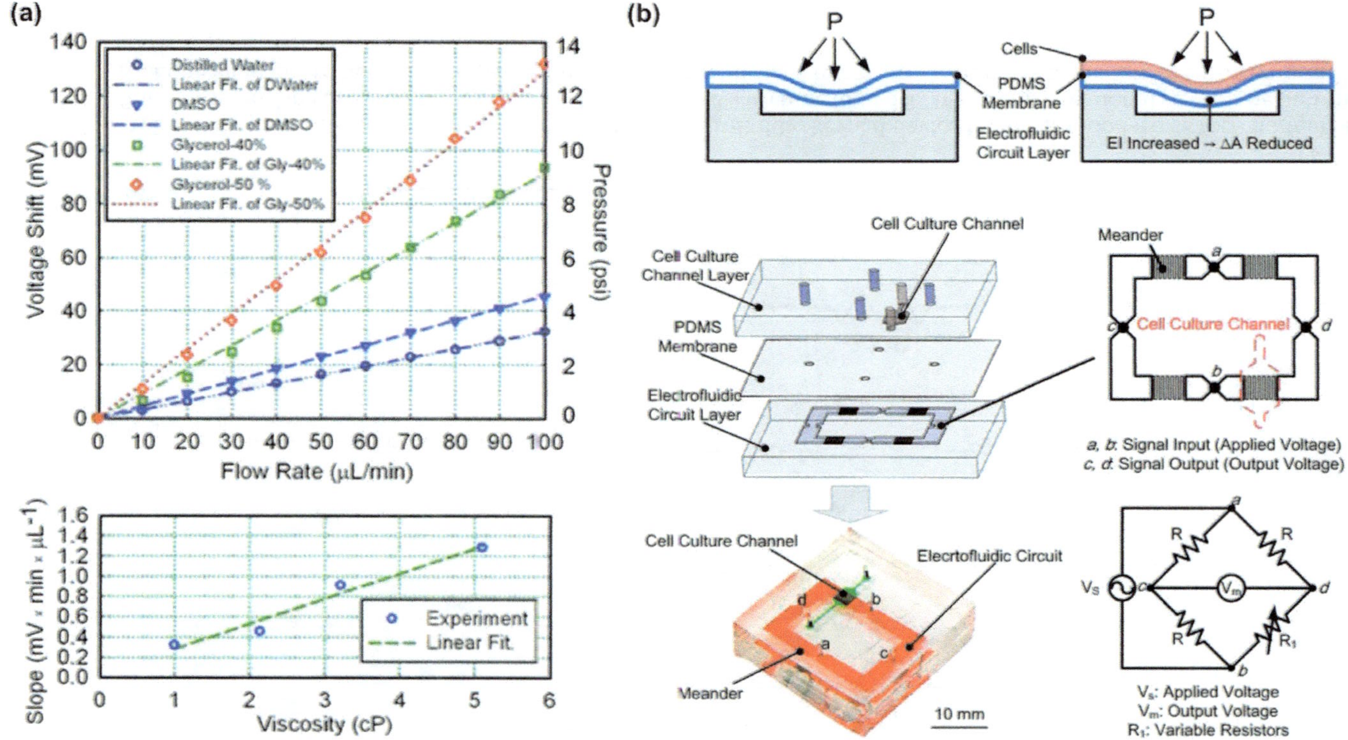

Figure 11.14 (a) Hydrostatic pressure measurement results of liquids with four different viscosities injected into an ionic liquid pressure sensor-integrated microfluidic device. The slope (sensitivity, the ratio of the output gate voltage to the flow rate) can be exploited to estimate the viscosity. (b) Operation principle of the microfluidic device capable of measuring in-plane elasticity of cell layers, and schematic and an equivalent electrical circuit of the microfluidic device with an embedded electrofluidic pressure sensor and a photograph of the fabricated device filled with colored food dyes. (a) Reproduced from ref. 21 with permission from the Royal Society of Chemistry. (b) Reprinted by permission from Macmillan Publishers Ltd: Scientific Reports (ref. 39), copyright 2016.

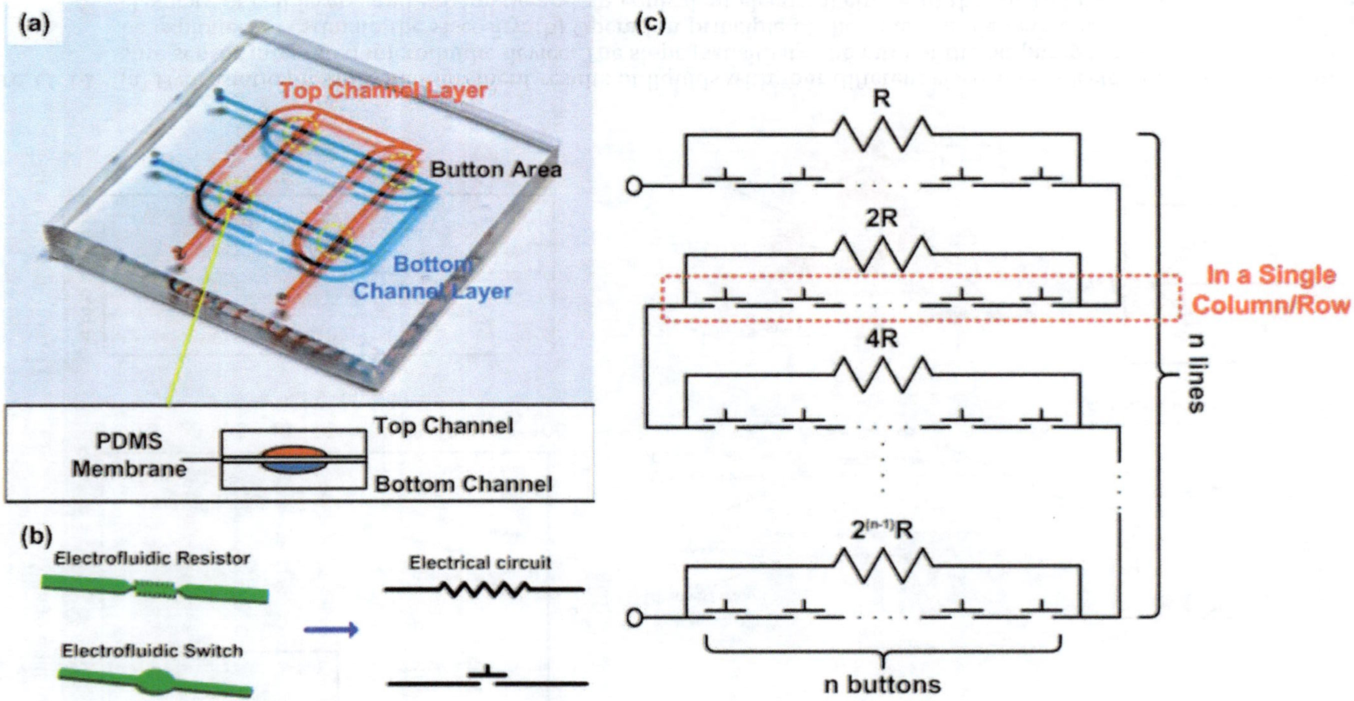

Figure 11.15 (a) Photograph of the fabricated ionic liquid-based electrofluidic keypad device filled with food dyes. Top channel layer is filled with red food dye, and bottom channel layer is filled with blue food dye. Inset: cross-sectional view of the button area. (b) Equivalent electrical circuit of electrofluidic components used to construct the circuit, and (c) the entire keypad circuit.

When an electrofluidic switch is pressed, the channel will be deformed to fully seal the channel and result in an open circuit. The rounded cross-sectional shape is capable of assuring the channel is fully closed. The device was fabricated using a similar process to that described in the previous section. Figure 11.15(a) shows a photograph of the fabricated 2 × 2 electrofluidic keypad device filled with food dyes.

The deice circuit is composed of a series of parallel arrangements of electrofluidic resistors and a series of switches. When one of the switches in the same column or row is pressed, the equivalent resistance of the circuit equals the value of the resistor placed in parallel. As a result, the column and row numbers on which the button is pressed can be determined by measuring the resultant resistance values of the electrofluidic circuits on top and bottom channel layers, respectively. Figure 11.16 shows the resistance

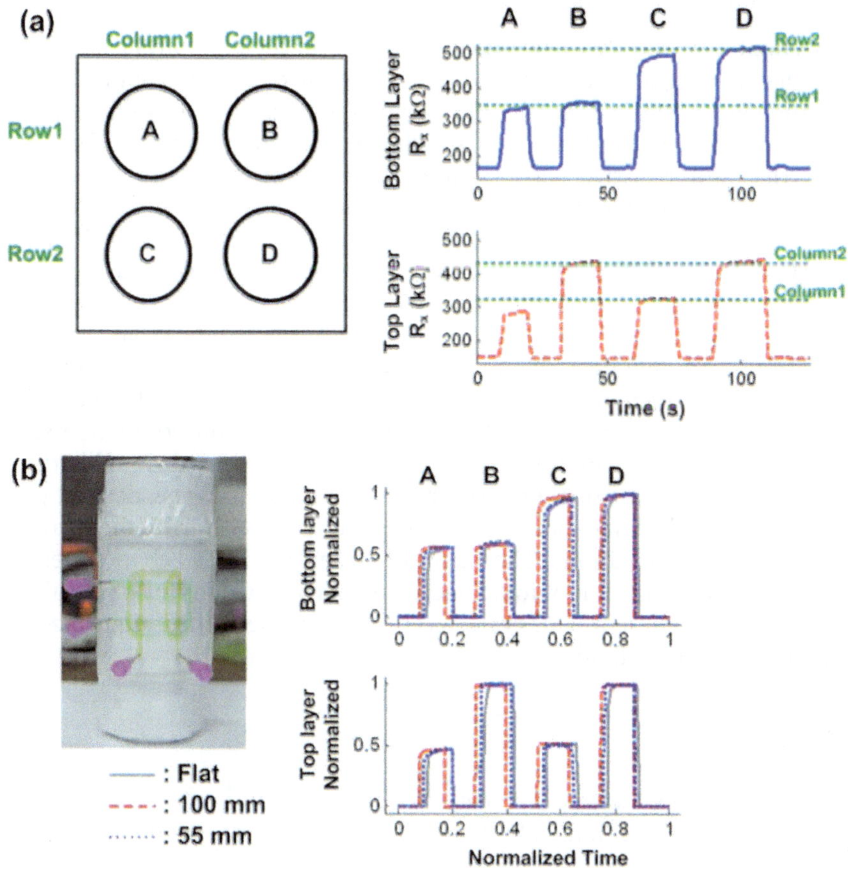

Figure 11.16 (a) Keypad layout and the measured resistance values of electrofluidic circuits at the top and bottom channel layers when different buttons are pressed. (b) Measured resistance values when different buttons are pressed when the device is bent with different radii of curvature.

measurement results when pressing different buttons on the keypad device. The bottom layer circuit measures on which row the button is pressed, and the top layer circuit measures on which column the button is pressed. Combining the measurements from both layers, the pressed button can be determined. Figure 11.16 further demonstrates the measurement results when the device is bent with different radii of curvature. The results suggest that the device can also be successfully operated under deformation. The developed ionic liquid-based keypad provides a promising technique to construct a bendable, optically transparent, and stable human input device with low cost and minimal fabrication requirements. The great stability of the device makes it suitable in the construction of various wearable devices for real world applications.

11.4 Conclusion

This chapter introduces new types of ionic-liquid-based physical sensors. The key-enabling breakthrough is the use of non-volatile, mechanically flexible, and electrically conductive ionic liquids to construct sensing elements. Ionic-liquid-based physical sensors with two types of sensing mechanisms, capacitive and resistive sensors, and their applications are introduced. For the capacitive sensors, the high capacitance at electrical double layers between an ionic liquid and electrodes, and the mechanical flexibility of ionic liquids, provide relatively high sensitivity for the sensors. Furthermore, on integration with transparent and deformable elastomeric substrates, the entire sensor has a rapid response and optical transparency that are useful for various wearable devices and biomedical applications. For the resistive sensors, a new concept of electrofluidic circuit, an ionic liquid filled fluidic channel network, is introduced. By arranging electrofluidic components, not only physical sensors but also circuit functions can be integrated within a single device without further assembly. The sensors possess desired great long-term and thermal stability, and they can be used for biomedical and daily practical applications. The examples described in this chapter demonstrate that the utilization of ionic liquids opens a new door for physical sensor development and provides various advantages due to the unique material properties of ionic liquids.

References

1. K. E. Johnson, *Electrochem. Soc. Interface*, 2007, 38.
2. G. A. Baker, S. N. Baker, S. Pandey and F. V. Bright, *Analyst*, 2005, **130**, 800.
3. J.-F. Liu, J. A. Jonsson and G.-B. Jiang, *Trends Anal. Chem.*, 2005, **24**(1), 20.
4. S. Pandey, *Anal. Chim. Acta*, 2006, **556**, 38.
5. N. V. Plechkova and K. R. Seddon, *Chem. Soc. Rev.*, 2008, **37**, 123.

6. S. Sowmiah, V. Srinivasadesikan, M.-C. Tseng and Y.-H. Chu, *Molecules*, 2009, **14**, 3780.
7. P. Sun and D. W. Armstrong, *Anal. Chim. Acta*, 2010, **661**, 1.
8. J. S. Wilkes and M. J. Zaworotko, *J. Chem. Soc., Chem. Commun.*, 1992, 965.
9. B. Nie, S. Xing, J. D. Brandt and T. Pan, *Lab Chip*, 2012, **12**, 1110.
10. J. C. Yeo and C. T. Lim, *Microsyst. Nanoeng.*, 2016, **2**, 16043.
11. P. Wang and Q. Liu, *Biomedical Sensors and Measurement*, Springer, 2011, p. 51.
12. A. Pantelopoulos and N. G. Bourbakis, *IEEE Trans. Syst. Man Cybern., Syst.*, 2010, **40**(1), 1.
13. A. Koh, D. Kang, Y. Xue, S. Lee, R. M. Pielak, J. Kim, T. Hwang, S. Min, A. Banks, P. Bastien, M. C. Manco, L. Wang, K. R. Ammann, K.-I. Jang, P. Won, S. Han, R. Ghaffari, U. Paik, M. J. Slepian, G. Balooch, Y. Huang and J. A. Rogers, *Sci. Transl. Med.*, 2016, **8**, 366ra165.
14. J. Yan and T. Pan, *IEEE Int. Conf. Micro Electro Mech. Syst., 24th*, 2011, 557.
15. M. A. Unger, H. P. Chou, T. Thorsen, A. Scherer and S. R. Quake, *Science*, 2000, **288**, 113.
16. B. Nie, R. Li, J. D. Brandt and T. Pan, *Lab Chip*, 2014, **14**, 1107.
17. Y. Z. Ding, E. Huang, K. S. Lam and T. Pan, *Lab Chip*, 2013, **13**, 1902.
18. G. M. Whitesides, *Nature*, 2006, **442**, 368.
19. D. Janasek, J. Franzke and A. Manz, *Nature*, 2006, **442**, 374.
20. Y.-C. Tung, Y. Torisawa, N. Futai and S. Takayama, *Lab Chip*, 2007, **7**, 1497.
21. C.-Y. Wu, W.-H. Liao and Y.-C. Tung, *Lab Chip*, 2011, **11**, 1740.
22. S. P. Timoshenko and S. Woinowsky-Krieger, *Theory of Plates and Shells*, McGraw-Hill, New York, 2nd edn, 1959.
23. Y.-C. Tung and C.-Y. Wu, *SPIE Newsroom*, 2012, DOI: 10.1117/2.1201207.004343.
24. A. Gunther and K. F. Jensen, *Lab Chip*, 2006, **6**, 1487.
25. R. L. Hartman and K. F. Jensen, *Lab Chip*, 2009, **9**, 2495.
26. H. A. Stone, A. D. Stroock and A. Ajdari, *Annu. Rev. Fluid Mech.*, 2004, **36**, 381.
27. T. M. Squires and S. R. Quake, *Rev. Mod. Phys.*, 2005, **77**, 977.
28. B. Mosadegh, C. H. Kuo, Y. C. Tung, Y. S. Torisawa, T. Bersano-Begey, H. Tavana and S. Takayama, *Nat. Phys.*, 2010, **6**, 433.
29. T. Thorsen, S. J. Maerkl and S. R. Quake, *Science*, 2002, **298**, 580.
30. J. G. Shackman, G. M. Dahlgren, J. L. Peters and R. T. Kennedy, *Lab Chip*, 2005, **5**, 56.
31. K. H. Chung, M. M. Crane and H. Lu, *Nat. Methods*, 2008, **5**, 637.
32. C. Fiitterer, N. Minc, V. Bormuth, J. H. Codarbox, P. Laval, J. Rossier and J. L. Viovy, *Lab Chip*, 2004, **4**, 351.
33. S. Zappe, M. Fish, M. P. Scott and O. Solgaard, *Lab Chip*, 2006, **6**, 1012.
34. D. W. Lee and Y. H. Cho, *Lab Chip*, 2009, **9**, 1681.

35. B. R. Schudel, C. J. Choi, B. T. Cunningham and P. J. A. Kenis, *Lab Chip*, 2009, **9**, 1676.
36. H. M. Xia, Z. P. Wang, Y. X. Koh and K. T. May, *Lab Chip*, 2010, **10**, 1712.
37. M. Abkarian, M. Faivre and H. A. Stone, *Proc. Natl. Acad. Sci. U. S. A.*, 2006, **103**, 538.
38. C.-Y. Wu, J.-C. Lu, M.-C. Liu and Y.-C. Tung, *Lab Chip*, 2012, **12**, 3943.
39. C.-H. Lin, C.-K. Wang, Y.-A. Chen, C.-C. Peng, W.-H. Liao and Y.-C. Tung, *Sci. Rep.*, 2016, **6**, 36425.
40. J.-C. Lu, Y.-J. Cheng, W.-H. Liao, and Y.-C. Tung, *The 5th International Conference on Optofluidics*, 2015.

CHAPTER 12

Aspects of Recent Advances in Smart Ionic Liquid Based Sensors

B. NATESH KUMAR, G. G. REDHI* AND C. RAJASEKHAR

Electroanalytical Laboratories, Department of Chemistry, Durban University of Technology, Durban, South Africa-4000
*E-mail: redhigg@dut.ac.za, nateshkumar786@gmail.com

12.1 An Overview on Ionic Liquids

Ionic liquids (ILs) are generally a diverse group of salts, which are liquids at ambient temperatures. While early work in this field tended to presume that ionic liquids had very similar properties as a class, it is now widely recognized that, in fact, they offer a very wide range of properties and that one of the only properties that can be thought of as ubiquitous among ionic liquids is ion conductivity.[1] ILs are a new class of purely ionic, salt-like materials that are liquid at unusually low temperatures. ILs can also be defined, as "Ionic Liquids are ionic compounds which are liquid below 100 °C". Especially, salts that are liquids at room temperature are called room-temperature ionic liquids (RTILs). RTILs are otherwise termed as molten organic liquids or fused salts with low melting points. The modern definition of a RTIL is any salt that has a melting point lower than ambient temperature. We begin with a brief overview of the types of cations and anions that are involved in the formation of ILs.

Smart Materials No. 28
Ionic Liquid Devices
Edited by Ali Eftekhari

Published by the Royal Society of Chemistry, www.rsc.org

12.1.1 Cations

The ionic liquid cation is generally an organic structure of low symmetry. The cationic centre most often involves a positively charged nitrogen or phosphorus. Those described thus far are based on ammonium, sulphonium, phosphonium, imidazolium, pyridinium, picolinium, pyrrolidinium, thiazolium, oxazlium or pyrazolium cations, usually completely substituted. The more recent research has mainly focused on room temperature ionic liquids composed of asymmetric dialkylimidazolium cations associated with a variety of anions. Through modification of the cation, the properties of the liquid, notably the melting point and liquid range,[2] viscosity[3] and miscibility with other solvents,[4] can be altered. On the basis of the cation, ILs are broadly divided into three categories: (1) five- and six-membered heterocyclic cations, (2) benzo-fused heterocyclic cations, (3) ammonium, phosphonium and sulphonium based cations.

12.1.1.1 Five-membered Heterocyclic Cations

The five-membered cations include imidazolium, pyrazolium, triazolium, thiazolium and oxazolium (Figure 12.1). While the halide salts are, usually, solids at room temperature there are many anions that lower the melting points of the salts to below room temperature. It is generally assumed that non-symmetrical N,N′-alkyl imidazolium cations yield salts having the lowest melting points; however, dibutyl-, dioctyl-, dinonyl- and dodecyl-imidazolium hexafluorophosphates are also liquids at room temperature.[5] The halide triazolium salts are solids at room temperature, but metathesis to salts, such as bis(trifluoromethanesulphonyl)amide, triflate or tetrafluoroborate, lowers the melting point to near or below room temperature.

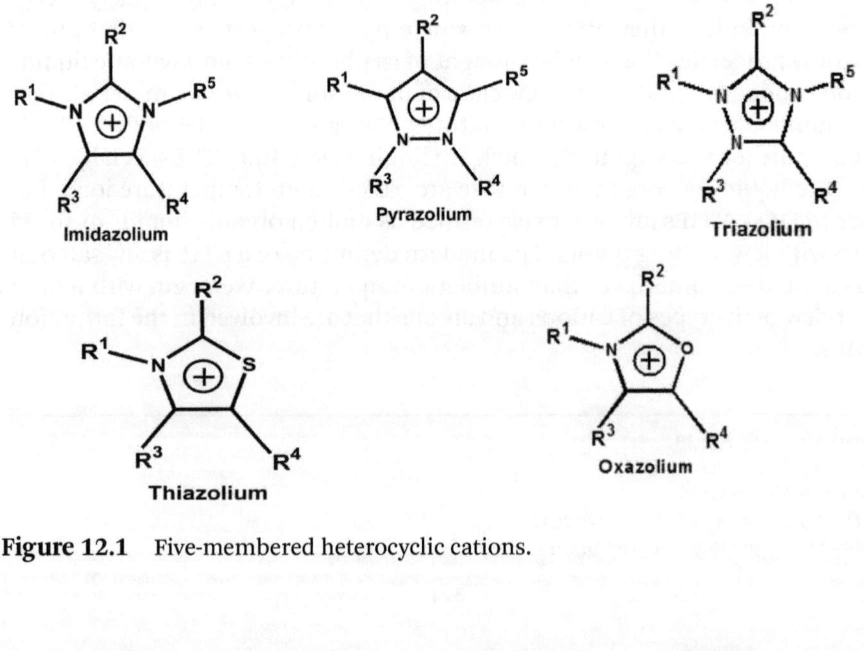

Figure 12.1 Five-membered heterocyclic cations.

12.1.1.2 *Six-membered Heterocyclic Cations*

Figure 12.2 shows the pyridinium, benzotriazolium, isoquinolinium and viologen-type cations. These salts have been known for quite some time, but interest in them has been far less intense than that of the imidazolium family. This is likely due to their more limited stability in the presence of nucleophiles and the toxicity of pyridine. Gordon *et al.*[6] have reported a series of pyridinium hexafluorophosphate salts with long alkyl chains (C_{12}–C_{18}), some of which melt below 100 °C. A more recent area of focus is the viologen family of ionic liquids. While most viologen-type ILs are very high melting solids, there is a handful that do exhibit much lower melting points, though not quite at room temperature. Benzotriazolium based ILs are often good solvents for aromatic species.[7]

12.1.1.3 *Ammonium, Phosphonium and Sulphonium Based Cations*

Tetraalkylammonium (Figure 12.3) salts have been known for centuries. In terms of their use as RTILs, earlier studies have led to the conclusion that longer alkyl chains are necessary required to obtain room temperature melting points. These are typically prepared by alkylation of the parent amine. To obtain low melting points, typically, at least two or three different alkyl groups

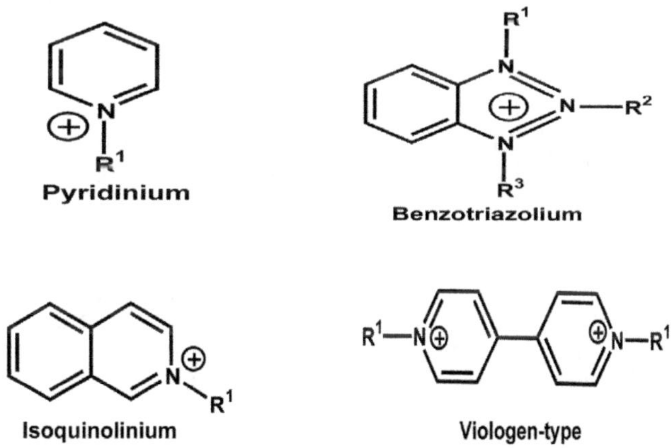

Figure 12.2 Six-membered and benzo fused heterocyclic cations.

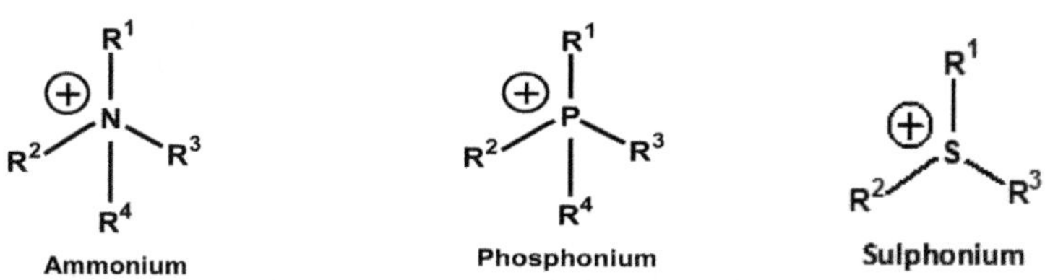

Figure 12.3 Ammonium, phosphonium and sulphonium based cations.

are required to create crystal-packing constraints; this usually requires several steps of alkylation. Although phosphonium RTILs are certainly known and finding growing applications in organic syntheses and other areas,[8] there is surprisingly little in the way of published data regarding their physical properties.[9] The hydrogen sulphate salt of the tetrabutylphosphonium cation has a melting point of 122–124 °C, while the hydrogen sulphate salt of the tributyl(decyl) phosphonium cation is a room temperature liquid.[10] In terms of other physical properties, the viscosity of these phosphonium RTILs is generally somewhat higher than their ammonium counterparts, but decreases very rapidly with increasing temperature. Phosphonium salts are generally more thermally stable than ammonium salts.[9] Phosphonium salts are typically made by alkylation of the parent phosphine.[11] As will be discussed further below, for the larger phosphonium cations this is straightforward; however, the pyrophoric nature of the lighter alkyl phosphines makes this a more difficult matter. One of the least studied types of RTILs is those based on the trialkyl-sulphonium cation.

As might be expected, the melting point and density typically decrease as the size of the cation increases. The viscosity, however, reaches a minimum with the triethyl compound and then often increases significantly for the tributyl compound.

12.1.2 Anions

Anions that form room temperature ionic liquids are usually weakly basic inorganic or organic compounds that have a diffuse or protected negative charge. On the basis of the anion, ILs may be divided into six groups: (1) ILs based on $AlCl_3$ and organic salts;[12] (2) ILs based on anions like PF_6^-,[5,13] BF_4^-,[14,15] and SbF_6^-;[13] (3) ILs based on anions like those in Figure 12.4 (a),[16,17] (b),[17,18] (c)[7] and (d);[18] (4) ILs based on anions like alkylsulphates,[19]

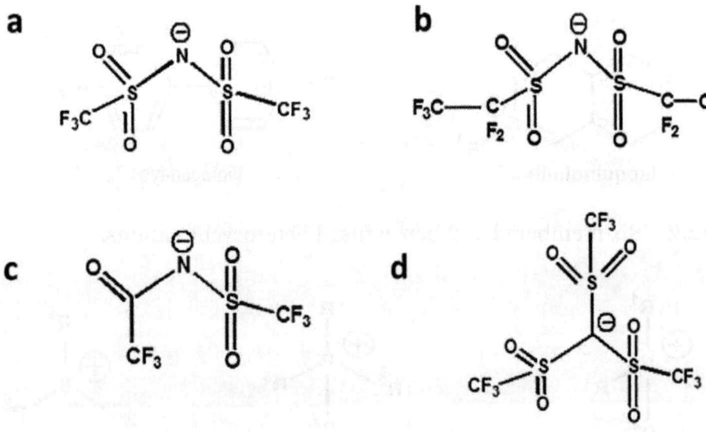

Figure 12.4 a) Bis(trifluoromethanesulonyl)amide, b) bis(perfluoroethyalsufonyl) amide, c) 2,2,2,-trifluoro-N-(trifluoromethanesulfonyl)acetamide, and d) tris(trifluoromethaneslfonyl)methanide.

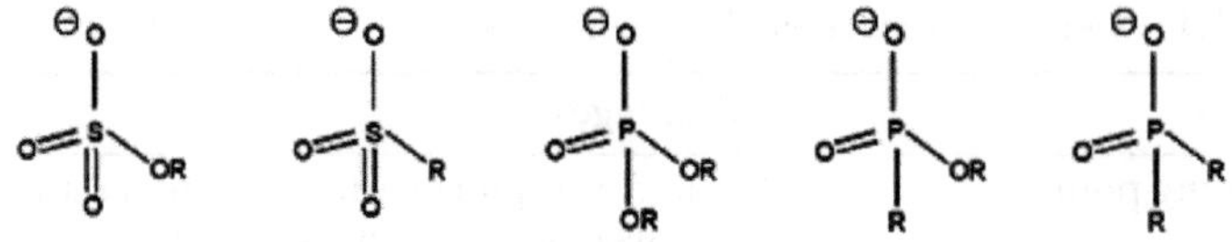

Figure 12.5 Sulphate, sulphonate, phosphate, phosphonate and phosphinate, anions.

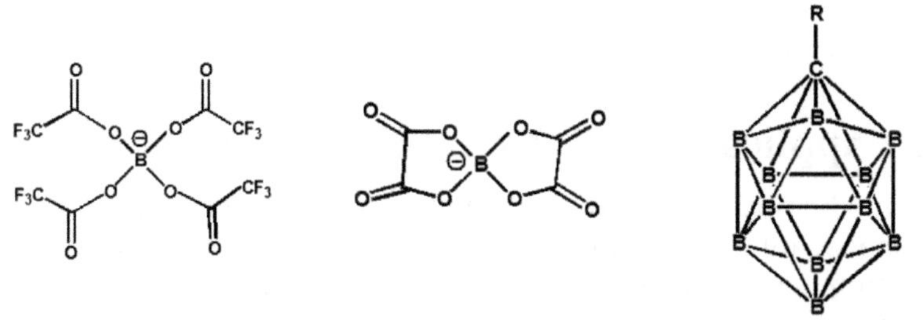

Figure 12.6 Borate and borane anions.

alkylsulphonates,[20] alkylphosphates,[9] alkylphosphonates[9] and alkylphosphinates[9] (Figure 12.5); (5) ILs based on anions such as mesylate,[21,22] tosylate ($CH_3PhSO_3^-$),[23] trifluoroacetate ($CF_3CO_2^-$),[3] acetate ($CH_3CO_2^-$),[7] SCN^-,[24] triflate ($CF_3SO_3^-$)[13,21,25] and dicyanamide [$N(CN)_2^-$];[26,27] (6) ILs based on anions such as the borates[28] and carboranes (Figure 12.6).[29]

12.1.3 General Properties of ILs

The unique properties of ionic liquids are summarized in Table 12.1. Note that these are entirely different properties from those of ordinary molecular liquids. In addition, every ionic liquid does not always show these properties. For electrochemical usage, the most important properties are both non-volatility and high ion conductivity. These are essentially the properties of advanced electrolyte solutions that are critical to energy devices for outdoor use. Safety is a more important issue than performance these days, and it has been taken into account in the materials developed for practical and commercial use. Thus, more developments in ionic liquids are expected in the future. The non-volatile electrolyte solution will change the shape and performance of electronic and ionic devices. These devices will become safer and have longer operational lives. They are composed of organic ions, and these organic compounds have unlimited structural variations because of the easy preparation of many different components. There are unlimited possibilities open to the relatively new field of ionic liquids. The most compelling idea is that ionic liquids are "fine-tunable." Therefore, we can easily expect extensive developments in fields using these remarkable materials.

Table 12.1 General properties of ionic liquids.

Properties	Advantages
Low melting point	Treated as liquid at ambient temperature Wide usable temperature range
Non-volatility	Thermal stability Flame retardancy
Composed by ions	High ion density High ion conductivity
Organic ions	Various kinds of salts Designable Unlimited combinations

12.2 Applications of ILs in the Sensory Field

An electrochemical biosensor is an analytical device that converts a biological response into an electrical signal. In general, it is difficult to exchange the electron between an enzyme and solid surface of the electrodes directly. This is because of the inaccessibility of its redox centre and loss of bioactivity of the enzyme due to the conformational changes arising through adsorption on the electrode surface. ILs have shown good compatibility with biomolecules and enzymes, and even whole cells. Thus, ILs can be used in electrochemical biosensors (Figure 12.7) typically as both binder and conductor. Common ILs advantages observed when they are incorporated into biosensors include higher conductivity, good biocatalytic ability,[29] long-term stability, superior sensitivity, improved linearity, better selectivity, and the ability to fabricate third-generation biosensors with direct electron transfer between protein and protein substrate electrode.

Applications of RTILs in chemo-sensing are recently being explored by many researchers. In one of the earlier reports, Dai and co-workers have developed and evaluated the performance of a quartz crystal microbalance device employing RTILs as the sensing materials for organic vapours.[30] The sensing mechanism was shown to be based on the rapid decrease in the viscosity of RTIL membrane due to solubilization of analytes in RTILs. It was further observed that this change in viscosity was specific to the gaseous chemical species as well as to the type of RTIL. The sensor was shown to have a rapid response (<2 s) with excellent reversibility due to the fast diffusion of analytes in RTILs. In a separate investigation, Compton and co-workers have demonstrated the use of RTILs in gas sensor design.[31] While comparing the attainable steady-state limiting currents and time responses of membrane-covered and membrane independent gas sensors, constituted of a variety of electrodes and electrolyte materials, Compton and co-workers considered a new design of a membrane-free microelectrode, modified with a thin layer of a RTIL. It was also concluded that the use of RTIL as an electrolyte eliminates the need for a membrane and supporting electrolyte; however, slower time responses resulted due to the higher viscosities of RTILs. It was suggested that these types of RTIL-based sensors might operate better under extreme operating conditions (*e.g.*, high temperature and pressure) due to

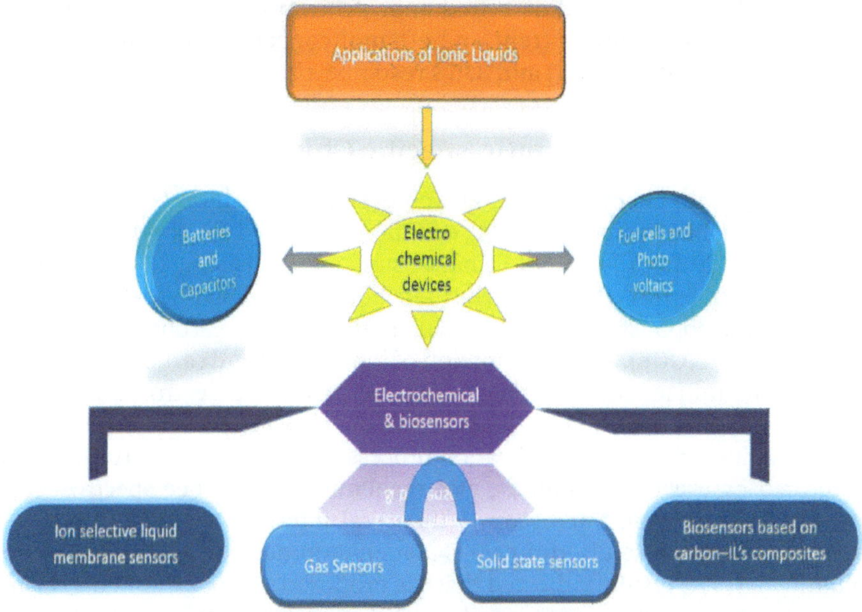

Figure 12.7 Schematic representation of electrochemical sensors applications of ionic liquids.

the thermal stability and extreme low volatility of RTILs. Towards more specific gas sensing applications, a solid-state amperometric O_2 sensor was built by using a supported hydrophobic bmimPF_6 porous polyethylene membrane as a solid-state electrolyte.[32] This RTIL-based O_2 gas sensor showed a wide detection range and high stability. In a similar investigation, the same group utilized a different hydrophilic RTIL, 1-ethyl-3-methylimidazolium tetrafluoroborate, to fabricate a solid-state amperometric sensor for O_2 detection.[33] Bennett and Leo have successfully shown the application of RTILs as stable solvents for ionic polymer transducers.[34] They explored the replacement of water with highly stable RTILs to saturate Nafion™ membranes, which are known to operate as electromechanical actuators and sensors. Their experimental results demonstrated that Nafion™ transducers solvated with the RTIL 1-ethyl-3-methylimidazolium trifluoromethyl-sulfonate have improved stability when operated in air, as compared to the same materials solvated with water. However, the reduction in the speed of the response as compared to water was also noteworthy. Finally, a critical evaluation of RTILs containing the lithium methyl sulfonyl group in ethanol sensing has been reported by Lee and Chou.[35] These RTILs were prepared from the precursor poly(propylene glycol)-*block*-(ethylene glycol)-*block*-(propylene glycol)-bis(2-aminopropyl ether) with different molecular weights. To establish the sensitivity of an ethanol sensor by using these RTILs, they were subjected to sequential electrochemical tests with Ni electrodes which performed at high sensitivity for the ethanol sensor. However, it was reported that only the derivative with low molecular weight could detect ethanol. A linear relationship was established between response current and ethanol concentration, with a detection limit around 0.13% (v/v) and 336 s response time.

12.2.1 IL Mediated Enantio-specific Chiral Sensor for DOPA

A sensitive electrochemical method was explored for chiral recognition of 3,4-dihydroxyphenylalanine (DOPA) enantiomers based on the immobilization of β-cyclodextrin (β-CD) onto a nanocomposite, consisting of multiwalled carbon nanotubes (MWCNTs) and the ionic liquid (IL) 1-butyl-3-methylimidazolium hexafluorophosphate (BIMIMPF6), modified glassy carbon electrode (β-CD/MWCNTs-IL/GCE). The combination of ionic liquid and carbon nanotubes could create unique nanomaterials, which facilitated the electron transfer. In addition, β-CD was brought in to act as the chiral selector for DOPA enantiomer recognition.[36] The immobilization process of β-CD/MWCNTs-IL/GCE was monitored by cyclic voltammetry, electrochemical impedance spectroscopy, and scanning electron microscopy. The chiral interface (β-CD/MWCNTs-IL/GCE) was employed to discriminate DOPA enantiomers *via* differential pulse voltammetry. The results indicated that the proposed sensor exhibited a stronger electrochemical response toward D-DOPA over the linear range from 4.0×10^{-3} to 4.0×10^{-9} mol L^{-1} and the detection limit was 1.2×10^{-9} mol L^{-1} ($S/N = 3$). The present findings give an available platform for enantioselective recognition of DOPA enantiomers based on the nanocomposite. β-CD/MWCNTs-IL/GCE exhibited a higher enantio-specificity for D-DOPA. The proposed method was sensitive over a wide linearity range and with a low detection limit. These features may provide an alternative method for the simple and sensitive recognition of DOPA enantiomers in chiral pharmaceutics.

12.2.2 IL Based Strain Sensor for Tendon Measurements

A robust and low cost ionic liquid based strain sensor has been fabricated for high strain measurements in biomedical applications (Figure 12.8). A tubular 5 mm long silicone microchannel with an inner diameter of 310 μm and an outer diameter of 650 μm is filled with an ionic liquid. Three ionic liquids have been investigated: 1-butyl-1-methylpyrrolidinium bis(trifluoromethylsulfonyl)imide, ethylammonium nitrate and cholinium ethanoate.[37] When the channel is axially stretched, geometrical deformations change the electrical impedance of the liquid channel. The sensors display a linear response and low hysteresis with an average gauge factor of 1.99 for strains up to 40%. Additionally, to fix the sensor by surgical stitching to soft biological tissue, a sensor with tube clamps consisting of photo-patternable SU-8 epoxy-based resin is proposed (Figure 12.9).

12.2.3 IL-based Actuator as a Humidity Sensor

Ionic electromechanically active polymer actuators with ionic liquid electrolytes working in air also respond to ambient relative humidity (RH). The physicochemical parameters, especially viscosity and conductivity, of ionic

Figure 12.8 Ionic liquid based strain sensor. Reproduced from *Biomedical Microdevices*, An ionic liquid based strain sensor for large displacement measurement, **19** (1), 2016, G. Keulemans, F. Ceyssens, R. Puers, © Springer Science+Business Media New York 2016 with permission of Springer.[37]

liquids are strongly dependent on temperature as well as on their water content. By measuring the impedance between the actuator electrodes, the relative humidity value can be extracted from the signal. The sensing actuator has the ability to convert electric energy into mechanical work and additionally to sense the parameters of the surrounding environment.

An iEAP laminate actuator can sense the ambient humidity around it. The iEAP laminate consists of highly hygroscopic constituents and starts immediately absorbing ambient water vapour when exposed to a humid environment. The impedance measurements show that the sorption of water causes the diminishing of internal resistance of the laminate up to as much as tenfold. The electrical double layer capacitance of a saturated laminate is increased by up to 30% compared to the dried condition.[38] The iEAP laminate was modelled as a series connection of resistance, capacitance and constant phase element, which gave a good fit in the frequency range corresponding to the typical range the actuator is used. The described effect turns an iEAP laminate made of TiC-CDC, IL and Nafion into a multi-sensing actuator. In addition to the ability to sense its position or motion and temperature, also information about the ambient humidity can be extracted by measuring the impedance. It is even possible to find an impedance parameter that even shows a linear response to the ambient RH value; for example, it is demonstrated that the phase angle measured at 1 Hz harmonic input signal shows linear correlation with RH.

12.2.4 Biosensor for Bisphenol-A

A novel surface of modified magnetite nanoparticles of Fe_3O_4 coated with cyclodextrin-functionalized ionic liquid (Fe_3O_4–CD-IL) has been obtained *via* the co-precipitation method in alkaline salt medium. Cyclodextrin-functionalized-ionic liquid was effectively coated onto the surface of Fe_3O_4 magnetite nanoparticles. The methods and instruments used to investigate the

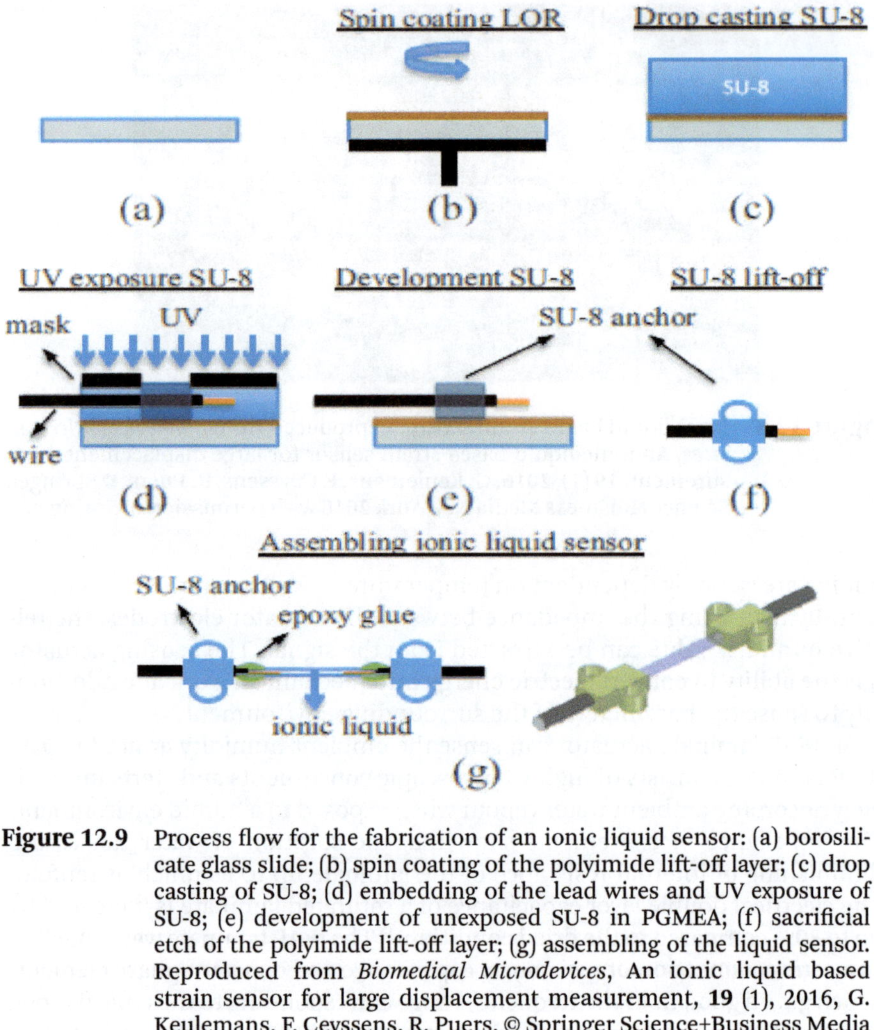

Figure 12.9 Process flow for the fabrication of an ionic liquid sensor: (a) borosilicate glass slide; (b) spin coating of the polyimide lift-off layer; (c) drop casting of SU-8; (d) embedding of the lead wires and UV exposure of SU-8; (e) development of unexposed SU-8 in PGMEA; (f) sacrificial etch of the polyimide lift-off layer; (g) assembling of the liquid sensor. Reproduced from *Biomedical Microdevices*, An ionic liquid based strain sensor for large displacement measurement, **19** (1), 2016, G. Keulemans, F. Ceyssens, R. Puers, © Springer Science+Business Media New York 2016 with permission of Springer.[37]

architecture are Fourier transform infrared spectroscopy, X-ray powder diffraction, electron microscope-energy dispersive X-ray spectrometry, transmission electron microscope, field emission scanning electron microscope, vibrating sample magnetometer and Brunauer–Emmett–Teller isotherm. A vibration sample magnetometer analysis verified that the Fe_3O_4–CD-IL attained excellent magnetic properties. The analysis by high-resolution transmission electron microscope shows that the Fe_3O_4–CD-IL produced had monodispersed particles with minimal aggregation.[39] Moreover, electrochemical studies have revealed that this new material showed outstanding ability to recognize Bisphenol A with lower electrochemical potential at 0.5 V than other comparative materials, as well as exhibiting a higher detection current. Thus, this material has promising potential as a new electrode material in sensor applications.

12.2.5 Simultaneous Determination of Bioactive Compounds

Novel poly(ionic liquid) functionalized polypyrrole nanotubes (PILs/PPyNTs) were successfully synthesized by Hui *et al.*, in 2015 (Figure 12.10) 1-Vinyl-3-ethylimidazole bromide (VEIB) was polymerized on the surface of novel polymerizable vinyl imidazolium-type IL modified PPyNTs prepared by a covalent method. Due to the modification of PILs, the dispersibility of PILs/PPyNTs in aqueous solution was significantly improved and their surface charge properties were obviously changed to electropositivity. Because of the synergetic effects of conductive PPyNTs and biocompatible PILs, excellent electrochemical catalytic activities towards dopamine (DA) and ascorbic acid (AA) were achieved using a PILs/PPyNTs modified glassy carbon electrode (GCE), which gave a potential difference large enough to readily distinguish DA from AA, with excellent sensitivity and good stability, for the simultaneous detection of DA and AA.[40] The existence of PILs effectively improved the transmission mode of electrons of DA and AA oxidation on the electrode and resulted in their different electrocatalytic performance.

12.2.6 Electrochemical Sensor for Metal Ion Detection

Fluorescence studies in acetonitrile solution proved a strong interaction between *N*′-(1-oxoacenaphthalen-2(1*H*)-ylidene) furan-2-carbohydrazide (L) and Yb^{3+} ions in comparison with other lanthanide ions. Thus, L can be used as a suitable ionophore in the construction of a Yb^{3+} potentiometric sensor. To have an electrode with enhanced mechanical resistance in real samples, L was used as a sensing material in new composite carbon paste electrodes.[41] The carbon paste was made based on a new nanocomposite that included multiwall carbon nanotubes (MWCNTs) and the room temperature ionic liquid 1-*n*-butyl-3-methylimidazolium tetrafluoroborate [bmim] BF_4. The nanocomposite electrode showed better sensitivity, selectivity, response time, response stability and lifetime in comparison with typical carbon paste electrodes. The best performance for the nanocomposite sensor was obtained with an electrode composition of 25% L, 25% [bmim] BF_4, 45% graphite powder, and 5% MWCNT. The new electrode exhibited a Nernstian response (19.6 ± 0.3 mV per decade) toward Yb^{3+} ions in the range of 1.0×10^{-8}–1.0×10^{-2} mol L^{-}.[1]

In this work, an Yb^{3+} nanocomposite carbon paste electrode based on MWCNT and ionic liquid is introduced. The performance of Yb^{3+} nanocomposite carbon paste sensor can be greatly improved by using a RTIL instead of a mineral oil (paraffin), and also by using MWCNTs as enhanced signal transducers. The modified CPEs show better potentiometric response than typical CPEs in terms of sensitivity, Nernstian slope, linear range, and response stability. The proposed sensor exhibits a longer lifetime.

12.2.7 Optoelectronic Sensor for Chemical Detection

An ionic liquid (IL)-based colorimetric sensor array for the detection and identification of chemicals in both the aqueous and vapour phases was developed by Waduge and co-workers in 2014. These facile and inexpensive

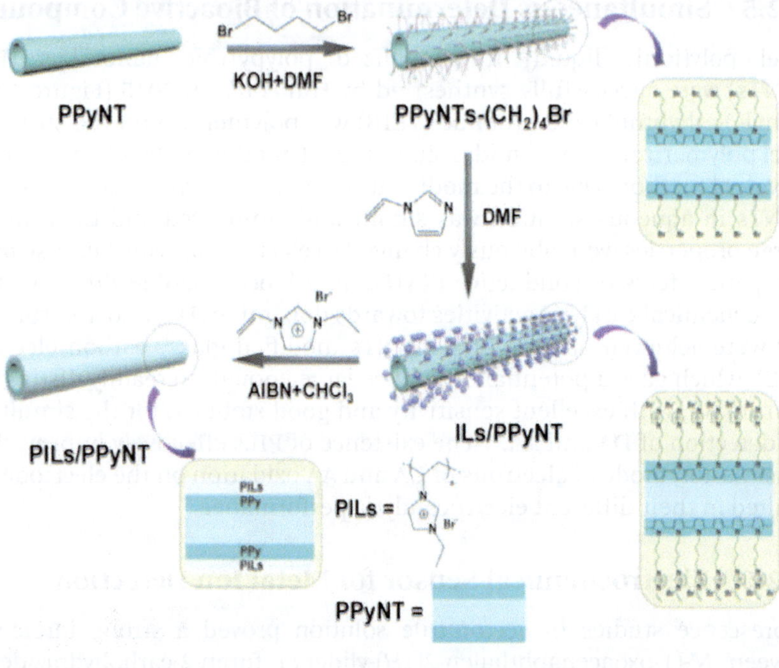

Figure 12.10 Reaction procedure for the preparation of PILs/PPyNTs. Reproduced from ref. 40 with permission from The Royal Society of Chemistry.

optoelectronic sensors were fabricated by using ionic liquids (ILs) derived from readily available pH indicator dyes. A series of 12 different chemosensory ILs were synthesized by pairing anionic pH indicator dyes with trihexyl (tetradecyl) phosphonium ([P_{66614}]) cation *via* an ion exchange reaction. The incorporation of the [P_{66614}] cation imparted hydrophobic characteristics to these ILs, and this induced hydrophobicity led to their desired low solubility in aqueous solutions, and also eliminated the need for a specialized hydrophobic matrix/substrate for immobilization. In the present investigation four different matrices, *i.e.*, glass microfiber filter papers, cotton threads, silica thin layer chromatography (TLC) plates, and alumina TLC plates, were employed for fabrication of sensor arrays. These sensor arrays were used to analyse pH values of aqueous solutions as well as for detection of acidic and basic vapors.[42] To further prove the applicability of these IL sensor arrays as tools to sense closely related complex materials, the arrays were applied to successful discrimination of aqueous solutions of smoke from three commercially available cigarette brands. The digital data generated from these sensor arrays were used in developing predictive models for accurately identifying various analytes. Two approaches were used for developing the models, and two methods were applied for assessing the predictive accuracy of the models. The use of cotton threads as a matrix led to the development of a more flexible, low volume, and lightweight array to estimate pH and detect a variety of vapours. These wearable arrays may possibly be incorporated into bandages, sweatbands, diapers, and similar systems. Overall, these IL-based sensor arrays should provide a new research direction in the development of advanced colorimetric sensor arrays for detection and identification of a range of analytes relevant to many different applications.

A facile, inexpensive, and disposable photonic IL-based sensor array that is sensitive to pH values, as well as acidic and basic vapours were successfully developed. The authors fabricated four different sensor arrays by using four different matrices, all of which are highly compatible for using in aqueous media. They also demonstrated that this sensor array allows discrimination between closely related complex mixtures in aqueous medium. Predictive models were used for successful discrimination and identification of a range of analytes. By use of a variety of matrices, we have demonstrated that these chemosensory ILs do not require any specialized hydrophobic matrix for fabrication of a sensor array. Thus, depending on the particular application, a broadly available, low-cost material can be used as a matrix when fabricating a sensor array. The fabrication of a wearable personal sensor arrays by using cotton threads has also been successfully demonstrated. These wearable sensor arrays can be easily fabricated by manipulating IL-stained cotton threads through use of common household tools such as sewing machines. These thread-based IL sensor arrays can be manufactured on a large scale, allowing applications in bandages, sweatbands, diapers, and other similar systems. Although it needs to be fully evaluated, an IL-based sensor is expected to exhibit a higher sensitivity than a traditional dye-based sensor because of easy accessibility of all dye molecules to the analytes. Finally, we believe that by proper choice of ILs and matrices, these arrays can be constructed to sense a wide range of analytes in a variety of applications, *e.g.* military applications, medical diagnosis, environmental monitoring, water quality analysis, and food safety.

12.2.8 IL-based Ethylene Sensor for Fruit and Vegetable Monitoring

Ethylene is a plant hormone and an indicator for the ripeness of various fruit and vegetables. Measuring the ethylene concentration is therefore a powerful method for monitoring and controlling the ripening process. However, current ethylene detection equipment is bulky and expensive or lacking sufficient sensitivity and selectivity. At the Holst Centre, a novel electrochemical ethylene sensor has been developed, in which a thin ionic-liquid film is used as an electrolyte. This enables the development of a small, low power and low cost ethylene sensor.[43] Presently, the detection limit of the sensor is less than 1 ppm. This is close to the specification needed for monitoring the ripening process during transport, and the storage of fruit and vegetables. Moreover, fabrication of the sensor on a flexible substrate has been demonstrated. This opens up possibilities for integrating the sensor in packaging materials, so-called smart packaging. Here, the sensor technology is explained and recent developments are presented.

The use of IL electrolytes holds great promise for electrochemical gas sensors in general, and for ethylene detection in particular. Our results demonstrate that ionic liquids enable the fabrication of small, low cost and low power ethylene sensors with single digit ppm sensitivity. Thus, IL-based electrochemical ethylene sensors are promising candidates for monitoring and controlling the ripening of fruits and vegetables during transport and storage. Furthermore, we have demonstrated the feasibility of fabricating our ethylene sensor on a flexible substrate. This is not only beneficial for reducing fabrication costs, it also opens up possibilities to integrate ethylene sensors in smart packaging materials.

12.2.9 IL-based Electrochemical Sensor for Ascorbic Acid in Foods and Pharmaceuticals

Bhajanthri and coworkers, in 2016, developed a high performance electrochemical sensor based on 1-butyl-3-methylimidazolium tetrafluoroborate ionic liquid (IL), boron nitride (BN) and a magnetite nanoparticle (Fe_3O_4NPs) based nanocomposite (IL-BN-Fe_3O_4-NPs) for the determination of ascorbic acid (AA) in various foods and pharmaceuticals.[44] The electrochemical sensing behaviour of AA with the bare GCE, MWCNTs, Fe_3O_4NPs and IL-BN-Fe_3O_4-NPs was monitored by using cyclic voltammetry. Differential pulse voltammetric determination of AA was performed in the concentration range 1–12 μM. The calibration plot showed good linearity in the increase of current responses with an R^2 of 0.998. The limit of detection (LOD) and limit of quantification (LOQ) of AA were calculated to be 0.042 and 0.139 μM, respectively. This indicates the excellent electrochemical sensing nature of the fabricated IL-BN-Fe_3O_4-NPs sensor over reported literature, with good sensitivity and selectivity. The sensor does not exhibit any remarkable change in the presence of interferents. The proposed sensor also showed good stability, repeatability and reproducibility. The developed sensor was successfully used for the determination AA in various real samples like fruit juices, vegetable juices and pharmaceuticals as well.

12.3 Conclusions

A rapidly emerging field in analytical research involves the development of sensors, electrodes, and diagnostic devices centred on ILs as alternatives to molecular solvents and conventional materials. In this chapter, we have reviewed and discussed some recent research results for various sensor formats that derive significant advantage from the incorporation of an IL component. Examples from optoelectrodes to actuators, including modified enzyme electrodes and lab-on-chip efforts are discussed. To date, work has focused mainly on ILs as drop-in substitute active medium, membrane, electrolyte, or coating in an existing analytical method or sensor construct. Additionally, ILs are just beginning to open the door to novel analytical methods that are not possible using classical solvent systems. Although the molecular basis for the design of ILs with desired properties lags somewhat behind these efforts, ultimately the term task specific IL will find meaningful relevance in the analytical sciences.

Acknowledgements

The authors gratefully acknowledge the Durban University of Technology, Durban, South Africa for financial support.

References

1. P. Nockemann, K. Binnemans and K. Driesen, *Chem. Phys. Lett.*, 2005, **415**, 131–136.
2. K. R. Seddon, *Kinet. Catal.*, 1996, **37**, 693.
3. P. Bonhote, A. P. Dias, N. Papageorgiou, K. Kalyanasundaram and M. Gratzel, *Inorg. Chem.*, 1996, **35**, 1168–1178.
4. J. G. Huddleston, H. D. Willauer, R. P. Swatloski, A. E. Visser and R. D. Rogers, *Chem. Commun.*, 1998, 1765–1766.
5. S. V. Dzyuba and R. A. Bartsch, *Chem. Commun.*, 2001, 1466–1467.
6. C. M. Gordon, J. D. Holbrey, A. R. Kennedy and K. R. Seddon, *J. Mater. Chem.*, 1998, **8**, 2627–2636.
7. S. A. Forsyth, J. M. Pringle and D. R. MacFarlane, *Aust. J. Chem.*, 2004, **57**, 113–119.
8. S. T. Handy, *Curr. Org. Chem.*, 2005, **9**, 959–988.
9. C. J. Bradaric, A. Downard, C. Kennedy, A. J. Robertson and Y. H. Zhou, *Green Chem.*, 2003, **5**, 143–152.
10. M. Degiorgi, D. Landini, A. Maia and M. Penso, *Synth. Commun.*, 1987, **17**, 521–533.
11. C. J. Bradaric, A. Downard, C. Kennedy, A. J. Robertson and Y. H. Zhou, *Abstr. Pap. Am. Chem. Soc.*, 2002, **224**, 611.
12. C. Chiappe and D. Pieraccini, *J. Phys. Org. Chem.*, 2005, **18**, 275–297.
13. P. Wasserscheid, C. Hilgers and A. Boesmann, Solvent Innovation Gmbh, EP1182196–A1, 2002.
14. J. D. Holbrey and K. R. Seddon, *J. Chem. Soc., Dalton Trans.*, 1999, 2133–2139.
15. S. Forsyth, J. Golding, D. R. MacFarlane and M. Forsyth, *Electrochim. Acta*, 2001, **46**, 1753–1757.
16. D. R. MacFarlane, P. Meakin, J. Sun, N. Amini and M. Forsyth, *J. Phys. Chem. B*, 1999, **103**, 4164–4170.
17. J. L. Goldman and A. B. McEwen, *Electrochem. Solid-State Lett.*, 1999, **2**, 501–503.
18. A. E. Visser, J. D. Holbrey and R. D. Rogers, *Chem. Commun.*, 2001, **23**, 2484–2485.
19. J. D. Holbrey, W. M. Reichert, R. P. Swatloski, G. A. Broker, W. R. Pitner, K. R. Seddon and R. D. Rogers, *Green Chem.*, 2002, **4**, 407–413.
20. L. Brinchi, R. Germani and G. Savelli, *Tetrahedron Lett.*, 2003, **44**, 2027–2029.
21. T. Kitazume and G. Tanaka, *J. Fluorine Chem.*, 2000, **106**, 211.
22. J. Golding, S. Forsyth, D. R. MacFarlane, M. Forsyth and G. B. Deacon, *Green Chem.*, 2002, **4**, 223–229.

23. J. M. Pringle, J. Golding, C. M. Forsyth, G. B. Deacon, M. Forsyth and D. R. MacFarlane, *J. Mater. Chem.*, 2002, **12**, 3475.
24. J. M. Leveque, J. L. Luche, C. Petrier, R. Roux and W. Bonrath, *Green Chem.*, 2002, **4**, 357.
25. D. R. MacFarlane, J. Golding, S. Forsyth, M. Forsyth and G. B. Deacon, *Chem. Commun.*, 2001, 1430–1431.
26. D. R. MacFarlane, S. A. Forsyth, J. Golding and G. B. Deacon, *Green Chem.*, 2002, **4**, 444–448.
27. W. Xu, L. M. Wang, R. A. Nieman and C. A. Angell, *J. Phys. Chem. B*, 2003, **107**, 11749.
28. A. S. Larsen, J. D. Holbrey, F. S. Tham and C. A. Reed, *J. Am. Chem. Soc.*, 2000, **122**, 7264.
29. S. Park and R. J. Kazlauskas, *Curr. Opin. Biotechnol.*, 2003, **14**, 432–437.
30. C. Liang, C. Y. Yuan, R. J. Warmack, C. E. Barnes and S. Dai, *Anal. Chem.*, 2002, **74**, 2172–2176.
31. M. C. Buzzeo, C. Hardacre and R. G. Compton, *Anal. Chem.*, 2004, **76**, 4583–4588.
32. R. Wang, S. Hoyano and T. Ohsaka, *Chem. Lett.*, 2004, **33**, 6–7.
33. R. Wang, T. Okajima, F. Kitamura and T. Ohsaka, *Electroanalysis*, 2004, **16**, 66–72.
34. M. D. Bennett and D. J. Leo, *Sens. Actuators, A*, 2004, **115**, 79–90.
35. Y. G. Lee and T. C. Chou, *Biosens. Bioelectron.*, 2004, **20**, 33–40.
36. C. Ya, H. Yihan, G. Dongmei, C. Cui, W. Qinghong and F. J. Yingzi, *Solid State Electrochem.*, 2014, **18**, 3463–3469.
37. K. Grim, C. Frederik and P. Robert, *Biomed. Microdevices*, 2017, **19**, 1.
38. M. Indrek, J. Urmas, K. Friedrich, P. Inga, P. Andres and A. Alvo, *IEEE/ASME International Conference on Advanced Intelligent Mechatronics (AIM) Wollongong, Australia*, 2013, pp. 9–12.
39. S. Subathra, M. Sharifah and S. A. M. Ninie, *Appl. Surf. Sci.*, 2015, **357**, 543–550.
40. M. Hui, Z. Haifeng, L. Jiachen, L. Daliang, W. Shuyao, Z. Yu, Z. Yuyang, W. Qiuhua, Z. Guolin and M. S. Xi-, *J. Mater. Chem. B*, 2015, **3**, 5310–5317.
41. F. Farnoush, R. G. Mohammad, P. H. Morteza and N. Parviz, *Int. J. Electrochem. Sci.*, 2010, **5**, 1103–1112.
42. I. S. G. Waduge, S. M. Kevin, L. D. R. Sergio, P. R. Bishnu, D. Susmita, H. Farhana, T. Attres and M. W. Isiah, *RSC Adv.*, 2014, **4**, 7225.
43. K. Wout, A. G. Z. Marcel, H. B. Sywert and C. C. Mercedes, *IMCS (2012) – The 14th International Meeting on Chemical Sensors*, 2012.
44. N. K. Bhajanthri, V. K. Arumugam, R. Chokkareddy and G. G. Redhi, *J. Mol. Liq.*, 2016, **222**, 370–376.

CHAPTER 13

Smart Ionic Liquids-based Gas Sensors

CHUAN ZHAO*, RICHARD GONDOSISWANTO AND D. BRYNN HIBBERT

School of Chemistry, UNSW Australia, Sydney, NSW 2052, Australia
*E-mail: chuan.zhao@unsw.edu.au

13.1 Introduction

Gas detection and measurement plays an important role in many areas, ranging from food safety, clinical aspects, through to environmental monitoring.[1] The atmosphere we live in contains various kinds of gases, some of which are vital to our daily life and many others that are hazardous at certain concentrations.[2] Vital gases such as oxygen and humidity should be kept at suitable levels and others such as hazardous emission gases from industrial processes should be controlled at safe levels. Such control is achievable when the concentrations of the gases can be measured in real time. Moreover, with increasing need for greater precision and accuracy following tightening of regulations for industrial gas emissions, fire protection systems and the requirements to detect several gases of clinical importance, there is an impetus to develop new and smarter gas monitoring systems.[1–3] Ionic liquids have desirable properties that make them promising candidates as components in gas sensing. This chapter provides an update on the latest developments in using ionic liquids for smart gas sensors, which has been reviewed previously,[4] and also highlights several future prospects in this rapidly evolving field.

Smart Materials No. 28
Ionic Liquid Devices
Edited by Ali Eftekhari

Published by the Royal Society of Chemistry, www.rsc.org

The market value of gas sensors in 2014 was US$1.9 billion and is expected to reach US$2.9 billion by 2022, growing at 5.4% over the forecast period.[5] Three particular gases: oxygen, carbon monoxide and carbon dioxide are responsible for half of the revenue from the gas sensor industry. Oxygen and carbon monoxide are mainly monitored in automotive and manufacturing industry in general, whereas carbon dioxide sensing is mainly required in food processing and storage.[6] Given that a chemical or physical property can be harnessed for detection of a gas, there are several key quantities that are used to characterize a sensor:

- Limit of detection (LOD) – the smallest concentration (amount fraction, volume fraction) that can be determined to exist with a certain probability of a false positive and a false negative decision. This is often calculated as three times the standard deviation of the baseline, which gives a probability of each error of around 95%.
- Sensitivity – the derivative of the measured quantity (*e.g.* current, absorbance, fluorescent intensity) with concentration of the gas. For linear calibrations it is the slope of the calibration line.
- Lower/upper limit of quantification (LLOQ, ULOQ) – the range of concentrations spanned by the calibration relation. The limits are usually determined when relative measurement uncertainty exceeds a given fraction (say 10%), or are dictated by the measurement problem (*e.g.* measurement of oxygen in ambient air).
- Response and recovery times (τ_f) – the time taken to establish a given fraction (*e.g.* $f = 50\%, 90\%$) of the steady state current for a given concentration, and the time taken to return to a given fraction of the baseline (*e.g.* $f = 10\%$) when the gas is removed.
- Selectivity – the degree to which a sensor can detect the target gas in the presence of other, interfering, gases. In potentiometric sensing, a selectivity coefficient is defined, but for the gas sensing literature selectivity is often semi-quantitatively discussed in relation to specific atmospheres.
- Lifetime of the sensor – the working life of a device.

Each of these factors will be discussed when comparing different detection systems. When considering an IL for use in a gas sensor, the ability to tailor the IL has the potential to improve each of the factors. However, all IL-based systems show great improvements in device lifetime.[6] Commercially available electrochemical gas sensors that are used at room temperature utilize liquid rather than solid electrolytes due to their high ionic conductivities and ease of use, but sacrifice lifetime.[7] In general, the lifetime of a liquid-state gas sensor is less than 3 years at typical operating conditions (25 °C, 60% RH), which is largely governed by the lifetime of its electrolyte.[8,9] Current liquid-state gas sensors utilize electrolytes such as conventional aqueous or organic solvents (*e.g.* H_2SO_4/H_2O, acetonitrile) which are prone to drying even under normal conditions of use and which perform badly under harsh conditions (*e.g.* high temperature and extreme humidity levels).[9] Membranes have been used to reduce solvent evaporation but they are unable to eliminate completely evaporative losses. In addition, the use of membranes increases the response time, a problem that will be discussed in a later section. Improvements in the design of gas sensor devices that can expand their lifetimes are anticipated to drive the remarkably competitive gas sensor market.

Room-temperature ionic liquids (RTILs) have received enormous interest in recent decades as electrolyte/solvent alternatives for gas sensors owing to their almost negligible volatility.[9,10] In addition, their superior electrochemical properties and thermal stability compared to conventional solvents and electrolytes make them ideal candidates for gas sensor electrolytes.[9] Another important feature of RTILs is their 'tunability'. By tailoring the cation and anion constituents of an RTIL, it is possible to alter the physical and chemical properties accordingly.[11] This is the motive behind the identification of ionic liquids as 'designer solvents' or 'task-specific' ILs in regard to the possibility to tailor them for specific applications. For example, it is conceivable to optimize a gas sensor device by designing an RTIL to adjust the conductivity, gas solubility, enhancing the sensitivity, selectivity and response time of the sensor. Owing to RTILs' superior properties, making them truly 'smart' materials, developing IL-based gas sensors has become a very attractive research field. Combining the many possibilities for innovative chemistry with the increased research activity in this field, we expect commercialization of the next generation of high-performance gas sensors with longer lifetimes.

During the past decade, there have been numerous research articles concerning the development of IL-based gas sensors. Earlier studies have been covered in numerous review journals and this book chapter will concentrate on the advancements in the field of ILs based gas sensors made from late-2011 to 2016.[8,11–13] The chapter is separated into sections according to the methodology of the detection mechanism, *i.e.* electrochemical, optical, and piezoelectric. Each section will start with an introduction describing the general detection mechanism and then IL-based sensors for different gases, including oxygen, carbon dioxide, ammonia, nitrogen oxides, and volatile organic compounds (VOCs), will be described.

13.2 Electrochemical Gas Sensors

13.2.1 Introduction

Electrochemical gas sensors can be classified based upon their detection mechanism, namely, potentiometric (measurement of voltage) and amperometric (measurement of current) gas sensors.[14] In potentiometric sensors, a local redox equilibrium is established at one electrode interface, and the response is the difference in potential between this electrode and a reference electrode.[13] The redox reaction often responds to a secondary species, such as H^+ (measuring pH changes in the presence of an acid or basic gas). Amperometric gas sensors measure current at a constant applied potential. The current arises from an electroactive species, often the gas molecule itself,

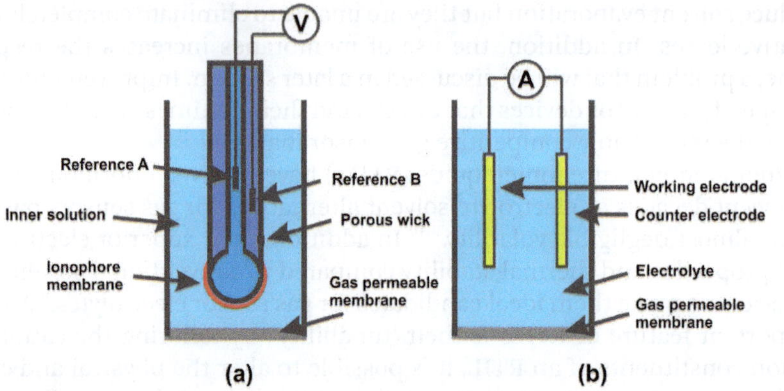

Figure 13.1 Schematic representation of (a) a potentiometric gas sensor and (b) an amperometric gas sensor.

that undergoes redox reactions at the electrode surface, and which can be related to the concentration of the gas at the outside of the membrane.[15]

A typical potentiometric gas sensor utilizes an ion-selective electrode (ISE), which is housed inside a solution and connected with the exterior gas by a thin membrane (Figure 13.1(a))[13] permeable to the target gas. The concentration of a species formed in the electrolyte from the gas (for example H^+ from the dissolution of CO_2 in aqueous electrolyte) is monitored by the ISE. While numerous IL-based potentiometric gas sensors have been reported,[16,17] there have been few recent publications regarding this type of gas sensor, possibly reflecting the superior characteristics of amperometric sensors and the greater complexity of working with H^+ in ILs.

Figure 13.1(b) displays a representation of a typical commercial amperometric gas sensor. The basic design consists of two electrodes (working and counter electrodes), electrolyte, and a gas-porous membrane. The electrolyte (*e.g.* H_2SO_4/H_2O) acts as a medium to capture gas from the membrane into the working electrode and for charge transport between electrodes. The gas membrane is designed to let gas pass through while providing a barrier against leakage or evaporation of liquid and fouling of electrodes.

13.2.2 IL-based Electrochemical Oxygen Sensors

Sensing dioxygen in air with high sensitivity and selectivity and fast-response times is the key to many processes and applications, including monitoring flammable or explosive gas mixtures, to control inert gas atmospheres in semiconductor fabrication, atmospheric monitoring in nuclear plants, food packaging, and mining.[18] The electrochemistry of oxygen is well studied and established and, as a result, oxygen has been the model system used to develop IL-based electrochemical gas sensors.

In aprotic ionic liquids, oxygen undergoes one-electron reduction to superoxide ($O_2^{\bullet-}$), which can be re-oxidized back to oxygen:[13,18]

$$O_2 + e^- \rightleftharpoons O_2^{\bullet-}$$

In protic ionic liquids, it has been reported that the O_2 formal standard potential for oxygen decreases due to fast reaction between superoxide and a source of protons (BH^+), following the initial one-electron reduction:[18]

$$O_2^{\bullet-} + BH^+ \rightarrow HO_2^{\bullet} + B$$

$$HO_2^{\bullet} + e^- \rightarrow HO_2^{\bullet-}$$

$$HO_2^{\bullet-} + BH^+ \rightarrow H_2O_2 + B$$

To avoid this complication IL-based electrochemical gas sensors usually utilize aprotic hydrophobic ILs having high solubility of oxygen. The commonest ILs used in gas sensor design contain fluorine atoms in the anion constituent, for example 1-butyl-1-methylpyrrolidinium bis(trifluoromethylsulfonyl)imide, $[C_4C_1pyr][tf_2N]$, or 1-*N*-butyl-3-methylimidazolium hexafluorophosphate, $[C_4C_1Im][PF_6]$.[8,13,19]

For example, Toniolo *et al.* reported a membrane-less oxygen gas sensor using trihexyltetradecylphosphonium-9,10-anthraquinone-1-sulphonate (THTDP-AQS) dissolved in 1-butyl-3-methylimidazolium bis(trifluoromethylsulfonyl) imide $[C_4C_1Im][tf_2N]$.[18] The presence of quinone could promote oxygen reduction at potentials lower than its standard potential reduction (*i.e.* less energy used). The enhancement is feasible through an inner sphere electron transfer following covalent linkage between the two redox partners.[20,21] These authors used a platinum (Pt) disc working electrode (1 mm) with Pt coil as the counter electrode and Pt wire as a pseudo-reference electrode. It was reported that this gas sensor can detect oxygen from 200 to 10^6 ppm (volume fraction) with a limit of detection (LOD) of 140 ppm. Interestingly, at higher temperatures (100 °C), the LOD can detect oxygen at a volume fraction as low as 10 ppm. In 2013, Baltes *et al.* reported a similar gas sensor using [HMIM][FAP] which also operated at elevated temperatures.[22] It was reported that the response time (τ_{90}, time to reach 90% of full signal) of using this IL-based sensor was approximately 15–20 s depending on the concentration of the oxygen gas. The LOD was calculated to be 5 ppm at room temperature. It was shown that this IL-based gas sensor could detect oxygen gas even up to 220 °C, but it suffers from hysteresis and irreproducible behaviour at this temperature.

In an attempt to design a membrane-less and spill-less (*i.e.* one that can be operated at different orientations) oxygen gas sensor, Lee *et al.* reported a gas sensor using a screen-printed electrode (SPE) integrated with 1-ethyl-3-methylimidazolium bis(trifluoromethylsulfonyl)imide, $[C_2C_1Im][tf_2N]$ and poly(methyl methacrylate) (PMMA).[23] The addition of polymer, mass fraction of 40–50%, increased the viscosity of the IL sufficiently to create a stable droplet even when the SPE was positioned at different orientations. The addition of PMMA, however, decreased the sensitivity of the gas sensor as the IL became more viscous, but no change in the LOD was observed.

Despite the useful properties of ILs they generally have greater viscosities than conventional organic/aqueous solvents. This leads to smaller diffusion coefficients of dissolved molecules and therefore slower response times, which could limit the application of ILs as alternative gas sensor electrolytes. To address this issue, Gunawan *et al.* reported a new design of IL-based gas sensor employing a microarray of IL droplets fabricated using the soft lithography technique of microcontact printing (Figure 13.2).[24] The microdroplets have extremely small volumes (~pL) and high surface-to-volume ratio, which significantly reduces the distance that dissolved oxygen must diffuse to reach the electrode surface. Each microdroplet acts as an individual sensing unit, forming a microarray gas sensor.

The result of these factors is enhancement in the sensitivity of the gas sensor and an increase in current density by five to six times, compared to that obtained from an equivalent macroscopic electrode sensor. Furthermore, the τ (τ_{50} = 25 s and τ_{90} = 45 s) are approximately ten times shorter compared to a macroscopic electrode sensor. The microarray sensor was shown to be sensitive to low oxygen volume fraction (130–1450 ppm), where the macroscopic sensor failed to respond. This study demonstrated the potential of miniaturizing ILs systems to improve the overall performance of an IL-based gas sensor.

Another approach to fabricate a membrane-less and spill-less gas sensor was reported by Gondosiswanto *et al.*[25] In their design, ionic liquids (1-[8-mercaptooctyl]-3-methylimidazolium bromide ([MOMIM]Br) and 1-[8-mercaptooctyl]-3-methylimidazolium hexafluorophosphate ([MOMIM][PF_6])) functionalized with terminal thiol groups were synthesized and used in the gas sensor. A terminal thiol group (–SH) forms a strong bond with gold and is used to anchor ILs to the surface of gold. In this design attaching the cation $[MOMIM]^+$ to the surface provides great physical stability to the ionic liquid droplet (Figure 13.3(a)).

The mechanical stability of the designed gas sensor was tested at different orientations and stable currents for the reduction of dioxygen were achieved (Figure 13.3(b)). A robust gas sensor is important when it is integrated into mobile objects, for example personnel attire or a moving vehicle. The designed gas sensor can operate at elevated temperature up to 40 °C. Above this temperature, the quality of the microdroplets is exposed and could lead to a decrease in the efficiency of the gas sensor.

A list of IL-based amperometric O_2 sensors reported since 2011 is summarized in Table 13.1.

13.2.3 IL-based Electrochemical Nitrogen Oxides (NO_X) Sensors

Nitrogen oxides (NO_x) are binary compounds containing nitrogen and oxygen, which include nitrogen monoxide (NO), nitrogen dioxide (NO_2), nitrous oxide (N_2O), dinitrogen trioxide (N_2O_3), dinitrogen tetroxide (N_2O_4), and

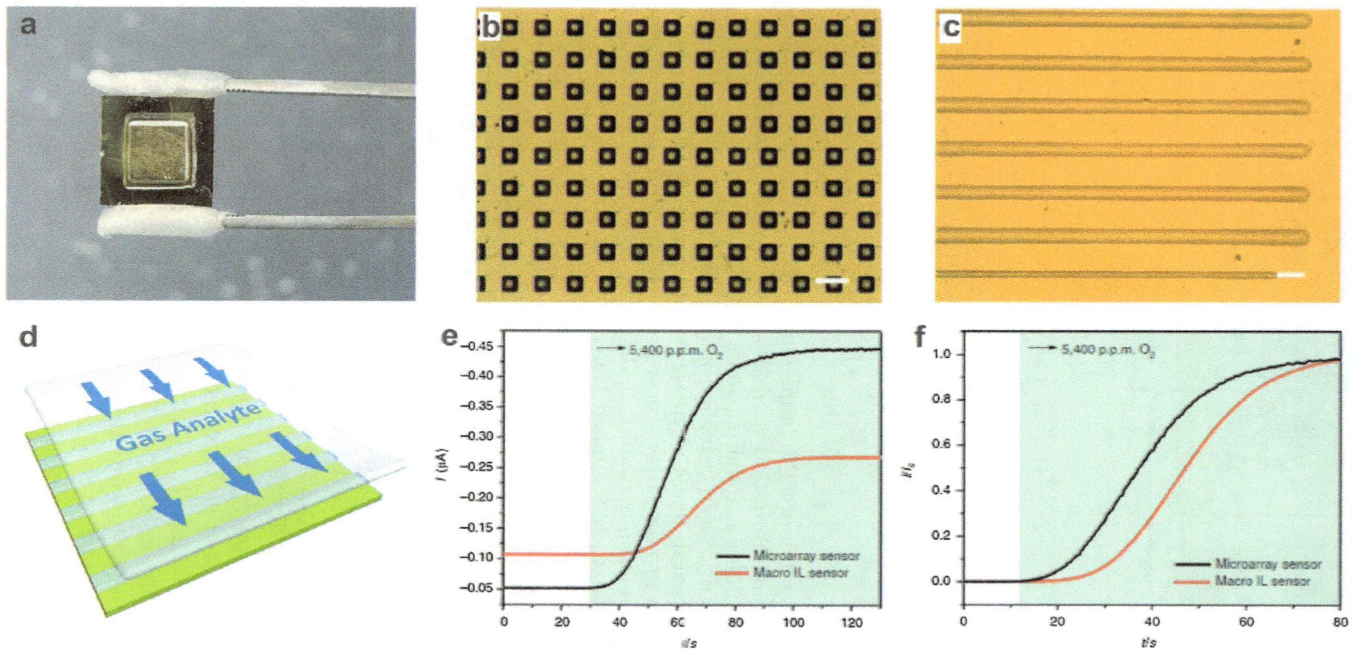

Figure 13.2 (a) Photograph of an ionic liquid microarray on a gold surface; (b) and (c) reflection micrographs of the ionic liquid with square and line patterns; (d) 3D schematic representation of IL droplet microarray sensor; (e) amperometric current–time (I–t) and (f) normalized (I/I_s) current–time responses of an IL microarray gas sensor *versus* a macrodisc electrode-based IL sensor, on exposure to 5400 ppm. O_2 as indicated by the green background. Reprinted by permission from Macmillan Publishers Ltd: Nature Communications (ref. 24), Copyright 2014.

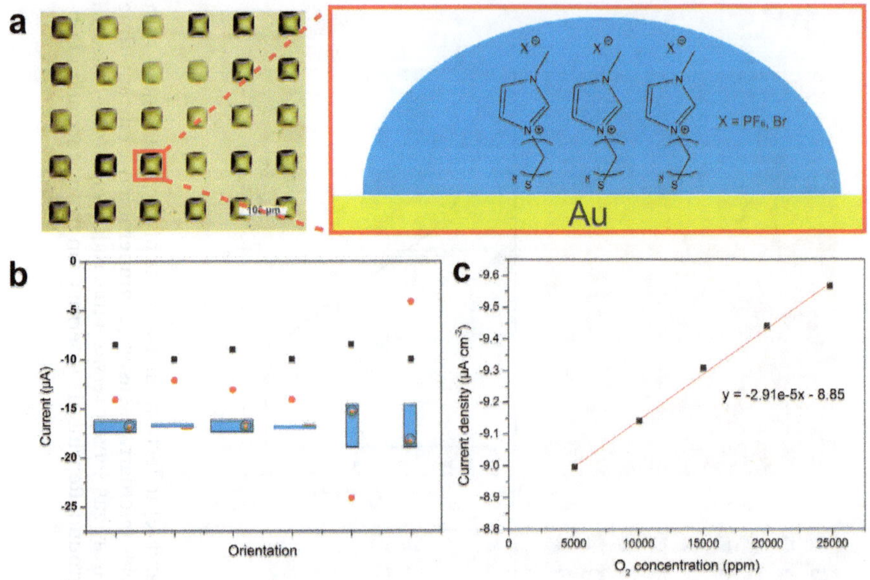

Figure 13.3 (a) Micrograph of a thiol-functionalized ionic liquid micropatterned with 50 μm fabrication *via* microcontact printing. Inset: illustration of the thiol-functionalized ionic liquid microdroplets. (b) Current variation of oxygen reduction in $[C_4C_1Im][Tf_2N]$ (red) and [MOMIM]Br (black) at different electrode orientation. (c) Calibration curve for [MOMIM]Br microarray sensor for dioxygen. Reprinted with permission from Gondosiswanto, R.; Gunawan, C. A.; Hibeert, D. B.; Harper, J. B.; Zhao, C. ACS Applied Materials & Interfaces. Copyright (2016) American Chemical Society.

dinitrogen pentoxide (N_2O_5). In the past five years, reported IL-based nitrogen gas sensors have concentrated on NO and NO_2. These gases are important components in the formation of photochemical smog, NO_2 being the brown haze seen in many cities, and which causes adverse effects on human health and environment.[38] A safe volume fraction limit of 20 ppb (nL L^{-1}, nmol mol^{-1}) was set by the European Union and 53 ppb by the Environmental Protection Agency of USA in 2010.[39] Measurement of NO_x volume fraction down to 1 ppb is therefore desirable.

Several groups have reported amperometric IL-based nitrogen oxide gas sensors. In 2012, Toniolo *et al.* reported the use of platinum disk electrode using four different RTILs ($[C_4C_1Im][tf_2N]$, 1-butyl-3-methylimidazolium trifluoromethanesulfonate ($[C_4C_1Im][TFMS]$), $[C_4C_1pyr][tf_2N]$, and 1-octyl-3-methylimidazolium bis(trifluoromethylsulfonyl)imide ($[C_8C_1Im][tf_2N]$)) for a membrane-less NO_x sensor.[39] A very small droplet of 0.5 μL IL of 70 μm thickness was formed on the platinum electrode. NO_2 predominantly occurs as its more stable form, N_2O_4, and must undergo dissociation before reversible oxidation to nitronium cation, NO_2^+. A detection limit of 0.96 ppb at 20 °C and

Table 13.1 Amperometric O_2 sensing using ionic liquids (reported since 2011); τ_ϕ is response time to reach *f*% of steady state, and LOD is limit of detection.

Working electrode	Ionic liquids	LOD	Sensitivity	τ_f	Ref.
Pt	$[C_2C_1Im][TCB]$ $[C_3C_1Im][Tf_2N]$	---	---	---	26
GC modified with CNTs-IL	$[C_4C_1Im][PF_6]$	126 μg L^{-1}	---	---	27
Pt	$[C_4C_1Im][Tf_2N]$, $[P_{6,6,6,14}][AQS]$	140 ppm	660 μA ppm^{-1}	---	18
Au	$[C_4C_1pyr][Tf_2N]$	0.08%	0.48 μA $[\%O_2]^{-1}$	---	28
Au macrodisc Pt microband	$[C_3C_1Im][Tf_2N]$	---	---	---	29
Pt microdisc Pt/Cu annular microband	$[C_3C_1Im][Tf_2N]$	0.5%	---	---	30
SPE (Au, Pt, Ag and C)	$[C_4C_1pyr][Tf_2N]$ $[C_2C_1Im][Tf_2N]$	1.5%	~1.93 μA $[\%O_2]^{-1}$	---	31
Porous Au arrays	$[C_4C_1Im][PF_6]$	0.0075%	6.1 μA$[\%O_2]^{-1}$	<10 s	32
Pt	$[C_6C_1Im][FAP]$	5 ppm	3.8 nA ppm^{-1}	15–20 s	33
Pt microdisc electrode	$[P_{6,6,6,14}][Tf_2N]$	---	---	---	34
Au/Pt microdisc electrode	$[P_{6,6,6,14}][Tf_2N]$, $[C_4C_1pyr][Tf_2N]$	---	---	---	35
Au microelectrode arrays	$[C_4C_1Im][Tf_2N]$	30 ppm	1 nA cm^{-2}/ppm	t_{90}: 45 s	25
Au macro disc		1060 ppm	0.07 nA cm^{-2}/ppm	55 s	
Pt microdisc	$[P_{6,6,6,14}][Tf_2N]$ $[P_{6,6,6,14}][FAP]$	---	---	---	36
Pt microdisc	$[P_{6,6,6,14}][FAP]$	---	9.8 nA bar^{-1}	---	37

0.55 ppb at 100 °C was measured using $[C_4C_1Im][tf_2N]$ and $[C_4C_1pyr][tf_2N]$ with τ_{90} of 10–15 s. However, since the oxidation potentials of NO and NO_2 are very close, it is very difficult to differentiate the concentration of individual NO_x in a mixture.

Research interest in NO_x sensors then moved towards IL/polymer-based sensors by creating a solid-state electrolyte that can operate at room temperature.[40–42] For example, Kuberský *et al.* reported an amperometric NO_2 sensor using IL/polymer composite on rigid and flexible SPE (ceramic and poly(ethylene terephthalate)).[41] The composite, containing IL, $[C_4C_1Im][tf_2N]$, poly(vinylidene fluoride) (PVDF), and 1-methyl-2-pyrrolidone in a ratio of 1 : 1 : 3, was drop-casted onto the SPE. This sensor is robust and gives a linear response to NO_2 with volume fraction between 0 and 10 ppm and LOD of 0.06 ppm. Moreover, the gas sensor was shown to be stable during 8 months of operation.

A list of IL-based amperometric NO_x sensors reported since 2011 is summarized in Table 13.2.

Table 13.2 IL-based amperometric NO_x sensors reported since 2011.

Working electrode	Ionic liquids	LOD/ppb	sensitivity (nA ppm^{-1})	t_{90}/t_{10}	Ref.
Au-minigrid	$[C_4C_1Im][PF_6]$	40	72.3 ± 1.6	15 s/33 s	40
	$[C_2C_1Im][Tf_2N]$	60	188.8 ± 6.2	11 s/21 s	
Pt wire	$[C_4C_1Im][Tf_2N]$	0.96	2	10–15 s	39
Porous carbon	$[C_4C_1Im][TFMS]$	60	600	60 s/120 s	43
Porous carbon	$[C_2C_1Im][Tf_2N]$	60	660	60 s/100 s	41
Graphite	$[C_2C_1Im][Tf_2N]$	---	190	Of order of 10 s	44
Graphite	$[C_2C_1Im][Tf_2N]$	6	590	70 s/60 s	42

13.2.4 IL-based Electrochemical Volatile Organic Compounds (VOCs) Sensors

Organic compounds that have high vapour pressure in standard conditions are prone to evaporation and so can be a hazard to health. Usually, VOCs are found in chemical-based industries, for example pesticides, paints, building materials, and so on. Moreover, detection of explosive VOCs, for example trinitrotoluene (TNT), is important for public safety. Several electrochemical VOCs gas sensors using ILs as electrolytes have been developed and reported.

In 2012, Dossi *et al.* reported a membrane-less gas sensor for detecting thiol-containing gases, such as 1-butanetiol.[45] In their gas sensor design, they fabricated a three-electrode cell on a paper-based substrate screen-printed with carbon ink. Several ILs ($[C_4C_1Im][PF_6]$, $[C_4C_1Im][tf_2N]$, and $[C_4C_1Im][BF_4]$ (1-butyl-3-methylimidazolium tetrafluoroborate)) were employed as electrolyte. It was reported that the oxidation potential of 1-butanethiol is very close to the IL's potential window (1.7 V *vs.* carbon ink). To shift the potential, the complex cobalt(II) phthalocyanine was included in the working electrode to act as an electrocatalyst. The gas sensor was able to detect 1-butanethiol within 6 s with LOD of 0.5 μM. A similar sensor design was used by Toniolo *et al.* to detect phenol and 1-butanethiol vapour.[46] Here, the electrolyte was an acetate-based RTIL, for example $[C_4C_1Im][Ac]$. It was claimed that such ILs can significantly reduce the oxidation potential of the gases without the need for an electrocatalyst due to stronger Brønsted acidity. This work highlights the ability of tuning the amphiphilic and/or acid–base properties of 'task specific' ILs for developing gas sensors.

A range of VOCs gas sensors reported since 2012 have been summarized in Table 13.3.

13.3 Optical Gas Sensors

13.3.1 Introduction

Gas sensors based on electromagnetic absorption, emission, luminescence or fluorescence have several advantages, including fast response time (less than 1 s is possible), high gas specificity and minimal drift.[49] In addition, the

Table 13.3 IL-based amperometric VOC sensors since 2012.

Analyte	Working electrode	Ionic liquids	LOD	Sensitivity	τ_f	Ref.
1-Butanethiol, phenol	GC disc electrode modified with Co(II) phthalocyanine	$[C_4C_1Im][Tf_2N]$	0.5 μM	54.42 nA μM^{-1}	<6 s	45
1-Butanethiol phenol	GC disc electrode,	$[C_4C_1Im][Tf_2N]$ $[C_4C_1Im][Ac]$	0.29 μM	1.09 mA μM^{-1}	<6 s	46
Benzaldehyde	Au layer	$[C_8C_1Im][BF_4]$	1 ppm	200 nA ppm^{-1}	---	47
Methyl-benzoate		$[C_4C_1Im][Tf_2N]$				
Acetophenone		$[C_4C_1Im][dca]$				
Methylamine	Pt micro-electrode	$[C_4C_1pyr][Tf_2N]$	33 ppm	1.24 pA ppm^{-1}	---	48
		$[C_2C_1Im][Tf_2N]$	34 ppm	2.13 pA ppm^{-1}		

vast range of available dyes and ease of fabrication, handling, and measurement means that optical arrays for detection of multiple analytes are possible – all ideal for studying complex samples such as odours.[50] Furthermore, real-time and *in situ* direct measurements can be conducted without consuming the gas sample, which can be important in process control.[51]

The general principle of optical gas sensors is the observation of spectral properties of a target molecule, or of its products of reaction with a dye which can be detected by the naked eye and/or quantified using spectroscopic measurements. The principle of measurement is usually based upon absorption or fluorescence spectroscopy, but these can also be combined with other techniques such as infrared or Raman spectroscopy. The reaction of the chemical dye with the analyte gas producing a change in colour (or fluorescence) may be due to several interactions, including dipole–dipole, π–π interaction, chemical reaction, *etc.* Optical gas sensor designs usually use a porous substrate, matrices, or hydrophobic membranes to immobilize the chemical dyes.[52–54] For example, Salinas *et al.* reported incorporation of various dyes into several inorganic supports (such as silica, alumina, and UVM-7) to visually monitor the aging process of chicken meat.[55]

The fabrication of an IL-based optical gas sensor was proposed by Oter *et al.*[56] The use of IL in the design is to dissolve the chemical dye while being immobilized inside porous materials or membranes. Moreover, an IL also acts as a medium to capture the gas from the atmosphere to interact with the chemical dye. This section will discuss recent trends in the design of IL-based optical gas sensors, addressed by individual gases.

13.3.2 IL-based Optical Oxygen Sensors

Numerous studies using dyes based on pyrene and its derivatives have been reported owing to their long fluorescence lifetimes.[57–61] However, pyrene-based dyes are unstable when dissolved in aqueous media or immobilized inside porous matrices, mainly due to evaporation and sublimation. Recently, Oter *et al.* designed an IL-based oxygen sensor by incorporating a pyrene-based dye and imidazolium-based IL into electrospun polymers (Figure 13.4).[62] The ILs, including $[C_4C_1Im][PF_6]$, $[C_4C_1Im][BF_4]$, $[C_4C_1Im][SCN]$ (1-butyl-3-methylimidazolium thiocyanate), or $[C_2C_1Im][BF_4]$, the polymer ethyl cellulose and the plasticizer dioctyl phthalate were used in the fabrication of the gas sensor. These materials, including the pyrene dye, were subsequently electrospun to form a porous structure. The gas sensor design was then tested for oxygen detection in an $O_2:N_2$ mixture. Oxygen quenches the fluorescence of the pyrene dye which results in an inverse relationship between fluorescence intensity and oxygen concentration. The LOD was measured to be 65 ppm with τ_{90} values of 6 to 20 s (varies with oxygen concentrations) and recovery time of 7 s. No cross-sensitivity was detected in the presence of carbon dioxide. The lifetime of the gas sensor was at least 100 days of operation; this is mainly due to the enhanced photostability of the pyrene-based dye in the IL and the negligible volatility of the IL itself.

Similar work was conducted by Ongun *et al.*, who used a fluorescent dicationic tris(2,2′-bipyridyl)ruthenium(II) chloride complex for optically sensing dioxygen.[64] Even though this dye is non-volatile, it can suffer from unstable photostability. However, when incorporated into an IL such as $[C_4C_1Im][PF_6]$ or $[C_4C_1Im][BF_4]$ and electrospun into an ethyl cellulose matrix, only a minor diminution was observed in the fluorescence intensity even after operating

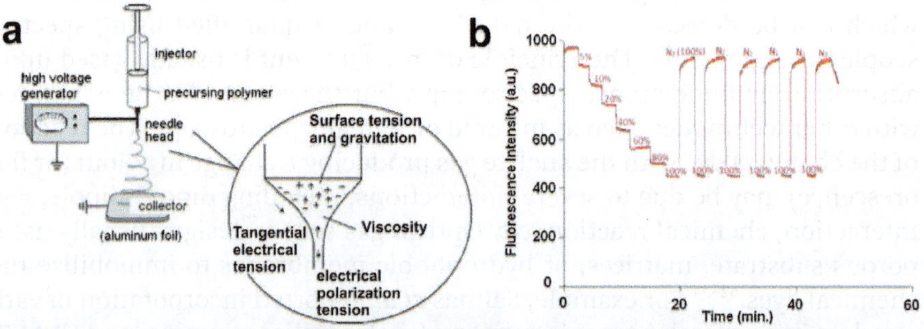

Figure 13.4 (a) Schematic representation of the electrospinning apparatus. Reproduced from ref. 63. © Springer Science+Business Media, LLC 2010, with permission of Springer. (b) Fluorescence emission response of the fabricated membrane at different O_2 concentrations. Reprinted from *Sensors and Actuators B: Chemical*, **183**, M. Z. Ongun, O. Oter, G. Sabanci, K. Ertekin, E. Celik, Enhanced stability of ruthenium complex in ionic liquid doped electrospun fibers, 11–19, Copyright 2013 with permission from Elsevier.[64]

for 3 years. A τ_{90} of 5–21 s was observed depending on the concentration of the oxygen gas, and the recovery time was about 5 s.

13.3.3 IL-based Optical Carbon Dioxide Sensors

Continuous monitoring of carbon dioxide (CO_2) with high accuracy is very important in environmental monitoring, seawater analysis, medicine and biological processes.[63] Optical CO_2 sensors based on absorbance or fluorescence changes of pH indicators or dye have been studied and developed since they were reported two decades ago.[56,65–71] Recently, Ertekin and co-workers investigated an optical CO_2 gas sensor using nano-scale electrospun fibre materials (with a similar design to the oxygen optical sensor).[63] In their work, a cocktail of polymers (ethyl cellulose [EC] or poly(methyl methacrylate) [PMMA]), plasticizer (dioctyl phthalate), IL ([C_2C_1Im][BF_4]), and 8-hydroxypyrene-1,3,6-trisulfonic acid (HPTS) trisodium salt as the dye, were paired with tetraoctylammonium (TOA^+) counterion to fabricate the nano-fibre. The counterion TOA^+ is used for stabilizing the anionic (deprotonated) form of HPTS in the lipophilic matrix. The gas sensing mechanism is divided into two processes: first diffusion of CO_2 through the membrane into the sensing region which contains a trace of water (from the formation of ion pair) forming carbonic acid. Interaction between carbonic acid and HPTS decreases the fluorescence of the dye through protolysis. The IL prevents dye evaporating and also acts as a buffer ([C_2C_1Im]OH), neutralizing carbonic acid and thus enhancing the long-term stability of the indicator. The change in fluorescence intensity with CO_2 concentration is hyperbolic, where greater sensitivity was observed at low concentrations of CO_2. The τ_{90} response time in EC and PMMA is in the range 15–20 s on exposure of 0–20% CO_2 and the recovery time is 2 min. The long-term stability of the sensor was tested over 7 months with no signal drift observed.

Chen *et al.* developed a new approach to the design of an IL-based optical carbon dioxide sensor using electrochemiluminescence (ECL) detection.[72] In ECL, electron-transfer between a molecule and the electrode surface generates excited states that emit light that can be quantified.[73] In a typical ECL sensor, the supporting electrolyte is used to increase the conductivities of the solvents and minimize uncompensated resistance, and also to establish a well-defined double layer, so that electrochemical reactions can occur.[74] Chen *et al.* exploited this idea for a CO_2 gas sensor using a co-solution of an amine and a luminophore (luminol). CO_2 gas, which itself not capable of producing electroluminescence, has been reported to react readily with substituted-amines to form a carbamate ionic liquid (CIL), resulting in an increase of ionic strength and viscosity.[72,75,76] Light emission produced by the luminophore on formation of a CIL can be detected and quantified.

Figure 13.5(b) displays the mechanism of the ECL gas sensor proposed by Chen *et al.* Dipropylamine (IL precursor) and dimethylformamide (solvent) were used as the co-solutions and luminol as the ECL luminophore which reacts with dissolved superoxide ion from dioxygen. In the absence of CO_2, no

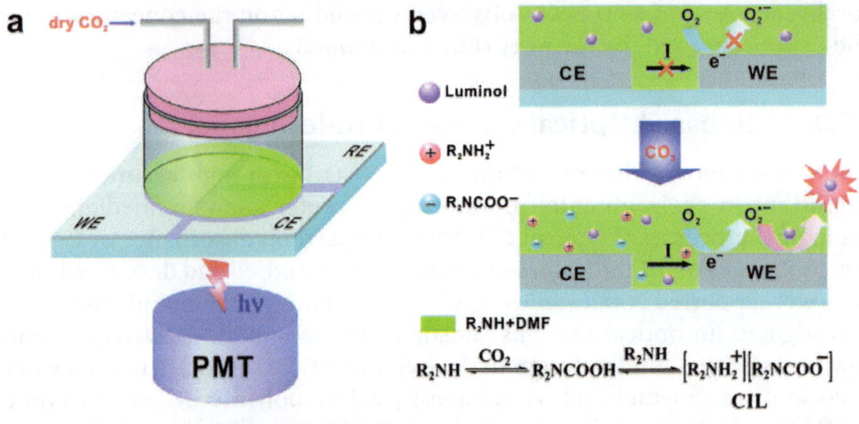

Figure 13.5 (a) Schematic representation of the ECL based sensor for CO_2 and (b) the principle of the CO_2 gas sensor based on ionic liquid induced ECL. Reprinted with permission from ref. 72. Copyright (2011) American Chemical Society.

emission can be observed as the electrolyte is not conductive (high resistance, no generation of superoxide ion). As CO_2 gas is introduced into the sensor chamber, the IL is formed and conductivity increases, dissolved oxygen is electro-reduced and ECL emission is produced.[77,78] This type of sensor has a good linear response range with a LOD of 80 ppm. In addition, selectivity to CO_2 gas was observed when exposed to a mixture of CO_2, SO_2, NO_2, and CO gases.

13.3.4 IL-based Optical Ammonia Gas Sensors

The ability to monitor ammonia and volatile amines in the environment is of importance in many fields, including leak detection in fertilizer factories and refrigeration systems, environmental gas analysis, breath analysis for medical diagnosis, air quality monitoring, and food processing.[79] Tailoring the cation or anion of an IL for the specific task of detecting ammonia was demonstrated by Yung *et al.*[80] They successfully synthesized a chemosensory ionic liquid with non-functional cation trihexyl(tetradecyl) phosphonium ($[P_{6,6,6,14}]^+$) and photoacidic trianion 8-hydroxypyrene-1,3,6-trisulfonate (HPTS, pyranine) to detect several amine vapours, such as ammonia (NH_3), methylamine ($MeNH_2$), ethylamine ($EtNH_2$), propylamine ($PrNH_2$), diethylamine (Et_2NH), and triethylamine (Et_3N). Thin films of $[P_{6,6,6,14}]$[HPTS] on glass slides were exposed to ammonia and amines resulting in the emission of different visible colours (Figure 13.6(a)) under a UV lamp ($\lambda = 365$ nm). The detection was quantified with fluorescence spectroscopy and the τ_{90} for a typical ammonia signal was about 51/2 min with the LOD varying from 2 to 200 ppm for different amines.

An optical ammonia gas sensor coupled with electrochemiluminescence detection was also fabricated by Chen *et al.*[79] In this design, tris(2,2′-bipyridine)

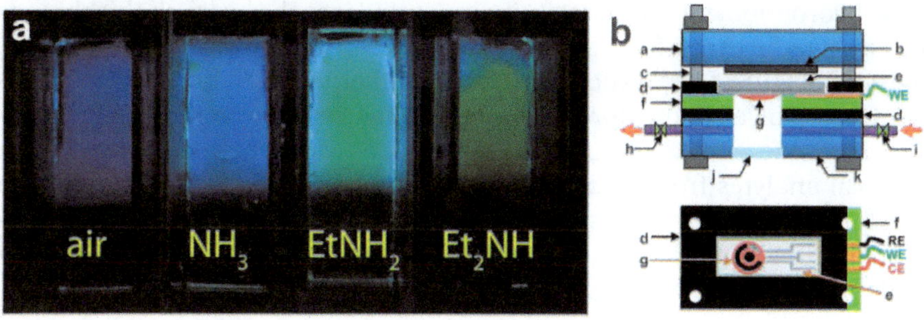

Figure 13.6 (a) Spin-casted thin films of $[P_{6,6,6,14}]$[HPTS] with different colour in response to amine vapours exposure under UV lamp with excitation laser at 365 nm. Adapted from ref. 80 with permission from the Royal Society of Chemistry. (b) Schematic representation of IL-mediated ECL sensor: a: top opaque PTFE plastic block, b: sponge, c: screw, d: air-tight rubber, e: screen-printed carbon electrode, f: PTFE plastic sheet with three Cu foil leads, g: IL-$Ru(bpy)_3^{2+}$ based sensing film, h: gas outlet, i: gas inlet; j: quartz optical window, k: bottom opaque PTFE plastic block. Adapted from ref. 79 with permission from the Royal Society of Chemistry.

ruthenium(II) ($Ru(bpy)_3^{2+}$) was used as the luminophore embedded inside a hydrophobic imidazole-based IL ($[C_4C_1Im][PF_6]$). Strong ECL emission was observed during reaction between $Ru(bpy)_3^{2+}$ and ammonia gas. $[C_4C_1Im][PF_6]$ was chosen as the supporting electrolyte for several reasons: it has a wide potential window (3.1 V for platinum electrodes and 4.0 V for gold electrodes)[81] and thus no interfering co-reactant can be generated from the ionic liquid breakdown; neither the cation nor the anion interferes with the ECL; and low water content in the ionic liquid prevents the ECL reaction between $Ru(bpy)_3^{2+}$ and OH^- dissociated from H_2O. Figure 13.6(b) shows a schematic representation of their gas sensor design. $Ru(bpy)_3^{2+}$-NH_3 is detected both with electrochemistry and luminescence. The film was 20 μm thick, and τ_{90} was 3 s with the luminescence reaching its peak intensity after 4 min. The LOD of this gas sensor is 10 ppt (pL L^{-1}) with a signal-to-noise ratio of 3. Selectivity to NH_3 was tested against other gases (*e.g.* SO_2, CO, CO_2, O_2, NO, and H_2S) and no ECL, other than that in response to NH_3, was observed.

13.3.5 IL-based Optical Volatile Organic Compound Sensors

There has been great interest in using cotton thread as the substrate for optical gas sensors due to its low cost, facile fabrication, low-volume, and exceptional ability to show changes in colour.[82,83] However, cotton threads are naturally hydrophobic and thus require expensive pretreatment such as plasma oxidation before they can be stained by aqueous dyes.[82] Recently, Galpothdeniya *et al.* designed an IL-based optoelectronic sensor array using several hydrophobic ILs with the cation $[P_{6,6,6,14}]^+$ and commercial dyes as the anion.

These hydrophobic ILs were easily incorporated into the hydrophobic threads and could reduce interference from humidity while allowing aqueous sample analysis.[84] Figure 13.7(a) shows the staining method of each thread using 10 μM of IL dissolved in dichloromethane. The sensor array (Figure 13.7(c)) contained 12 different ILs for analysis. The sensor was demonstrated to detect chemical analytes from cigarettes in aqueous solutions, and could also be used for sensing the vapours of different volatile compounds, *e.g.* hydrochloric acid, trifluoroacetic acid, formic acid and acetic acid. The analytes could be identified by colour mapping of an array.

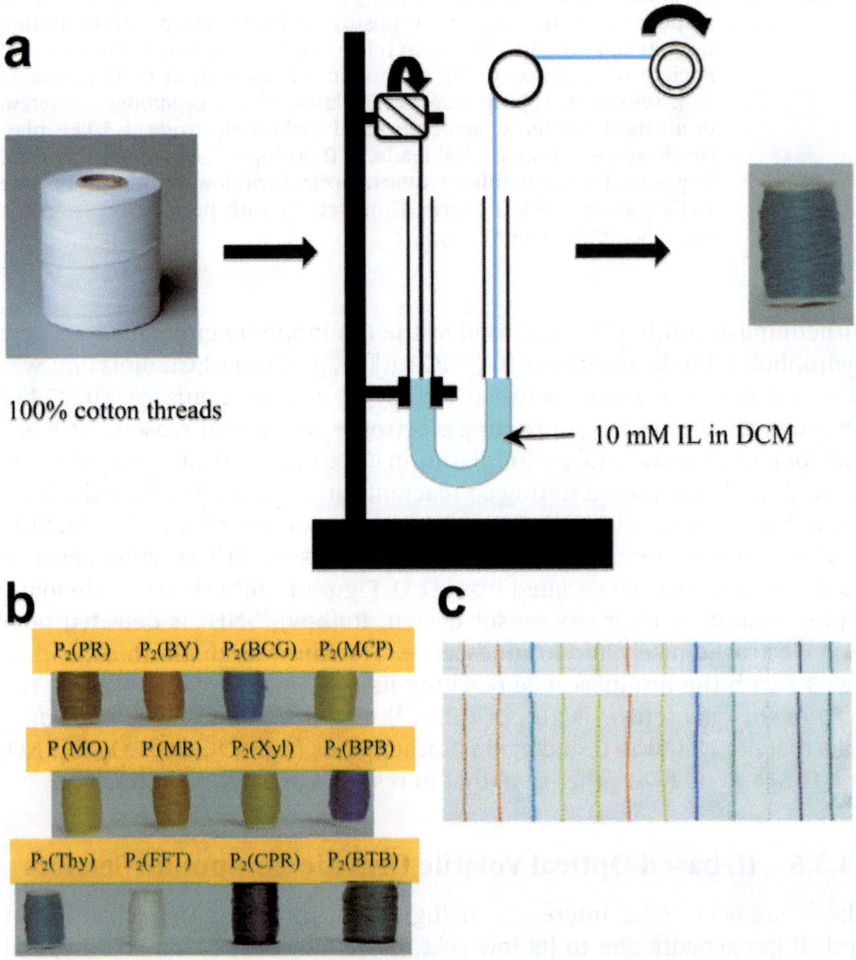

Figure 13.7 (a) Illustration of the staining procedure of the cotton threads with ILs in DCM. (b) Cotton thread spools stained with chemosensory ILs (P refers to $[P_{6,6,6,14}]^+$ cation). (c) Sensor array fabricated using IL-stained threads sewed onto a cloth. Adapted from ref. 84 with permission from the Royal Society of Chemistry.

13.4 Piezoelectric Gas Sensors

Piezoelectric crystals were first developed by Sauerbrey and lay the foundation of modern piezoelectric sensors.[85] A piezoelectric sensor employs a change in resonant frequency (Δf) that is directly proportional to a change in mass of the surface of the sensor.[86] Generally, piezoelectric gas sensors are classified into two different types: quartz crystal microbalance (QCM) and surface acoustic waves (SAW) devices. Both utilize the change in mass of gas adsorbed on the surface to detect the analyte.[87–90]

13.4.1 IL-based Quartz-crystal Microbalance Sensors

Quartz-crystal microbalance (QCM) sensors consist of a resonating polymer-coated disk with metal electrodes, in which each side is connected to a lead wire and the device resonates at a characteristic frequency due to excitation by a specific oscillating signal.[12] The adsorption of analytes on the electrode results in an increase in mass at the oscillating surface, and thus shifts the frequency. The frequency change in QCM is proportional to the mass. Quartz crystal microbalances can respond to 1 ng mass changes,[12] but need a material at the surface capable of capturing the gas molecules.

ILs have received great interest for mediating the capture of the gas molecules at QCM sensors. The ability to tailor the cation or anion of the ILs can be exploited to enhance (or reduce) incorporation of, and interaction with, the gas molecules.[91] For example, Chu *et al.* demonstrated an IL-based QCM sensor by tailoring the IL to selectively react with explosive azide-based gas analytes.[92] Figure 13.8(a) shows task-specific ionic liquids TIL 1 and TIL 2 with the cation functionalized with an alkyne group to undergo click-chemistry with azide-containing gases. The irreversible reaction integrates trace amounts of the gas at the QCM surface. TIL 1 has a cyclooctyne fused with cyclopropane ring which severely distorts the ideal linear geometry and thus raises its ground state energy, resulting in lower reaction barriers for cycloaddition (relative to TIL 2).[93,94] Figure 13.8(b) shows that TIL 1 has a good response to butyl azide gas, whereas TIL 2 had virtually no response under the same conditions. Selectivity was also demonstrated over several other volatile organic compounds such as acetone, methanol, ethyl acetate, hexane and moisture which gave virtually no response.

In 2016, Chu *et al.* synthesized four functionalized ionic liquids (FILs) that react with alkene-based gases by the Diels–Alder reaction.[95] This reaction is well understood, and has been a foundation for many of total syntheses of natural products. The general mechanism of the reaction is the cycloaddition of a conjugated diene with a substituted alkene to form a substituted cyclohexene system. For example, Figure 13.8(c) displays the response curve of a QCM sensor with a thin layer of FIL on the electrode surface. Several volatile solvents were tested and the FIL was observed to selectively give response only to vapours having an alkene moiety (1-pentene, cyclopentene, methyl acrylate, acrolein, and acryloyl chloride).

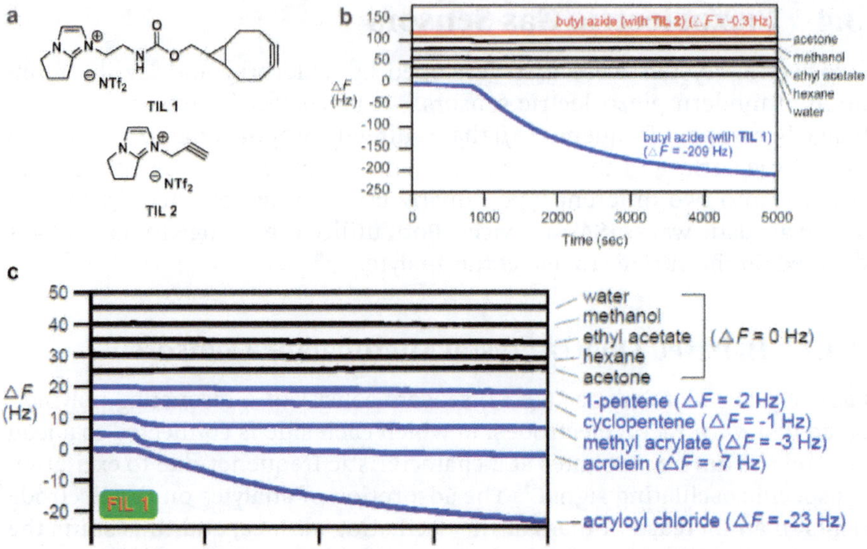

Figure 13.8 (a) Structure of tailored ILs TIL 1 and TIL 2. (b) Response curves of butyl azide, acetone, methanol, ethyl acetate, hexane, and water gases (660 ppb each) using N_2 as carrier gas; detected using TIL films on a QCM electrode. Adapted with permission from ref. 92. Copyright (2014) American Chemical Society. (c) QCM response curves of water, methanol, ethyl acetate, hexane, acetone, 1-pentene, cyclopentene, methyl acrylate, acrolein and acryloyl chloride gases (1.51 ppm each) with FIL 1. Adapted with permission from ref. 95. Copyright (2016) American Chemical Society.

Another approach for an IL-based QCM sensor was demonstrated by Speller *et al.* who described a so-called IL-based virtual sensor array (VSA).[96] VSA is a promising approach that allows generation of analyte-specific patterns from a single physical sensor that can produce multiple responses.[97] Thin films of IL, such as $[C_8C_1Im][Br]$ and $[C_8C_1Im][SCN]$, on QCM chips were exposed to gas mixtures (*e.g.* alcohols, chloromethane, aromatic hydrocarbon and nitriles). Frequency shifts on exposing the sensors to different vapours were examined using multiple harmonics on the basis of different decay lengths. Results obtained from the harmonics were then further analysed using principal component analysis and quadratic discriminant analysis as an input for models that could distinguish within each set of analogous gas analytes. Using this method, two similar complex mixtures could be distinguished correctly in 100% of the cases studied. This level of accuracy is very promising for analysing complex mixtures and chemically distinct analytes.

13.4.2 Surface Acoustic Wave Sensors

Surface acoustic wave (SAW) resonator technology has been established to detect mass change with high sensitivity (down to 1 pg).[98–102] Recently, Murakawa *et al.* designed a wireless SAW-based sensor network to detect hydrogen sulfide

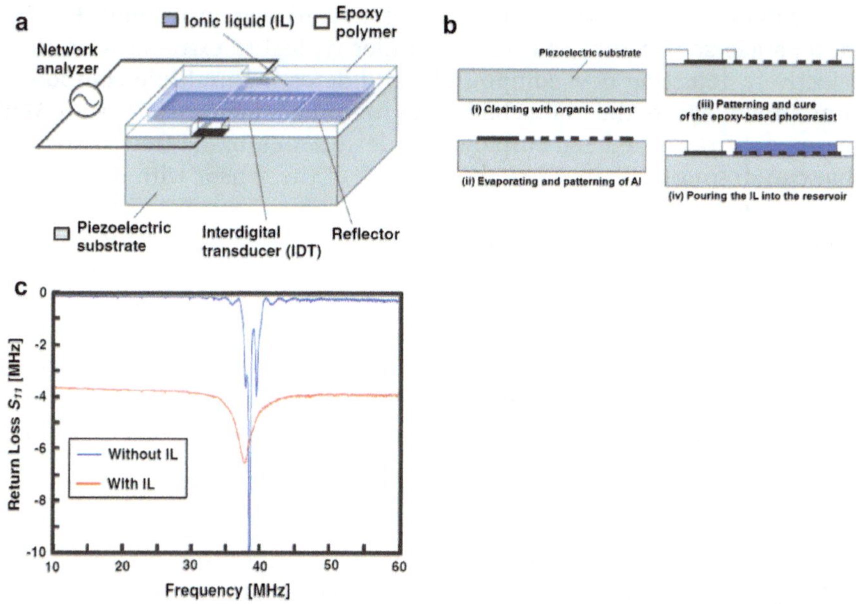

Figure 13.9 (a) Illustration of the H_2S SAW-based sensor. (b) Fabrication steps for the gas sensor. (c) Frequency characteristic of the sensor. Adapted with permission from ref. 103. Copyright 2013, American Chemical Society.

gas (H_2S), using an IL to capture the gas selectively.[103] H_2S is a common, highly-toxic gas in our environment, typically generated by bacteria from organic matter containing sulfur. H_2S and organic sulfides are generated in anaerobic environments such as sewage lines, paper and fish factories, and garbage dumps. Moreover, volcanic activity also produces H_2S. Figure 13.9(a) shows the schematic representation of the proposed gas sensor with a step-by-step fabrication method of the device (Figure 13.9(b)). Generally, the gas adsorbent in a typical SAW gas sensor is a solid, which inevitably gives a noisy signal when it is integrated on the interdigital transducer (detector). On the other hand, using an IL as the absorbent reduces noise since the surface is always smooth.

The resonant characteristic of the sensor with and without IL was investigated using a network analyser. Figure 13.9(c) shows that upon applying IL the resonance frequency shifts from 38.6 to 37.9 MHz with attenuated signal and no background noise. For H_2S sensing, the device was integrated inside a vacuum chamber and the air inside was evacuated before being purged with N_2 gas. Lastly, H_2S gas was introduced (~6.8 ppm) into the gas chamber.

13.5 Trends and Future Directions

13.5.1 Overview

Real-time detection and monitoring of gas molecules in the air that we breathe is of great importance in modern society. There is a continuing task to design robust gas sensors with improved lifetimes and better overall performance (sensitivity, limit of detection and response time). Furthermore, a gas sensor that can operate under typical to extreme conditions, selectively detecting new compounds, and detecting multiple compounds simultaneously are the main challenges for today's gas sensor design. With these goals, ILs have been employed to design membrane-free gas sensors that can dramatically increase the lifetime of the sensor with a simplified sensor design.

Piezoelectric-based gas sensors that utilize ILs represent the smallest subsection in this chapter in terms of references, suggesting that there is no significant upward trend in research, at least for the past five years. Despite being an attractive area of research, piezoelectric-based gas sensors have the major drawback of high sensitivity of the detector to surrounding vibrational noise. Here, ILs are clearly advantageous in allowing facile fabrication of the absorbent substrate with introduced functionality. Ultimately, while there might be niche applications of these sensors, helped by the use of ILs, this area is likely to remain quiet in the next few years.

The number of reports of IL-based colorimetric gas sensors is also relatively low, likely due to the fact that it is difficult to quantify the concentration of gas in a system without further instrumentation. Furthermore, task-specific ILs based upon colorimetric molecules tend to have high melting points, such that the resulting ionic salts may not even be considered ionic liquids. Specific, one-use devices with direct colour change could find use for particular applications.

IL-based electrochemical gas sensors represent the largest subsection in this chapter. As such, a wide range of research covering IL-based electrochemical gas sensors for detecting numerous gas analytes with different sensing approaches has been reported just in the past four years. Research into IL-based electrochemical gas sensors is significant, with strong growth in innovation and interest. The intrinsically-conductive nature of ILs lends them to electrochemical systems.

13.5.2 Extreme Environments and Environmental Sensitivity

ILs have basically no volatility and have excellent thermal, chemical and electrochemical stability, all of which are advantageous in developing robust gas sensors. In particular, recent research has focused upon investigating ILs in low humidity environments and at high temperatures; here ILs are obviously superior to aqueous and/or organic solvents.

Another issue that is beginning to be explored is the performance of IL-based sensors in response to changing ambient conditions, when response signals can fluctuate depending on changes in temperature and humidity, due to variable diffusion, viscosity and solubility values. However, these issues are becoming well understood and can now be corrected, for example 1,2-diferrocenylethylene has been demonstrated to be a voltammetric dual thermometer and humidity meter when dissolved in ILs, and is compatible with simultaneous oxygen gas detection.[26]

13.5.3 Designer Solvents and Task-specific Ionic Liquids

There are numerous possibilities in tailoring the cations and anions to make ILs. As such, there is arguably more scope for picking an IL for specific physicochemical properties than with conventional organic solvents. This is clearly apparent for gas sensing applications, as ILs have been chosen and then modified for their gas solubility, hydrophilicity, hydrophobicity, viscosity, *etc*. A further direct outcome are so-called 'task-specific' functionalized ILs, where the structure of the IL deliberately incorporates a motif that performs a specific task and thus enables or improves a gas sensing methodology. The click chemistry of the QCM sensor for detecting explosive azides described in Section 13.4.1 is an example of this. There is clearly an increasing trend, with the application of ILs becoming more sophisticated and targeted to specific processes (*e.g.* CO_2 capture).

13.5.4 ILs on Screen-printed (Disposable) Substrates

Significant research interest on the incorporation of ILs into cellulosic (paper) and screen-printed electrode sensing systems is growing. This is likely due to the relatively easy and cheap manufacturing processes, and the substrate is, in theory, disposable. Additionally, only a small volume of IL needs to be incorporated onto the substrate, and as the IL cannot dry out the shelf life of such systems is significantly extended. However, the cost of ILs and the environmental fate of ILs require further work before such systems can be widely commercialized.

13.5.5 Miniaturized Gas Sensors and Arrays of Gas Sensors

The relatively high viscosities of ILs are clearly correlated with slow response times. One way to address this issue is by applying an extremely thin layer of IL, which can dramatically enhance response time by reducing the distance travelled by the target molecule. Furthermore, the ability to arrange ILs into geometrically dense but stable microarrays can further enhance performance, for example a microelectrode array can significantly enhance the current density with improved signal-to-noise values. The application of potentially hundreds of sensing nodes also reduces the consequences from the failure of an individual sensing unit. The ease with which such liquid systems can be deployed on a surface is in marked contrast to the use of solids.

13.6 Conclusions

Gas sensor technology has developed quite significantly from canaries in coalmines to the present state of the art. This chapter establishes some of the more recent trends in the application of 'smart' ILs in gas sensor devices. Superior properties of ILs designed for specific characteristics have been proven to improve the sensitivity, response time, selectivity, stability, life-times, and the overall performance of gas sensors. Electrochemical IL-based gas sensors combined with miniaturized devices is currently the dominant trend.

References

1. G. Sberveglieri, *Gas Sensors: Principles, Operation and Developments*, Springer Science & Business Media, 2012.
2. D. K. Aswal and S. K. Gupta, *Science and Technology of Chemiresistor Gas Sensors*, Nova Publishers, 2007.
3. A. E. Cote, *Operation of Fire Protection Systems*, Jones & Bartlett Learning, 2003.
4. C. Zhao, C. Gunawan, M. Ge, R. Gondosiswanto and L. Aldous, Recent Advances in Ionic Liquid-based Gas Sensors, *Analytical Applications of Ionic Liquids*, 2016, p. 287.
5. Insight, G. M., Gas Sensors Market Size by Technology (Electrochemical, Semiconductor, Pid, Solid State, Catalytic, Infrared), by Product (Oxygen, Carbon Dioxide, NOx, Carbon Monoxide), by End-use (Medical, Environmental, Petrochemical, Building Automation & Domestic Appliances, Automotive, Industrial), Industry Analysis Report, Regional Outlook, Application Potential, Price Trend, Competitive Market Share & Forecast, 2015–2022, https://www.gminsights.com/industry-analysis/gas-sensors-market-size.
6. R. W. Seifert, B. F. Leleux and C. L. Tucci, *Nurturing Science-based Ventures: An International Case Perspective*, Springer Science & Business Media, 2008.
7. M. J. Tierney and H. O. L. Kim, Electrochemical Gas Sensor with Extremely Fast Response Times, *Anal. Chem.*, 1993, **65**, 3435–3440.
8. E. I. Rogers, A. M. O'Mahony, L. Aldous and R. G. Compton, (Invited) Amperometric Gas Detection Using Room Temperature Ionic Liquid Solvents, *ECS Trans.*, 2010, **33**, 473–502.
9. M. C. Buzzeo, C. Hardacre and R. G. Compton, Use of Room Temperature Ionic Liquids in Gas Sensor Design, *Anal. Chem.*, 2004, **76**, 4583–4588.
10. M. Freemantle, *An Introduction to Ionic Liquids*, Royal Society of Chemistry, Cambridge, 2010.
11. D. S. Silvester, Recent Advances in the Use of Ionic Liquids for Electrochemical Sensing, *Analyst*, 2011, **136**, 4871–4882.
12. S. K. Pandey, K.-H. Kim and K.-T. Tang, A Review of Sensor-based Methods for Monitoring Hydrogen Sulfide, *Trends Anal. Chem.*, 2012, **32**, 87–99.
13. L. Xiong and R. G. Compton, Amperometric Gas Detection: A Review, *Int. J. Electrochem. Sci.*, 2014, **9**, 7152–7181.
14. J. R. Stetter, W. R. Penrose and S. Yao, Sensors, Chemical Sensors, Electrochemical Sensors, and Ecs, *J. Electrochem. Soc.*, 2003, **150**, S11–S16.
15. J. R. Stetter and J. Li, Amperometric Gas Sensors a Review, *Chem. Rev.*, 2008, **108**, 352–366.
16. D. S. Silvester and L. Aldous, Electrochemical Detection Using Ionic Liquids, *Electrochemical Strategies in Detection Science*, The Royal Society of Chemistry, Cambridge, 2016, Ch. 10, pp. 341–386.

17. D. Wei and A. Ivaska, Applications of Ionic Liquids in Electrochemical Sensors, *Anal. Chim. Acta*, 2008, **607**, 126–135.
18. R. Toniolo, N. Dossi, A. Pizzariello, A. P. Doherty, S. Susmel and G. Bontempelli, An Oxygen Amperometric Gas Sensor Based on Its Electrocatalytic Reduction in Room Temperature Ionic Liquids, *J. Electroanal. Chem.*, 2012, **670**, 23–29.
19. R. G. Evans, O. V. Klymenko, S. A. Saddoughi, C. Hardacre and R. G. Compton, Electroreduction of Oxygen in a Series of Room Temperature Ionic Liquids Composed of Group 15-Centered Cations and Anions, *J. Phys. Chem. B*, 2004, **108**, 7878–7886.
20. S. M. Hubig, R. Rathore and J. K. Kochi, Steric Control of Electron Transfer. Changeover from Outer-Sphere to Inner-Sphere Mechanisms in Arene/Quinone Redox Pairs, *J. Am. Chem. Soc.*, 1999, **121**, 617–626.
21. A. Houmam, Electron Transfer Initiated Reactions: Bond Formation and Bond Dissociation, *Chem. Rev.*, 2008, **108**, 2180–2237.
22. N. Baltes, F. Beyle, S. Freiner, F. Geier, M. Joos, K. Pinkwart and P. Rabenecker, Trace Detection of Oxygen–Ionic Liquids in Gas Sensor Design, *Talanta*, 2013, **116**, 474–481.
23. J. Lee, G. Du Plessis, D. W. Arrigan and D. S. Silvester, Towards Improving the Robustness of Electrochemical Gas Sensors: Impact of Pmma Addition on the Sensing of Oxygen in an Ionic Liquid, *Anal. Methods*, 2015, **7**.
24. C. A. Gunawan, M. Ge and C. Zhao, Robust and Versatile Ionic Liquid Microarrays Achieved by Microcontact Printing, *Nat. Commun.*, 2014, **5**, 3744.
25. R. Gondosiswanto, C. A. Gunawan, D. B. Hibbert, J. B. Harper and C. Zhao, Microcontact Printing of Thiol-Functionalized Ionic Liquids Microarrays for "Membrane-Less" and "Spill-Less" Gas Sensors, *ACS Appl. Mater. Interfaces*, 2016, **8**, 31374.
26. L. H. J. Xiong, A. M. Fletcher, S. Ernst, S. G. Davies and R. G. Compton, An Electrochemical Thermometer: Voltammetric Measurement of Temperature and Its Application to Amperometric Gas Sensing, *Analyst*, 2012, **137**, 2567–2573.
27. L. Bai, D. Wen, J. Y. Yin, L. Deng, C. Z. Zhu and S. J. Dong, Carbon Nanotubes-Ionic Liquid Nanocomposites Sensing Platform for Nadh Oxidation and Oxygen, Glucose Detection in Blood, *Talanta*, 2012, **91**, 110–115.
28. X. Y. Mu, Z. Wang, X. Q. Zeng and A. J. Mason, A Robust Flexible Electrochemical Gas Sensor Using Room Temperature Ionic Liquid, *IEEE Sens. J.*, 2013, **13**, 3976–3981.
29. L. H. J. Xiong, D. Lowinsohn, K. R. Ward and R. G. Compton, Fabrication of Disposable Gold Macrodisc and Platinum Microband Electrodes for Use in Room-Temperature Ionic Liquids, *Analyst*, 2013, **138**, 5444–5452.
30. L. H. J. Xiong, P. Goodrich, C. Hardacre and R. G. Compton, Evaluation of a Simple Disposable Microband Electrode Device for Amperometric Gas Sensing, *Sens. Actuators, B*, 2013, **188**, 978–987.

31. J. Lee, K. Murugappan, D. W. M. Arrigan and D. S. Silvester, Oxygen Reduction Voltammetry on Platinum Macrodisk and Screen-Printed Electrodes in Ionic Liquids: Reaction of the Electrogenerated Superoxide Species with Compounds Used in the Paste of Pt Screen-Printed Electrodes?, *Electrochim. Acta*, 2013, **101**, 158–168.
32. C. G. Hu, X. Y. Bai, Y. K. Wang, W. Jin, X. Zhang and S. S. Hu, Inkjet Printing of Nanoporous Gold Electrode Arrays on Cellulose Membranes for High-Sensitive Paper-Like Electrochemical Oxygen Sensors Using Ionic Liquid Electrolytes, *Anal. Chem.*, 2012, **84**, 3745–3750.
33. N. Baltes, F. Beyle, S. Freiner, F. Geier, M. Joos, K. Pinkwart and P. Rabenecker, Trace Detection of Oxygen - Ionic Liquids in Gas Sensor Design, *Talanta*, 2013, **116**, 474–481.
34. L. H. J. Xiong, E. O. Barnes and R. G. Compton, Amperometric Detection of Oxygen under Humid Conditions: The Use of a Chemically Reactive Room Temperature Ionic Liquid to 'Trap' Superoxide Ions and Ensure a Simple One Electron Reduction, *Sens. Actuators, B*, 2014, **200**, 157–166.
35. E. E. L. Tanner, L. H. J. Xiong, E. O. Barnes and R. G. Compton, One Electron Oxygen Reduction in Room Temperature Ionic Liquids: A Comparative Study of Butler–Volmer and Symmetric Marcus-Hush Theories Using Microdisc Electrodes, *J. Electroanal. Chem.*, 2014, **727**, 59–68.
36. P. L. Li, E. O. Barnes, C. Hardacre and R. G. Compton, Microelectrode Voltammetry of Dioxygen Reduction in a Phosphonium Cation-Based Room-Temperature Ionic Liquid: Quantitative Studies, *J. Phys. Chem. C*, 2015, **119**, 2716–2726.
37. P. L. Li and R. G. Compton, Electrochemical High Concentration Oxygen Sensing Using a Phosphonium Cation Based Room Temperature Ionic Liquid: Analytical Studies, *Electroanalysis*, 2015, **27**, 1550–1555.
38. G. Whitten, The Chemistry of Smog Formation: A Review of Current Knowledge, *Environ. Int.*, 1983, **9**, 447–463.
39. R. Toniolo, N. Dossi, A. Pizzariello, A. P. Doherty and G. Bontempelli, A Membrane Free Amperometric Gas Sensor Based on Room Temperature Ionic Liquids for the Selective Monitoring of Nox, *Electroanalysis*, 2012, **24**, 865–871.
40. M. Nádherná, F. Opekar, J. Reiter and K. Štulík, A Planar, Solid-State Amperometric Sensor for Nitrogen Dioxide, Employing an Ionic Liquid Electrolyte Contained in a Polymeric Matrix, *Sens. Actuators, B*, 2012, **161**, 811–817.
41. P. Kuberský, A. Hamacek, S. Nespurek, R. Soukup and R. Vik, Effect of the Geometry of a Working Electrode on the Behavior of a Planar Amperometric NO_2 Sensor Based on Solid Polymer Electrolyte, *Sens. Actuators, B*, 2013, **187**, 546–552.
42. P. Kuberský, T. Syrový, A. Hamáček, S. Nešpůrek and L. Syrová, Towards a Fully Printed Electrochemical NO_2 Sensor on a Flexible Substrate Using Ionic Liquid Based Polymer Electrolyte, *Sens. Actuators, B*, 2015, **209**, 1084–1090.

43. P. Kuberský, A. Hamáček, M. Kroupa, J. Štulík and V. Zwiefelhofer, *Potentiostat Solution for Electrochemical Amperometric Gas Sensor, Electronics Technology (ISSE), 2012 35th International Spring Seminar on, IEEE*, 2012, pp. 388–393.
44. P. Kuberský, P. Sedlak, A. Hamacek, S. Nespurek, T. Kuparowitz, J. Sikula, J. Majzner, V. Sedlakova, L. Grmela and T. Syrovy, Quantitative Fluctuation-Enhanced Sensing in Amperometric NO_2 Sensors, *Chem. Phys.*, 2015, **456**, 111–117.
45. N. Dossi, R. Toniolo, A. Pizzariello, E. Carrilho, E. Piccin, S. Battiston and G. Bontempelli, An Electrochemical Gas Sensor Based on Paper Supported Room Temperature Ionic Liquids, *Lab Chip*, 2012, **12**, 153–158.
46. R. Toniolo, N. Dossi, A. Pizzariello, A. Casagrande and G. Bontempelli, Electrochemical Gas Sensors Based on Paper-Supported Room-Temperature Ionic Liquids for Improved Analysis of Acid Vapours, *Anal. Bioanal. Chem.*, 2013, **405**, 3571–3577.
47. J. Gębicki, A. Kloskowski and W. Chrzanowski, Prototype of Electrochemical Sensor for Measurements of Volatile Organic Compounds in Gases, *Sens. Actuators, B*, 2013, **177**, 1173–1179.
48. K. Murugappan, C. Kang and D. S. Silvester, Electrochemical Oxidation and Sensing of Methylamine Gas in Room Temperature Ionic Liquids, *J. Phys. Chem. C*, 2014, **118**, 19232–19237.
49. J. Hodgkinson and R. P. Tatam, Optical Gas Sensing: A Review, *Meas. Sci. Technol.*, 2012, **24**, 012004.
50. N. A. Rakow and K. S. Suslick, A Colorimetric Sensor Array for Odour Visualization, *Nature*, 2000, **406**, 710–713.
51. M. Lackner, Tunable Diode Laser Absorption Spectroscopy (Tdlas) in the Process Industries–a Review, *Rev. Chem. Eng.*, 2007, **23**, 65–147.
52. K. S. Suslick, An Optoelectronic Nose:"Seeing" Smells by Means of Colorimetric Sensor Arrays, *MRS Bull.*, 2004, **29**, 720–725.
53. C. Zhang and K. S. Suslick, A Colorimetric Sensor Array for Organics in Water, *J. Am. Chem. Soc.*, 2005, **127**, 11548–11549.
54. L. Feng, C. J. Musto, J. W. Kemling, S. H. Lim and K. S. Suslick, A Colorimetric Sensor Array for Identification of Toxic Gases Below Permissible Exposure Limits, *Chemi. Commun.*, 2010, **46**, 2037–2039.
55. Y. Salinas, J. V. Ros-Lis, J.-L. Vivancos, R. Martinez-Manez, M. D. Marcos, S. Aucejo, N. Herranz and I. Lorente, Monitoring of Chicken Meat Freshness by Means of a Colorimetric Sensor Array, *Analyst*, 2012, **137**, 3635–3643.
56. O. Oter, K. Ertekin, D. Topkaya and S. Alp, Emission-Based Optical Carbon Dioxide Sensing with Hpts in Green Chemistry Reagents: Room-Temperature Ionic Liquids, *Anal. Bioanal. Chem.*, 2006, **386**, 1225–1234.
57. Y. Fujiwara and Y. Amao, An Oxygen Sensor Based on the Fluorescence Quenching of Pyrene Chemisorbed Layer onto Alumina Plates, *Sens. Actuators, B*, 2003, **89**, 187–191.
58. Y. Fujiwara and Y. Amao, 1-Pyrenedecanoic Acid Chemisorption Film as a Novel Oxygen Sensing Material, *Sens. Actuators, B*, 2002, **85**, 175–178.

59. T. Ishiji and M. Kaneko, Photoluminescence of Pyrenebutyric Acid Incorporated into Silicone Film as a Technique in Luminescent Oxygen Sensing, *Analyst*, 1995, **120**, 1633–1638.
60. D. W. Lübbers and N. Opitz, Opticl Fluorescence Sensors for Continuous Measurement of Chemical Concentrations in Biological Systems, *Sensors and Actuators*, 1983, **4**, 641–654.
61. O. Oter and A. C. Ribou, Quenching of Long Lifetime Emitting Fluorophores with Paramagnetic Molecules, *J. Fluoresc.*, 2009, **19**, 389–397.
62. O. Oter and S. Sahin, Gaseous and Dissolved Oxygen Sensing with Stabilized Pyrene in Ionic Liquid Modified Electrospun Slides, *Turk. J. Chem.*, 2015, 395–411.
63. S. Aydogdu, K. Ertekin, A. Suslu, M. Ozdemir, E. Celik and U. Cocen, Optical CO_2 Sensing with Ionic Liquid Doped Electrospun Nanofibers, *J. Fluoresc.*, 2011, **21**, 607–613.
64. M. Z. Ongun, O. Oter, G. Sabancı, K. Ertekin and E. Celik, Enhanced Stability of Ruthenium Complex in Ionic Liquid Doped Electrospun Fibers, *Sens. Actuators, B*, 2013, **183**, 11–19.
65. D. Wencel, J. Moore, N. Stevenson and C. McDonagh, Ratiometric Fluorescence-Based Dissolved Carbon Dioxide Sensor for Use in Environmental Monitoring Applications, *Anal. Bioanal. Chem.*, 2010, **398**, 1899–1907.
66. B. H. Weigl and O. S. Wolfbeis, New Hydrophobic Materials for Optical Carbon Dioxide Sensors Based on Ion Pairing, *Anal. Chim. Acta*, 1995, **302**, 249–254.
67. B. Muller and P. C. Hauser, Fluorescence Optical Sensor for Low Concentrations of Dissolved Carbon Dioxide, *Analyst*, 1996, **121**, 339–343.
68. G. Neurauter, I. Klimant and O. S. Wolfbeis, Microsecond Lifetime-Based Optical Carbon Dioxide Sensor Using Luminescence Resonance Energy Transfer, *Anal. Chim. Acta*, 1999, **382**, 67–75.
69. O. Oter, K. Ertekin, D. Topkaya and S. Alp, Room Temperature Ionic Liquids as Optical Sensor Matrix Materials for Gaseous and Dissolved CO_2, *Sens. Actuators, B*, 2006, **117**, 295–301.
70. K. Ertekin and S. Alp, Enhanced Emission Based Optical Carbon Dioxide Sensing in Presence of Perfluorochemicals (Pfcs), *Sens. Actuators, B*, 2006, **115**, 672–677.
71. O. Oter, K. Ertekin and S. Derinkuyu, Ratiometric Sensing of CO_2 in Ionic Liquid Modified Ethyl Cellulose Matrix, *Talanta*, 2008, **76**, 557–563.
72. L. Chen, D. Huang, S. Ren, Y. Chi and G. Chen, Carbon Dioxide Gas Sensor Based on Ionic Liquid-Induced Electrochemiluminescence, *Anal. Chem.*, 2011, **83**, 6862–6867.
73. M. M. Richter, Electrochemiluminescence (Ecl), *Chem. Rev.*, 2004, **104**, 3003–3036.
74. A. M. Bond and S. W. Feldberg, Analysis of Simulated Reversible Cyclic Voltammetric Responses for a Charged Redox Species in the Absence of Added Electrolyte, *J. Phys. Chem. B*, 1998, **102**, 9966–9974.
75. P. G. Jessop, D. J. Heldebrant, X. Li, C. A. Eckert and C. L. Liotta, Green Chemistry: Reversible Nonpolar-to-Polar Solvent, *Nature*, 2005, **436**, 1102.

76. L. Phan, J. R. Andreatta, L. K. Horvey, C. F. Edie, A.-L. Luco, A. Mirchandani, D. J. Darensbourg and P. G. Jessop, Switchable-Polarity Solvents Prepared with a Single Liquid Component, *J. Org. Chem.*, 2008, **73**, 127–132.
77. H. Cui, Y. Xu and Z.-F. Zhang, Multichannel Electrochemiluminescence of Luminol in Neutral and Alkaline Aqueous Solutions on a Gold Nanoparticle Self-Assembled Electrode, *Anal. Chem.*, 2004, **76**, 4002–4010.
78. P. J. Welford, B. A. Brookes, J. D. Wadhawan, H. B. McPeak, C. E. W. Hahn and R. G. Compton, The Electro-Reduction of Carbon Dioxide in Dimethyl Sulfoxide at Gold Microdisk Electrodes: Current | Voltage Waveshape Analysis, *J. Phys. Chem. B*, 2001, **105**, 5253–5261.
79. L. Chen, D. Huang, Y. Zhang, T. Dong, C. Zhou, S. Ren, Y. Chi and G. Chen, Ultrasensitive Gaseous NH_3 Sensor Based on Ionic Liquid-Mediated Signal-on Electrochemiluminescence, *Analyst*, 2012, **137**, 3514–3519.
80. K. Y. Yung, A. J. Schadock-Hewitt, N. P. Hunter, F. V. Bright and G. A. Baker, 'Liquid Litmus': Chemosensory Ph-Responsive Photonic Ionic Liquids, *Chem. Commun.*, 2011, **47**, 4775–4777.
81. V. M. Hultgren, A. W. Mariotti, A. M. Bond and A. G. Wedd, Reference Potential Calibration and Voltammetry at Macrodisk Electrodes of Metallocene Derivatives in the Ionic Liquid [Bmim][PF_6], *Anal. Chem.*, 2002, **74**, 3151–3156.
82. M. Reches, K. A. Mirica, R. Dasgupta, M. D. Dickey, M. J. Butte and G. M. Whitesides, Thread as a Matrix for Biomedical Assays, *ACS Appl. Mater. Interfaces*, 2010, **2**, 1722–1728.
83. E. B. Anderson and T. E. Long, Imidazole- and Imidazolium-Containing Polymers for Biology and Material Science Applications, *Polymer*, 2010, **51**, 2447.
84. W. I. S. Galpothdeniya, K. S. McCarter, S. L. De Rooy, B. P. Regmi, S. Das, F. Hasan, A. Tagge and I. M. Warner, Ionic Liquid-Based Optoelectronic Sensor Arrays for Chemical Detection, *RSC Adv.*, 2014, **4**, 7225–7234.
85. G. Sauerbrey, Use of Quartz Vibration for Weighing Thin Films on a Microbalance, *J. Phys.*, 1959, **155**, 206–212.
86. J. J. McCallum, Piezoelectric Devices for Mass and Chemical Measurements: An Update. A Review, *Analyst*, 1989, **114**, 1173–1189.
87. T. B. Pollard, T. D. Kenny, J. F. Vetelino and M. P. da Cunha, Pure Sh-Saw Propagation, Transduction and Measurements on Knbo/Sub 3, *IEEE Trans. Sonics Ultrason.*, 2006, **53**, 199–208.
88. O. Tigli and M. E. Zaghloul, A Novel Saw Device in Cmos: Design, Modeling, and Fabrication, *IEEE Sens. J.*, 2007, **7**, 219–227.
89. A. W. Czanderna and T. M. Thomas, A Quartz-Crystal Microbalance Apparatus for Water Sorption by Polymers, *J. Vac. Sci. Technol., A*, 1987, **5**, 2412–2416.
90. Z.-H. Sheng, L. Shao, J.-J. Chen, W.-J. Bao, F.-B. Wang and X.-H. Xia, Catalyst-Free Synthesis of Nitrogen-Doped Graphene *Via* Thermal Annealing Graphite Oxide with Melamine and Its Excellent Electrocatalysis, *ACS Nano*, 2011, **5**, 4350–4358.

91. A. Rehman, A. Hamilton, A. Chung, G. A. Baker, Z. Wang and X. Zeng, Differential Solute Gas Response in Ionic-Liquid-Based Qcm Arrays: Elucidating Design Factors Responsible for Discriminative Explosive Gas Sensing, *Anal. Chem.*, 2011, **83**, 7823–7833.
92. M.-C. Tseng and Y.-H. Chu, Reaction-Based Azide Gas Sensing with Tailored Ionic Liquids Measured by Quartz Crystal Microbalance, *Anal. Chem.*, 2014, **86**, 1949–1952.
93. D. Heber, P. Rösner and W. Tochtermann, Cyclooctyne and 4-Cyclooctyn-1-Ol – Versatile Building Blocks in Organic Synthesis, *Eur. J. Org. Chem.*, 2005, **2005**, 4231–4247.
94. M. F. Debets, S. S. van Berkel, J. Dommerholt, A. J. Dirks, F. P. J. T. Rutjes and F. L. van Delft, Bioconjugation with Strained Alkenes and Alkynes, *Acc. Chem. Res.*, 2011, **44**, 805–815.
95. T.-H. Hsu, S.-J. Chiang and Y.-H. Chu, Quartz Crystal Microbalance Analysis of Diels–Alder Reactions of Alkene Gases to Functional Ionic Liquids on Chips, *Anal. Chem.*, 2016, **88**, 10837–10841.
96. N. C. Speller, N. Siraj, B. P. Regmi, H. Marzoughi, C. Neal and I. M. Warner, Rational Design of Qcm-D Virtual Sensor Arrays Based on Film Thickness, Viscoelasticity, and Harmonics for Vapor Discrimination, *Anal. Chem.*, 2015, **87**, 5156–5166.
97. A. Schutze, A. Gramm and T. Ruhl, Identification of Organic Solvents by a Virtual Multisensor System with Hierarchical Classification, *IEEE Sens. J.*, 2004, **4**, 857–863.
98. J. Beckley, V. Kalinin, M. Lee and K. Voliansky, Non-contact Torque Sensors Based on Saw Resonators, *Frequency Control Symposium and PDA Exhibition, IEEE International*, 2002, pp. 202–213.
99. B. Donohoe, D. Geraghty and G. E. O'Donnell, Wireless Calibration of a Surface Acoustic Wave Resonator as a Strain Sensor, *IEEE Sens. J.*, 2011, **11**, 1026–1032.
100. M. Jungwirth, H. Scherr and R. Weigel, Micromechanical Precision Pressure Sensor Incorporating Saw Delay Lines, *Acta Mech.*, 2002, **158**, 227–252.
101. Y. Nakai, T. Nakao, K. Nishiyama and M. Kadota, Surface Acoustic Wave Duplexer Composed of SiO_2/Cu Electrode/Linbo 3 Structure Having Convex and Concave Portions, *Jpn. J. Appl. Phys.*, 2009, **48**, 07GG02.
102. V. Kalinin, *Passive Wireless Strain and Temperature Sensors Based on Saw Devices, Radio and Wireless Conference, 2004 IEEE, 19–22 Sept. 2004*, 2004, pp. 187–190.
103. Y. Murakawa, M. Hara, H. Oguchi, Y. Hamate and H. Kuwano, Surface Acoustic Wave Based Sensors Employing Ionic Liquid for Hydrogen Sulfide Gas Detection, *Microsyst Technol.*, 2013, **19**, 1255–1259.

CHAPTER 14

Design and New Energy Application of Ionic Liquids

SHIRO SEKI*[a,b], SHIMPEI ONO[b], NOBUYUKI SERIZAWA[b], YASUHIRO UMEBAYASHI[c], SEIJI TSUZUKI[d], KAZUHIDE UENO[e] AND MASAYOSHI WATANABE[e]

[a]Kogakuin University, Department of Environmental Chemistry and Chemical Engineering, School of Advanced Engineering, 2665-1 Nakano-machi, Hachioji, Tokyo 192-0015, Japan; [b]Central Research Institute of Electric Power Industry, Materials Science Research Laboratory, 2-6-1 Nagasaka, Yokosuka City, Kanagawa 240-0196, Japan; [c]Niigata University, Graduate School of Science and Technology, 8050 Ikarashi, 2-no-cho, Nishi-ku, Niigata City, Niigata 950-2181, Japan; [d]National Institute of Advanced Industrial Science and Technology (AIST), 1-1-1, Umezono, Tsukuba, Ibaraki 305-8568, Japan; [e]Yokohama National University, Department of Chemistry and Biotechnology, 79-5 Tokiwadai, Hodogaya-ku, Yokohama City, Kanagawa 240-8501, Japan
*E-mail: shiro-seki@cc.kogakuin.ac.jp

14.1 Introduction

Room-temperature ionic liquids (ILs) are liquid salts consisting entirely of cations (positive ions) and anions (negative ions), which show interesting physical properties such as low flammability and low volatility (negligible vapor pressure), high ionic conductivity, and high thermal and electrochemical stability.[1] ILs are known as "designer's solvents," because their properties

Smart Materials No. 28
Ionic Liquid Devices
Edited by Ali Eftekhari

Published by the Royal Society of Chemistry, www.rsc.org

can be easily tuned by varying the chemical structure and combination of cations and anions. It has already been pointed out that the number of ILs resulting from the combinations of cation and anion exceeds 1 million.

In materials chemistry research fields, IL have attracted considerable attention due to promising physicochemical properties for various potential applications. For instance, studies on ionic aggregation[2–6] in ILs and on hybrid materials with ILs and polymers,[7–10] glasses,[11] and inorganic materials[12] have been reported. Research and development of ILs has also looked at their performance as solvents for synthesis,[13,14] and as catalysts[15–17] in reactions. For electrochemical purposes IL components have found uses in energy storage–conversion systems as new task-specific electrolytes for devices such as lithium secondary batteries, electric double-layer capacitors,[18–20] dye-sensitized solar cells,[21,22] fuel cells,[23–26] electrochromic devices,[27] and field-effect transistors.[28] In particular, the number of studies on lithium secondary battery electrolytes has increased in recent years owing to strong demand for storage of renewable energy.

In this chapter, we present our research and development of IL systems and discuss how fundamental properties (general characteristics, special characteristics) of ILs can affect their applications in electrochemical fields (lithium batteries, electrochemical double layer capacitors, transistors, *etc.*).

14.2 General Physical Properties of ILs

14.2.1 Volume Property

The relatively high viscosity of ILs is often a serious problem when using them. Therefore, to design a low viscosity IL, it is important to elucidate the quantitative relationship between the viscosity and the molecular structure of an IL. To clarify the molecular origin of the macroscopic physical properties of ILs, many spectroscopic and theoretical investigations of the physical properties of ILs have been performed. However, there are few reports on the detailed analysis of the density of IL and its temperature dependency (degree of expansion).[3,4] It is well known that the dynamic properties of liquids (*e.g.* diffusion, flowability and viscosity) strongly correlate with the molar volume of liquid (solution, IL) as a thermodynamic quantity.[29–31] However, the molar volume should be an intrinsically static thermodynamic quantity,[32,33] thus, more experimental and theoretical effort are needed to understand dynamic physical properties at the molecular level. For these purposes, it is important to obtain experimental molar volume (density) and viscosity data for a wide range of IL samples over a wide temperature range. We have performed precise density measurements of 12 kinds of ILs (Table 14.1) with six different cations (chemical structures shown in Figure 14.1) and seven different anions ($N(SO_2C_2F_5)_2$, $N(SO_2CF_3)_2$, $N(SO_2F)_2$, SO_3CF_3, BF_4, SCN, $N(CN)_2$). Relationships between the molecular structure of the IL components and the IL physical properties were examined.

Figure 14.2(a) shows the relationship between the molecular weight (M_w) of the 12 types of IL and their molar volumes ($V_{measure}$) obtained from

Table 14.1 Measurement matrix for cations and anions of ILs.

Cation/ anion	$N(SO_2C_2F_5)_2$	$N(SO_2CF_3)_2$	$N(SO_2F)_2$	SO_3CF_3	BF_4	SCN	$N(CN)_2$
EMIm	A	B	C	D	E	F	G
EDMIm		H					
P_{13}		I					
PP_{13}		J					
ETMP		K					
TMPA		L					

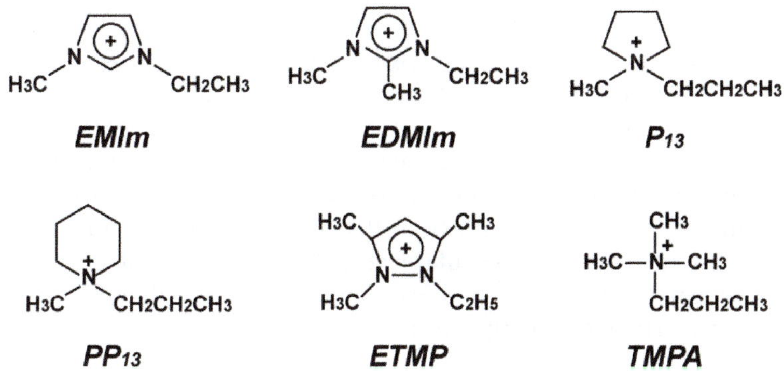

Figure 14.1 Chemical structures of cation species of measured ILs.

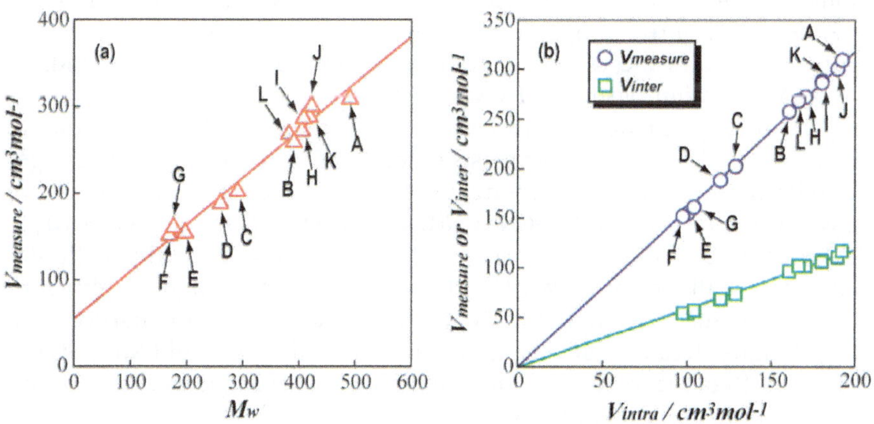

Figure 14.2 Relationship between molecular weight (M_w) of ILs and measured volume per mole ($V_{measure}$) obtained from ρ (a), and relationship between V_{intra} (molecular frame volume) and $V_{measure}$ and between V_{intra} and free volume (V_{inter}) (b) at 303.15 K.

density measurements. A relatively good linear fit to the data ($R^2 = 0.979$) was obtained with a non-zero intercept. *Ab initio* calculations of interactions in ion pairs in the gas phase show that electrostatic interactions are the major source of the attraction, but the dispersion interactions are also important for the attraction in the ionic liquids.[34] In fact, the relative importance of van der Waals interactions in the liquids state has been shown by molecular dynamics simulations.[35,36] Figure 14.2(b) shows the relationships between V_{intra}, and $V_{measure}$, and V_{inter}. Both $V_{measure}$ and V_{inter} exhibit clear linear relationships with V_{intra}, and the calculated gradients were 1.593 ($V_{measure}$) and 0.594 (V_{inter}). Therefore, the ILs examined in this study have a free (void) volume of approximately 37%.[37] Notably, a simple but significant linear relationship between V_{intra} and V_{inter} holds for the studied conventional 12 kinds of ILs, containing cations of many molecular frame structures and anions of diverse numbers of fluorine and oxygen as interaction sites. If you plot $V_{measure}$ and V_{inter} for a molecular liquid, it is worth mentioning that a linear relationship will be found as well as for ILs. Recent molecular dynamics simulations based on reliable force fields can quantitatively reproduce the experimental values of liquid physical properties such as density, viscosity, and self-diffusion coefficients for both ILs and molecular liquids, making a clear understanding at the molecular level more possible.[38] However, further efforts both experimental and theoretical are necessary to determine general structure–property relationships for ILs.

14.2.2 Refractive Index

The refractive index is also an important physicochemical property. We attempted to clarify the relationship between the refractive index and polarizability of constituent molecules. The refractive indices of the several ILs were measured in the temperature range 283.15–353.15 K at 5 K intervals. Figure 14.3 shows the temperature dependences of the refractive indices of the EMIm-cation based RTILs with various anion species. A clear linear temperature dependence was observed. The range of the obtained refractive indices was 1.35–1.60, which suggests that the refractive index has a stronger dependence on the anion component than that of the cation. The refractive indices decreased in the order SCN $\gg$ $N(CN)_2$ > SO_3OH > SO_3CH_3 > $SO_3OC_2H_5$ $\gg$ $N(SO_2F)_2$ > TCB $\gg$ $N(SO_2CF_3)_2$ > BF_4 $\gg$ FAP (tris(pentafluoroethyl)trifluorophosphate) at all temperatures. The IL EMIm-SCN had the highest refractive index (1.55 at 303.15 K) in this study, which was close to that of quartz crystal (1.54), a high-refraction index material. The refractive index of EMIm-SCN is also higher than that of liquid immersion oil (1.51), which is used to increase the resolution of optical microscopes. This high refractive index suggests many potential applications of ILs as optical materials. Furthermore, the observed density and refractive index values agreed with previous reports (for EMIm-$N(SO_2CF_3)_2$: EMIm-TFSA).[39] The relationship between the refractive indices of ILs and other physicochemical properties of ILs has attracted much attention. Shimizu *et al.* reported on the relationship between the

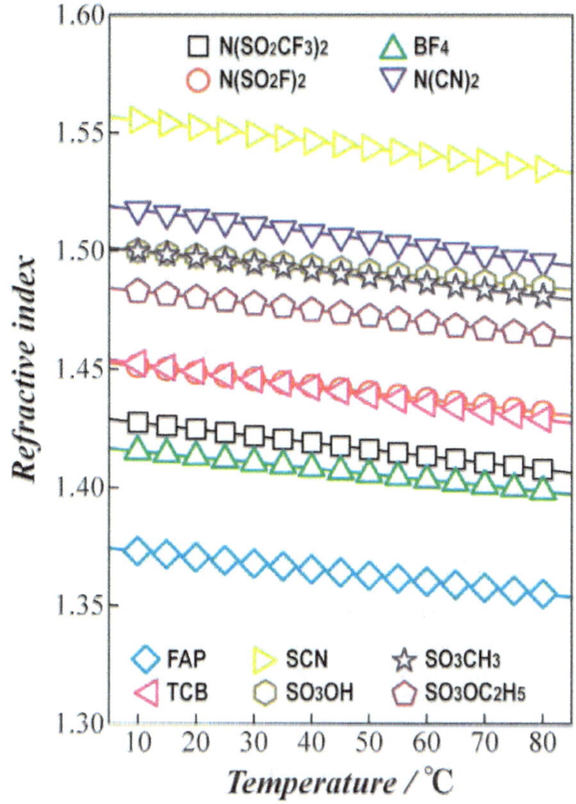

Figure 14.3 Temperature dependence of refractive indices for EMIm-cation-based ILs (353.15–283.15 K).

molecular refractive indices and the cohesive molecular internal energy.[40] Many researchers has since reported relationships between the refractive indices of ILs with various physicochemical parameters, such as density, viscosity, surface tension, and heat capacity.[41–45] The refractive index indicates the magnitude of dielectric response of a material to an electrical field induced by electromagnetic waves (light). Refractive index can be thought of as a response to electronic polarization within an instantaneous time scale as a first order approximation. Thus, the refractive index may be predicted from knowledge of the polarizabilities of constituent cations and anions. Therefore, in our study of the refractive index of ILs, the molecular polarizabilities of ions were calculated. The calculations show that large ions have generally large molecular polarizabilities. The molecular polarization is in turn determined by the number of constituent atoms and their polarization. Therefore, molecular polarizability can be approximately estimated as the sum of atomic polarizabilities. Thus, the refractive index should be proportional to the polarizability of a material per volume. We investigated the relationship between the refractive index of ILs and the calculated polarizability

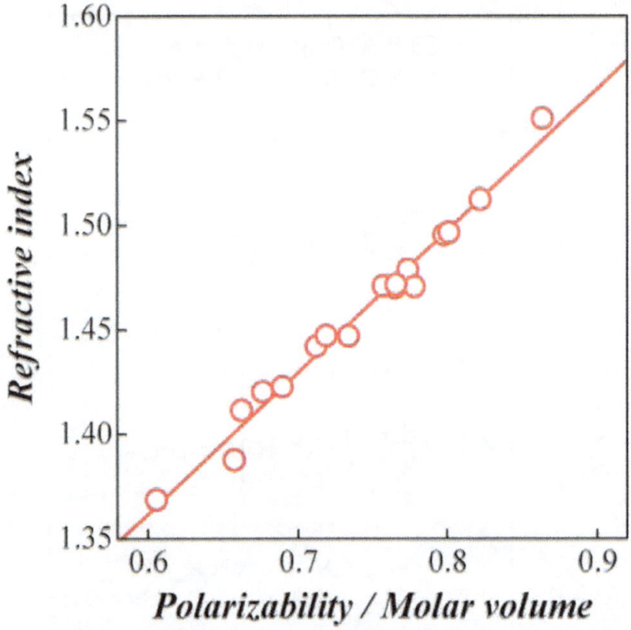

Figure 14.4 Relationship between polarizability/molecular volume and refractive index for ILs at 303.15 K.

of ion pair/molecular volume (Figure 14.4). The polarizability of ion pairs was the sum of the calculated polarizabilities of isolated cations and anions. A high correlation ($R^2 > 0.98$) was obtained for the 17 types of ILs, suggesting that good refractive index estimation is possible from the calculation of polarizability according to the Lorentz–Lorenz equation. No simple estimation method for the refractive index of ILs has yet been reported. However, this initial observation, based on analysis of the optical properties of ILs and the molecular polarizabilities of constituent cations and anions, may guide future work in this area. We found a linear relationship between the refractive index of ILs and the molecular polarizability of ion pairs,[46] which suggests that ILs can be designed to have even higher refractive indices based on calculations of the polarizabilities for each ionic species.

14.3 Special Physical Properties of ILs

We observed an unusual ionic conduction phenomenon related to a phase transition of a novel phosphonium-cation-based IL. Upon cooling, a clear increase in ionic conductivity was found as temperature was decreased, which differs from the behavior of conventional ILs. This was the first report of an abnormal correlation between temperature and ionic conduction for IL electrolytes.[47] Figure 14.5 shows Arrhenius plots of σ calculated from the impedance spectra of neat TEMMP-TFSA (cation structure shown

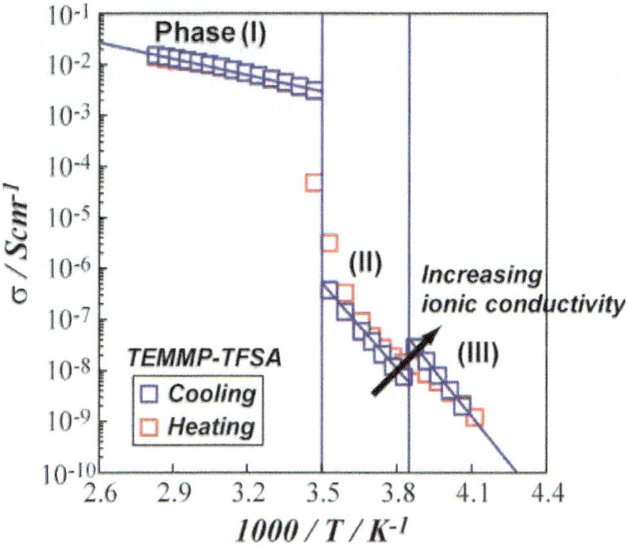

Figure 14.5 Arrhenius-type plots of ionic conductivity (σ) for TEMMP-TFSA (blue: cooling; red: heating).

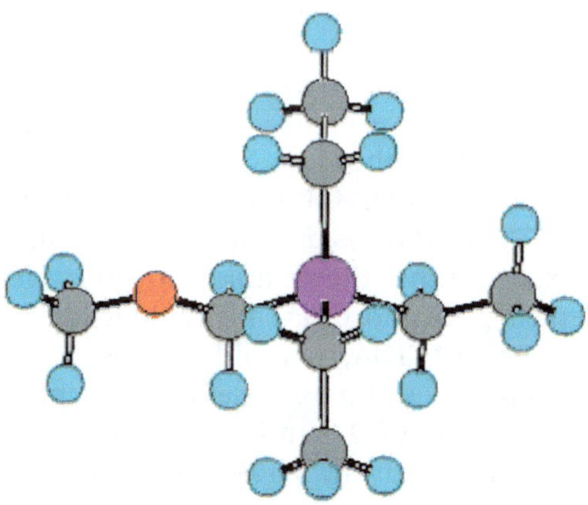

Figure 14.6 Most stable structure of TEMMP cation calculated from relative energies of rotamers at the MP2/6-311G**//HF/6-311G** level (pink: phosphorus; red: oxygen; grey: carbon; light blue: hydrogen).

in Figure 14.6). During the cooling process, the temperature dependence of σ for neat TEMMP-TFSA could be classified into three clear temperature regions. We assumed that the three types of ionic conduction behavior are associated with phase changes [denoted as phases (I)–(III)]. For most conventional ionic conductors (*e.g.*, non-aqueous electrolytes, polymer electrolytes, inorganic electrolytes, ILs and plastic crystals) there have been no reports of more than two types of clear conduction behaviors that show temperature dependence. The temperature changes of phase (I) to (II), and phase (II) to (III) occurred 283.15–288.15 K and from 258.15 to 261.15 K, respectively. In phase (I), a relatively high σ of 5.3×10^{-3} S cm^{-1} at 303.15 K and a low activation energy (E_a) of approximately 21 kJ mol^{-1} were obtained. In phase (II), the behavior changed and σ rapidly decreased by four orders of magnitude (3.8×10^{-7} S cm^{-1} at 283.15 K) compared with that at 288.15 K (3.1×10^{-3} S cm^{-1}). The apparent E_a of this phase change was approximately 109 kJ mol^{-1}, which is five times greater than that of phase (I). The ionic conductivity σ is generally expressed as the product of the number of carrier ions, the valence of the ion species, and the mobility of the ions. In this study, the relationships between temperature and σ were shown to be linked with the mobility of ions. In other words, a marked viscosity change (transition from the solid-state phase) was observed at 283.15 K. In addition, as the temperature decreased in the phase (II) region, an increase in σ was found until the transition from phase (II) to phase (III) (261.15 K: 7.5×10^{-9} S cm^{-1}; 258.15 K: 2.9×10^{-8} S cm^{-1}; apparent E_a: 118 kJ mol^{-1}). To the best of our knowledge, phenomena involving an increase of σ with decreasing temperature have not been reported to date. At present, we attribute the accelerated ionic conduction to a reorientation of the ionic conduction path. The apparent E_a of phase (III) was 118 kJ mol^{-1}, which was similar to that of phase (II). The limiting temperature was 243.15 K, where σ could hardly be measured. Although the effects of the heating process corresponded well with those observed during cooling, the change in σ at the transition from phase (III) to (II) was smaller compared with that occurring during cooling. However, no such phase changes were found with triethyl(2-ethoxymethyl)phosphonium TFSA (TEMEP-TFSA), which has a similar chemical structure to TEMMP-TFSA, suggesting that the position of the oxygen atom in TEMMP plays an important role in the phase change. In particular, we expected that the flexible and short methoxymethyl base connecting the central phosphorus atom should lead to these phenomena. In the case of bulk TEMMP-TFSA, no rapid decrease in σ was observed, and the temperature dependences of σ were classified at approximately 268.15 K as two apparent E_a values of 27 and 126 kJ mol^{-1} for high- and low-temperature regions, respectively. Thus, typical plastic crystal behavior may be expected, which could lead to the development of a solid-state high Li ionic conductors.

14.4 Electrochemical Properties of ILs

14.4.1 Lithium Batteries

Many ILs reported to date (*e.g.*, imidazolium-cation-based ILs) lack electrochemical stability at the reduction potential of Li ions. Thus, various battery systems with different negative materials have been proposed. Over the last decade, Nakagawa *et al.*[48] and Garcia *et al.*[49] developed [$LiCoO_2$ | imidazolium-based IL | $Li_4Ti_5O_{12}$] cells to address the problem of the reduction limit of imidazolium cations. Sato *et al.* also reported a battery using a carbon anode on which a solid electrolyte interphase was formed with the addition of vinylene carbonate and ethylene carbonate to ILs.[50] To increase the reduction potential of ILs, Sakaebe and Matsumoto proposed an IL mixture of *N*-methyl-*N*-propylpiperidinium TFSA (PP_{13}-TFSA) containing LiTFSA, and prepared [$LiCoO_2$ positive electrode | PP_{13}-TFSA/LiTFSA | Li metal negative electrode] cells.[51] In this paper, 30 charge–discharge cycles with a coulombic efficiency higher than 97% was demonstrated. Matsumoto *et al.* reported a mixture of an RTIL based on an alkyl ammonium cation having an ether structure and a Li salt for use as a Li battery electrolyte, which showed a wide electrochemical window particularly in terms of its reduction potential.[52] We have also reported advanced batteries with a long charge/discharge cycle life of over 100 cycles, which use quaternary ammonium cations based on (*N*,*N*-diethyl-*N*-methyl-*N*-(2-methoxyethyl)ammonium) (DEME-TFSA)[53,54] and imidazolium cations based on 1,2-dimethyl-3-propylimidazolium bis(trifluoromethylsulfonyl)amide (DMPIm-TFSA).[55,56] Molecular design of ILs is an effective approach to improve the performance of Li secondary batteries. We have summarized relationships between the molecular structure of ILs (imidazolium and pyrazolium[57]) and battery performance (capacity retention at 50 cycles), as shown in Figure 14.7. In imidazolium cations, the introduction of an electron-donating substituent (methyl base) at the second position of the imidazolium cation and a long alkyl chain at the third position can promote charge delocalization over the imidazolium cation and improve the reduction stability of the IL. However, for pyrazolium-based ILs, the performance was decreased by this modification for unclear reasons. Nevertheless, chemical structure changes to the IL were effective and guided molecular design

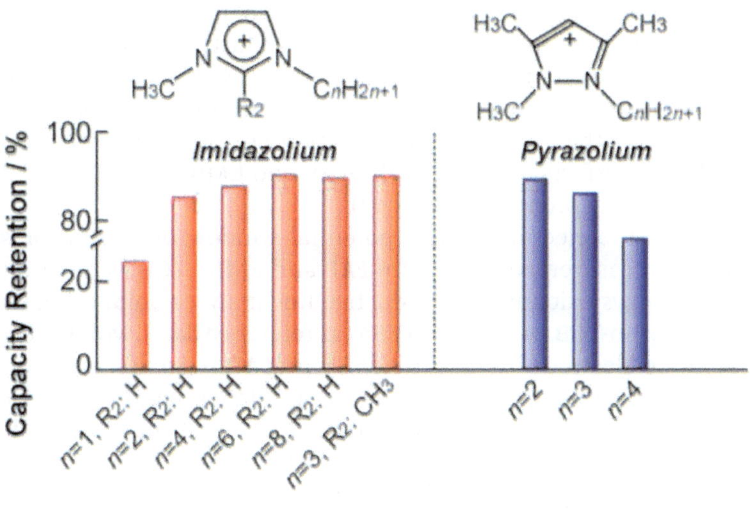

Figure 14.7 Relationships between chemical structures of cation and capacity retention at 50 cycles of [$LiCoO_2$ positive electrode | IL electrolyte | Li metal negative electrode] cells.

of ILs shows promise for achieving effective high-performance Li secondary batteries.

In Li secondary batteries that make practical use of ILs, the application of a graphite negative electrode results in incomplete interfacial formation. Thus, the use of other negative electrode materials, such as Li metal and $Li_4Ti_5O_{12}$, has been reported. Ishikawa *et al.*[58] and Guerfi *et al.*[59] reported favorable charge/discharge operation of [graphite electrode | IL electrolyte | Li metal electrode] cells using bis(fluorosulfonyl)amide (FSA) anion-based ILs. The FSA anion-based ILs had as very low viscosity compared with those based on typical imide anions (*e.g.*, TFSA) and are also promising ILs from the viewpoint of realizing batteries with high-rate output characteristics.[60] We have also confirmed the formation of a suitable interfacial layer (solid electrolyte interphase) between a graphite electrode and FSA-based IL electrolyte by the application of electrochemical calorimetric measurements during charge/discharge operations.[61]

14.4.2 Electrochemical Double Layer Capacitors

Electric double-layer capacitors (EDLCs) have a high rate capability and semi-permanent long cycle life, which suit the requirements of smart grids and energy harvesting devices. The favorable performance of EDLCs is attributed to the fast charge/discharge mechanism of activated carbon electrodes, which involves an electric double layer and does not depend on Faradic reaction (electrochemical reaction). For these applications, ILs with high ionic conductivity and electrochemical stability are strongly desired as electrolyte materials for EDLCs. Tris(pentafluoroethyl)trifluorophosphate (FAP)[62] and tetracyanoborate (TCB)[63] anions have been reported as new anion species. The FAP anion consists of three electrochemically stable fluoroethyl groups and has a high-molecular-weight (M_w: 445.01). TCB contains a boron atom substituted by four low-molecular-weight CN bases (M_w: 114.88), and has a low viscosity owing to the small inter/intra-friction of its molecules. We have investigated the electrochemical properties (*i.e.*, electrochemical window and EDLC performances) of EMIm-FAP and EMIm-TCB ILs, and clarify the electrochemical properties of each IL.

To estimate the electrochemical potential window of ILs, linear sweep voltammetry measurements of EMIm-FAP and EMIm-TCB were performed. Figure 14.8 shows oxidative and reductive linear sweep voltammograms of EMIm-FAP, EMIm-TCB, and EMIm-TFSA (as reference data), with Pt electrode as the working electrode. In the case of EMIm-FAP, reduction and oxidation currents started to flow with gas evolution occurring near −2.6 and +2.4 V *vs* Ag/Ag(I), respectively. Thus, EMIm-FAP is electrochemically stable at a potential of approximately 5.0 V. Compared with the well-known EMIm-TFSA (−2.7 to +2.0 V *vs* Ag/Ag(I)), EMIm-FAP showed a high oxidation resistivity and achieved a large stable potential window. However, in the EMIm-TCB system, although the oxidative current increased slightly near 1.6 V, no significant current attributable to oxidative decomposition was observed even at potentials beyond 5.0 V *vs* Ag/Ag(I). An inactive layer formed on the Pt working

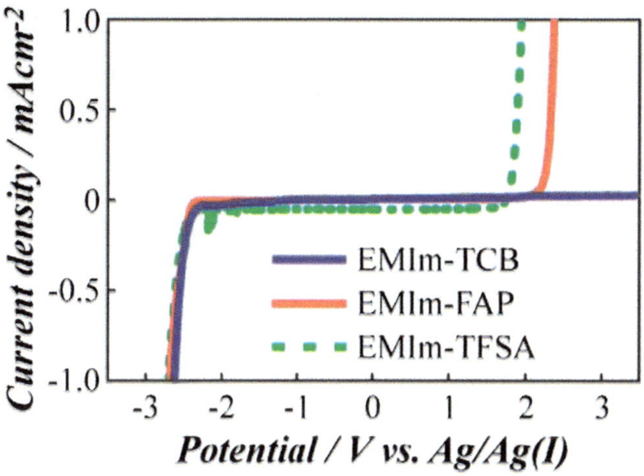

Figure 14.8 Linear sweep voltammograms of EMIm-FAP, EMIm-TCB, and EMIm-TFSA using platinum electrodes.

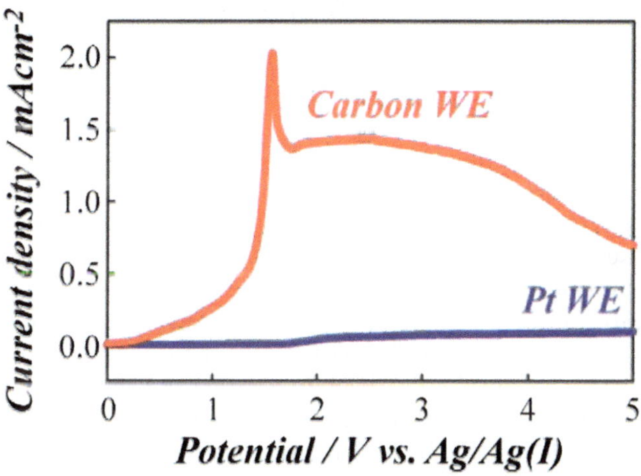

Figure 14.9 Linear sweep voltammograms of EMIm-TCB using platinum and carbon working electrodes.

electrode surface, which might have suppressed decomposition. To investigate film formation at the Pt electrode, another electrode was tested. Figure 14.9 shows the linear sweep voltammograms of EMIm-TCB measured with a C working electrode instead of a Pt working electrode. Switching the electrodes allowed measurement of an oxidation current at potentials lower than 1 V and a brown decomposition product was observed at higher oxidation potentials. The reason for the oxidation current observed in Figure 14.9 cannot be determined; however, although EMIm-TCB showed a high oxidation resistance on a Pt electrode, an oxidation current was observed at a relatively low potential when a C working electrode was used. These findings indicate

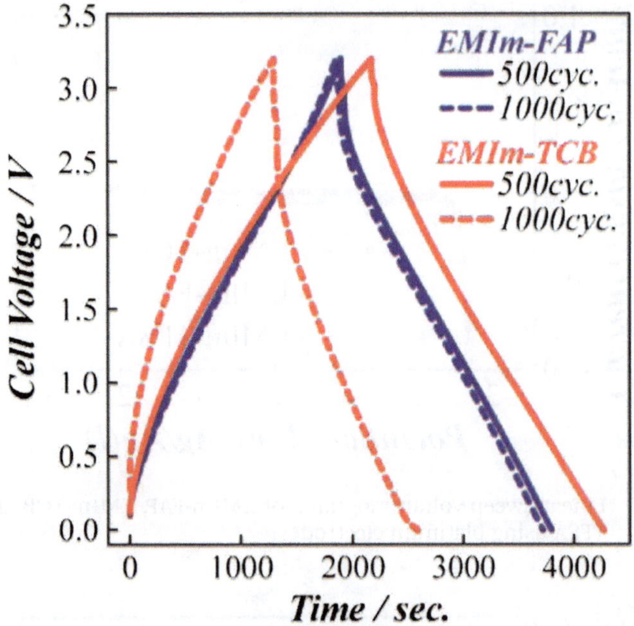

Figure 14.10 Charge/discharge curves for electric double-layer capacitors using EMIm-FAP and EMIm-TCB.

the importance of selecting a suitable electrode material when using EMIm-TCB as an electrolyte in electrochemical devices. Furthermore, the wide potential window of about 5.0 V suggests the possibility of using EMIm-FAP in EDLCs operating at high-voltage exceeding 3 V. The reductive potential for ILs is generally limited by the cation species.[60] Thus, these three ILs with the same EMIm cation showed similar reduction decomposition behavior.

Figure 14.10 shows the charge–discharge curves of EDLCs based on EMIm-FAP and EMIm-TCB at the 500th and 1000th cycles. The observed discharge capacitances (2.0–1.0 V)[64] of the EMIm-FAP electrolyte EDLC were 28.3 (at the 500th cycle) and 28.1 F g^{-1} (at the 1000th cycle). On the other hand, those of the EMIm-TCB electrolyte EDLC were 29.6 (at the 500th cycle) and 18.3 F g^{-1} (at the 1000th cycle). Although the initial charge–discharge capacitances of the EMIm-TCB system were slightly higher than those of the EMIm-FAP system, the EMIm-FAP system showed better reversibility than that of the EMIm-TCB system. The ionic mobility parameters (related to ionic radius, and double-layer capacitance), such as the inverse of the viscosity, conductivity, and ion diffusion coefficients of EMIm-TCB, were larger than those of EMIm-FAP. However, the wider electrochemical window of EMIm-FAP with a C electrode than that of EMIm-TCB suggested good cycle performance. Although EDLCs using ILs have been reported to operate mainly at less than 3.0 V, EMIm-FAP based EDLCs are capable of higher voltage operation owing to their wide electrochemical window (Figure 14.8).[65] High voltage operation of an EDLC can lead to greater amounts of stored charge. Extension of the cycle life by establishing relationships between voltage and cycle-life and high charge and discharge current operated EDLCs are important areas of future work.

14.4.3 New Applications for Electrical Devices

Electric field control of charge carrier density has long been a key technology to tune the physical properties of condensed matter, explored by the modern semiconductor industry. One of the big challenges is to increase the maximum attainable carrier density; however, it is limited by the quality of gate dielectrics. Recently, the electric double layers (EDLs) gating technique using electrolyte opens a new avenue not only to novel physics of solid-state materials but also to potential applications for low consumption devices. The advantage of electrolyte gating using the EDLs is that much higher-density carriers are accumulated at the interface of semiconductor channels with low gate-bias voltages because the gate electric field is sustained only at typically 1 nm thick EDLs. Therefore, such devices realize excellent current amplification even at small gate voltage (V_G) of less than 1 V, with the benefit of the large capacitance of the EDLs. Several groups have reported that the EDL gating using ILs is indeed useful for organic field-effect transistors (OFETs),[66–69] as they have shown to enable a very fast response (compared to other types of electrolytes) to changes in the gate voltage. In addition, we recently found that by choosing proper ILs, the mobility of 12.4 $cm^2\ V^{-1}\ s^{-1}$ can be achieved for rubrene single crystal transistors,[69] which is only a fraction of the value of intrinsic material property, demonstrating an excellent field-effect switching operation. As the result, sufficiently large ON current is achieved even with the gate voltage less than 0.5 V, with the benefit of the large capacitance of the EDLs.

This method can be applicable not only to organic semiconductors, but also inorganic semiconductors such as ZnO,[70] VO_2,[71] InGaZnO,[72] and so on. Especially, in metal–insulator–semiconductor field-effect transistors based on VO_2 (vanadium dioxide), electrostatic charging at the surface induces a three-dimensional metallic ground state[71] extending the concept of electric-field control to macroscopic phase control. Not only electric states, but also magnetic properties can be modified with EDLs with ILs; for example, the paramagnetic-to-ferromagnetic transition temperature of Co ultra-thin films at room temperature by an EDL formation at the interface between a cobalt layer and a polymer film containing an ionic liquid.[73] This low-voltage (a few Volts) control of magnetism at room temperature is a significant step towards the realization of future low-power magnetic applications and opens up a new dimension for the research field of spintronics. Recently, several teams incorporated EDL gating in transistors by replacing the solid gate dielectric with electrolyte and have successfully observed surprising effects including a 10% modulation of the resistance in Au film,[74] an electric field induced superconductivity in $SrTiO_3$,[75] ZrNCl[76] and high-Tc cuprate.[77] All these achievements strongly indicate that the EDL gating is a powerful disorder-free technique for controlling electronic phases simply by electric field, for inducing novel electronic states that are impossible to realize by a chemical route, and for future applications in dissipationless devices and nanodevices.

14.5 New ILs: Solvate ILs

It has been reported that equimolar mixtures of glymes (oligo-ethers) and certain Li salts such as Li[TFSA] exist in a liquid state at room temperature and exhibit high thermal stability, low volatility, and low flammability, which are attractive features for electrolytes in lithium secondary batteries.[78,79] These molten complexes behave like ILs because the glymes complex with Li cations to form a complex cation, $[Li(glyme)]^+$, with a neglectable amount of free glyme molecules. Moreover, the complex cations and TFSA anions form an IL-like liquid electrolyte. These electrolyte materials feature many similarities to ILs with strongly solvated Li ions and glyme molecules (behaving as cations) and dissociated anions (denoted as Li solvate ionic liquids, SILs). SILs can be used as electrolytes for 4 V class lithium batteries. Moreover, SILs have weak coordinating ability, which allows for use in innovative Li-sulfur batteries and give a long cycle life (over 400 cycles, Figure 14.11) by suppressing the dissolution of the polysulfide into the electrolyte.[80] More specifically, it is becoming clear that SILs have important benefits in terms of performance (low Lewis acidity and basicity) in addition to safety advantages.

We investigated the battery performance of SILs using a $LiCoO_2$ positive electrode, and confirmed that the battery worked over 200 cycles in a triglyme (G3) system ($[Li(G3)_1]$TFSA).[81] To further improve the battery performance, we switched to a $LiNi_{1/3}Mn_{1/3}Co_{1/3}O_2$-based electrode and investigated the long-cycle performance above 4.0 V. Figure 14.12(a) shows the charge/discharge profiles obtained in cycle tests (voltage range: 2.7–4.2 V, current: C/8) of the [$LiNi_{1/3}Mn_{1/3}Co_{1/3}O_2$ positive electrode | SILs ($[Li(G3)_1]$TFSA) electrolyte | Li metal negative electrode] cell at 303.15 K. A relatively high discharge capacity (approximately 150 mA h g^{-1} per positive electrode) was obtained in the first cycle, which corresponded to approximately 52% of the theoretical capacity of $Li_xNi_{1/3}Mn_{1/3}Co_{1/3}O_2$ (283.7 mA h g^{-1}, $0 < x < 1$), although the initial irreversible capacity was larger (1st Coulombic efficiency: 87.5%). The electrode material $LiNi_{1/3}Mn_{1/3}Co_{1/3}O_2$ has been reported to have a low initial Coulombic efficiency (approximately 90%, electrolyte: EC/DMC 1M $LiPF_6$).[82] The average discharge voltage was 3.74 V and the calculated energy density was approximately 553 W h kg^{-1} at the first cycle. Although the polarization increased with the charge/discharge cycle number (attributed to the increased resistance of the cells), stable charge/discharge operation continued up to 400 cycles.[83] We have also reported stable charge/discharge operation of a 4V-class $LiCoO_2$ positive electrode after 200 cycles,[81] which showed a similar capacity decreases for both $LiCoO_2$ and $LiNi_{1/3}Mn_{1/3}Co_{1/3}O_2$ cells. These results demonstrate the favorable features of the SILs for long cycle life, which can be attributed to the combination of the

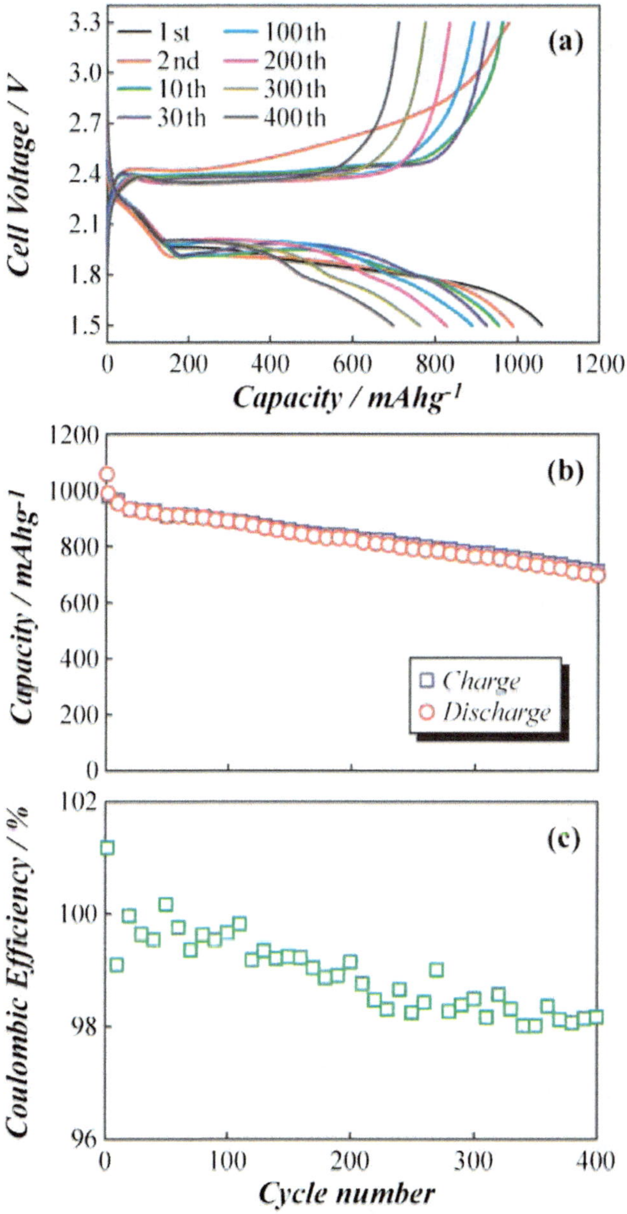

Figure 14.11 Charge/discharge profiles of [sulfur positive electrode | $[Li(G3)_1]$TFSA electrolyte | Li metal negative electrode] cell at the 1st, 2nd,10th, 30th, 100th, 200th, 300th, and 400th cycles ((a) voltage range: 3.3–1.5 V, current: C/18), cycle number dependences of charge/discharge capacities per positive electrode (b), and Coulombic efficiencies (c) at 303.15 K.

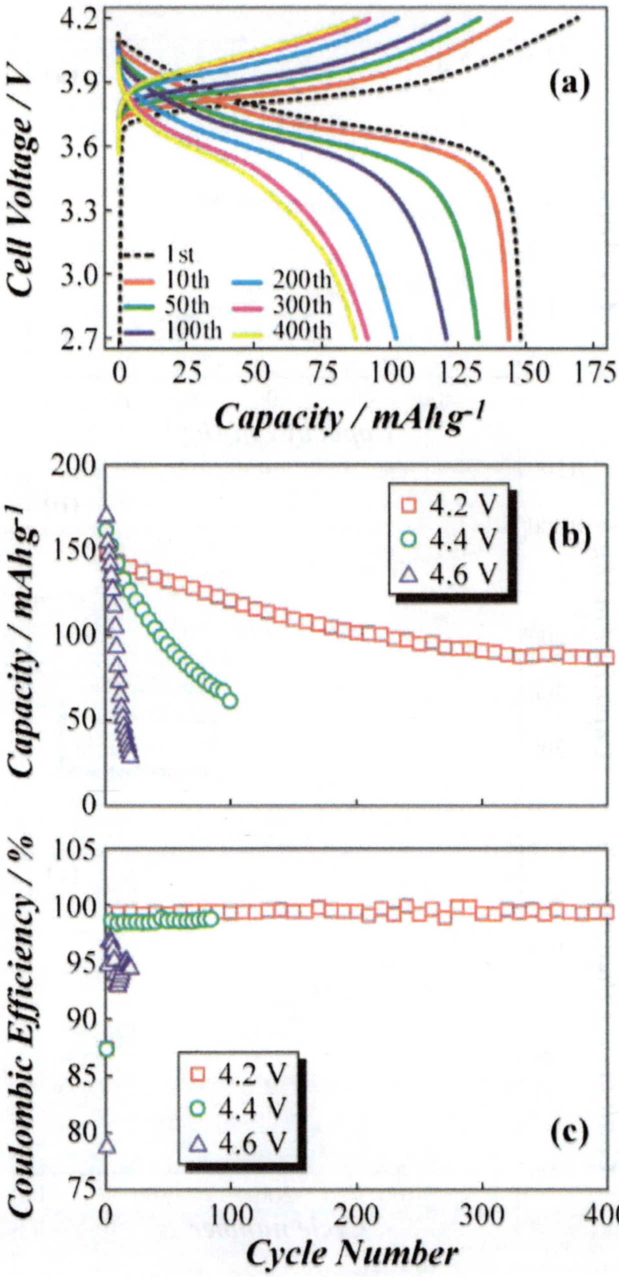

Figure 14.12 Charge/discharge profiles of [$LiNi_{1/3}Mn_{1/3}Co_{1/3}O_2$ positive electrode | $[Li(G3)_1]$TFSA electrolyte | Li metal negative electrode] cell at the 1st, 10th, 50th, 100th, 200th, 300th, and 400th cycles ((a), voltage range: 2.7–4.2 V, current: C/8), cycle number dependences of discharge capacities per positive electrode (b), and Coulombic efficiencies (c) for different upper charge cutoff voltages (4.2, 4.4, and 4.6 V) at 303.15 K.

highly-oxidation-resistant SILs and the $LiNi_{1/3}Mn_{1/3}Co_{1/3}O_2$ positive electrode. To clarify the upper charge cutoff voltage and cycle properties of the prepared cells, the cycle number dependence on the charge and discharge capacities (Figure 14.12(b)) and Coulombic efficiencies (Figure 14.12(c)) were also plotted by changing the upper charge cutoff voltages. When the upper charge cutoff voltage was relatively low (4.2 V), a sufficient capacity retention was achievable for more than 400 cycles (approximately 60% of the first discharge capacity). However, when the upper charge cutoff voltage was increased, the charge/discharge reversibility decreased, particularly at 4.6 V, even though the initial discharge capacity increased. This tendency correlates with previously reported linear sweep voltammetry studies on the oxidative stability of such devices.[79] A similar tendency for coulombic efficiency was observed with increasing upper charge cutoff voltage. Considering the stable interfacial behavior of SILs ($[Li(G3)_1]TFSA$) and electrolyte/Li metal electrode, we believe that the decrease of capacity likely occurs through degradation at the $LiNi_{1/3}Mn_{1/3}Co_{1/3}O_2$ positive electrode/SILs ($[Li(G3)_1]TFSA$) electrolyte interface, owing to instability of the electrode at high voltage (*e.g.*, release of oxygen). However, the cell in which the upper charge cutoff voltage was set to 4.2 V exhibited almost 100% coulombic efficiency; this result is consistent with the slight capacity decrease with increasing charge/discharge cycle number. As mentioned earlier, SILs electrolytes have the potential to be used as new electrolyte materials for safe Li secondary batteries under suitable operation voltage conditions. In particular, the molar ratio of glyme and Li cations at the electrode interfaces (for both the positive and negative electrodes) may shift during charge/discharge. If the molar ratio of glyme and Li cation is important to the battery performance (electrode stability), we considered the concentration profile during electrochemical reactions was a key factor for investigation.

The oxidative stability of SILs system relies on the strong coordination of glyme molecule to Li cation. The fraction of free glyme was estimated to be smaller than 3% in the equimolar mixture of LiTFSA and G3 or tetraglyme (G4) using Raman spectroscopy.[84] However, the concentration of electroactive species changes near the electrode/electrolyte interface during electrochemical (charge/discharge) reactions to form the concentration gradient layer. In the case of Li secondary battery using a SILs, Li^+ is consumed at the interface during charge of negative electrode or discharge of positive electrode, leading to the decrease of local concentration of Li^+ around the electrode interface. Therefore, glyme molecules may liberate transiently when the rate of diffusion of the glyme molecules and the rate of ligand exchange reaction are slower than those of the electrode reaction.[85] The transient generation of free glyme molecule may influence the oxidative stability of SILs especially at the initial stage of charging process at a positive electrode. The viscosity of SILs strongly depends on the composition ratio between glyme and Li salt.[86,87] Therefore, one can investigate the transient liberation of glyme under reactions from the *in situ* measurement of local viscosity near the electrode interface.

The electrochemical quartz crystal microbalance (EQCM) technique is widely used as an *in situ* monitoring tool of the electrode mass based on the piezoelectric properties of a quartz crystal.[88] The mass change of quartz crystal electrode can be estimated from the change in resonance resistance based on the Sauerbrey equation.[89] Moreover, an impedance technique EQCM (also called EQCM-A) can measure the resonance resistance of a quartz crystal (R_{res}). R_{res} has been reported to reflect the product of the viscosity (η_L) and the density (ρ_L) of the media in contact with the quartz crystal, and to be proportional to $(\eta_L\rho_L)^{0.5}$:

$$R_{res} = \frac{A(2\pi f_0 \eta_L \rho_L)}{k^2} \tag{14.1}$$

where f_0 is the fundamental resonance frequency, A is the surface area of the electrode, and k is an electromechanical coupling factor.[90] The local change in $\eta_L\rho_L$ value corresponding to the concentration distribution of Ag^+ was observed during deposition and dissolution of Ag in an amide-type IL using the EQCM-A technique.[91] Therefore, it is informative to conduct EQCM-A measurements in SILs to understand the transient coordination of glyme to Li^+.

The deposition and dissolution of Li metal on a Ni-coated quartz crystal (f_0: 9.08 MHz) was investigated in SILs ($[Li(G3)_1]TFSA$) using EQCM-A because Ni does not react with Li to form an intermetallic compound.[87] The local changes in $\eta_L\rho_L$ value during galvanostatic measurement at different current densities is shown in Figure 14.13. The estimated value is the average $\eta_L\rho_L$ at a few micrometers from the electrode because the amplitude of a quartz crystal decays exponentially in a liquid media. The change in local viscosity seems to be dominant during measurement because the density change showed relatively small dependence on the molar ratio. During the deposition of Li, the local viscosity decreased with

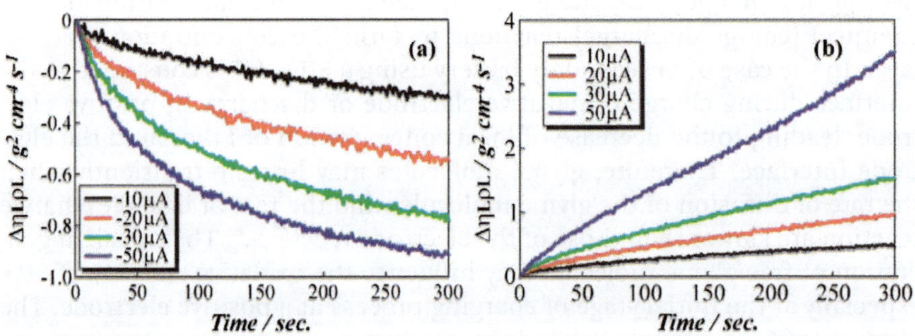

Figure 14.13 Changes in $\eta_L\rho_L$ value during the galvanostatic (a) deposition and (b) dissolution of lithium with various currents at 298.15 K.

reaction, suggesting that glyme liberated near the electrode. The current dependence indicates the distribution of the composition ratio between glyme and Li salt increased with an increase in the rate of electrode reaction. On the other hand, the $\eta_I \rho_L$ value largely increased during dissolution of Li due to the increase of local concentration of Li^+ and formation of such bulky anion as $[Li(TFSA)_2^-]$.[92,93] In the same manner, the local viscosity may increase during extraction of Li^+ at a positive electrode in charging or negative electrode in discharging. This dynamic change in physical properties of an electrolyte possibly affects the diffusion phenomena of Li^+ and rate performance of batteries.

The amount of free glyme near the electrode/electrolyte interface may be reduced by adding excess LiTFSA. In fact, the free glyme mole fraction to total glyme in the SIL has been estimated to be less than 2.3% in the SIL system with a 1 : 1 glyme : LiTFSA ratio as previously reported by Raman speciation analysis.[84] Hence, to improve the electrochemical stability of SILs, it is necessary to form long-lived robust complexes between oxygen atoms in the electrolyte and Li cations. We proposed a new-type of non-stoichiometric SILs, which can be prepared by adding excess Li salt to achieve a higher Li salt concentration than in a 1 : 1 equimolar complex electrolyte. Although this addition may increase the viscosity of electrolyte, the excess Li salt is necessary to compensate Li cations close to the electrodes during the charge/discharge process.[94]

To investigate the electrochemical performances of the three types of $[Li(G3)_x]TFSA$ electrolytes, we performed charge/discharge cycle tests. Figure 14.14 shows (a) the discharge profiles (1st and 30th cycles) and the cycle number dependencies of (b) the average discharge capacities and (c) average coulombic efficiencies obtained in the cycle tests. The initial discharge profiles for each of the cells showed values close to the theoretical capacity for Li_yCoO_2 (150 mA h g^{-1}; $0.45 < y < 1$), although we also observed slight polarizations caused by the increment in the Li salt concentration. The changes in the average discharge voltage with the cycle number depended on the LiTFSA : G3 ratio. Considerable increases in polarization and decreases in discharge capacity were observed at the lower Li salt concentration, $[Li(G3)_{1.25}]TFSA$. The discharge capacities retained after 30 cycles were 66.1% ($[Li(G3)_{1.25}]TFSA$), 74.6% ($[Li(G3)_1]TFSA$), and 84.4% ($[Li(G3)_{0.8}]TFSA$). A similar improvement in coulombic efficiency (associated with the suppression of side reactions) was observed with increasing Li salt concentration. Therefore, we suggest that these three effects are related to the composition ratio of G3 and the Li cations. We have confirmed that excess Li salts in glyme solvent (non-equimolar SILs) can compensate ionic species for the oxidation/reduction (dissolution/deposition) process. On the basis of this approach the electrochemical stability at the electrolyte/electrode ($LiCoO_2$, positive electrode side) interface can be improved with respect to the charge/discharge reaction, leading to longer cycle life of the batteries.

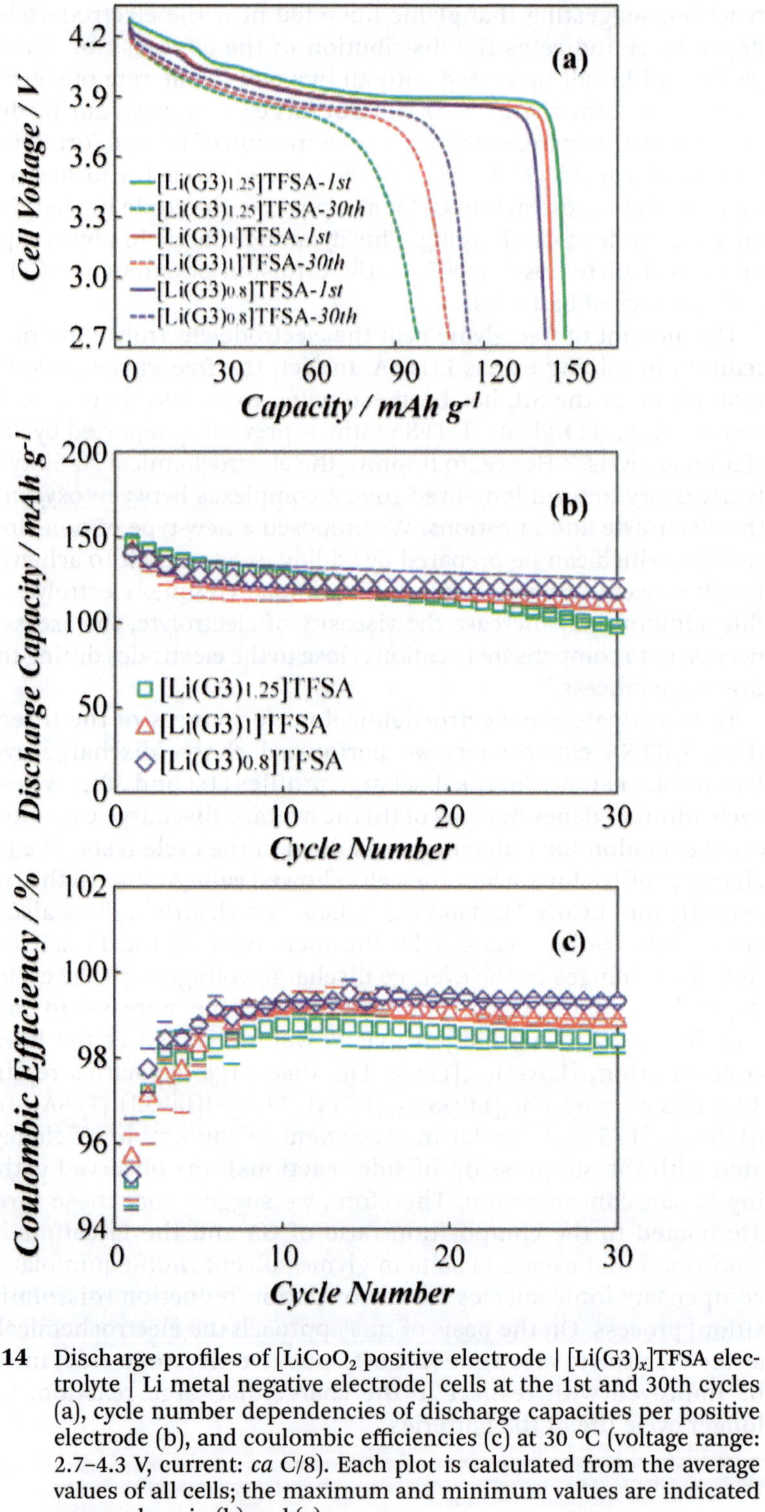

Figure 14.14 Discharge profiles of [$LiCoO_2$ positive electrode | [Li(G3)$_x$]TFSA electrolyte | Li metal negative electrode] cells at the 1st and 30th cycles (a), cycle number dependencies of discharge capacities per positive electrode (b), and coulombic efficiencies (c) at 30 °C (voltage range: 2.7–4.3 V, current: *ca* C/8). Each plot is calculated from the average values of all cells; the maximum and minimum values are indicated as error bars in (b) and (c).

14.6 Conclusion

In this chapter, we have introduced our research and development of ILs as new electrolyte materials for various electrochemical applications. The deeper understanding that we have developed concerning the relationships between molecular structure and physicochemical properties of IL will help advance research into new electrolytes and increase the possibilities of IL based systems.

Acknowledgements

The authors acknowledge Ms N. Yoshioka (Yokohama National University) for technical and editorial support in preparing the manuscript. The references to our works were partially supported by the Advanced Low Carbon Technology Research and Development Program (ALCA) of the Japan Science and Technology Agency (JST) and KAKENHI from the Japan Society for the Promotion of Science (JSPS), Japan.

References

1. K. E. Johnson, *Electrochem. Soc. Interface*, 2007, **16**, 38.
2. A. Noda, K. Hayamizu and M. Watanabe, *J. Phys. Chem. B*, 2001, **105**, 4603.
3. H. Tokuda, K. Hayamizu, K. Ishii, M. A. B. H. Susan and M. Watanabe, *J. Phys. Chem. B*, 2004, **108**, 16593.
4. H. Tokuda, K. Hayamizu, K. Ishii, M. A. B. H. Susan and M. Watanabe, *J. Phys. Chem. B*, 2005, **109**, 6103.
5. H. Tokuda, K. Ishii, M. A. B. H. Susan, S. Tsuzuki, K. Hayamizu and M. Watanabe, *J. Phys. Chem. B*, 2006, **110**, 2833.
6. H. Tokuda, S. Tsuzuki, M. A. B. H. Susan, K. Hayamizu and M. Watanabe, *J. Phys. Chem. B*, 2006, **110**, 19593.
7. A. Noda and M. Watanabe, *Electrochim. Acta*, 2000, **45**, 1265.
8. S. Washiro, M. Yoshizawa, H. Nakajima and H. Ohno, *Polymer*, 2004, **45**, 1577.
9. M. A. B. H. Susan, T. Kaneko, A. Noda and M. Watanabe, *J. Am. Chem. Soc.*, 2005, **127**, 4976.
10. T. Fukushima, A. Kosaka, Y. Ishimura, T. Yamamoto, T. Takigawa, N. Ishii and T. Aida, *Science*, 2003, **300**, 2072.
11. A. Hayashi, M. Yoshizawa, C. A. Angell, F. Mizuno, T. Minami and M. Tatsumisago, *Electrochem. Solid-State Lett.*, 2003, **6**, E19.
12. Y. Tominaga, S. Asai, M. Sumita, S. Panero and B. Scrosati, *Electrochem. Solid-State Lett.*, 2005, **8**, A22.
13. K. R. Seddon, A. Stark and M. J. Torres, *Pure Appl. Chem.*, 2000, **72**, 2275.
14. J. Dupont, R. F. De Souza and P. A. Z. Suarez, *Chem. Rev.*, 2002, **102**, 3667.
15. P. Wasserscheid and W. Keim, *Angew. Chem., Int. Ed.*, 2000, **39**, 3773.
16. T. Welton, *Coord. Chem. Rev.*, 2004, **248**, 2459.

17. R. A. Sheldon, R. Madeira Lau, M. J. Sorgedrager, F. Van Rantwijk and K. R. Seddon, *Green Chem.*, 2002, **4**, 147.
18. A. B. McEwen, H. L. Ngo, K. LeCompte and J. L. Goldman, *J. Electrochem. Soc.*, 1999, **146**, 1687.
19. M. Ue, M. Takeda, T. Takahashi and M. Takehara, *Electrochem. Solid-State Lett.*, 2002, **5**, A119.
20. M. Ue, M. Takeda, A. Toriumi, A. Kominato, R. Hagiwara and Y. Ito, *J. Electrochem. Soc.*, 2003, **150**, A499.
21. D. Kuang, P. Wang, S. Ito, S. M. Zakeeruddin and M. Grätzel, *J. Am. Chem. Soc.*, 2006, **128**, 7732.
22. T. Kato, T. Kado, S. Tanaka, A. Okazaki and S. Hayase, *J. Electrochem. Soc.*, 2006, **153**, A626.
23. M. Doyle, S. K. Choi and G. Proulx, *J. Electrochem. Soc.*, 2000, **147**, 34.
24. A. Noda, M. A. B. H. Susan, K. Kudo, S. Mitsushima, K. Hayamizu and M. Watanabe, *J. Phys. Chem. B*, 2003, **107**, 4024.
25. R. Hagiwara, T. Nohira, K. Matsumoto and Y. Tamba, *Electrochem. Solid-State Lett.*, 2005, **8**, A231.
26. M. A. Navarra, S. Panero and B. Scrosati, *Electrochem. Solid-State Lett.*, 2005, **8**, A324.
27. W. Lu, A. G. Fadeev, B. Qi and B. R. Mattes, *J. Electrochem. Soc.*, 2004, **151**, H33.
28. S. Ono, S. Seki, R. Hirahara, Y. Tominari and J. Takeya, *Appl. Phys. Lett.*, 2008, **92**, 103313.
29. A. P. Abbott, *ChemPhysChem*, 2004, **5**, 1242.
30. J. M. Slattery, C. Daguenet, P. J. Dyson, T. J. S. Schubert and I. Krossing, *Angew. Chem., Int. Ed.*, 2007, **46**, 5384.
31. A. S. Pensado, A. A. H. Padua, M. J. P. Comunas and J. Fernandez, *J. Phys. Chem. B*, 2008, **112**, 5563.
32. J. Palomar, V. R. Ferro, J. S. Torrecilla and F. Rodriguez, *Ind. Eng. Chem. Res.*, 2007, **46**, 6041.
33. J. Jacquemin, R. Ge, P. Nancarrow, D. W. Rooney, M. F. C. Gomes, A. A. H. Padua and C. Hardacre, *J. Chem. Eng. Data*, 2008, **53**, 716.
34. S. Tsuzuki, H. Tokuda, K. Hayamizu and M. Watanabe, *J. Phys. Chem. B*, 2005, **109**, 16474.
35. O. Borodin and G. D. Smith, *J. Phys. Chem. B*, 2006, **110**, 11481.
36. K. Fujii, Y. Soejima, Y. Kyoshoin, S. Fukuda, R. Kanzaki, Y. Umebayashi, T. Yamaguchi, S. Ishiguro and T. Takamuku, *J. Phys. Chem. B*, 2008, **112**, 4329.
37. S. Seki, T. Kobayashi, Y. Kobayashi, K. Takei, H. Miyashiro, K. Hayamizu, S. Tsuzuki and Y. Umebayashi, *J. Mol. Liq.*, 2010, **152**, 9.
38. S. Tsuzuki, W. Shinoda, H. Saito, M. Mikami, H. Tokuda and M. Watanabe, *J. Phys. Chem. B*, 2009, **113**, 10641.
39. M. Tariq, P. A. S. Forte, M. F. C. Gomes, J. N. C. Lopes and L. P. N. Rebelo, *J. Chem. Thermodyn.*, 2009, **41**, 790.
40. K. Shimizu, M. Tariq, M. F. C. Gomes, L. P. N. Rebelo and J. N. C. Lopes, *J. Phys. Chem. B*, 2010, **114**, 5831.

41. E. Gómez, B. González, N. Calvar, E. Tojo and Á. Domínguez, *J. Chem. Eng. Data*, 2006, **51**, 2096.
42. H. Jin, B. O'Hare, J. Dong, S. Arzhantsev, G. A. Baker, J. F. Wishart, A. J. Benesi and M. Maroncelli, *J. Phys. Chem. B*, 2008, **112**, 81.
43. K.-S. Kim, B.-K. Shin, H. Lee and F. Ziegler, *Fluid Phase Equilib.*, 2004, **218**, 215.
44. A. B. Pereiro, J. L. Legido and A. Rodríguez, *J. Chem. Thermodyn.*, 2007, **39**, 1168.
45. A. B. Pereiro, P. Verdía, E. Tojo and A. Rodríguez, *J. Chem. Eng. Data*, 2007, **52**, 377.
46. S. Seki, S. Tsuzuki, K. Hayamizu, Y. Umebayashi, N. Serizawa, K. Takei and H. Miyashiro, *J. Chem. Eng. Data*, 2012, **57**, 2211.
47. S. Seki, Y. Umebayashi, S. Tsuzuki, K. Hayamizu, Y. Kobayashi, Y. Ohno, T. Kobayashi, Y. Mita, H. Miyashiro, N. Terada and S. Ishiguro, *Chem. Commun.*, 2008, 5541.
48. H. Nakagawa, S. Izuchi, K. Kuwana, T. Nukuda and Y. Aihara, *J. Electrochem. Soc.*, 2003, **150**, A695.
49. B. Garcia, S. Lavallée, G. Perron, C. Michot and M. Armand, *Electrochim. Acta*, 2004, **49**, 4583.
50. T. Sato, T. Maruo, S. Marukane and K. Takagi, *J. Power Sources*, 2004, **138**, 253.
51. H. Sakaebe and H. Matsumoto, *Electrochem. Commun.*, 2004, **5**, 594.
52. H. Matsumoto, M. Yanagida, K. Tanimoto, M. Nomura, Y. Kitagawa and Y. Miyazaki, *Chem. Lett.*, 2000, **29**, 922.
53. S. Seki, Y. Kobayashi, H. Miyashiro, Y. Ohno, Y. Mita, A. Usami, N. Terada and M. Watanabe, *Electrochem. Solid-State Lett.*, 2005, **8**, A577.
54. S. Seki, Y. Ohno, H. Miyashiro, Y. Kobayashi, A. Usami, Y. Mita, N. Terada, K. Hayamizu, S. Tsuzuki and M. Watanabe, *J. Electrochem. Soc.*, 2008, **155**, A421.
55. S. Seki, Y. Kobayashi, H. Miyashiro, Y. Ohno, A. Usami, Y. Mita, N. Kihira, M. Watanabe and N. Terada, *J. Phys. Chem. B*, 2006, **110**, 10228.
56. S. Seki, Y. Ohno, Y. Kobayashi, H. Miyashiro, A. Usami, Y. Mita, H. Tokuda, M. Watanabe, K. Hayamizu, S. Tsuzuki, M. Hattori and N. Terada, *J. Electrochem. Soc.*, 2007, **154**, A173.
57. S. Seki, T. Kobayashi, N. Serizawa, Y. Kobayashi, K. Takei, H. Miyashiro, K. Hayamizu, S. Tsuzuki, T. Mitsugi, Y. Umebayashi and M. Watanabe, *J. Power Sources*, 2010, **195**, 6207.
58. M. Ishikawa, T. Sugimoto, M. Kikuta, E. Ishiko and M. Kono, *J. Power Sources*, 2006, **162**, 658.
59. A. Guerfi, S. Duchesne, Y. Kobayashi, A. Vijh and K. Zaghib, *J. Power Sources*, 2008, **175**, 866.
60. H. Matsumoto, H. Sakaebe and K. Tatsumi, *J. Power Sources*, 2005, **146**, 45.
61. S. Seki, Y. Kobayashi, H. Miyashiro, Y. Ohno, Y. Mita, N. Terada, P. Charest, A. Guerfi and K. Zaghib, *J. Phys. Chem. C*, 2008, **112**, 16708.
62. N. V. Ignat'ev, U. Welz-Biermann, A. Kucheryna, G. Bissky and H. Willner, *J. Fluorine Chem.*, 2005, **126**, 1150.

63. D. Williams, B. Pleune, J. Kouvetakis, M. D. Williams and R. A. Andersen, *J. Am. Chem. Soc.*, 2000, **122**, 7735.
64. N. Handa, T. Sugimoto, M. Yamagata, M. Kikuta, M. Kono and M. Ishikawa, *J. Power Sources*, 2008, **185**, 1585.
65. S. Seki, N. Serizawa, K. Hayamizu, S. Tsuzuki, Y. Umebayashi, K. Takei and H. Miyashiro, *J. Electrochem. Soc.*, 2012, **159**, A967.
66. S. Ono, S. Seki, R. Hirahara, Y. Tominari and J. Takeya, *Appl. Phys. Lett.*, 2008, **92**, 103313.
67. T. Uemura, R. Hirahara, Y. Tominari, S. Ono, S. Seki and J. Takeya, *Appl. Phys. Lett.*, 2008, **93**, 263305.
68. Y. Yomogida, J. Pu, H. Shimotani, S. Ono, S. Hotta, Y. Iwasa and T. Takenobu, *Adv. Mater.*, 2012, **24**, 263301.
69. S. Ono, K. Miwa and S. Seki, *Appl. Phys. Lett.*, 2016, **108**, 063301.
70. H. Shimotani, H. Asanuma, A. Tsukazaki, A. Ohtomo, M. Kawasaki and Y. Iwasa, *Appl. Phys. Lett.*, 2007, **91**, 082106.
71. M. Nakano, K. Shibuya, D. Okuyama, T. Hatano, S. Ono, M. Kawasaki, Y. Iwasa and Y. Tokura, *Nature*, 2012, **487**, 459.
72. M. N. Fujii, Y. Ishikawa, K. Miwa, H. Okada, Y. Uraoka and S. Ono, *Sci. Rep.*, 2015, **5**, 18168.
73. K. Shimamura, D. Chiba, S. Ono, S. Fukami, N. Ishiwata, M. Kawaguchi, K. Kobayashi and T. Ono, *Appl. Phys. Lett.*, 2012, **100**, 122402.
74. D. Daghero, F. Paolucci, A. Sola, M. Tortello, G. A. Ummarino, M. Agosto, R. S. Gonnelli, J. R. Nair and C. Gerbaldi, *Phys. Rev. Lett.*, 2012, **108**, 066807.
75. K. Ueno, S. Nakamura, H. Shimotani, A. Ohtomo, N. Kimura, T. Nojima, H. Aoki, Y. Iwasa and M. Kawasaki, *Nat. Mater.*, 2008, **7**, 855.
76. J. T. Ye, S. Inoue, K. Kobayashi, Y. Kasahara, H. T. Yuan, H. Shimotani and Y. Iwasa, *Nat. Mater.*, 2010, **9**, 125.
77. A. T. Bollinger, G. Dubuis, J. Yoon, D. Pavuna, J. Misewich and I. Bozovic, *Nature*, 2011, **472**, 458.
78. T. Tamura, T. Hachida, K. Yoshida, N. Tachikawa, K. Dokko and M. Watanabe, *J. Power Sources*, 2010, **195**, 6095.
79. T. Tamura, K. Yoshida, T. Hachida, M. Tsuchiya, M. Nakamura, Y. Kazue, N. Tachikawa, K. Dokko and M. Watanabe, *Chem. Lett.*, 2010, **39**, 753.
80. K. Dokko, N. Tachikawa, K. Yamauchi, M. Tsuchiya, A. Yamazaki, E. Takashima, J. Park, K. Ueno, S. Seki, N. Serizawa and M. Watanabe, *J. Electrochem. Soc.*, 2013, **160**, A1304.
81. K. Yoshida, M. Nakamura, Y. Kazue, N. Tachikawa, S. Tsuzuki, S. Seki, K. Dokko and M. Watanabe, *J. Am. Chem. Soc.*, 2011, **133**, 13121.
82. T. Ohzuku and Y. Makimura, *Chem. Lett.*, 2001, **30**, 642.
83. S. Seki, N. Serizawa, K. Takei, K. Dokko and M. Watanabe, *J. Power Sources*, 2013, **243**, 323.
84. K. Ueno, R. Tatara, S. Tsuzuki, S. Saito, H. Doi, K. Yoshida, T. Mandai, M. Matsugami, Y. Umebayashi, K. Dokko and M. Watanabe, *Phys. Chem. Chem. Phys.*, 2015, **17**, 8248.

85. K. Yoshida, M. Tsuchiya, N. Tachikawa, K. Dokko and M. Watanabe, *J. Electrochem. Soc.*, 2012, **159**, A1005.
86. K. Yoshida, M. Tsuchiya, N. Tachikawa, K. Dokko and M. Watanabe, *J. Phys. Chem. C*, 2011, **115**, 18384.
87. N. Serizawa, S. Seki, K. Takei, H. Miyashiro, K. Yoshida, K. Ueno, N. Tachikawa, K. Dokko, Y. Katayama, M. Watanabe and T. Miura, *J. Electrochem. Soc.*, 2013, **160**, A1529.
88. M. Thompson, A. L. Kipling, W. C. Duncan-Hewitt, L. V. Rajaković and B. A. Čavić-Vlasak, *Analyst*, 1991, **116**, 881.
89. G. Z. Sauerbrey, *Z. Phys.*, 1959, **155**, 206.
90. H. Muramatsu, E. Tamiya and I. Karube, *Anal. Chem.*, 1988, **60**, 2142.
91. N. Serizawa, Y. Katayama and T. Miura, *J. Electrochem. Soc.*, 2009, **156**, D503.
92. J.-C. Lassègues, J. Grondin and D. Talaga, *Phys. Chem. Chem. Phys.*, 2006, **8**, 5629.
93. Y. Umebayashi, T. Mitsugi, S. Fukuda, T. Fujimori, K. Fujii, R. Kanzaki, M. Takeuchi and S.-I. Ishiguro, *J. Phys. Chem. B*, 2007, **111**, 13028.
94. S. Seki, N. Serizawa, K. Takei, S. Tsuzuki, Y. Umebayashi, Y. Katayama, T. Miura, K. Dokko and M. Watanabe, *RSC Adv.*, 2016, **6**, 33043.

CHAPTER 15

Ionic Liquid Based Nanocarriers for Topical and Transdermal Drug Delivery

M. MONIRUZZAMAN[a,b], H. MAHMOOD[a] AND M. GOTO*[c,d]

[a]Universiti Teknologi PETRONAS, Chemical Engineering Department, Seri Iskandar 32610, Perak, Malaysia; [b]Universiti Teknologi PETRONAS, Centre of Research in Ionic Liquids (CORIL), Seri Iskandar 32610, Perak, Malaysia; [c]Kyushu University, Department of Applied Chemistry, Graduate School of Engineering, 744 Moto-oka, Fukuoka 819-0395, Japan; [d]Kyushu University, Center for Future Chemistry, Fukuoka 819-0395, Japan
*E-mail: m-goto@mail.cstm.kyushu-u.ac.jp

15.1 Introduction

The tremendous potential of microemulsions (MEs) as modern colloidal nanocarriers for topical and transdermal drug delivery is well recognized.[1] MEs, which are systems composed of water, oil, and a surfactant/mixture of surfactants, offer a variety of advantages in pharmaceutical applications, including nanometer size, long-term stability, nontoxicity, easy preparation, and high drug solubilization capacity and improved drug delivery.[2–4] Generally, water-soluble drugs can be formulated in water-in-oil (w/o) MEs, whereas oil-soluble drugs are better suited for oil-in-water (o/w) MEs. These versatile colloidal systems are of great technological and scientific interest to researchers because of their potential to incorporate a wide range of both hydrophilic

Smart Materials No. 28
Ionic Liquid Devices
Edited by Ali Eftekhari

Published by the Royal Society of Chemistry, www.rsc.org

and hydrophobic drug molecules owing to the presence of both lipophilic and hydrophilic domains. However, many drug molecules are insoluble or sparingly soluble in water and most pharmaceutically acceptable organic solvents. It is well recognized in modern medicine that many discovered drugs never enter the formulation stage because of their poor solubility in conventional solvents and difficulties in delivery. These limitations demand the provision of alternative solvents that can be used to make a dispersed phase in MEs for solubilization of sparingly soluble drug molecules. Taking these points into account, as well as considering the diverse properties of room temperature ILs, ILs are a good choice because they are capable of dissolving various poorly soluble compounds, including drug molecules.[5–7]

In the past few years, ILs – pairs of organic cations and organic/inorganic anions (Figure 15.1) – have been increasingly used as solvents or co-solvents in a wide range of applications, including biological applications. These include biotechnology (*e.g.*, biocatalysis, biomolecule purification, and biofuel production),[8] pharmaceutics (drug delivery and drug formulation),[9] and antimicrobial agents.[10] The technological utility of drug molecules, particularly sparingly soluble compounds, can be significantly enhanced by their use in ILs rather than in conventional organic solvents or their natural medium (water).[11,12] This advantage comes from the favorable combination of the IL's green properties with tailor-made physicochemical and biological properties. These properties of ILs enable the dissolution of many sparingly soluble

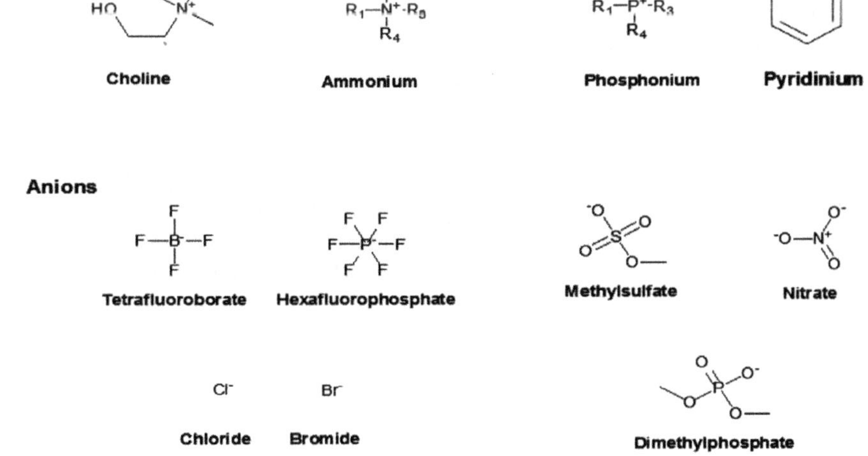

Figure 15.1 Commonly used ions in ILs.

drug molecules at room temperature.[7,8] However, IL-borne drugs cannot penetrate into or across skin owing to the hydrophobic barrier function of skin.[13] In addition, encapsulation of drug molecules in a particulate carrier can protect them against degradation *in vitro* and *in vivo*, and the release can be controlled with specific targeting. Formation of IL droplets in a continuous oil phase using a nontoxic surfactant or a blend of surfactants could produce efficient carriers for sparingly soluble drugs, which are dissolved in ILs. In fact, the IL in IL-in-oil (IL/o) nonaqueous MEs can load drug molecules in the core and the oil phase can be used to pass the barrier of the skin.

In the last decade, IL-based nonaqueous MEs have attracted considerable attention owing to the favorable physical and chemical properties of ILs. Han and co-workers reported formation of IL/o MEs using the nonionic surfactant Triton X-100.[14] Subsequently, the size, shape, formation mechanism, and interior polarity of such MEs have been extensively studied.[15–17] However, the Triton X-100 surfactant is not an acceptable pharmaceutical component. Our group developed a IL/o ME with a blend of the two nontoxic surfactants sorbitan monooleate (Tween 80) and sorbitan laurate (Span 20) in an isopropyl myristate (IPM) continuous phase.[6] Such MEs can significantly solubilize sparingly soluble drugs (*e.g.*, acyclovir (ACV), methotrexate, and 1-[(5-(*p*-nitrophenyl)furfurylidene)amino]hydantoin sodium) in the IL droplets. Preliminary results suggested that IL-based MEs could be potentially used for topical/transdermal administration of drug molecules, particularly sparingly soluble drugs. This new IL/o system needs to be characterized in terms of the drug solubility and delivery capacity. More importantly, the safety profile of the IL-assisted ME should be determined to promote its application in pharmaceuticals.

In this chapter, we describe a novel IL/o ME that is able to dissolve pharmaceutical molecules that are insoluble or sparingly soluble in both water and most pharmaceutical grade organic liquids. Nanometer-sized IL droplets in IPM are formed with a combination of the nonionic surfactants Tween 80 and Span 20. These two surfactants were selected because of their diverse interfacial properties that are beneficial for formation of IL-supported MEs. In addition, Tween 80 has hydrophilic poly(ethylene oxide) (PEO) groups, which have a strong affinity for the imidazolium cation of ILs. Dynamic light scattering (DLS) was used to characterize the systems. The effect of different Tween 80/Span 20 weight ratios on the solubility of ACV as a model sparingly soluble drug and the permeation of ACV into and across Yucatan hairless micropig (YMP) porcine skin was investigated. Finally, the safety profile of the new carrier *in vitro* was evaluated by a cytotoxicity assay using the LabCyte EPI-MODEL, which is a commercially available reconstructed human-cultured epidermal model.

15.2 Experimental

15.2.1 Materials

Tween 80 and Span 20 were obtained from Sigma-Aldrich Chemical Co. (St. Louis, MO) and used without further purification. The IL [C_1mim][$(MeO)_2PO_2$] (dimethylimidazolium dimethyl phosphate) was purchased from Solvent Innovation GmbH (Germany), and it was used as received. Frozen (−80 °C) porcine skin samples (excised from a 5 month old female YMP) were obtained from Charles River Japan Inc. (Tokyo, Japan). ACV was obtained as a generous gift from ASPION pharmaceuticals (Kobe, Japan). IPM was purchased from Tokyo Chemical Industries Co. Ltd. (Tokyo, Japan). Dulbecco's phosphate-buffered saline (D-PBS) was obtained from Invitrogen. The reconstructed human epidermal model LabCyte EPI-MODEL 12 was obtained from Japan Tissue Engineering Co., Ltd. (Aichi, Japan). The 3-[4,5-dimethylthiazol-2-yl]-2,5-diphenyl-tetrazolium bromide (MTT) medium was purchased from Dojindo Co., Kumamoto, Japan. An ACV topical cream was purchased from Taisho Pharmaceutical Co., Ltd. (Tokyo, Japan). All of the other reagents used in the experiments were of analytical grade.

15.2.2 Preparation of IL/o MEs

To prepare the MEs described in our previous study,[6] Tween 80 was blended with Span 20 at various fixed weight ratios (Table 15.1). A 20 wt% mixture of the surfactants was then thoroughly mixed with IPM until a clear and optically transparent solution was obtained. The IL was then added to produce IL/o MEs by vigorously vortexing the resultant mixture at 25 °C. The drug-loaded MEs were obtained by injecting an appropriate amount of the IL in which the drug (10–20 wt%) was solubilized in advance.

15.2.3 Determination of the Drug Solubility in the MEs

The drug solubility in each ME was investigated as follows. The drug was added in excess to MEs prepared as described in Section 15.2.2. The sample was stirred for 24 h at 25 °C. The precipitated drug was removed by filtration through a 0.45 μm Millipore Millex-LG filter. The amount of ACV in the resulting clear filtrate was determined using an ultraviolet (UV) spectrophotometer at 252 nm after appropriate dilution with methanol, retaining the respective blank.

Table 15.1 Effect of surfactant concentration on the solubility of ACV in IL/o ME.[a]

Formulations	Tween80 : Span-20 (mass ratio/molar ratio)	HLB[b]	Viscosity (mPa s)	Solubility of ACV (mg mL^{-1} of ME)
ME1	3 : 1/1 : 1.3	13.4	32.5	7.26
ME2	3 : 2/1 : 2.5	12.4	28.8	7.88
ME3	1 : 1/1 : 3.8	11.7	26.7	4.86
ME4	1 : 2/1 : 7.6	10.6	24.2	3.65
ME5	1 : 3/1 : 11.3	10.1	22.6	3.09

[a]Microemulsions prepared with 20 wt% surfactant and 4 wt% IL in IPM.
[b]HLB values of surfactant mixture (HLB_{mix}) were calculated by using the relation: $HLB_{mix} = f_A HLB_A + f_B HLB_B$, where HLB_A, HLB_B are the HLB values, and f_A, f_B are the weight fractions of Tween-80 and Span-20, respectively.

15.2.4 Characterization of the ME Systems

The sizes and size distributions of ME droplets with and without the drug were determined by the DLS method using a computerized inspection system (Zetasizer Nano ZS, Malvern Instruments, Malvern, UK). All of the samples were equilibrated for more than 5 h before the measurements, and no macroscopic heterogeneity was visible. The experiments were performed at 25 ± 0.1 °C. The samples were equilibrated for 15 min before data collection. The reported mean diameter of the ME droplets is the average of five replicated experiments.

15.2.5 Physical Stability of the MEs

The physical stability of the MEs with ACV was investigated by clarity and phase separation observation and droplet size determination by periodic inspection over 2 months at 25 °C. In addition, a centrifuge test was performed to assess the physical stability of the MEs, involving centrifuging the MEs at 15 000 rpm for 30 min.[18]

15.2.6 Skin Permeation Studies

To investigate the performance of the new ME systems, drug penetration into the skin was evaluated *in vitro*. Penetration of the drug from the MEs into and across full thickness skin pieces of YMP was investigated using hand-made Franz diffusion cells consisting of a donor and a receiver compartment. The skin piece prepared according to our previously study[19] was clamped between the donor and receiver compartments of the diffusion cell. The stratum corneum faced the donor side of the cell while the dermal side of the skin was allowed to be in contact with PBS solution (pH 7.4), which has maintained at 32.5 ± 0.1 °C using a thermostatically controlled water bath and magnetically stirred at 600 rpm throughout the experiments. The skin pieces were hydrated with PBS for 1 h before the permeation experiments and then 0.5 mL of the formulation containing the drug was placed over the skin. After a predetermined time, the receptor chambers were emptied and the solutions were analyzed for the drug content. The donor solutions were removed and the skin was washed twice with 0.1 M HCl before unclamping the diffusion cells. Finally, the tested skin pieces were processed to determine the drug content inside the skin.[19]

15.2.7 Analysis Assay

The drug contents in the receptor solutions and skin were determined by high-performance liquid chromatography. Column: Shiseido CAPCELL PAK C18 MG (4.6 × 250); mobile phase: 0.1 wt% trifluoroacetic acid : acetonitrile = 1 : 99; flow rate = 0.5 mL min^{-1}; detection: UV at 252 nm.

15.2.8 Data Analysis

To determine the transdermal flux, the cumulative amount of ACV that permeated through the YMP skin was plotted as a function of time. The slope and intercept of the linear portion of the plot were derived by regression. The transdermal flux (J, μg cm^{-2} h^{-1}) was calculated as the slope divided by the skin surface area. The skin permeability P was calculated from the transdermal flux using the equation $P = J/\Delta C$, where ΔC is the concentration difference across the skin.

15.2.9 *In vitro* Cytotoxicity Studies

The MTT cell viability assay was performed using the reconstructed human epidermal model LabCyte EPI-MODEL 12, as reported in the literature with some modifications.[20] In brief, the tissues in 24-well plates (BD Biosciences, San Jose, CA, USA) with assay medium (1 mL) were incubated for 12 h at 37 °C in a 5% CO_2 environment. The formulations (100 μL) were placed on the tissue surface and the cultures were incubated for 24 h (37 °C, 5% CO_2, humidified atmosphere). After the incubation period, the tissues were carefully rinsed 10–12 times with D-PBS to remove any remaining MEs from the tissue surface. After removing the culture medium, the treated and control tissues were postincubated for 12 h at the same conditions as previously mentioned. Freshly prepared MTT medium (1 mL, 0.5 mg mL^{-1}) was added to the wells and the tissues were incubated for 3 h. The tissue fragment was then transferred to a microtube containing 0.5 mL of 2-propanol, completely immersing the tissue, and extracted for over 2 h at 25 °C in the dark. Subsequently, the extracted solution was transferred to 96-well plates and the optical density was measured at 570 nm and 650 nm as reference absorbance using 2-propanol as a blank. The tissue viability was determined as the percentage relative to the viability of the negative control of D-PBS.

15.3 Results and Discussion

15.3.1 Selection of the Components for ME Formation

The choice of the components for pharmaceutical grade MEs is very important. The components should be nontoxic and able to form a stable nanocarrier with enhanced properties, including high drug solubility and permeation rate. In the present study, a mixture of Tween 80 (with a high hydrophilic–lipophilic balance (HLB)) and Span 20 (with a low HLB) was selected as the emulsifying agent and IPM was used as the oil phase, because these components are considered to be safe and they are widely used solvents/materials to prepare MEs for pharmaceutical use (Figure 15.2).[21,22] The IL $[C_1mim][(MeO)_2PO_2]$ was used as the dispersed phase because of its high solubility efficiency for ACV[7] and ability to form MEs. ILs containing coordinating anions that are strong hydrogen bond acceptors (*e.g.*, $(CH_3O)_2PO_2^-$ and CH_3COO^-)[21] have been found

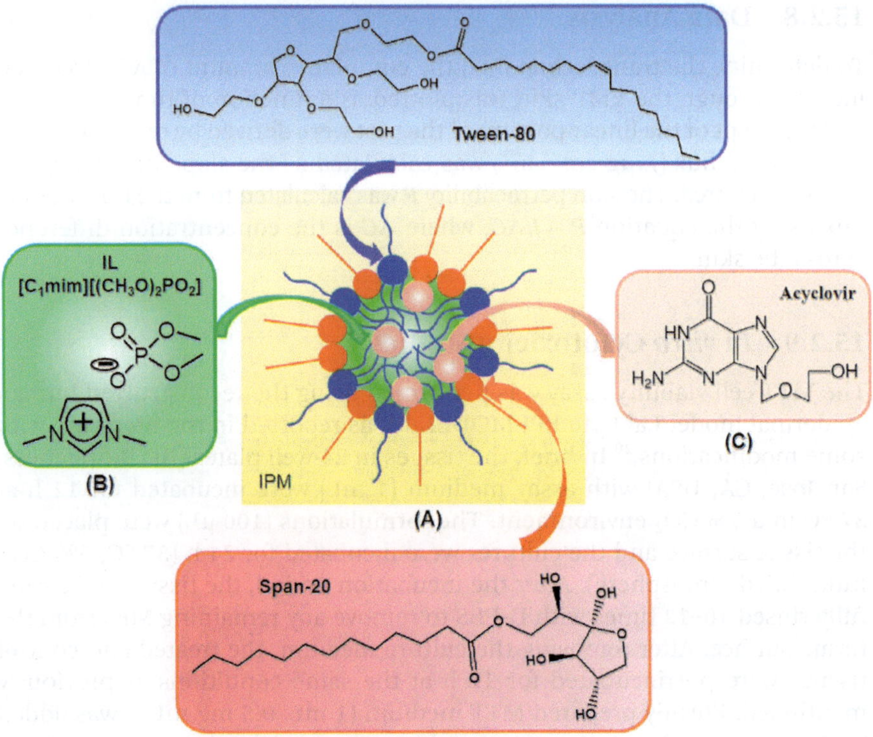

Figure 15.2 Schematic diagram of IL/o MEs.

to be very effective for forming the stable disperse phase in bulk IPM owing to formation of hydrogen bonds between the anions of the IL and the hydroxyl groups provided by both surfactants.[21] In addition, Tween 80 has hydrophilic ethylene oxide (EO) groups, which have a strong affinity for the imidazolium cation in ILs through electrostatic interactions.[16] Because a low content of surfactant favors dermal/transdermal administration, 20 wt% surfactant mixtures were used. According to the phase diagram (data not shown), MEs prepared with this surfactant concentration can solubilize a large amount of IL. In the ME, the IL phase is the internal or dispersed phase, whereas the IPM makes up the external or continuous phase (see Figure 15.2).

15.3.2 Effect of the Surfactant on the Solubility of ACV in the IL/o ME

The solubility of ACV in IL/o MEs prepared with different Tween 80 : Span 20 ratios was investigated because the drug solubility in MEs is highly dependent on the content of each individual surfactant.[23] The solubility of ACV in IL-free Tween 80/Span 20/IPM micelles was found to be very low (about 0.1 mg mL^{-1}).[24] In the presence of IL to form IL/o MEs, a significant amount of ACV can be stabilized in the MEs by solubilization in the IL dispersed phase (Table 15.1). In a previous study, we found that the IL [C_1mim][$(MeO)_2PO_2$] can solubilize more than 25 wt% ACV at room temperature.[7] Interestingly, the solubility of ACV was found to depend on the Tween 80 : Span 20 weight ratio. Higher Tween 80 : Span 20 ratios led to solubilization of ACV in the formulations. In particular, when the Tween 80 : Span 20 weight ratio was above 1 : 1, the solubility of ACV was significantly reduced. The large increase in the ACV solubility in ME1 and ME2 (Table 15.1) is most likely related to formation of stable ME droplets with a larger interface than the others MEs.[6] Because the Tween 80 : Span 20 ratio of 3 : 2 shows the maximum solubility of ACV, this surfactant ratio was used for the subsequent experiments reported in this chapter.

15.3.3 Characterization of MEs Loaded with ACV

To investigate the effectiveness of MEs as drug delivery carriers, their microstructure needs to be characterized. First, the variation in the sizes of MEs at different IL concentrations (defined as the molar ratio of IL to surfactants) was investigated. It was found that with increasing IL content in the MEs, the size of the MEs increases (data not shown), which is a common phenomenon for IL/o or other conventional MEs. This confirmed that the IL is in the core, and hence additional IL causes the core of the MEs to expand. This observation is very common in conventional MEs.[3] We also found that the shape of the newly developed MEs is spherical.[6,7] To check the effect of encapsulated drug on the size of the ME droplets, the sizes and size distributions of droplets of IL/o ME2 with various drug concentrations were characterized by DLS at 25 °C. It was found that incorporation of ACV into the MEs (0–5 mg mL^{-1}) decreases the mean diameter (Figure 15.3), and this is more pronounced at higher ACV concentration. This trend is consistent with loading of drug molecules in w/o and o/w ME systems.[25,26] It is not clear why the droplet size of surfactant aggregates decreases with increasing drug content. Further study is required to determine the exact reason. However, a possible reason is that a certain amount of undissolved drug could deposit at the interface of the ME, which may increase the interfacial area. Consequently, the size of the drug-loaded IL droplets decreases. The polydispersity index values are about 0.2, which means that a narrow size distribution and homogeneous distribution of surfactant aggregates were produced in the MEs.

15.3.4 Physical Stability of the MEs

The stability of the carrier containing the drug molecules is very important in the design of MEs for drug delivery. Therefore, the stability of ACV-loaded ME1, ME2, and ME3 was evaluated by clarity and phase separation observations and droplet size determination over 2 months at 25 °C. These properties of ME1 and ME2 did not significantly change during the observed time. However, ME3 became turbid after 45 days. This is possibly because of the lower amount of the

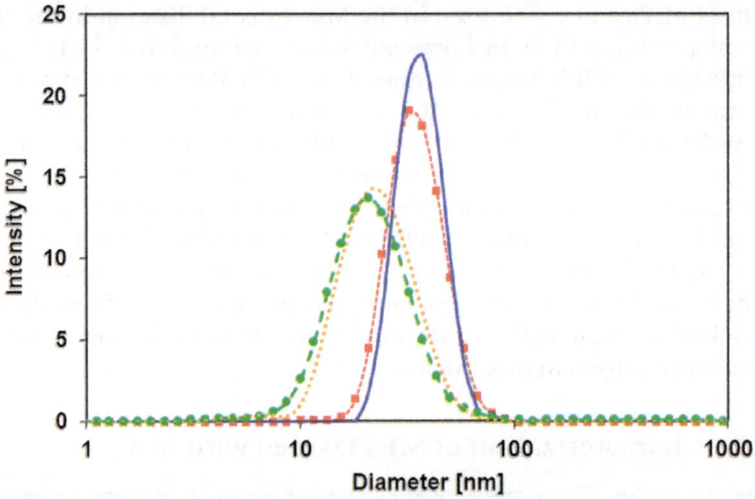

Figure 15.3 Size distributions of droplets in IL-in-IPM MEs with various ACV concentrations at 25 °C. The overall weight fraction of surfactants is 20 wt% (Tween 80:Span 20 = 3:2 (w:w)) and [IL] = 4 wt%. —— ACV =0, ··■·· 1 mg mL^{-1}, ····· 3 mg mL^{-1}and ··●·· 5 mg mL^{-1}.

Tween 80 surfactant because it has been reported that higher Tween 80 content in the surfactant mixture favors formation of stable MEs.[6] The physical stability of the MEs was also investigated by centrifugation.[18] The results showed that the formulations have excellent physical stability. In addition, identical samples of ME2 containing ACV were stored at different temperatures (4, 25, and 37 °C) to investigate the effect of the storage temperature on the stability of the MEs. These samples were then subjected to visual inspection and DLS monitoring at determined time intervals. The droplet sizes and size distributions did not significantly change with time up to 45 days (data not shown). The results show that the storage temperature has a negligible effect on the long-term stability of drug-containing IL/o MEs.

15.3.5 Skin Permeation Studies

The main goal of this study is to prove that IL/o MEs can be used for transdermal and topical drug delivery. The permeation capability of ACV from IL/o MEs into and across YMP skin was investigated *in vitro* and compared with other formulations as control experiments (IL, IPM, surfactants–IPM, w/o MEs, and commercially available ACV cream). Notably, ACV in IPM, surfactants–IPM, and w/o ME formulations is in a dispersed state whereas ACV in IL and IL/o is completely solubilized. As shown in Figure 15.4, the permeation of ACV into the skin improved by several orders of magnitude compared with conventional w/o MEs when IL/o MEs were used. Interestingly, application of IL/o MEs induced significant transdermal permeation of ACV, whereas the

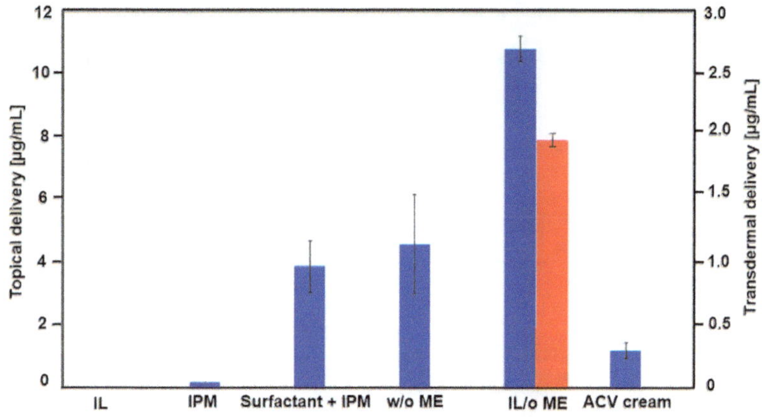

Figure 15.4 ACV penetration into and across the skin using various formulations at 32.4 °C. Blue and red columns represent topical and transdermal delivery, respectively.

other formulations exhibited almost zero permeation (*i.e.*, below the detection level). This is expected to be because of the advantages of conventional MEs as drug carriers, including their high solubilization capacity, favorable drug transportation mechanism, and so forth.[3,4,21,22] Drugs are only able to penetrate the skin in the dissolved form, and IL/o MEs can solubilize a large amount of ACV in the IL droplets, which can act as a drug reservoir. In addition, the lipophilic parts of the IL/o MEs can readily obstruct the stratum corneum (SC) lipid structure so that nanometer size IL droplets containing solubilized drug can move through the skin. During *in vitro* absorption, the drug can easily diffuse into the external phase from the IL droplets. For the IL–ACV formulation, although ACV is completely solubilized in the formulation, the hydrophobic barrier function of the skin offers significant resistance for penetration of ACV, which is provided by the highly ordered structure of the SC, because of the extremely hydrophilic characteristics of ILs.[22] Furthermore, in the formulations with a high concentration of IPM, which is a well-known enhancer to overcome the barrier of the SC, the presence of a large amount of suspended ACV could disrupt the skin.[27] Considering all of the factors discussed above, it can be concluded that the excellent skin permeation rate achieved using IL/o MEs is mainly because of two factors: ACV is completely dissolved in nanometer-sized IL droplets and the presence of the surfactants and IPM decrease the resistance of the SC lipid structure.

Because the molar ratio of the individual surfactant components in the formulation defines the physicochemical properties of MEs and has a significant effect on its drug delivery behavior, ACV permeation was investigated with various ratios of the two different surfactants and a fixed overall surfactant concentration of 20 wt%. The transdermal permeation profiles (Figure 15.5) show the cumulative mass of ACV diffused through the skin as a function of time. These results clearly indicate that skin penetration of ACV considerably

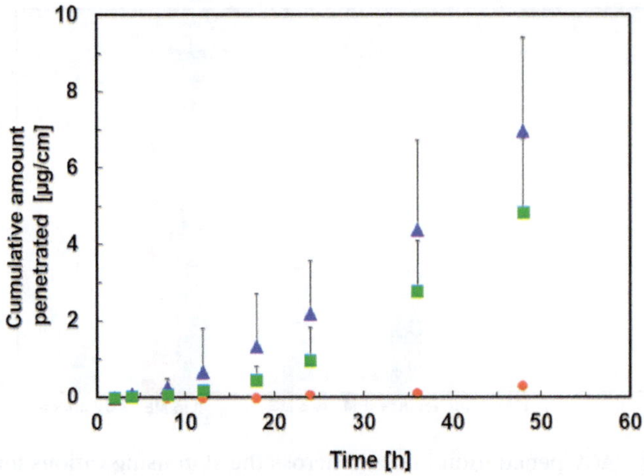

Figure 15.5 Permeation profile of ACV across the skin from MEs containing different weight ratios of Tween 80 to Span 20 (n = 3). [Overall surfactant] = 20 wt% and [IL] = 4 wt%. Symbols: (●) 1:1, (▲) 3:1, and (■) 3:2. Reprinted from *International Journal of Pharmaceutics*, **400** (1–2), M. Moniruzzaman, M. Tamura, Y. Tahara, N. Kamiya, M. Goto, 243–250, Copyright (2010) with permission from Elsevier.[24]

decreased when ME3 was used as a carrier. The skin permeabilities to ACV were estimated based on the transdermal fluxes. The flux values obtained from the transdermal penetration profiles of ACV for various MEs (data not shown) indicate that MEs prepared with mass ratios of 3:1 (ME1) and 3:2 (ME2) are efficient for transdermal permeation of ACV based on the skin permeation parameters. However, transdermal delivery considerably decreased when the MEs (ME3) were prepared with a high concentration of Span 80 (ME3). The interfacial area of the IL droplets may be responsible for this difference. Generally, a larger stable interfacial area of the dispersed phase at a specific surfactant concentration promotes both transdermal and topical delivery.[28,29] We have measured the mean diameter of MEs in the absence of drug molecules and found 26.5 nm, 25.0 nm and 16.3 nm for ME1, ME2 and ME3, respectively. These results clearly indicated that ME1 and ME2 have larger interfacial areas than ME3, which could be reflected in the increased transdermal permeation. The HLB values of the surfactant mixtures indicate that the thermodynamic activities of ME1 and ME2 in IPM are higher than those of ME3 because the combined solubilities of the surfactants of ME1 and ME2 in IPM are less than those of ME3. The drug permeation capabilities of the various MEs through the skin could be affected by the differences in their thermodynamic activities.[30] ME2 showed the best performance as a transdermal drug carrier. These findings confirm that an optimized amount of Tween 80 is useful to regulate the IL droplet interfacial area required for preparing efficient MEs for drug delivery, which is a common characteristic observed in MEs.

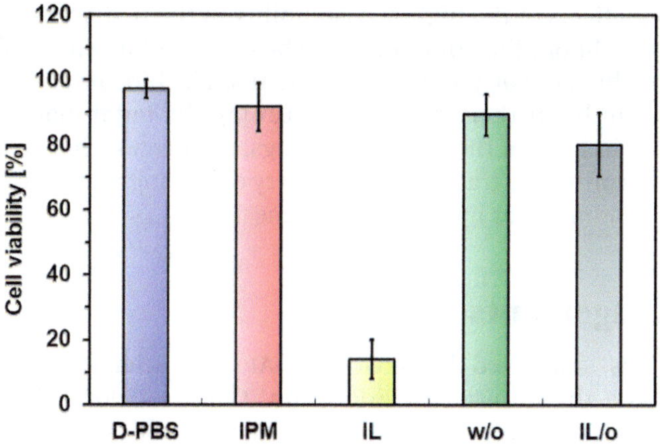

Figure 15.6 Cytotoxicity of various formulations according to the MTT assay (n = 3). The MEs contained 20 wt% surfactants (Tween 80 : Span 20 = 3 : 2), whereas 4 wt% IL and water were used in the IL/o and w/o MEs, respectively. Reprinted from *International Journal of Pharmaceutics*, **400** (1–2), M. Moniruzzaman, M. Tamura, Y. Tahara, N. Kamiya, M. Goto, 243–250, Copyright (2010) with permission from Elsevier.[24]

15.3.6 *In vitro* Cytotoxicity Studies

The cytotoxic effects of IL/Span 20/Tween 80/IPM (IL/o) MEs were compared with those of water/Tween 80/Span 20/IPM (w/o) MEs and D-PBS to investigate their relative safety. Notably, substitution of 4 wt% IL by 4 wt% water is the only difference between IL/o and w/o MEs. The cell viability was the lowest when pure IL was used (Figure 15.6). However, the MEs with 4 wt% IL showed more than 80% cell viability compared with the control samples. The trend is consistent with many necessary and useful chemicals being used in the pharmaceutical industry. For example, dimethyl sulfoxide (DMSO), which is a common chemical enhancer for transdermal drug delivery, is believed to be safe at low concentrations (up to 10%), although high concentrations of DMSO considerably increase the system cytotoxicity, which generates erythema and wheals of the SC.[30] Notably, the cell viability of the IL/o MEs is comparable with those of w/o ME and IPM. Based on these cytotoxicity measurements, we suggest that this newly developed IL/o ME is a safe transdermal drug delivery system, especially for sparingly soluble drugs.

15.4 Conclusion

We have shown the potential of IL/o MEs for effective transdermal delivery of drugs that are insoluble or poorly soluble in water and other conventional organic solvents. MEs prepared with a 3 : 2 ratio of Tween 80 to Span 20 showed high solubility of ACV and high permeation into YMP skin. The improvement of ACV diffusion using this new IL/o ME could be because of the combined effects of the high ACV solubility in the IL disperse phase and the presence of lipophilic components. The results of a cell viability study confirmed the biocompatibility of the IL/o MEs. The biocompatibility of the formulation can be further enhanced by tuning the tailor-made attributes of ILs. Finally, in light of the present findings, we suggest that IL/o MEs are promising for controlled transdermal delivery of sparingly soluble drugs that are difficult to administer by other approaches.

Acknowledgements

This work was supported by a Grant-in-Aid for Scientific Research (S) 16H06369 from the Ministry of Education, Culture, Sports, Science, and Technology of Japan. H. Mahmood would like to thank Universiti Teknologi PETRONAS, Malaysia for financial support through the graduate assistantship scheme.

References

1. A. Rösler, G. W. Vandermeulen and H.-A. Klok, *Adv. Drug Delivery Rev.*, 2012, **64**, 270–279.
2. S. Heuschkel, A. Goebel and R. H. Neubert, *J. Pharm. Sci.*, 2008, **97**, 603–631.
3. A. Kogan and N. Garti, *Adv. Colloid Interface Sci.*, 2006, **123**, 369–385.
4. S. Gupta and S. Moulik, *J. Pharm. Sci.*, 2008, **97**, 22–45.
5. V. Jaitely, A. Karatas and A. T. Florence, *Int. J. Pharm.*, 2008, **354**, 168–173.
6. M. Moniruzzaman, N. Kamiya and M. Goto, *J. Colloid Interface Sci.*, 2010, **352**, 136–142.
7. M. Moniruzzaman, Y. Tahara, M. Tamura, N. Kamiya and M. Goto, *Chem. Commun.*, 2010, **46**, 1452–1454.
8. M. S. Rajabi, M. Moniruzzaman, H. Mahmood, M. Sivapragasam and M. A. Bustam, *J. Mol. Liq.*, 2017, **227**, 15–20.
9. J. L. Shamshina, P. S. Barber and R. D. Rogers, *Expert Opin. Drug Delivery*, 2013, **10**, 1367–1381.
10. J. Pernak, K. Sobaszkiewicz and I. Mirska, *Green Chem.*, 2003, **5**, 52–56.
11. M. Sivapragasam, M. Moniruzzaman and M. Goto, *Biotechnol. J.*, 2016, **11**, 1000–1013.
12. N. Adawiyah, M. Moniruzzaman, S. Hawatulaila and M. Goto, *MedChemComm*, 2016, **7**, 1881–1897.
13. G. Cevc, *Adv. Drug Delivery Rev.*, 2004, **56**, 675–711.
14. H. Gao, J. Li, B. Han, W. Chen, J. Zhang, R. Zhang and D. Yan, *Phys. Chem. Chem. Phys.*, 2004, **6**, 2914–2916.
15. D. Chakrabarty, D. Seth, A. Chakraborty and N. Sarkar, *J. Phy. Chem. B*, 2005, **109**, 5753–5758.
16. Y. a. Gao, J. Zhang, H. Xu, X. Zhao, L. Zheng, X. Li and L. Yu, *ChemPhysChem*, 2006, **7**, 1554–1561.

17. J. Eastoe, S. Gold, S. E. Rogers, A. Paul, T. Welton, R. K. Heenan and I. Grillo, *J. Am. Chem. Soc.*, 2005, **127**, 7302–7303.
18. H. Chen, X. Chang, T. Weng, X. Zhao, Z. Gao, Y. Yang, H. Xu and X. Yang, *J. Controlled Release*, 2004, **98**, 427–436.
19. Y. Tahara, S. Honda, N. Kamiya, H. Piao, A. Hirata, E. Hayakawa, T. Fujii and M. Goto, *J. Controlled Release*, 2008, **131**, 14–18.
20. M. Katoh, F. Hamajima, T. Ogasawara and K.-i. Hata, *J. Toxicol. Sci*, 2009, **34**, 327–334.
21. R. H. Neubert, U. Schmalfuß, R. Wolf and W. A. Wohlrab, *J. Pharm. Sci.*, 2005, **94**, 821–827.
22. U. Schmalfuss, R. Neubert and W. Wohlrab, *J. Controlled Release*, 1997, **46**, 279–285.
23. S. K. Kim, E. H. Lee, B. Vaishali, S. Lee, Y.-k. Lee, C.-Y. Kim, H. T. Moon and Y. Byun, *J. Controlled Release*, 2005, **105**, 32–42.
24. M. Moniruzzaman, M. Tamura, Y. Tahara, N. Kamiya and M. Goto, *Int. J. Pharm.*, 2010, **400**, 243–250.
25. K.-M. Park and C.-K. Kim, *Int. J. Pharm.*, 1999, **181**, 173–179.
26. G. Kantarcı, I. Özgüney, H. Y. Karasulu, S. Arzık and T. Güneri, *AAPS PharmSciTech*, 2007, **8**, 75–81.
27. F. Pattarino, M. Carlotti and M. R. Gasco, *Pharmazie*, 1994, **49**, 72.
28. A. C. Sintov and L. Shapiro, *J. Controlled Release*, 2004, **95**, 173–183.
29. G. Ktistis and I. Niopas, *J. Pharm. Pharmacol.*, 1998, **50**, 413–418.
30. M. I. Afouna, S. C. Mehta, A. H. Ghanem, W. I. Higuchi, E. R. Kern, E. De Clercq and H. H. El-Shattawy, *J. Pharm. Sci.*, 1998, **87**, 917–922.

CHAPTER 16

Bioactivity of Ionic Liquids

RICARDO FERRAZ*[a,b], CÁTIA TEIXEIRA[b], PAULA GOMES[b] AND CRISTINA PRUDÊNCIO[a,c]

[a]Ciências Químicas e das Biomoléculas/Centro de Investigação em Saúde e Ambiente (CISA), ESS—Escola Superior de Saúde do Porto do Instituto Politécnico do Porto, Rua Dr. António Bernardino de Almeida, 400 4200-072, Porto, Portugal; [b]LAQV-REQUIMTE, Departamento de Química e Bioquímica Faculdade de Ciências, Universidade do Porto, Rua do Campo Alegre 687, P-4169-007 Porto, Portugal; [c]I3S – Instituto de Investigação e Inovação em Saúde, Universidade do Porto Rua Alfredo Allen 208, 4200-135 Porto, Portugal
*E-mail: ricardoferraz@eu.ipp.pt

16.1 Introduction

> *"The sky is the limit: Ionic liquids are well-known as modern solvents, but they can be so much more!"*

Professor Ralf Giernoth in "Task-specific ionic liquids"[1]

Ionic liquids (ILs) were discovered a century ago, but it was only three decades ago that they started to catch the attention of the scientific community. ILs are defined as organic salts composed of an organic cation and an organic or inorganic anion[2,3] that are stable below their melting point. These salts are known by their remarkable physical and chemical properties[4–6] and

Smart Materials No. 28
Ionic Liquid Devices
Edited by Ali Eftekhari

Published by the Royal Society of Chemistry, www.rsc.org

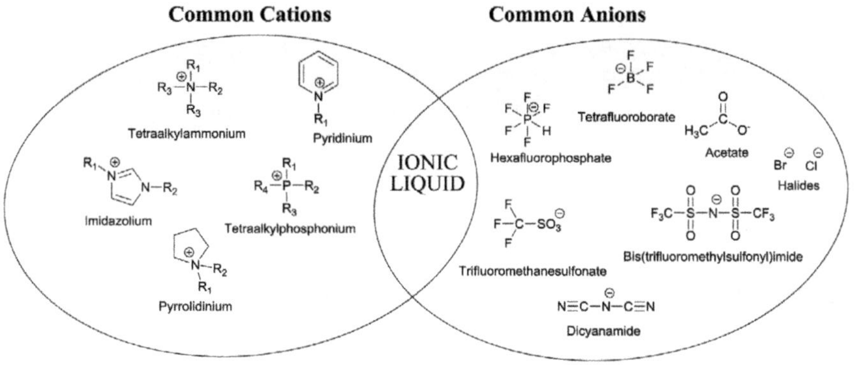

Figure 16.1 Some examples of cations and anions commonly used to produce ILs.

by their tuneability, in the sense that they can be tailored to exert the desired property by choosing the right set of cations and anions.[4]

Typical organic cations used in ILs are imidazolium, pyridinium, ammonium, phosphonium and pyrrolidinium[7–10] (Figure 16.1) while the most typical organic and inorganic anions are halides, tetrafluoroborate, hexafluorophosphate, bis(trifluoromethylsulfonyl)imide, acetate and dicyanamide (Figure 16.1).[4,11] It is therefore easy to perceive that different combinations of the broad variety of cations and anions available may lead to a huge number of different ILs waiting to be tested in diverse applications.[7,11]

ILs discovery occurred when Paul Walden reported the physical properties of ethylammonium nitrate $[EtNH_3][NO_3]$ (m.p. 13–14 °C). This compound was produced by the neutralization of ethylamine with concentrated nitric acid.[12] Paul Walden made a very detailed description of this compound and clearly reported it as a new class of liquids, but his work did not attract much attention and interest at the time.[12] However, nowadays, this discovery is widely recognized as the cornerstone on which ILs-based research was built, leaving an important legacy for the scientific community. Actually, it is from Walden's pioneering work that emerged the most well-known definition of ILs: "ILs are most commonly defined as materials that are composed of cations and anions which melt at or below 100 °C". Although this temperature value does not have any physical or chemical significance, it has endured until today, and it was only recently that it started to be questioned.[13–16]

With the advance of the frontiers of knowledge in this area, modern ILs have been produced for a variety of applications. In addition, as ILs were associated early on with several important properties, the scientific community has been witnessing a steady rise in the number of publications and patents in this field.[3,12,17–19] It is widely known that one of the first applications of ILs was as green-solvents, which contributed to the increase of performance and security in a wide range of chemical procedures.[3,12,18–23]

Up to now, and based on their properties, three different generations of ILs are known (Figure 16.2):[4,24,25] (i) the first (1st) generation is based on ILs

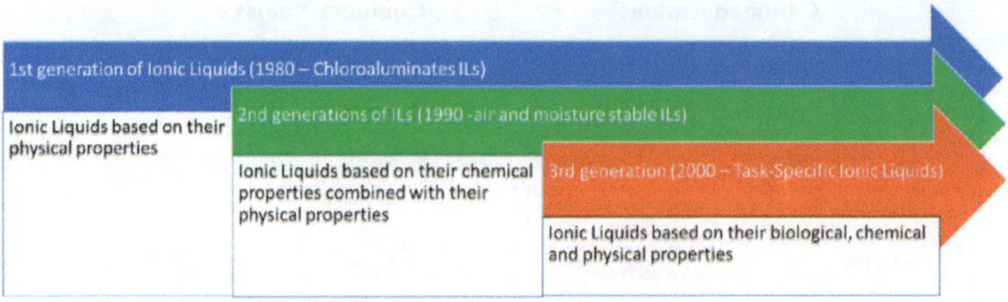

Figure 16.2 The three generations of ILs established thus far.

physical properties like melting point, density, viscosity, thermal stability, conductivity and hydrophobicity;[4,24,25] (ii) the second (2nd) generation is characterized by the possibility to combine physical and chemical properties (chemical reactivity, coordination, and solvation);[4,24,25] and (iii) around the beginning of this century, the third (3rd) generation appeared, corresponding to the so-called task-specific ILs, which combine biological with physical and chemical properties, and sometimes include active pharmaceutical ingredients (APIs) in their composition.[4,24,25]

From the above, one can easily predict that a wide panoply of task-specific ILs can be designed for diverse applications in health and life sciences. In this connection, promising studies on bioactive ILs have been emerging over the past few years, some of which are highlighted next.

16.2 Ionic Liquids as Antimicrobial Agents

One of the earliest types of bioactivity ascribed to ILs regards their antimicrobial potential. Interestingly, this potential was originally discovered in the course of studies aimed at establishing whether ILs were "really green", by assessing their toxicity towards several types of microorganisms or other cells; several of these studies have enabled the discovery of ILs capable of displaying excellent antimicrobial activity.[26–29]

Antimicrobial agents are a broad range of compounds able to kill or inhibit the growth of microorganisms. The understanding and development of antimicrobials has increased since 1800s, with antibiosis, disinfection, preservation, sterilization and environmental control. Unfortunately, since 1970s, the unreasonable use of antibiotics led to the growth of resistance profiles and selection towards multidrug resistance, threatening the end of the "antibiotic era".[30,31] Therefore, alternative ways to fight bacterial infections are urgently needed, which makes the discovery of antimicrobial ILs particularly important. It has been observed that solutions of antimicrobial ILs retain potent antimicrobial activity, indicating that the mechanistic basis for their biocidal properties could be attributable to one or both ions. Hence, the dual nature of ILs can be exploited by synthesizing ILs that contain both cation and anion with inherent antimicrobial activity.[27,30]

Antimicrobial ILs in aqueous media share structural and mechanistic analogies with established cationic biocides and surfactants such as quaternary ammonium compounds (QACs, quaternary salts or "quats").[30,32,33] The common features of an inherently charged hydrophilic head group with one or more hydrophobic "tails", observed both in common ILs and cationic surfactants, such as benzalkonium chloride (BAC, a mixture of alkylbenzyl(dimethyl)ammoniums), cetylpyridinium chloride/bromide (CPC/B) or cetrimonium bromide (cetyltrimethylammonium bromide, CTAB), suggest a common mode of action. These similarities extend towards a tendency for ILs to aggregate in solution, forming amphiphilic micelles and displaying surface activity as characteristically dictated by increasing lipophilicity.[30,32–34]

ILs antimicrobial activities have been studied for decades using standard dilution tests.[30,35] Since then, typical characteristics and factors that influence antimicrobial activity have been identified. Generally, ILs can display broad activity spectra, affecting Gram positive and Gram negative bacteria as well as mycobacteria and fungi.[30,35] However, antimicrobial activity is governed by a balance between key factors, which are defined by the molecular architecture of each IL.[30]

In general, antimicrobial activity of ILs increases with increasing lipophilicity. This effect has been quantified for pyridinium salts by demonstrating that calculated ($\log P$) and experimental ($\log k_w$) octanol–water partition coefficients, well-known lipophilic descriptors, showed highly significant correlation ($P < 0.001$ for a selection of Gram positive and Gram negative organisms) with antimicrobial activity following quantitative structure–activity relationship (QSAR) analysis.[30,36]

A study performed by Yu and Nie[37] demonstrated that the growth of some bacteria (*Escherichia coli*, *Staphylococcus aureus* and *Bacillus subtilis*) may be inhibited in the presence of ILs containing a halide (Figure 16.3). The observed inhibition of bacterial growth was related to ILs structure, as toxicity increased along with the length of the alkyl chain.

Some classic cations like imidazolium, pyridinium, pyrrolidinium, piperidinium, and ammonium were shown to inhibit the growth of pathogenic and nonpathogenic bacteria and fungi.[17,30] From the medical and pharmaceutical perspectives, drug resistance is a pressing issue with an urgent need for antimicrobial innovation.[17,38–40] In this context, targeted attempts to develop antimicrobial ILs for medical purposes have been made.[41] Indeed,

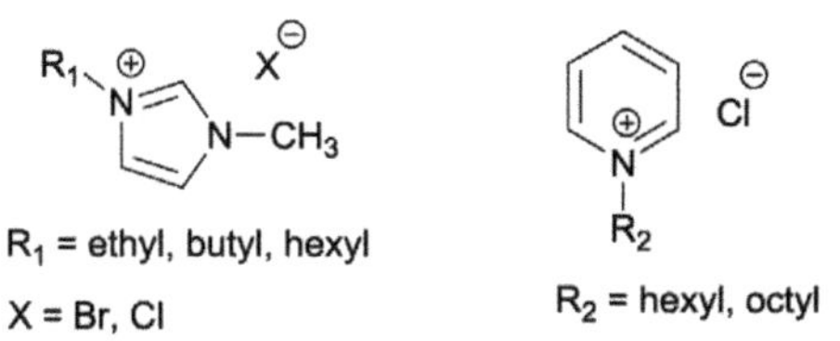

Figure 16.3 Structure of the antimicrobial ILs studied by Yu and Nie. For further reading please see ref. 37.

comparison of activity of ampicillin-carrying ILs with that of sodium ampicillin clearly showed possible advantages of using ILs as antimicrobial agents.[41]

Another example is the case of 1-alkyl-3-methylimidazolium fumarates that showed antimicrobial potential against Gram-positive *Bacillus subtilis*, Gram-negative *Escherichia coli*, and yeast *Saccharomyces cerevisiae*[42] suggesting that the ILs may have potential applications in modern biotechnology.[17,42] The antimicrobial properties of phosphonium and ammonium ILs have also been studied.[17] For example, ammonium cations combined with azolate anions, and hydroxylammonium-based ILs, were found to have antibacterial and antifungal properties, with the latter actually presenting activity against relevant human pathogens such as *Staphylococcus aureus*, *Salmonella typhi*, and *Vibrio cholera*.[17,43,44] Several studies on bioactive phosphonium-based ILs have also been reported, such as the work by Santos and co-workers focused on the toxicity of those ILs within the context of the food industry.[45] In addition, it was found that triphenylamine phosphonium ILs can spontaneously adopt nanostructures displaying activity against Gram positive bacteria like *S. aureus*.[17,46] Moreover, some studies on diphosphonium-based ILs showed these to exhibit a broad spectrum of antimicrobial activity against ocular pathogens.[17,47]

16.2.1 Anti-biofilm Ionic Liquids

A biofilm can be described as a community of microbial cells irreversibly bound to a substrate, to an interface, or even to each other when embedded in a matrix of extracellular polymeric substances; when integrated in a biofilm, each microbial species produces and exhibits an altered phenotype with respect to gene transcription and growth rate, as compared to planktonic cells of the same species.[28–30,48–50] The distinct physiological properties of microbial cells in biofilms makes them particularly hard to fight against, given their considerably high growth rate and resistance to the action of antimicrobial agents.[28–30,35,51] As such, health risks are increased for biofilm-related infections, as treatments are more difficult to apply and have lowered success rates.[26,28–30,48–50,52] Indeed, treatments are sometimes useless because the action of the antimicrobial agents is prevented due to the barrier formed by the *stratum corneum* and the biofilm.[51]

To avoid some of these issues posed by biofilm-related infections, the application of ILs as anti-biofilm agents has been investigated.[28–30,48,51,52] This led to discovery of ILs showing interesting inhibitory results against resistant biofilm-forming microorganisms.[17,28,29] For instance, low concentrations of [C16MIM][Cl] inhibited growth of multidrug-resistant *Candida tropicalis*,[17,48] whereas 1-alkyl-3-methylimidazolium chlorides, bromides, and iodide have shown activity against both planktonic and biofilm bacteria and fungi.[17,28,29,52]

Based on two human bacterial pathogens, *Pseudomonas aeruginosa* and *Salmonella enterica*, Zakrewsky *et al.* have analyzed the effect of ILs on biofilm disruption and drug action (transport); of the 12 ILs examined, the choline-geranate was revealed as a multi-purpose IL, with antimicrobial activity, decreased toxicity to epithelia and skin, and enhanced effective permeation into the biofilm.[51]

The interest in ILs as anti-biofilm agents goes well beyond health sciences. Biofouling, *i.e.*, contamination of wet surfaces by living organisms, is not only a threat to health but also to industry and economy, from food processing or paper production to the marine industry. For example, marine infrastructures can be seriously damaged by the corrosion of surfaces caused by microorganisms, whereas fouling organisms on ship hulls cause an increase of frictional resistance and, consequently, an increase in expenditures for fuel (that can increase by about 45%).[53] In view of this, Seter *et al.*[54] determined the antifouling properties of the organic salt cetrimonium nalidixate, based on the concept of ILs; this salt was incorporated into a polyurethane coating and complete inhibition of microbial activity was observed, which supports the potential of ILs-based coatings for inhibition of microbial-related corrosion in marine environments.[53]

16.3 Ionic Liquids as Anti-tumor Agents

Cancer is one of the leading causes of mortality and morbidity worldwide, demanding new and alternative treatments that could both avoid some of the unpleasant side effects associated with conventional therapies and be more effective.[4,25,55]

In connection with this, the toxicity of different ILs against a colon adenocarcinoma (CaCo-2) cell line has been reported.[8,9,56–58] Overall, the results demonstrated that ILs could represent a new entry in antitumoral approches[2,25] as innovative task-specific ILs, with tailored properties, might be obtained by the right combination of anion and cation.[25,59–61] This capacity for tuning the physical, chemical and biological properties of ILs is a major advantage, since it enables the tailoring of the cytotoxic profile, as well as of the physicochemical and pharmacological properties necessary to have the desired therapeutic effect *in vivo*.[62]

As summarized by Dias *et al.*[25] (Table 16.1), the research into ILs as anti-tumor agents is a recent topic.[7–10,25,56,57,60,61,63] The studies compiled in Table 16.1 demonstrate a new perspective in cancer therapy, based on application of ILs.[25] Several reported studies highlighted the potential antitumor activity of ILs, and examined their selective toxicity against different human cancer cell lines.[4,25,40,60,61] To this end, the main method used *in vitro* was the MTT assay,[25] which consists of measuring the activity of cellular enzymes that reduce the tetrazolium dye MTT or the similar WST dyes.[64,65] Such *in vitro* assays are very useful as they can be employed for a fast and simple screening of ILs cytotoxicity, enabling prediction of possible effects on humans.[64] Still, the discovery of antitumoral ILs is an emerging field, so complementary studies are needed to establish the value of ILs as anti-neoplastic agents; for example, studies focused on identification of basal cellular and subcellular targets of ILs in human cancer cell lines are of chief importance to unravel the ILs mode of antitumoral action and, hence, to improve the design of next-generation anti-cancer ILs.[25,65]

Table 16.1 Examples of literature reports on *in vitro* assays of ILs against human cancer cell lines; reproduced from *ChemMedChem*, The Anticancer Potential of Ionic Liquids, A. R. Dias, J. Costa-Rodrigues, M. H. Fernandes, R. Ferraz, C. Prudêncio, John Wiley and Sons, © 2017 Wiley-VCH Verlag GmbH & Co. KGaA, Weinheim.[25]

Ref.	Publication year	Cell lines	Cell viability assay	Ionic liquids tested
57	2007	Colon carcinoma	MTT assay	Imidazolium based cations with different side chains, choline, 1-butyl-2,3-dimethylimidazolium cation, tetrahexyldimethyl-guanidinium cation, trihexyl-tetradecylphosphonium cation and trioctyl-methylammonium cation combined with different type of anions: BF_4, PF_6, dicyanoamide, acesulfame, saccharine and NTf_2
66	2007	Cervical cancer	MTT assay	Ethyl-, butyl-, octyl-, benzyl-, and allyl-substituted 1-alkyl-3-methylimidazolium, alkylpyridinium, *N*-alkyl-*N*,*N*-dimethyl-*N*-(2-hydroxyethyl)ammonium (choline derivatives) and alkyl-triethylammonium salts
67	2009	Breast cancer	MTT assay	Pyridinium, piperidinium, pyrrolidinium or imidazolium cations with various alkyl chain lengths, anions bromide, bis(trifluoro-methanesulfone)imide (NTf2), trifluoromethylsulfonate (TfO), or nonafluoromethylsulfonate (NfO) Three additional new hydrophobic and task-specific ILs: 1-(3-cyanopropyl)-1-methylpiperidinium $[(CF_3SO_2)_2N]$, 1-methyl-1-[4,5-bis(methylthio)-pentyl]piperidinium] $[(CF_3SO_2)_2N]$ and 1-methyl-1-[4,5-bis(methylthio)-pentyl]pyrrolidinium $[(CF_3SO_2)_2N]$
61	2009	Leukemia, non-small lung cancer, CNS cancer, melanoma, breast cancer, ovarian cancer, renal cancer, prostate cancer cells	MTT assay	Seven phosphonium and ammonium based ILs
56	2009	Colon carcinoma	MTT assay	Eighty ILs from different classes of cations and anions based on imidazolium, guanidinium, ammonium, phosphonium, pyridinium and pyrrolidinium

60	2010	Leukemia, non-small lung cancer, CNS cancer, melanoma, breast cancer, ovarian cancer, renal cancer, prostate cancer cells	MTT assay	Imidazolium based ILs with varying side chain length/functionalized side chains and different anions: $[(CF_3SO_2)_2N]$, BF_4, PF_6, trifluorotris(1,1,2,2,2-pentafluoroehyl)phosphate, octyl sulfate and chloride
62	2012	Brain cancer cells, and human embryonic kidney	MTT assay	Four ammonium and imidazolium based ILs
68	2012	Melanoma, neuroblastoma, breast adenocarcinoma, and epidermoid carcinoma	MTT assay	ILs based on betulinic acid with cations as choline, and benzalkonium
69	2013	Hepatocarcinoma	WST-1 assay	1-Methyl-3-octyimidazolium chloride
63	2015	Breast cancer, prostate cancer, colon cancer, Hepatocellular carcinoma, osteosarcoma	MTT assay	Ampicillin salts as ILs based on cations tetraethylammonium, trihexyltetradecylphosphonium, cetylpyridinium, choline, 1-ethyl-3-methylimidazolium, and 1-hydroxy-ethyl-3-methylimidazolium
70	2015	Hepatocarcinoma	WST-8 assay	1-Methyl-3-octyimidazolium bromide
71	2015	Colon carcinoma	MTT assay	Salicylic salts as ILs based on imidazolium cations

16.4 Modes of Action and Structure–Activity Relationships of Bioactive Ionic Liquids

16.4.1 Modes of Antimicrobial Action

Identification of biological pathways that may be the targets of ILs is one of the major challenges for the development of ILs within the context of the pharmaceutical industry.[13] Although the mechanisms of action (MOA) through which ILs exert their biological activities are not yet understood,[30] data gathered from several studies permitted the observation that some structural features are determinant for ILs biological activities, such as[25] (i) the size of the side chain of the cations; (ii) the presence or absence of a functional group; (iii) the nature of the cation; and (iv) the nature of the anion.[25,64] Based on structure–activity relationships (SAR) established for some bioactive ILs, putative MOAs have been advanced, as follows.

It has been mentioned previously that the shared structure between some antimicrobial ILs and cationic biocides and surfactants[30,32] corresponds to a charged hydrophilic head group and hydrophobic tails, which confer to ILs the tendency to form micelles[30,32] and to display surface activity.[30,34,72] These properties are both related to the increase of lipophilicity,[30,73] which could be indicative of a common mechanism of action.[30] Based on these similarities between antimicrobial ILs, surfactants and cationic biocides, some mechanisms of action have been proposed, such as the possible disruption of cell membrane integrity.[30] Yet, results regarding electrostatic properties are not coherent; for example, Pernak and co-workers carried out studies to understand the effect of electrostatic activity of ILs. One study was on choline-based antimicrobial ILs[30,74] and the other was on phosphonium-based ILs;[75] the results were not coherent among each other.[30,75] Another property that has been related to the biological activity of ILs is the surface tension as described by Jungnickel and collaborators.[72,73] Their work with 1-alkyl-3-methylimidazolium-based ILs showed that the "chain length" and "anion effects" upon the surface tension are directly proportional to the ILs antimicrobial efficacy.[30,72,73]

Additional MOAs for antimicrobial ILs have been proposed. For instance, quaternary ammonium compounds have been claimed to exert their action by coagulating cytoplasmic constituents, and, consequently, lead to denaturation of enzymes; one example is that of cetrimonium bromide, which has been identified as an inhibitor of membrane-bound enzymes.[30] A different perspective was presented by Ferraz *et al.*[63] for ILs based on ampicillin as the anion.[63] In this work, the authors showed the importance of pairing the classic antibiotic with the right cation, as stronger and more stable cation–anion interactions lead to more efficient ion pairing, which permitted the final pharmaceutical salt to act as lipophilic drug carrier, while having improved water solubility, permeability, and bioavailability (Figure 16.4). In other words, in this particular report, the enhanced antimicrobial properties

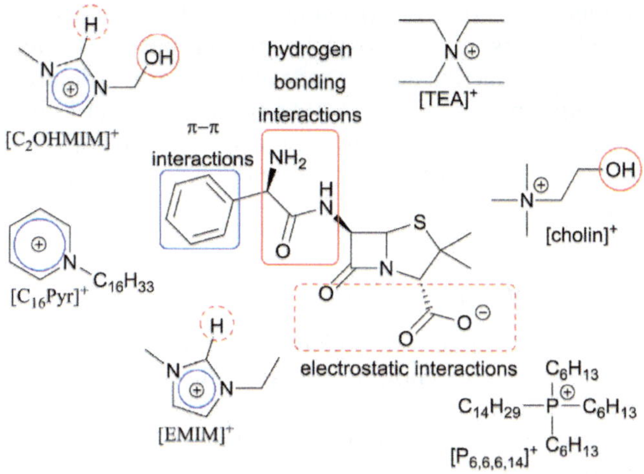

Figure 16.4 Structures of ampicillin salts and their possible cation–anion interactions. Reproduced with permission from R. Ferraz *et al.*, Antitumor Activity of Ionic Liquids Based on Ampicillin, ChemMedChem, John Wiley and Sons, © 2015 Wiley-VCH Verlag GmbH & Co. KGaA, Weinheim.[63]

of the ampicillin-based salts described may be more related to improved pharmacokinetics, rather than to any kind of MOA different from that of ampicillin itself.

16.4.2 Structure–Activity Relationships

16.4.2.1 Effect of Cation Side-chain Length and Functionalization

Numerous studies have been performed to establish SARs and understand how ILs act in biological processes. One of the characteristics that has been repeatedly observed is the length of the cation side chain, since this may be related to the toxic effects exerted by several ILs.[25] Indeed, several halides of imidazolium, ammonium, phosphonium, and pyridinium as head groups linked to alkyl "tails" ranging from 1 to 18 carbons displayed toxicity towards microbial cells that increased along with the length of the alkyl side chain.[9,25,65]

The length of the cation's side chain has also been reported to influence the antitumoral activity of ILs.[10,25] For example, the pioneering studies of the Malhotra and Kumar groups on the antitumor activity of imidazolium-based ILs[25,60,61] showed that anti-cancer activity was significantly improved with the increase of the alkyl chain length, ranging from C_8 to C_{12}, while the cytotoxicity towards healthy human cell lines remained low in most of the cases.[60]

Besides the length of the cation's side chain, the presence or absence of polar functional groups, like hydroxyl, has been reported to affect the bioactivity of ILs.[9,25] A study carried out by Frade and co-workers demonstrated that the hydroxyl group present in C_2OHMIM and C_5OHMIM (1-(2-hydroxyethyl)-3-methyimidazolium)-based ILs, having tetrafluoroborate (BF_4^-) or hexafluorophosphate (PF_6^-) as counter-ions, contributes to enhancing the viability of HT-29 cells.[57]

16.4.2.2 *Effects of Specific Cations or Anions*

Wang and co-workers studied the activity of different cations (imidazolium, ammonium, phosphonium, choline and pyridinium) in HeLa cells,[25,66] showing that hydrophobic and aromatic cations (pyridinium and imidazolium) have more activity than the highly polar choline- and ammonium-based ILs.[25,66] Still, studies by Kumar and co-workers, among other similar ones, suggest that the effect of changing the cationic head group on biological and toxicological properties of ILs is smaller than that of the side chain length.[25,64,65,67]

Based on literature reports, it is more difficult to understand the effects of the anion on the biological activity of ILs.[25,64] For example, Frade *et al.*[9,56,57] investigated different classes of anions, and demonstrated that, regardless of the cation used, the effects on toxicity for both HT-29 and Caco2 cells were always lower when the $[(CF_3SO_2)_2N]^-$ anion was used in comparison to BF_4^-, PF_6^- or dicyanoamide[9,56,57] (Table 16.1).

Kumar and co-workers investigated the activity of ILs where different anions were combined with different cationic head groups against human MCF-7 breast cancer cells;[60,61,65,67] they concluded that, for a same cation, the hydrophilic and small anions bromide, tetrafluoroborate and hexafluoroarsenate have less effect on cell viability than the highly fluorinated, bulky and hydrophobic Tf_2N, TfO or NfO (Table 16.1).

In contrast with the above reports, other authors carried out studies where it was demonstrated that the anion had no effects on the biological activity of imidazolium-based ILs.[25,64,65] Hence, overall, no common trend has been observed regarding the anion's influence on the cytotoxicity of ILs,[14,47] and so the matter remains under debate. Additional studies are therefore urgently needed to shed some light onto the SAR and MOA of bioactive ILs, which would greatly contribute to the development, design and synthesis of more efficient antimicrobial- and anticancer-specific ILs.[17]

16.5 Ionic Liquids as Active Pharmaceutical Ingredients (APIs)

Historically, ILs have been introduced to replace toxic organic solvents to help industry in general, and the pharmaceutical industry in particular, for the production of compounds in a more sustainable way.[20] Nevertheless, the toxicity and ILs safety profiles were sometimes inconsistent with the tightly regulated manufacturing practices to which pharma companies are subjected.[20] As such, although 3rd generation task-specific ILs represent a new entry in drug development, they remain widely put aside by the pharmaceutical industry.[20] Consequently, a better understanding of the therapeutic value of ILs is of utmost importance to induce a paradigm shift in this context.

Most APIs currently used are in the solid form, which may bring about polymorphism-related problems. The conversion of a drug into a different crystalline form may change relevant properties (melting point, solubility, bioavailability) leading to unpredictable changes in the drug's pharmacokinetics.[17,76–78] From a pharmaceutical point of view, and based on these properties, the formulation of APIs in the form of ILs (API-ILs) might eliminate liabilities related to polymorphism, while contributing to the improvement of solubility and/or bioavailability.[17,25,79] In addition, because the properties of ILs can be easily tuneable, tailoring their properties through the appropriate combination of cations and anions might improve their toxicity and lipophilicity, among other features.[4,13,17,24,27,80–82]

Another noteworthy aspect is the huge costs incurred by the pharmaceutical industries to support all steps of the drug development pipeline, from early pre-clinical research to clinical trials, patenting, licensing by the appropriate agencies, and marketing.[20,83] Also in this context, API-ILs can emerge as a cost-effective alternative, as most of the methods associated to the production of ILs are typically not very expensive.[84]

Efforts towards development of ILs suitable for the pharmaceutical industry started in the late-1990s, initially with the focus not on novel API-ILs but on the use of ILs as alternative formulations where one of the ions of the classical formulation was replaced by a safer or natural surrogate.[20] There are several examples, such as the work of Davis and collaborators who, in 1998, produced an alkylated derivative of miconazole (antifungal drug) that was combined with hexafluorophosphates, demonstrating that it was possible produce IL from a biologically active molecule.[20,85] In 2004, the same group produced ILs from saccharin and acesulfame (sweeteners), through combination with different organic cations, demonstrating that is possible to substitute a fluorine anion and that it is possible to have a biological characteristic coupling as IL.[20,86]

From 2007 onwards, we have witnessed the rise of API-ILs as a potential new class of therapeutic agents devoid of polymorphism issues, possibly displaying improved production costs, physico-chemical properties, delivery, release rates, handling, and efficacy.[4,20,24] This has been accompanied by a steady increase in the number of publications in this area. For instance, Hough *et al.* reported the production of lidocaine docusate, where the anesthetic and emollient properties of lidocaine and docusate, respectively, were combined.[24,87,88] Stoimenovski and co-workers have eliminated the polymorphism issue in propantheline bromide (antimuscarinic drug) through liquefaction of the drug by producing an IL based propantheline.[20,81,89] In 2012, Ferraz *et al.* demonstrated the possibility of reversing microbial resistance to classical antibiotics through their formulation of API-ILs;[41] in this case, ampicillin-based ILs were reported with improved solubility and lipophilicity,[80] while being active against ampicillin-resistant strains.[27] Moreover, the same API-ILs may be repurposed to act as antitumor agents.[63] More recently, the same authors have shown that API-ILs may be used for the "recycling" of classical antimalarials by synthesizing primaquine-based ILs that were active against three different stages of parasite.[90]

All these findings reinforce the importance of additional studies to contribute to a fuller understanding of how API-ILs can be tuned to exhibit high potency and drug-like properties. In addition, it will be necessary to update their regulation and, thus, facilitate their introduction into broad use in the market.[13]

16.6 Conclusions and Challenges

During recent years, we have witnessed the unveiling of the biological and therapeutic potential of ILs and, along with all the knowledge that is being thereof produced, more and more questions arise. Still, it is now well established that ILs, and specifically ionic interactions, are crucial for biological properties.

The use of ILs as bioactive agents naturally has advantages and disadvantages (Figure 16.5). Moreover, if a highly valuable API-IL candidate is successfully identified after all stages of clinical development, regulation is needed not only by the usual institutions, such as the European Medicines Agency or the Food and Drug Administration Agency, but also in regard to intellectual property protection and patenting. Aspects amenable to be

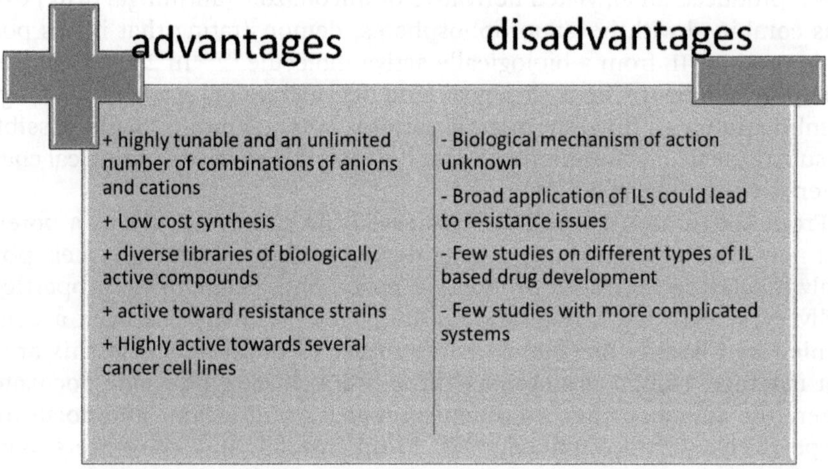

Figure 16.5 Advantages and disadvantages of the use of ILs as bioactive agents.

patented must be clearly identified, *i.e.*, it must be established what is actually patented, if the API-IL can be patentable as a new therapeutic agent or as a new formulation, or if only its clinical properties should be patented; moreover, since API-ILs are salts amenable to dissociation in solution, it must be defined whether it is the salt as the non-dissociated ion pair or/ and its ions that are patentable.[13] These regulations are fundamental for the pharmaceutical industry to analyze the cost/benefit of developing an API-IL as a new medicine.[13]

In terms of research on bioactive ILs a long road remains to be travelled. For instance, studies towards identification of mechanisms of biological action, or of synergistic effects of ILs ions in neutral, charged, and partially ionized forms are needed. The improvement of the properties of ILs to make them fully suited for biological and therapeutic applications can only arise from a collaborative effort between experts from distinct science and technology fields, who will certainly uncover even more striking possibilities in the near future. In fact, many *in vitro* and *in vivo* studies (with rodents) have started to appear,[17,91] but a major challenge resides in the establishment of reliable *in silico* methods, which are of chief importance in modern drug development. Although several *in silico* studies on physico-chemical properties of ILs have been reported,[92] works where molecular docking and molecular dynamics have been applied to the design of bioactive ILs are non-existent, apart from a few studies on the toxicity profile of ILs.[93]

The challenges posed by the search for useful bioactive ILs are not confined to the arena of research and development in health and life sciences. Besides reinforcing the knowledge of scholars and industrial experts in order to get a better understanding and higher familiarity with ILs, it is important to introduce the study of this interesting class of compounds, and of their applications, in the schedules of chemistry and related courses. In fact, a paradigm shift must occur in the way the formation of organic compounds is traditionally explained in many academic institutions, as it almost exclusively settles on the making and breaking of covalent bonds, whereas the formation and applications of organic salts, as ILs, is often neglected.[13]

Undoubtedly, ILs represent an extremely rich field which only now we are beginning to get acquainted with. In particular, the development of bioactive ILs for therapeutic, biological and biotechnological applications, although still in its infancy, has already been shown to enclose an enormous potential to be explored.

Acknowledgements

Thanks are due to Fundação para a Ciência e Tecnologia (FCT, Portugal) for funding research unit LAQV-REQUIMTE (ref. UID/QUI/50006/2013). Thanks are also due to "Comissão de Coordenação e Desenvolvimento Regional do Norte (CCDR-N)/NORTE2020/Portugal 2020" for funding through project DESignBIOtechHealth (ref. Norte-01-0145-FEDER-000024).

References

1. R. Giernoth, *Angew. Chem., Int. Ed.*, 2010, **49**, 2834–2839.
2. I. M. Marrucho, L. C. Branco and L. P. N. Rebelo, *Annu. Rev. Chem. Biomol. Eng.*, 2014, **5**, 527–546.
3. H. Mizuuchi, V. Jaitely, S. Murdan and A. T. Florence, *Eur. J. of Pharm. Sci.*, 2008, **33**, 326–331.
4. R. Ferraz, L. C. Branco, C. Prudencio, J. P. Noronha and Z. Petrovski, *ChemMedChem*, 2011, **6**, 975–985.
5. M. E. Heckenbach, F. N. Romero, M. D. Green and R. U. Halden, *Chemosphere*, 2016, **150**, 266–274.
6. A. A. Tietze, P. Heimer, A. Stark and D. Imhof, *Molecules*, 2012, **17**, 4158–4185.
7. L. C. Branco, G. V. S. M. Carrera, J. Aires-de-Sousa, I. L. Martin, R. Frade and C. A. M. Afonso, in *Ionic Liquids: Theory, Properties, New Approaches*, ed. P. A. Kokorin, InTech, 2011, ch. 3.
8. R. F. M. Frade, S. Simeonov, A. A. Rosatella, F. Siopa and C. A. M. Afonso, *Chemosphere*, 2013, **92**, 100–105.
9. R. F. M. Frade, A. Matias, L. C. Branco, N. M. T. Lourenco, J. N. Rosa, C. A. M. Afonso and C. M. M. Duarte, *Ionic Liquid Applications: Pharmaceuticals, Therapeutics, and Biotechnology*, 2010, vol. 1038, pp. 135–144.
10. R. F. M. Frade and C. A. M. Afonso, *Hum. Exp. Toxicol.*, 2010, **29**, 1038–1054.
11. M. C. Bubalo, K. Radosevic, I. R. Redovnikovic, J. Halambek and V. G. Srcek, *Ecotoxicol. Environ. Saf.*, 2014, **99**, 1–12.
12. N. V. Plechkova and K. R. Seddon, *Chem. Soc. Rev.*, 2008, **37**, 123–150.
13. J. L. Shamshina, S. P. Kelley, G. Gurau and R. D. Rogers, *Nature*, 2015, **528**, 188–189.
14. F. M. F. Vallana, R. P. Girling, H. Q. N. Gunaratne, L. A. M. Holland, P. M. McNamee, K. R. Seddon, J. R. Stonehouse and O. Todini, *New J. Chem.*, 2016, **40**, 9958–9967.
15. M. T. Clough, C. R. Crick, J. Grasvik, P. A. Hunt, H. Niedermeyer, T. Welton and O. P. Whitaker, *Chem. Sci.*, 2015, **6**, 1101–1114.
16. M. G. Freire, in *Ionic-liquid-based Aqueous Biphasic Systems: Fundamentals and Applications*, ed. M. G. Freire, Springer, Berlin, Heidelberg, 2016, pp. 1–25.
17. K. S. Egorova, E. G. Gordeev and V. P. Ananikov, *Chem. Rev.*, 2017, **117**, 7132–7189.
18. M. Petkovic, K. R. Seddon, L. P. N. Rebelo and C. S. Pereira, *Chem. Soc. Rev.*, 2011, **40**, 1383–1403.
19. D. R. MacFarlane and K. R. Seddon, *Aust. J. Chem.*, 2007, **60**, 3–5.
20. M. Smiglak, J. M. Pringle, X. Lu, L. Han, S. Zhang, H. Gao, D. R. MacFarlane and R. D. Rogers, *Chem. Commun.*, 2014, **50**, 9228–9250.
21. O. A. Cojocaru, J. L. Shamshina and R. D. Rogers, *Chimi. Oggi*, 2013, **31**, 24–28.
22. H. Rodriguez, K. Bica and R. D. Rogers, *Trop. J. Pharm. Res*, 2008, **7**, 1011–1012.

23. W. L. Hough and R. D. Rogers, *Bull. Chem. Soc. Jpn.*, 2007, **80**, 2262–2269.
24. W. L. Hough, M. Smiglak, H. Rodriguez, R. P. Swatloski, S. K. Spear, D. T. Daly, J. Pernak, J. E. Grisel, R. D. Carliss, M. D. Soutullo, J. H. Davis and R. D. Rogers, *New J. Chem.*, 2007, **31**, 1429–1436.
25. A. R. Dias, J. Costa-Rodrigues, M. H. Fernandes, R. Ferraz and C. Prudêncio, *ChemMedChem*, 2017, **12**, 11–18.
26. S. Y. Choi, H. Rodriguez, H. Q. N. Gunaratne, A. V. Puga, D. Gilpin, S. McGrath, J. S. Vyle, M. M. Tunney, R. D. Rogers and T. McNally, *RSC Adv.*, 2014, **4**, 8567–8581.
27. R. Ferraz, V. Teixeira, D. Rodrigues, R. Fernandes, C. Prudencio, J. P. Noronha, Z. Petrovski and L. C. Branco, *RSC Adv.*, 2014, **4**, 4301–4307.
28. A. Busetti, D. E. Crawford, M. J. Earle, M. A. Gilea, B. F. Gilmore, S. P. Gorman, G. Laverty, A. F. Lowry, M. McLaughlin and K. R. Seddon, *Green Chem.*, 2010, **12**, 420–425.
29. L. Carson, P. K. W. Chau, M. J. Earle, M. A. Gilea, B. F. Gilmore, S. P. Gorman, M. T. McCann and K. R. Seddon, *Green Chem.*, 2009, **11**, 492–497.
30. J. N. Pendleton and B. F. Gilmore, *Int. J. Antimicrob. Agents*, 2015, **46**(2), 131–139.
31. A. Cornellas, L. Perez, F. Comelles, I. Ribosa, A. Manresa and M. T. Garcia, *J. Colloid Interface Sci.*, 2011, **355**, 164–171.
32. M. Blesic, M. H. Marques, N. V. Plechkova, K. R. Seddon, L. P. N. Rebelo and A. Lopes, *Green Chem.*, 2007, **9**, 481–490.
33. J. Pernak, A. Skrzypezak, G. Lota and E. Frackowiak, *Chem.–Eur. J.*, 2007, **13**, 3106–3112.
34. J. Luczak, J. Hupka, J. Thoming and C. Jungnickel, *Colloids Surf.*, 2008, **329**, 125–133.
35. B. F. Gilmore, *Antimicrobial Ionic Liquids*, INTECH Open Access Publisher, 2011.
36. J. Pernak, J. Kalewska, H. Ksycinska and J. Cybulski, *Eur. J. Med. Chem.*, 2001, **36**, 899–907.
37. Y. Yu and Y. Nie, *J. Environ. Prot. Ecol.*, 2011, **2**, 298.
38. M. R. Cole, J. A. Hobden and I. M. Warner, *Molecules*, 2015, **20**, 6466–6487.
39. J. L. Martinez, *Drug Discovery Today*, 2014, **11**, 33–39.
40. K. S. Egorova and V. P. Ananikov, *ChemSusChem*, 2014, **7**, 336–360.
41. R. Ferraz, L. C. Branco, I. M. Marrucho, J. M. M. Araujo, L. P. N. Rebelo, M. N. da Ponte, C. Prudencio, J. P. Noronha and Z. Petrovski, *MedChemComm*, 2012, **3**, 494–497.
42. B. Y. He, G. N. Ou, C. Y. Zhou, M. H. Wang and S. Y. Chen, *J. Chem.*, 2013, 1–7.
43. M. I. Hossain, M. El-Harbawi, Y. A. Noaman, M. A. B. Bustam, N. B. M. Alitheen, N. A. Affandi, G. Hefter and C. Y. Yin, *Chemosphere*, 2011, **84**, 101–104.
44. F. Walkiewicz, K. Materna, A. Kropacz, A. Michalczyk, R. Gwiazdowski, T. Praczyk and J. Pernak, *New J. Chem.*, 2010, **34**, 2281–2289.
45. A. G. Santos, B. D. Ribeiro, D. S. Alviano and M. A. Z. Coelho, *RSC Adv.*, 2014, **4**, 37157–37163.
46. F. Brunel, C. Lautard, F. Garzino, S. Giorgio, J. M. Raimundo, J. M. Bolla and M. Camplo, *Bioorg. Med. Chem. Lett.*, 2016, **26**, 3770–3773.

47. G. A. O'Toole, M. Wathier, M. E. Zegans, R. M. Q. Shanks, R. Kowalski and M. W. Grinstaff, *Cornea*, 2012, **31**, 810–816.
48. V. Z. Bergamo, R. K. Donato, D. F. Dalla Lana, K. J. Z. Donato, G. G. Ortega, H. S. Schrekker and A. M. Fuentefria, *Lett. Appl. Microbiol.*, 2015, **60**, 66–71.
49. D. Y. Hu, X. Li, P. K. Sreenivasan and W. DeVizio, *Clin. Ther.*, 2009, **31**, 2540–2548.
50. M. E. Olson, H. Ceri, D. W. Morck, A. G. Buret and R. R. Read, *Can. J. Vet. Res.*, 2002, **66**, 86–92.
51. M. Zakrewsky, K. S. Lovejoy, T. L. Kern, T. E. Miller, V. Le, A. Nagy, A. M. Goumas, R. S. Iyer, R. E. Del Sesto, A. T. Koppisch, D. T. Fox and S. Mitragotri, *Proc. Natl. Acad. Sci. U. S. A.*, 2014, **111**, 13313–13318.
52. Y. V. Nancharaiah, G. K. K. Reddy, P. Lalithamanasa and V. P. Venugopalan, *Biofouling*, 2012, **28**, 1141–1149.
53. M. B. A. Rahman, K. Jumbri, M. Basri, E. Abdulmalek, K. Sirat and A. Salleh, *Molecules*, 2010, **15**, 2388–2397.
54. M. Seter, M. J. Thomson, J. Stoimenovski, D. R. MacFarlane and M. Forsyth, *Chem. Commun.*, 2012, **48**, 5983–5985.
55. T. Torimoto, T. Tsuda, K. Okazaki and S. Kuwabata, *Adv. Mater.*, 2010, **22**, 1196–1221.
56. R. F. M. Frade, A. A. Rosatella, C. S. Marques, L. C. Branco, P. S. Kulkarni, N. M. M. Mateus, C. A. M. Afonso and C. M. M. Duarte, *Green Chem.*, 2009, **11**, 1660–1665.
57. R. F. M. Frade, A. Matias, L. C. Branco, C. A. M. Afonso and C. M. M. Duarte, *Green Chem.*, 2007, **9**, 873–877.
58. C. Florindo, A. Costa, C. Matos, S. L. Nunes, A. N. Matias, C. M. M. Duarte, L. P. N. Rebelo, L. C. Branco and I. M. Marrucho, *Int. J. Pharm.*, 2014, **469**, 179–189.
59. V. Kumar and V. Malhotra Sanjay, in *Ionic Liquids Applications: Pharmaceutical Therapeutics and Biotechnology*, ed. S. V. Malhorta, American Chemical Society, Washington, DC, 2010, pp. 1–12.
60. S. V. Malhotra and V. Kumar, *Bioorg. Med. Chem. Lett.*, 2010, **20**, 581–585.
61. V. Kumar and S. V. Malhotra, *Bioorg. Med. Chem. Lett.*, 2009, **19**, 4643–4646.
62. N. K. Kaushik, P. Attri, N. Kaushik and E. H. Choi, *Molecules*, 2012, **17**, 13727–13739.
63. R. Ferraz, J. Costa-Rodrigues, M. H. Fernandes, M. M. Santos, I. M. Marrucho, L. P. N. Rebelo, C. Prudencio, J. P. Noronha, Z. Petrovski and L. C. Branco, *ChemMedChem*, 2015, **10**, 1480–1483.
64. T. P. T. Pham, C.-W. Cho and Y.-S. Yun, *Water Res.*, 2010, **44**, 352–372.
65. J. Arning and M. Matzke, *Curr. Org. Chem.*, 2011, **15**, 1905–1917.
66. X. F. Wang, C. A. Ohlin, Q. H. Lu, Z. F. Fei, J. Hu and P. J. Dyson, *Green Chem.*, 2007, **9**, 1191–1197.
67. R. A. Kumar, N. Papaiconomou, J. M. Lee, J. Salminen, D. S. Clark and J. M. Prausnitz, *Environ. Toxicol.*, 2009, **24**, 388–395.

68. C. Suresh, H. Zhao, A. Gumbs, C. S. Chetty and H. S. Bose, *Bioorg. Med. Chem. Lett.*, 2012, **22**, 1734–1738.
69. C. Q. Jing, X. Y. Li, J. H. Zhang and J. J. Wang, *J. Biochem. Mol. Toxicol.*, 2013, **27**, 330–336.
70. X. Y. Li, J. Ma and J. J. Wang, *Ecotoxicol. Environ. Saf.*, 2015, **120**, 342–348.
71. K. S. Egorova, M. M. Seitkalieva, A. V. Posvyatenko, V. N. Khrustalev and V. P. Ananikov, *ACS Med. Chem. Lett.*, 2015, **6**, 1099–1104.
72. J. Luczak, C. Jungnickel, I. Lacka, S. Stolle and J. Hupka, *Green Chem.*, 2010, **12**, 593–601.
73. C. Jungnickel, J. Luczak, J. Ranke, J. F. Fernandez, A. Muller and J. Thoming, *Colloids Surf., A*, 2008, **316**, 278–284.
74. J. Pernak, A. Syguda, I. Mirska, A. Pernak, J. Nawrot, A. Pradzynska, S. T. Griffin and R. D. Rogers, *Chem.–Eur. J.*, 2007, **13**, 6817–6827.
75. A. Cieniecka-Roslonkiewicz, J. Pernak, J. Kubis-Feder, A. Ramani, A. J. Robertson and K. R. Seddon, *Green Chem.*, 2005, **7**, 855–862.
76. J. Bernstein, *Polymorphism in Molecular Crystals*, Oxford University Press, 2002.
77. D. Singhal and W. Curatolo, *Adv. Drug Delivery Rev.*, 2004, **56**, 335–347.
78. R. Hilfiker, *Polymorphism in the Pharmaceutical Industry*, Wiley-VCH Verlag GmbH, Weinheim, 2006.
79. K. C. Waterman, R. C. Adami, K. M. Alsante, A. S. Antipas, D. R. Arenson, R. Carrier, J. Y. Hong, M. S. Landis, F. Lombardo, J. C. Shah, E. Shalaev, S. W. Smith and H. Wang, *Pharm. Dev. Technol.*, 2002, **7**, 113–146.
80. C. Florindo, J. M. M. Araujo, F. Alves, C. Matos, R. Ferraz, C. Prudencio, J. P. Noronha, Z. Petrovski, L. Branco, L. P. N. Rebelo and I. M. Marrucho, *Int. J. Pharm.*, 2013, **456**, 553–559.
81. J. Stoimenovski, D. R. MacFarlane, K. Bica and R. D. Rogers, *Pharm. Res.*, 2010, **27**, 521–526.
82. J. L. Shamshina, P. S. Barber and R. D. Rogers, *Expert Opin. Drug Delivery*, 2013, **10**, 1367–1381.
83. M. Moniruzzaman and M. Goto, *J. Chem. Eng. Jpn.*, 2011, **44**, 370–381.
84. R. Ferraz, J. Noronha, F. Murtinheira, F. Nogueira, M. Machado, M. Prudencio, S. Parapini, S. D'Alessandro, C. Teixeira, A. Gomes, C. Prudencio and P. Gomes, *RSC Adv.*, 2016, **6**, 56134–56138.
85. J. H. Davis, K. J. Forrester and T. Merrigan, *Tetrahedron Lett.*, 1998, **39**, 8955–8958.
86. E. B. Carter, S. L. Culver, P. A. Fox, R. D. Goode, I. Ntai, M. D. Tickell, R. K. Traylor, N. W. Hoffman and J. H. Davis, *Chem. Commun.*, 2004, 630–631.
87. W. L. Hough-Troutman, M. Smiglak, S. Griffin, W. M. Reichert, I. Mirska, J. Jodynis-Liebert, T. Adamska, J. Nawrot, M. Stasiewicz, R. D. Rogers and J. Pernak, *New J. Chem.*, 2009, **33**, 26–33.
88. J. Pernak, M. Smiglak, S. T. Griffin, W. L. Hough, T. B. Wilson, A. Pernak, J. Zabielska-Matejuk, A. Fojutowski, K. Kita and R. D. Rogers, *Green Chem.*, 2006, **8**, 798–806.

89. L. Borka and J. K. Haleblian, *Acta Pharm. Jugosl.*, 1990, **40**, 71–94.
90. C. Teixeira, N. Vale, B. Perez, A. Gomes, J. R. B. Gomes and P. Gomes, *Chem. Rev.*, 2014, **114**, 11164–11220.
91. S. Goindi, P. Arora, N. Kumar and A. Puri, *AAPS PharmSciTech*, 2014, **15**, 810–821.
92. M. Deetlefs, K. R. Seddon and M. Shara, *Phys. Chem. Chem. Phys.*, 2006, **8**, 642–649.
93. C. W. Cho, S. Stolte and Y. S. Yun, *Sci. Rep.*, 2016, **6**, 1–9.

CHAPTER 17

Functional DNA in Ionic Liquids

B. OLAVE[a] AND T. SCHÄFER*[a,b]

[a]Polymat, University of the Basque Country, Av. Tolosa 72, Donostia-San Sebastián, 20018, Spain; [b]Ikerbasque – Basque Foundation for Science, Bilbao, Spain
*E-mail: thomas.schafer@ehu.eus

17.1 Functional DNA

17.1.1 Introduction

Deoxyribonucleic acid (DNA) has long been associated only with the storage of genetic information, this information being transcribed into ribonucleic acid (RNA) sequences, and then being translated into amino-acid sequences, a multitude of which serves to build proteins. However, with DNA technology developing during the last two decades, a lot of applications have been found for different DNA based structures, such as their use as catalysts,[1] the design of logic gates,[2,3] the construction of nanomaterials[4] and the creation of diagnostic tools and sensors[5] among many others.[6] Evolution in nature has generated a multitude of functional macromolecules; however, with regard to nucleic acids we have just started to discover the vast possibilities that they can offer in biology, bionanotechnology and biophysics, among other fields. The significant role that DNA plays in life began to be understood in 1953 with the discovery of the double stranded (ds) DNA.[7] Exploring the use of nucleic acid probes began in the early 1960s[8] and application of nucleic acids outside of a purely biological context was discovered in the early 1980s by

Smart Materials No. 28
Ionic Liquid Devices
Edited by Ali Eftekhari

Published by the Royal Society of Chemistry, www.rsc.org

Nadrian Seeman[9] who proposed new ideas of exploiting the molecular recognition properties of nucleic acids for more sophisticated nanostructures.

Yet, DNA technology has hitherto almost exclusively been associated with liquid environments at or near physiological conditions. However, taking into account the tuneability of ionic liquids, they could indeed be a highly interesting alternative liquid medium for hosting functional DNA molecules as this would significantly increase the degree of freedom to design smart DNA-based nanodevices. Thus, after first introducing the main aspects of DNA structure and function, this chapter will present the potential of ionic liquids as a possible alternative liquid medium to host these smart biological building blocks.

The genesis of DNA technology would not have been possible without the fact that even the most complex nucleic acid structures are generated by interactions between their very simple building blocks, namely the four nucleotides (Figure 17.1). DNA nucleotides are made of three different components: a phosphate group, a five-carbon sugar group called deoxyribose and a nitrogenous base. Four different bases can be found in DNA nucleotides: adenine (A), thymine (T), guanine (G) and cytosine (C). The single stranded (ss) DNA turns into a double stranded (ds) DNA after the base pairing, or hybridization, which occurs between the two complementary nucleotides adenine and thymine, as well as guanine and cytosine, respectively. There are two types of base pairing, Watson–Crick and Hoogsteen base pairing, the former being the most common one. Apart from the formation of these duplexes, triplex and quadruplex forms of DNA also exist and lend DNA additional functions.

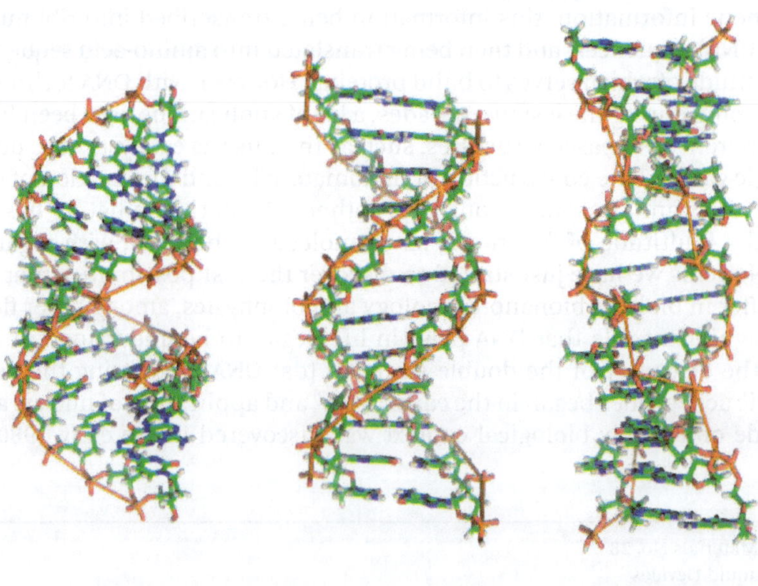

Figure 17.1 From left to right: PDB structures of the A-, B- and Z-conformations, respectively, that DNA assumes depending on the water and ions present in their environment.

Nowadays, there are mostly three different types of DNA based functional nucleic acids (FNAs): DNA-aptamers (or simply *aptamers*), deoxyribozymes (*DNAzymes*) and allosteric deoxyribozymes (deoxy*aptazymes*). Compared to antibodies or enzymes, FNAs have the advantage that they can be chemically synthesized at relatively low cost and high purity, apart from being more stable chemically.

17.1.2 Aptamers

Aptamers were discovered in the 1990s and the term "aptamer" was coined by the groups of Szostak and Ellington who reported on RNA molecules that bound to a small organic dye.[10] It is since then that we understand "aptamers" as single stranded nucleic acids (DNA or RNA) that can strongly bind to their target due to a molecular recognition event. Aptamers based on RNA outdo those based on DNA in various aspects, namely, they fold better into 3D structures and establish in general stronger interactions due to one more hydroxyl group in their ribose, resulting in higher binding affinities. However, the synthesis of RNA aptamers is much more expensive than that of DNA aptamers, and they are furthermore highly prone to degradation. Therefore, it is DNA aptamers that are preferably used as functional building blocks in nanodevices and which will consequently be focused on in this chapter.

The size of DNA aptamers commonly varies between 20 and 80 bases, which can arrange into different functional structures such us hairpins, quadruplexes or bulges, among others. These functional structures ensure fitting and high specificity to their targets through multiple molecular interactions as well as steric arrangements. Targets that are recognized can be highly varied and consist of inorganic or organic low molecular weight compounds, nucleotides, amino acids, peptides, proteins, and even whole cells (Figure 17.2).

Aptamer sequences that specifically recognise their targets are commonly obtained through *in vitro* selection protocols called "systematic evolution of ligands by exponential enrichment" (SELEX, Figure 17.3 [11,12]). The advantage of SELEX is that it is a rather straightforward method to identify specific aptamer sequences that bind to ligands with high affinity and selectivity and can therefore be applied to a variety of different targets. SELEX starts with a ssDNA library of about 10^{14} sequences containing two fixed regions of 10–20 nucleotides needed for library amplification and a central region containing a random sequence of 20–80 nucleotides for target recognition.

After incubation of the sequence library with the target, unbound nucleic acid sequences are separated from those that have created the complex with the target. The latter are subsequently amplified using polymerase chain reaction (PCR), and subjected to a new round of the SELEX procedure, reducing in this way gradually the variety of the DNA sequences and enriching the sequences with high affinity and specificity to the target. Checking on the binding affinity after reach round allows stopping of the procedure when after three rounds there is no further improvement of specificity.

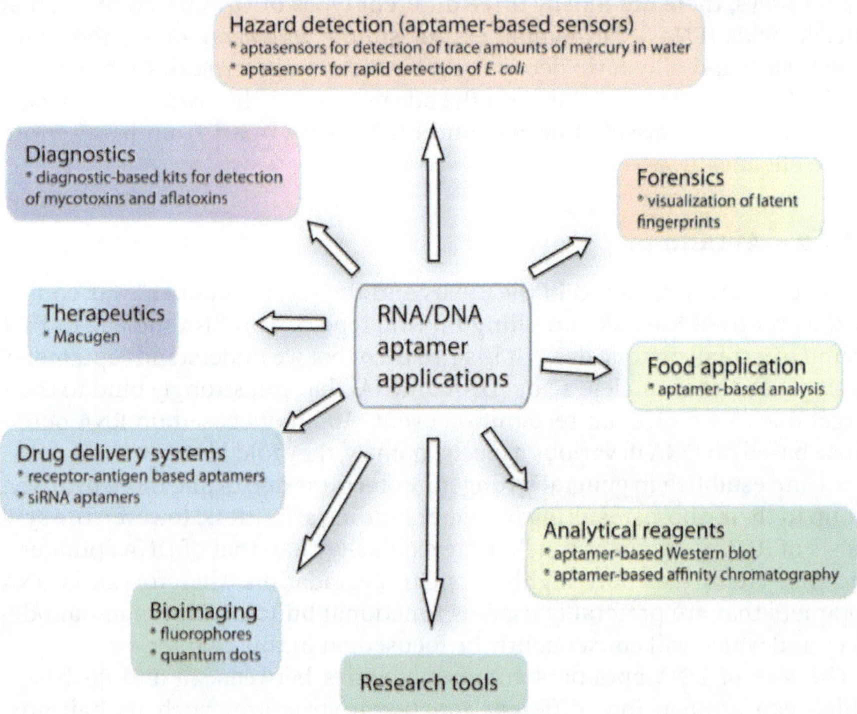

Figure 17.2 Fields of aptamer applications. Reprinted from *Biotechnology Advances*, 33 (6), M. Darmostuk, S. Rimpelova, H. Gbelcova, T. Ruml, Current approaches in SELEX: An update to aptamer selection technology, 1141–1161, Copyright 2015 with permission from Elsevier.[11]

Normally, after 6–20 rounds it is possible to proceed to the final step, which is the cloning and sequencing of the specific DNA aptamer. The SELEX procedure is constantly being improved,[13] with the aim of, for example, increasing the aptamer specificity and reducing the number of selection rounds.

Due to their low cost of production, high specificity and affinity to the target, DNA aptamers have been used in a wide variety of fields as biosensors in order to detect allergens, pathogens and toxins, amongst others, in food and water.[14] Owing to the conformational changes they might undergo upon target recognition, they can also be used as "gatekeepers" in nanoporous systems mimicking biological cell membranes,[15] or in biomedical therapeutic applications where they can activate or deactivate signalling pathways (differentiation, proliferation, anchoring) similar to antibodies.

17.1.3 DNAzymes

Besides the development of aptamers, the folding properties and inherent chirality of the double helical DNA have also been exploited to develop DNA-based catalysts. Here the link to biotransformations in ionic liquids is

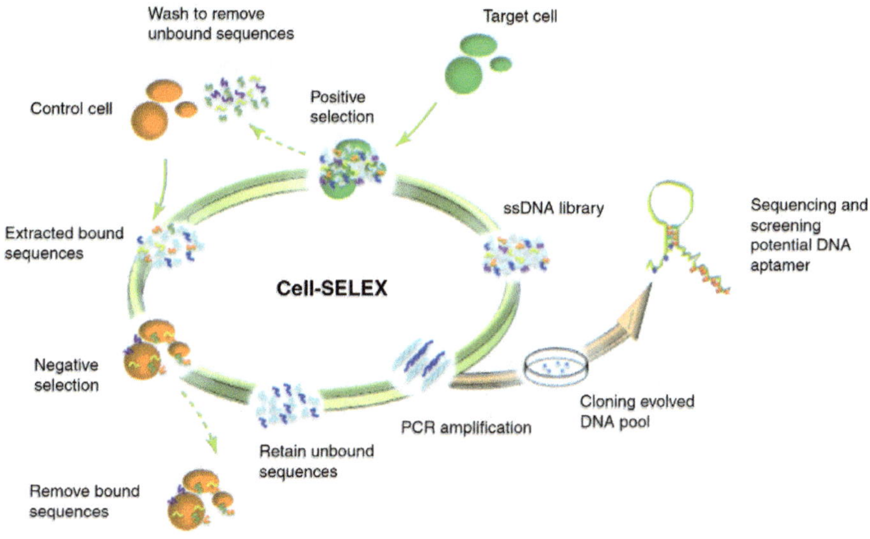

Figure 17.3 Representation of SELEX procedure using cells as targets, this procedure is very similarly used with other types of targets. Reprinted from *Trends in Biotechnology*, **29** (12), W. Tan, H. Wang, Y. Chen, X. Zhang, H. Zhu, C. Yang, R. Yang, C. Liu, Molecular aptamers for drug delivery, 634–640, Copyright 2011 with permission from Elsevier.[12]

even more evident. Catalytic DNA consists of supramolecular hybrids containing metal complexes and dsDNA molecules or ssDNA sequences (the DNAzymes), the latter being obtained by *in vitro* selection adapted from SELEX.[1]

In 1994 the first DNA enzyme (Figure 17.4) was created by Breaker and Joyce.[16] DNAzymes can be significantly more stable thermally than other types of enzymes. Furthermore, DNAzymes can incorporate both a catalytic and sensing/reporting function as opposed to conventional enzymes which need to be conjugated with respective labels.

DNAzymes are specifically characterized by their capability to cleave RNA, but they can also cleave DNA and other bonds.[1] Following the discovery of RNA-cleaving DNAzymes, molecular biologists began to investigate whether DNAzymes could be exploited as therapeutic agents, with their principal aim being antisense therapy: knowing the genetic sequence of a particular gene responsible for causing a particular disease, a specific DNAzyme can be synthesized which will bind to the messenger RNA produced by that gene and subsequently inactivate it, "turning off" the respective gene efficiently.

The ligand-binding and catalytic features of nucleic acids can be combined to generate aptazymes. An outstanding feature of the aptazymes is that their catalytic activity is increased or inhibited when they are bound to a specific target, a property which can be predicted and adequately

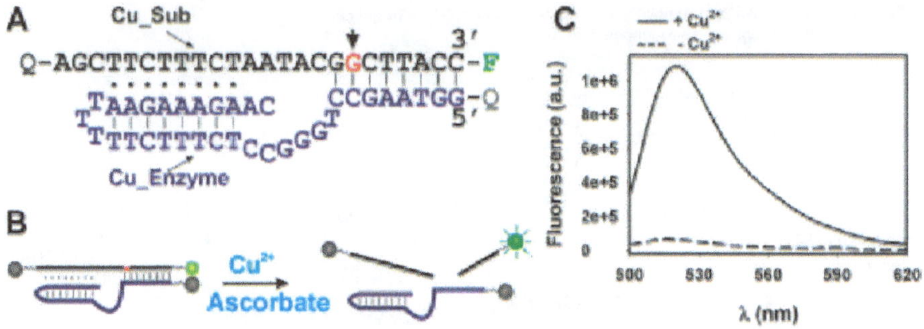

Figure 17.4 (A) Chemical structure of a DNAzyme which acts as Cu^{2+} sensor. (B) Mechanism of the signal generation due to the separation of the fluorophore and the quencher. (C) Fluorescence spectra of the sensor before and 10 min after the addition of 20 μm Cu^{2+}. Reprinted with permission from ref. 20. Copyright (2007) American Chemical Society.

designed based on the fundamental base-pairing concept. It has been shown that it is more difficult to design ligand-dependent repair of an inactive structure than perturb an active one, with the latter becoming therefore the most interesting function of aptazymes. In the 1990s the first aptazymes were already described: the addition of an oligonucleotide complementary to known inhibitory sequences was shown to activate a misfolded HDV ribozyme by over three orders of magnitude.[17] Nowadays, combining existing know-how on systematically designing aptamers and DNAzymes has resulted in smart solutions for applications such as sensors.[18]

17.1.4 Other Functional Nucleic Acids

Aptamers, DNAzymes and aptazymes can be modified after selection to obtain specific structures. One common transformation is their transformation into what is denoted "molecular beacons". An example of a molecular beacon is an ssDNA hairpin based on an aptamer, with its two ends labelled by a fluorophore and a fluorescence quencher (Figure 17.4). The proximity of the fluorophore/quencher pair results in FRET and hence no fluorescence is emitted when the hairpin is in its closed state. Since one leg of the hairpin loop sequence is the aptamer sequence recognizing a target, exposure to the molecular target or target DNA results in a binding event which forces the hairpin to open. As a consequence, the distance between fluorophore and quencher increases and so will the fluorescence. The ingenuity of this concept has resulted in molecular beacons being applied in many different fields[19–21] and they can also be conveniently being used as probes to explore the function of DNA in chemical environments different from conventional physiological conditions.

17.2 DNA Stability and Function in Non-conventional Environments

17.2.1 Aqueous Buffer Solutions

The structure and function of nucleic acids are a consequence of interactions with the molecules surrounding them. Privé *et al.*[22] stated in 1991 that "...DNA is built from five structural elements rather than three: the familiar bases, sugars and phosphate groups, but also ordered water molecules and bound counterions." Nucleic acids and other biomolecules mostly find their applications under physiological conditions in aqueous buffer solutions which maintain a constant pH of the solution, while only minor differences in the concentration of other compounds such as ions or target molecules may occur. One must therefore bear in mind that the ingenuity of nature is to a significant extent possible owing to the adaptation of a system to a liquid environment which undergoes only minor fluctuations in its physicochemical properties.

The most important feature of water as a solvent for biomolecules is its hydrogen bonding capability and its large dipole moment. Due to the latter, it is a very appropriate solvent for salts as cations interact favourably with the oxygen, and anions with the hydrogen atoms. Nucleic acids are highly charged polyanions with one negative charge per phosphate group which needs to be adequately compensated by cations in solution. Depending on how this compensation takes place in solution as regards both the presence of water and ions adjacent to the negatively charged DNA, dsDNA will assume different conformations.[23] It is well-known that at high temperature the dsDNA structure melts into an open coil, which is why the melting temperature related to this event is commonly used as the representative parameter to measure the stability of dsDNA. The thermal stability of DNA is mainly associated with the presence of the hydration shells on the DNA surface and as such related to the entropy and enthalpy of water. Hence, on increasing the temperature of the solution the number and strength of the water–DNA interactions is gradually reduced, which eventually results in the melting of dsDNA.[24]

17.2.2 Effect of Ions

High salt or metal ion concentrations can influence the water activity and often provoke a transition among polynucleotide conformations such as the B-Z or B-A transitions of DNA. During folding of nucleic acids cations such as sodium or magnesium are required to neutralize the negative charges on the backbones and reduce in this way the Coulombic interactions between the phosphates. This stabilization by higher ion concentration is more pronounced for dsDNA than for ssDNA because dsDNA is more negatively charged and this increases the bound ions required for compensation.

Monovalent cations such as Na^+, and except for Ag^+, appeared to undergo nonspecific interactions while, in contrast, divalent cations such as Mg^{2+} showed evidence of site-binding and hydrogen bond formation with an increased concentration of metal cations bound to phosphates, increasing in this way the melting temperature[25] (*e.g.*, 3×10^{-4} M Mg^{2+} stabilizes DNA as effectively as 5×10^{-2} M Na^+). In addition to stabilizing electrostatic interactions, alkaline earths and transition metals can destabilize the double helix. The destabilization occurs by metal ion binding to the bases, which means that the relative importance of destabilization in view of stabilization follows the preference of the ion for base over the phosphate. Cation size is also an important factor to be considered for DNA structure stabilization, as it has been suggested that larger cations are sterically hindered from approaching the DNA backbone as closely as densely packed or smaller cations, while additionally the decreased surface charge density of larger cations decreased also the strength of the Coulombic interactions with the phosphates.[26] Consequently, the number of cations located in the immediate vicinity of the dsDNA correlates inversely with the cation sizes studied. Ions not only affect DNA stability and function based on their interactions with different parts of the nucleotides, but they apparently also bind differently in the major and minor grooves of the dsDNA, with AT-rich sequences driving the binding of metallic cations in the minor groove as opposed to GC-rich sequences.

17.2.3 Organic Solvents as Non-conventional Environment

As much as water is the most conventional liquid environment for DNA, it is not the ideal solvent in some aspects. For example, when storing DNA in water hydrolytic reactions take place which cause the denaturation of DNA. In detail, depurination occurs involving the purine deoxyribonucleosides, deoxyadenosine and deoxyguanosine, and in which the β-*N*-glycosidic bond is hydrolytically cleaved to release a nucleic base, adenine or guanine, respectively. Depurination hinders the long-term storage of DNA as this reaction is even faster if the sample is not frozen. This brought about interest in studying the structure and stability of DNA in non-conventional media, first only in molecular organic solvents and subsequently also in ionic liquids.

A well-known example is the use of ethanol as a precipitation agent for DNA. It was observed that until 50% (v/v) of ethanol in solution dsDNA maintains its B conformation, but its melting temperature decreases. Beyond this concentration, dsDNA undergoes a conformational change, agglomerates and eventually precipitates.[27] Conversely, the duplex structure of a 21-mer synthetic DNA in 99% (v/v) glycerol and ethylene glycol maintains the double-helical structure and also its thermostability with a melting point of 30 and 50 °C lower than in water respectively; however, in other molecular solvents such as 99% (v/v) formamide, methanol or DMSO no duplex structure of DNA was detected.[28]

Two decades ago, in 1997, it was shown by Klibanov and co-workers that organic solvents could replace water in biotransformations using enzymes, increasing in this way tremendously the degree of freedom for processes involving enzyme-based catalysis.[29] They proved that an enzyme lyophilized from an aqueous solution maintains its conformation and, hence, activity when subsequently being introduced into an anhydrous organic solvent: in the absence of water, which would enable a conformational adaptation to the new environment, the enzyme remained "locked" in its original conformation and therefore function, although admittedly with less efficiency. In fact, in such a system the enzyme is rather dispersed in the solvent than solvated as it will not interact with the organic solvent.

As regards DNA, on the other hand, its use in organic solvents is still being studied, with the recognition capacity of DNA recently being reported in nine different solvents and for different salt concentrations by using a molecular beacon:[30] by varying the concentration of methanol, ethanol, isopropanol, acetonitrile, formamide, dimethylformamide (DMF), dimethyl sulfoxide (DMSO), ethylene glycol, and glycerol, respectively, up to 75% (v/v), the rate of hybridization of the molecular beacon was much faster compared with water up to a certain concentration depending on the solvent, improving also the specificity. This surprising phenomenon of an accelerated hybridization was attributed to a reduced activation energy barrier. A maximum increase was observed in 56% (v/v) ethanol where the molecular beacon showed a 70-fold rate enhancement of hybridization rate as compared to physiological conditions (Figure 17.5). Beyond this concentration, DNA was considered to lose its stabilizing hydration shell, resulting in precipitation.

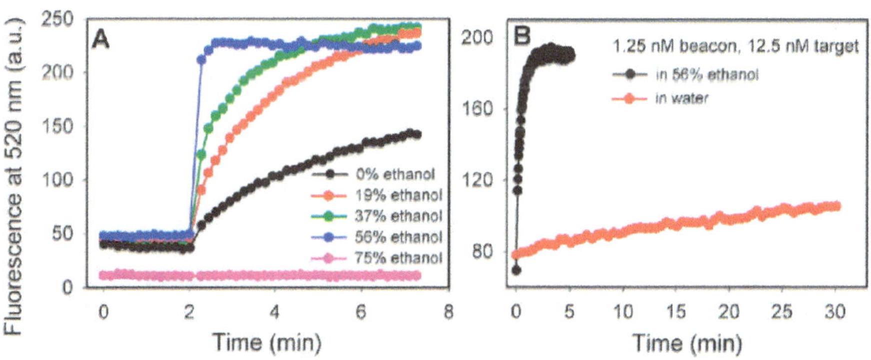

Figure 17.5 (A) Kinetics of molecular beacon hybridization after the addition of the target at minute 2; the recognition of the target was studied increasing the concentration of ethanol (% v/v) in a solution containing 50 nM of molecular beacon and 200 nM of target. (B) Comparison of the kinetics with the best ethanol concentration and pure buffer. Buffer conditions: 300 mM NaCl, 10 mM HEPES, pH 7.6. Reprinted with permission from ref. 30. Copyright (2010) American Chemical Society.

17.2.4 Polymers as Non-conventional Environment

The presence of high concentrations of macromolecules in solutions of nucleic acids is of utmost importance as it reflects in fact the actual cellular environment. Indeed, the intracellular liquid is considered to be highly "crowded", affecting significantly the structure and function of nucleic acids as compared to when they are in a dilute solution.[31] This heterogeneous and highly volume-occupied or crowded environment can considerably influence the reactivity and distribution of macromolecules that take part in the intracellular biochemical processes, including the structure and function of nucleic acids. The formation and the function of these structures are of interest because of their roles in important biological processes, such as aging, and their potential as therapeutic targets for cancer.[32] On the other hand, G-quadruplexes and i-motifs (guanine and cytosine rich quadruplexes, respectively[33]) are excellent candidates for the construction of a variety of novel nanomechanical devices with many potential applications in drug delivery, miniaturized biosensors, nanorobotics and dynamic nanomaterials[34] and their use in crowded environments is therefore currently subject to intense investigation. Poly(ethylene glycol) (PEG) is one of the most commonly used polymers in aqueous solution to mimic cellular environments (Figure 17.6[31]) and thus acts as a crowding agent, as is assumed that PEG does not interact chemically with nucleotides but merely serves as physically reducing the space, a role that might be subject to discussion.

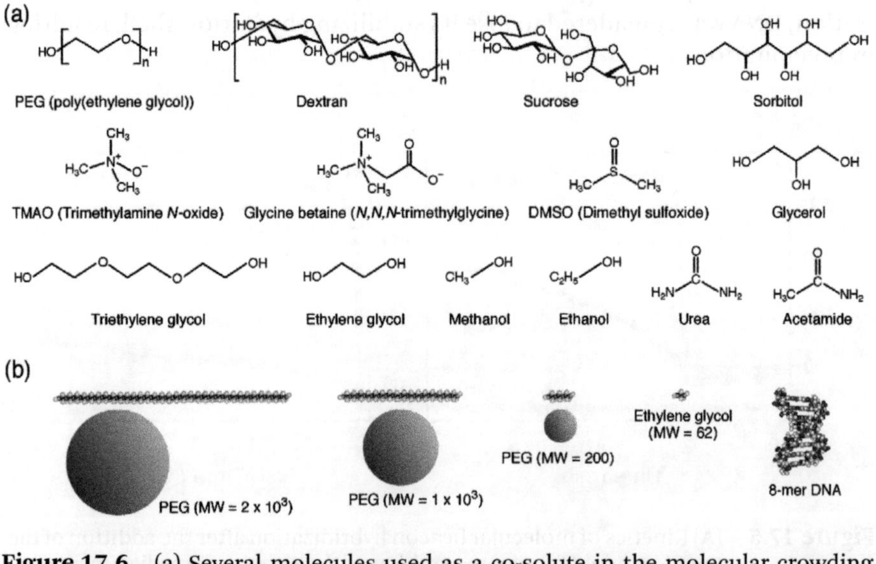

Figure 17.6 (a) Several molecules used as a co-solute in the molecular crowding studies of nucleic acids. (b) Comparison of the molecular sizes of PEGs of different average molecular weights, ethylene glycol, and a short DNA duplex. A sphere represents the gyration diameters of the PEG molecules. Reprinted with permission from ref. 31. Copyright (2014) American Chemical Society.

17.2.5 Ionic Liquids and Deep Eutectic Solvents as Non-conventional Environment

Ionic liquids and deep eutectic solvents (DESs) possess physico-chemical properties that make them versatile alternative solvents and distinct from both molecular organic solvents (uncharged) and regular salts (solid). This lends ILs and DES a high degree of tuneability of their physico-chemical properties, making them extremely attractive (co-)solvents. Yet, the solvation of DNA in these systems is complex and not fully understood at a molecular level, with the function of nucleic acids just beginning to be explored.

A DES is defined a mixture of (a) a metal chloride hydrate with a hydrogen bond donor or a quaternary ammonium salt with (b) a metal chloride, (c) metal chloride hydrate or (d) hydrogen bond donor. The physico-chemical properties of DESs are very similar to those of ILs and therefore were initially considered as a new class of ILs.[35] In some cases room temperature DESs can be cheaper than ILs, less toxic and more biodegradable. A DES can potentially solvate nucleic acids better than some ILs; very common room temperature DESs used with nucleic acids are the choline chloride mixtures with urea (Figure 17.7), glycerol or ethylene glycol at 1 : 2 molar ratios.[36,37]

ILs have been considered "green solvents" owing to their almost negligible vapour pressure; however, their impact on the environment is a subject of discussion. An ideal green solvent should have low toxicity[38] and be easily degradable in the environment,[39] but depending on the cation and the anion the impact of ILs will vary, as can be seen for example with the very commonly used imidazolium-based ILs containing BF_4^- and PF_6^- anions which degrade in the presence of water (protons). The synthesis of some ILs at industrial scale also needs to be taken into account as it can produce contaminants such as halogens or VOCs. However, the degree of tuneability of room temperature DESs is not comparable to that of RTILs, which makes the latter the best option for designing functional nanostructures in non-conventional solvents – much concerning the understanding and predicting the respective interactions remains challenging.

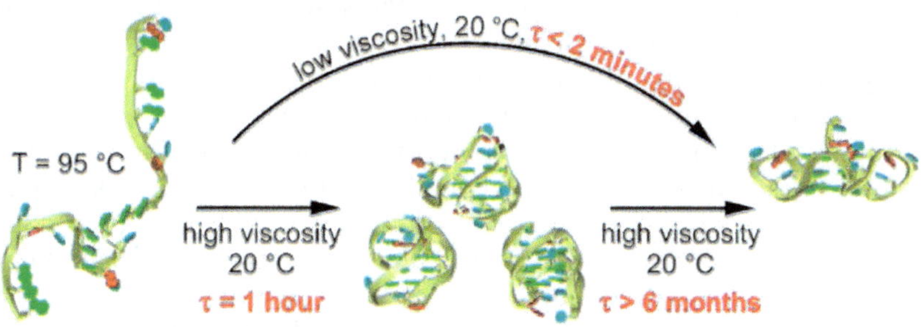

Figure 17.7 DNA refolding in viscous choline chloride/urea (1 : 2). Reprinted with permission from ref. 37. Copyright (2012) American Chemical Society.

It has been shown that nucleic acids can form secondary structures that reversibly denature under heating in a neat DES or IL, but this does not automatically mean that DNA may also lose its function: function may rely precisely on the capability of reversibly changing a structure. For example, in the previously described hairpin structure it is the reversible hybridization/denaturing of the stem sequence that lends the DNA its function, rather than a structural stability. Hence, the term "stability" refers in this case to maintaining a dynamic function rather than a static system. Nevertheless, although the static structural stability is not always related to the functionality, it is certainly necessary to maintain stability to the extent that DNA aggregation and precipitation is prevented. Being electrolytes, ILs possess in principle the capacity to act both as stabilizing electrolytes and solvents such that the function of DNA can be fine-tuned, which raises the question as to which are the most appropriate ILs or IL mixtures as (co-)solvents to do so and based on what physico-chemical properties. In this context, ILs must not be considered as a general group of solvents as is commonly found in the literature, because their interactions will vary depending entirely on their individual physico-chemical properties. Therefore, in the following the role of the physico-chemical properties of different ionic liquids for the static and dynamic stability of DNA is discussed.

17.3 Functional DNA in Ionic Liquids

17.3.1 Understanding the Molecular Interactions Between DNA and Ionic Liquids

ILs have been used for extraction and purification of nucleic acids and other biomolecules such as amino acids, proteins, and lipids, as well as cellulose representing bulk biomass. Since the functionality of the biomolecules is not a concern in these processes, they are not going to be discussed in this chapter but the interested reader is rather referred to an excellent critical review about this topic published recently (Figure 17.8 [40]). Contrary to what could be expected, most ILs employed in these extraction and purification processes are not highly hydrophobic water-immiscible ILs. Water-miscible ILs can rather be used in aqueous biphasic systems (ABS). Hereby, a salting-out agent is added to the water-miscible ILs to create a second liquid phase, with these systems consisting of hydrophilic ILs and inorganic salts in aqueous solution, or of conventional polymer-salt based ABS with ILs as adjuvants.[41]

The ability of ILs to maintain the structural stability and function of nucleic acids is essential for their role as potentially substituting aqueous solutions. The capacity of ILs to behave as salts and as solvents is due to their different interactions with DNA which involve dispersion forces, hydrogen bonding, steric effects and hydrophobic interactions. The water content must be carefully determined because the macroscopically observed effect of an IL on DNA will be drastically different depending on whether the IL is hydrated or not. Hereby, it is particularly relevant in terms of how far the hydration shell of DNA is affected by the presence of ILs. On one hand, cations of ILs screen

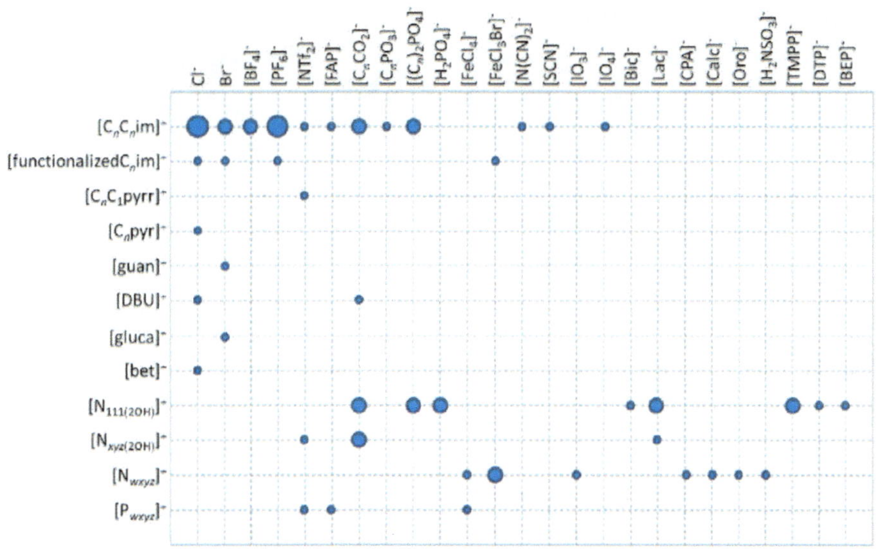

Figure 17.8 ILs used for the separation and purification of nucleic acids as a function of cation–anion combinations. The usage incidence (number of articles) is represented by the size of the circles, which proportionally increases as follows: 1 < 2 < 3. Reprinted with permission from ref. 40. Copyright (2017) American Chemical Society.

the DNA negative charge, presumably as a cation cloud, and increase the thermal stability; on the other hand, it is the IL anion that is not only considered essential for the solubility of the ionic liquid in water but also has been found to be determinant for the binding energy between the IL and the DNA.[42] In addition, the capacity of ionic liquids to form hydrogen bonds is important as anions and cations which are hydrogen bond donors or/and acceptors can participate in the solvation mechanism of DNA bases.[43] The hydrogen bond donor and acceptor properties have been predicted for many ILs using COSMO-RS,[44] which can be a useful first screening filter to select the ILs of interest.

17.3.2 Potential Application and Sustainability of Choline Based Ionic Liquids

Finding a medium in which DNA is soluble without losing its structure and being stable for long periods of time at room temperature is a challenge. While molecular solvents such as ethylene glycol and to some extent glycerol have been found to stabilize DNA, employing ILs for this purpose is an ongoing research topic.

Choline based ionic liquids with different anions have been at the centre of attention due to their biodegradability, biocompatibility, low toxicity and their capability of maintaining the stability of DNA.[45–50] Quaternary ammonium salts[51] have yielded satisfying stabilization of DNA structures, with two studies showing that A-T base pairs are more stable than G-C base pairs in the

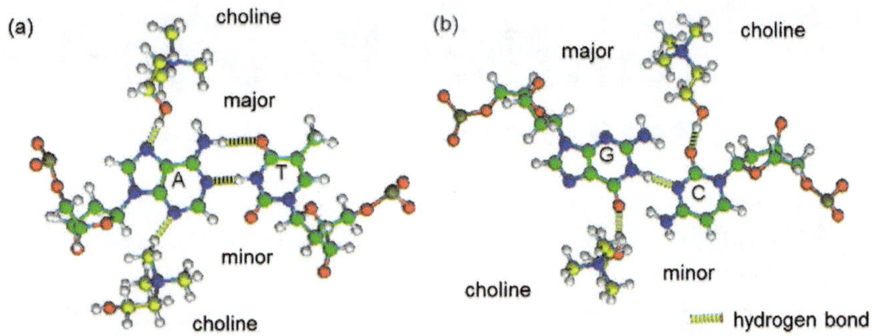

Figure 17.9 Representative alignments of choline ions bound to (a) an A–T base pair and (b) a G–C base pair. Phosphorus, oxygen, nitrogen, and hydrogen atoms are shown in gold, red, blue, and white, respectively. The carbons of choline ions are shown in yellow, and the carbons of DNA are in green. Reprinted with permission from ref. 53. Copyright (2014) American Chemical Society.

presence of a hydrated choline based ionic liquid: in one study[52] the stability of DNA base pairs at a concentration of 100 μM was evaluated in choline dihydrogen phosphate (CDHP) solutions in a concentration range from 0.1 to 4 M using UV spectroscopy to compare the experimental observation with the effect of NaCl. It was reported that in the presence of the latter the melting temperature decreased if the amount of A-T base pairs increased while in the case of CDHP the opposite trend was observed. It was hypothesized that choline ions stabilize an A-T rich DNA duplex by binding in the grooves, and destabilize a G-C rich DNA duplex by binding to guanines, favouring in this way the ssDNA form. In a follow-up study[53] these observations were further explored by molecular dynamics (Figure 17.9) showing that CDHP apparently resided closer to the dsDNA than did sodium ions, with the sodium ions interacting with the phosphate groups electrostatically whereas CDHP formed multiple hydrogen bonds with the DNA nucleotides. The latter was observed to occur in competition with the hydrogen bonds between G and C bases that form the duplex, therefore supposedly resulting in a destabilization of G-C base pairs and hence the entire DNA duplex.

Dynamic DNA systems which are related to the function of nucleic acids have not been deeply studied so far in the presence of ionic liquids although this may offer exciting possibilities and would foster new applications of DNA technology in non-conventional solvents such as ionic liquids. An example of such systems is discussed in the following.

17.3.3 Molecular DNA-aptamer Beacons as Molecular Recognition Sensors in Ionic Liquids

Machado *et al.*[19,54] reported on the ability of a DNA-aptamer molecular beacon to maintain not only its hairpin structure but also its function even at high concentrations of ionic liquids in buffer solutions. As mentioned above, molecular beacons have been widely employed in biosensing owing to the ease of detecting the fluorescence signal upon interaction with the target. Fluorescence markers have been employed in various structural studies involving DNA and ionic liquids. However, in the work of Machado it was shown that ionic liquids in solution can severely quench fluorescence which can potentially lead to erroneous interpretations of experimental data. In fact, the need to carefully use fluorophores as reporter molecules in this context is even more evident as ILs have even been employed to extract dyes from solutions of stained DNA.[55]

To understand how different ILs influence the stability and hence dynamics of the stem part of the aptamer hairpin structure, the hybridization kinetics and melting temperature of two representative complementary 12-mer ssDNA in solution were studied in the presence of these ILs. The sequence of the two complementary ssDNAs was identical to that of the stem of the respective molecular aptamer beacon which in this case was an aptamer beacon responsive to ATP. It was assumed that the ionic liquid could affect the molecular recognition of ATP in ionic liquids a threefold (Figure 17.10): (1) the stem which hybridizes/denatures reversibly and is responsible for the reversible opening/closing of the hairpin structure, (2) the linker between both stems based on polyethylene glycol which provided the DNA-aptamer beacon with a flexible loop warranting its reversible opening and closing function and (3) the recognition binding "pocket" formed by the binding sequence and which is the main part responsible for the recognition of the target. The ATP-aptamer was chosen as a probe for this study due to its previous full characterization and to the fact that the ATP or AMP target is a small molecule, which reduces the amount of possible interactions between the ionic liquid and the target and therefore allows better control over the experimental design.

The dissociation or binding constant (K_d) represents the sensitivity of the aptamer for recognizing the target and it was determined by increasing the concentrations of AMP target (Figure 17.11 [19,54]) both in pure buffer solutions and in presence of ILs. Furthermore, hybridization studies were conducted between the two complementary ssDNAs of the stem yielding the melting temperature of the respective stem DNA duplex, the hybridization rate and the amount of DNA that remained hybridized with the addition of increasing concentrations of ILs was calculated, as shown with the IL ethylammonium nitrate (EAN) (Figure 17.11 [19,54]). In this way, these studies covered both the structural stability of DNA as well as its functional stability.

Regarding the structural stability of the stem DNA duplex, it was observed that none of the ILs used had any detrimental effect up to 10% (v/v) when compared with pure TBS buffer solution. Beyond this concentration, the effects observed depended on the physico-chemical properties of the ILs. Choline-based ILs affected hybridization of the dsDNA stem only slightly when increasing the IL concentration beyond 10% (v/v). On the contrary, butyl imidazolium chloride (BMIM-Cl) and 1,1,3,3-tetramethylguanidinie lactate (TMG-L) were the two ILs that destabilized dsDNA strongest in the same concentration range. At a concentration of 40% (v/v) hybridization was no longer observed in the case of BMIM-Cl, and it decreased by almost 90%

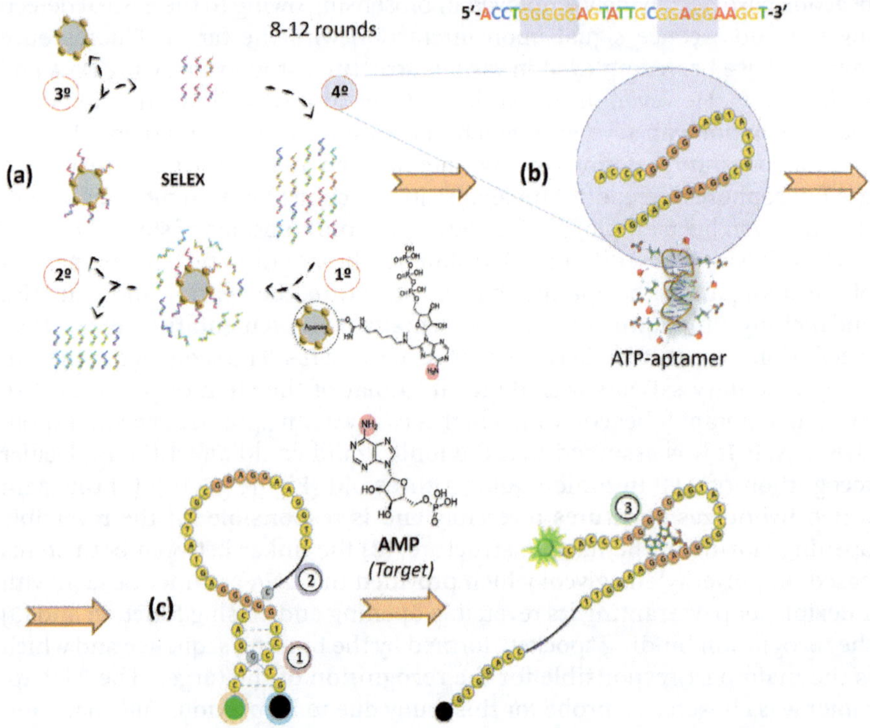

Figure 17.10 (a) Procedure of a single round of SELEX; (b) and (c): Aptamer and molecular beacon structures used. Adapted from I. Machado, V. Cengiz Özalp, E. Rezabal, T. Schäfer, DNA Aptamers are Functional Molecular Recognition Sensors in Protic Ionic Liquids, *Chemistry – A European Journal*, Copyright © 2014 Wiley-VCH Verlag GmbH & Co. KGaA, Weinheim.[19]

for TMG-L. It was concluded that mainly the choline-based ILs were suitable for maintaining the DNA functions as their cation could adequately shield the negative charge of the DNA phosphate groups. Conversely, $BMIM^+$ and TMG^+ interacted less with the phosphates of dsDNA and in this way destabilized the duplex. As regards EAN, a tendency in between these two extremes was observed. *Ab initio* calculations showed that hydrophobic interactions between the ILs and the DNA bases played a minor role compared to the strong electrostatic interactions occurring with the DNA phosphate groups. Hereby, it was shown that the distance between two negatively charged DNA-phosphates was too small to accommodate two compensating cations from BMIM-Cl or TMG-L, but provides enough space for two smaller cations such as choline or ethylammonium, in this way eventually resulting in a better stabilization of the dsDNA by the latter.

These results partially contradict what was reported by Chandran *et al.*[56] mainly based on molecular simulations, claiming that in an aqueous

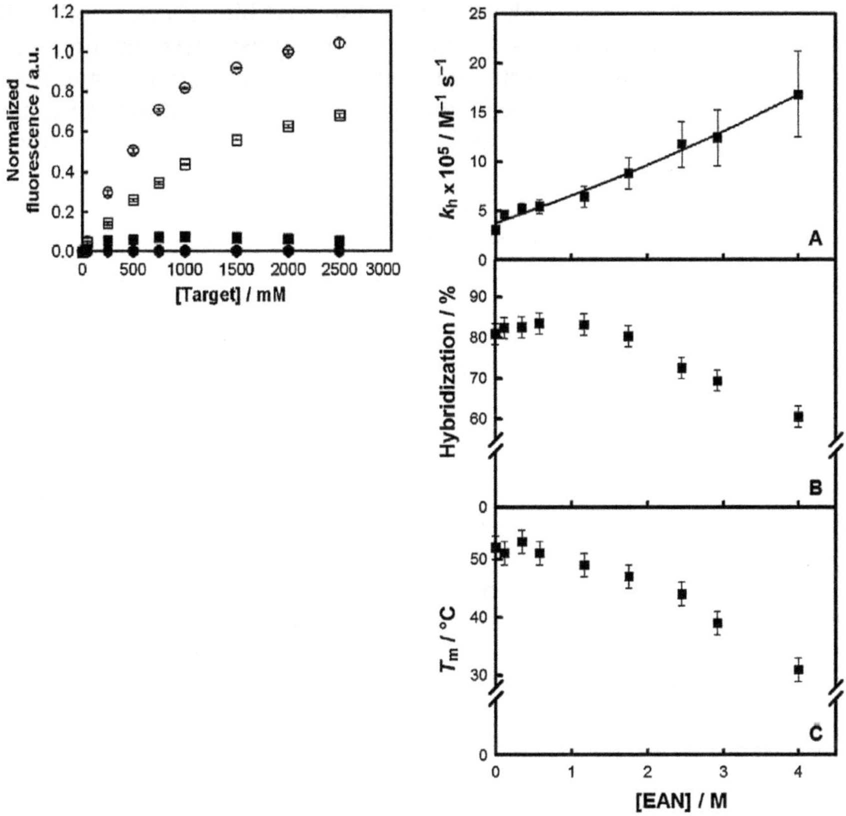

Figure 17.11 (Left) Normalized fluorescence obtained by the recognition of AMP molecular aptamer beacon of increasing concentrations of AMP in 0 M (open circles) and 2 M (open squares) of ethylammonium nitrate (EAN) with the respective error bars and GMP in 0 M (filled circles) or 2 M (filled squares) of EAN in TBS solution at 25 °C. (Right) Graphic representation of the: (A) hybridization rates, k_h, (B) amount of DNA hybridized (%), and (C) melting temperatures, T_m, as a function of the increase of EAN concentration in TBS solution. Reproduced from I. Machado, V. Cengiz Özalp, E. Rezabal, T. Schäfer, DNA Aptamers are Functional Molecular Recognition Sensors in Protic Ionic Liquids, *Chemistry – A European Journal*, Copyright © 2014 Wiley-VCH Verlag GmbH & Co. KGaA, Weinheim.[19]

solution of 58% (v/v) BMIM-Cl dsDNA retains its B-conformation. Moreover, they postulated that the partially hydrophobic BMIM-based ILs would stabilize DNA by intercalating in between its grooves due to their planar geometry, while hydrophilic choline-based ILs would not interact as well due to their tetrahedral geometry. These contradictory results between simulation and experiment reveal that simulation parameters of MD simulations may still not exactly reflect experimental conditions when dealing with such complex structures as DNA, and respective conclusions need to be analysed carefully.

The reason for this is that hitherto it is still impossible to tackle both bulk phenomena and complex DNA-dynamics at an atomistic level by molecular dynamics calculations, but both strongly determine the structure and function of DNA. As regards the recognition capacity of the aptamer molecular beacon, Machado *et al.*[19,54] observed a significant decrease in sensitivity but the recognition specificity of the ATP-aptamer molecular beacon was maintained against guanosine monophosphate (GMP) even at high concentrations of ILs and when compared with buffer solution. This observation was in full agreement with results from the hybridization kinetics of the two oligomers in the presence of ILs, indicating that it is indeed the stem part of DNA hairpin structures that is crucial for maintaining the reversible opening and closing upon a molecular recognition event in ionic liquids.

17.4 Conclusions

The use of functional nucleic acids based on DNA in a non-conventional environment has opened up new opportunities in nanotechnology. They can be used as biosensors, nanomaterials or for drug delivery and the specificity of these applications can be finely tuned using ionic liquids or deep eutectic solvents which represent a unique liquid environment for nucleic acids. Due to the vast amount of different ionic liquids and deep eutectic solvents and the respective differences in physico-chemical properties, it is not possible to generalize any experimental evidence obtained. Molecular simulations my help in understanding and designing smart systems based on DNA in ionic liquids, but it remains a challenge to tackle both at the same time bulk events and subtle molecular interactions within the DNA structure. Still, the first studies indicating that smart DNA-constructs may maintain their function even in the presence of large amounts of certain ionic liquids encourage a further systematic investigation of how ionic liquids may eventually replace aqueous environments for such purposes.

Acknowledgements

T. S. acknowledges ERC StG 209842-Matrix, ERC PoC 713641-ESSENSE, Basque Cooperative Research Project Elkartek-Lisol, as well as grants from the Spanish Ministry of Economy and Competitiveness CTQ2016-80375-P and CTQ2013-41113-R.

References

1. M. Hollenstein, DNA Catalysis: The Chemical Repertoire of DNAzymes, *Molecules*, 2015, **20**(11), 20777–20804.
2. I. Willner, B. Shlyahovsky, M. Zayats and B. Willner, DNAzymes for Sensing, Nanobiotechnology and Logic Gate Applications, *Chem. Soc. Rev.*, 2008, **37**(6), 1153–1165.

3. Y. Guo, L. Zhou, L. Xu, X. Zhou, J. Hu and R. Pei, Multiple Types of Logic Gates Based on a Single G-Quadruplex DNA Strand, *Sci. Rep.*, 2014, **4**, 7315.
4. M. R. Jones, N. C. Seeman and C. a. Mirkin, Programmable Materials and the Nature of the DNA Bond, *Science*, 2015, **347**(6224), 1260901.
5. A. P. F. Turner, Biosensors: Sense and Sensibility, *Chem. Soc. Rev.*, 2013, **42**(8), 3184–3196.
6. Y. Krishnan and F. C. Simmel, Nucleic Acid Based Molecular Devices, *Angew. Chem., Int. Ed.*, 2011, **50**(14), 3124–3156.
7. J. D. Watson and F. H. C. Crick, Molecular Structure of Nucleic Acids: A Structure for Deoxyribose Nucleic Acid (Reprinted from Nature, April 25, 1953), *Nature*, 1969, **224**(5218), 470–471.
8. B. D. Hall and S. Spiegelman, Sequence Complementarity of T2-DNA and T2-Specific RNA, *Proc. Natl. Acad. Sci.*, 1961, **47**(2), 137–146.
9. N. C. Seeman, Nucleic Acid Junctions and Lattices, *J. Theor. Biol.*, 1982, **99**(2), 237–247.
10. A. D. Ellington and J. W. Szostak, *In vitro* Selection of RNA Molecules That Bind Specific Ligands, *Nature*, 1990, **346**(6287), 818–822.
11. M. Darmostuk, S. Rimpelova, H. Gbelcova and T. Ruml, Current Approaches in SELEX: An Update to Aptamer Selection Technology, *Biotechnol. Adv.*, 2014, **33**(6), 1141–1161.
12. W. Tan, H. Wang, Y. Chen, X. Zhang, H. Zhu, C. Yang, R. Yang and C. Liu, Molecular aptamers for drug delivery, *Trends Biotechnol.*, 2011, **29**(12), 634–640.
13. L. Hernandez, I. Machado, T. Schafer and F. Hernandez, Aptamers Overview: Selection, Features and Applications, *Curr. Top. Med. Chem.*, 2015, **15**(12), 1066–1081.
14. C. L. A. Hamula, H. Zhang, F. Li, Z. Wang, X. Chris Le and X. F. Li, Selection and Analytical Applications of Aptamers Binding Microbial Pathogens, *TrAC, Trends Anal. Chem.*, 2011, **30**(10), 1587–1597.
15. M. B. Serrano-Santos, E. Llobet, V. C. Ozalp and T. Schafer, Characterization of Structural Changes in Aptamer Films for Controlled Release Nanodevices, *Chem. Commun.*, 2012, **48**(81), 10087–10089.
16. R. R. Breaker and G. F. Joyce, A DNA Enzyme That Cleaves RNA + R., *Chem. Biol.*, 1994, **1**, 223–229.
17. G. van der Horst, A. Christian and T. Inoue, Reconstitution of a Group I Intron Self-Splicing Reaction with an Activator RNA, *Proc. Natl. Acad. Sci. U. S. A.*, 1991, **88**(1), 184–188.
18. Z. Zhou, L. Xiao, Y. Xiang, J. Zhou and A. Tong, A General Approach for Rational Design of Fluorescent DNA Aptazyme Sensors Based on Target-Induced Unfolding of DNA Hairpins, *Anal. Chim. Acta*, 2015, **889**, 179–186.
19. I. Machado, V. C. Özalp, E. Rezabal and T. Schäfer, DNA Aptamers Are Functional Molecular Recognition Sensors in Protic Ionic Liquids, *Chem.–Eur. J.*, 2014, **20**(37), 11820–11825.

20. J. Liu and Y. Lu, A DNAzyme Catalytic Beacon Sensor for Paramagnetic Cu 2+ Ions in Aqueous Solution with High Sensitivity and Selectivity, *J. Am. Chem. Soc.*, 2007, **129**(32), 9838–9839.
21. J. Huang, Y. He, X. H. Yang, K. Quan and K. M. Wang, Inhibited Aptazyme-Based Catalytic Molecular Beacon for Amplified Detection of Adenosine, *Chin. Chem. Lett.*, 2014, **25**(9), 1211–1214.
22. G. G. Privé, K. Yanagi and R. E. Dickerson, *J. Mol. Biol.*, 1991, **217**, 177.
23. A. Ghosh and M. Bansal, A Glossary of DNA Structures from A to Z, *Acta Crystallogr., Sect. D: Biol. Crystallogr.*, 2003, **59**(4), 620–626.
24. K. Jumbri, M. B. Abdul Rahman, E. Abdulmalek, H. Ahmad and N. M. Micaelo, An Insight into Structure and Stability of DNA in Ionic Liquids from Molecular Dynamics Simulation and Experimental Studies, *Phys. Chem. Chem. Phys.*, 2014, **16**(27), 14036–14046.
25. Z.-J. Tan and S.-J. Chen, Nucleic Acid Helix Stability: Effects of Salt Concentration, Cation Valence and Size, and Chain Length, *Biophys. J.*, 2006, **90**(4), 1175–1190.
26. E. Stellwagen, J. M. Muse and N. C. Stellwagen, Monovalent Cation Size and DNA Conformational Stability, *Biochemistry*, 2011, **50**(15), 3084–3094.
27. J. Piškur and A. Rupprecht, Aggregated DNA in Ethanol Solution, *FEBS Lett.*, 1995, **375**(3), 174–178.
28. G. Bonner and A. M. Klibanov, Structural Stability of DNA in Nonaqueous Solvents, *Biotechnol. Bioeng.*, 2000, **68**(3), 339–344.
29. J. L. Schmitke, L. J. Stern and A. M. Klibanov, The Crystal Structure of Subtilisin Carlsberg in Anhydrous Dioxane and Its Comparison with Those in Water and Acetonitrile, *Proc. Natl. Acad. Sci. U. S. A.*, 1997, **94**, 4250–4255.
30. N. Dave and J. Liu, Fast Molecular Beacon Hybridization in Organic Solvents with Improved Target Specificity, *J. Phys. Chem. B*, 2010, **114**(47), 15694–15699.
31. S. Nakano, D. Miyoshi and N. Sugimoto, Effects of Molecular Crowding on the Structures, Interactions, and Functions of Nucleic Acids, *Chem. Rev.*, 2014, **114**, 2733–2758.
32. H. Han and L. H. Hurley, G-Quadruplex DNA: A Potential Target for Anti-Cancer Drug Design, *Trends Pharmacol. Sci.*, 2000, **21**(4), 136–142.
33. D. E. Gilbert and J. Feigon, Multistranded DNA Structures, *Curr. Opin. Struct. Biol.*, 1999, **9**, 305–314.
34. P. Alberti and J.-L. Mergny, DNA Duplex-Quadruplex Exchange as the Basis for a Nanomolecular Machine, *Proc. Natl. Acad. Sci. U. S. A.*, 2003, **100**(4), 1569–1573.
35. A. P. Abbott, T. J. Bell, S. Handa and B. Stoddart, Cationic Functionalisation of Cellulose Using a Choline Based Ionic Liquid Analogue, *Green Chem.*, 2006, **8**(9), 784–786.
36. D. Mondal, M. Sharma, C. Mukesh, V. Gupta and K. Prasad, Improved Solubility of DNA in Recyclable and Reusable Bio-Based Deep Eutectic Solvents with Long-Term Structural and Chemical Stability, *Chem. Commun.*, 2013, **49**(83), 9606.

37. F. M. Lannan, I. Mamajanov and N. V. Hud, Human Telomere Sequence DNA in Water-Free and High-Viscosity Solvents: G-Quadruplex Folding Governed by Kramers Rate Theory, *J. Am. Chem. Soc.*, 2012, **134**(37), 15324–15330.
38. C. Pretti, C. Chiappe, N. Gathergood and P. J. Scammells, Acute Toxicity of Ionic Liquids to the Zebrafish (Danio Rerio), *Green Chem.*, 2006, **8**(3), 238–240.
39. A. Jordan and N. Gathergood, Biodegradation of Ionic Liquids - a Critical Review, *Chem. Soc. Rev.*, 2015, **44**(22), 8200–8237.
40. S. P. M. Ventura, F. A. e Silva, M. V. Quental, D. Mondal, M. G. Freire and J. A. P. Coutinho, Ionic-Liquid-Mediated Extraction and Separation Processes for Bioactive Compounds: Past, Present, and Future Trends, *Chem. Rev.*, 2017, **117**, 6984–7052.
41. M. G. Freire, A. F. M. Cláudio, J. M. M. Araújo, J. A. P. Coutinho, I. M. Marrucho, J. N. C. Lopes and L. P. N. Rebelo, Aqueous Biphasic Systems: A Boost Brought about by Using Ionic Liquids, *Chem. Soc. Rev.*, 2012, **41**(14), 4966.
42. A. Benedetto and P. Ballone, Room Temperature Ionic Liquids Meet Biomolecules: A Microscopic View of Structure and Dynamics, *ACS Sustainable Chem. Eng.*, 2016, **4**(2), 392–412.
43. L. Cardoso and N. M. Micaelo, DNA Molecular Solvation in Neat Ionic Liquids, *ChemPhysChem*, 2011, **12**(2), 275–277.
44. T. Zhou, L. Chen, Y. Ye, L. Chen, Z. Qi, H. Freund and K. Sundmacher, An Overview of Mutual Solubility of Ionic Liquids and Water Predicted by COSMO-RS, *Ind. Eng. Chem. Res.*, 2012, **51**(17), 6256–6264.
45. M. Sharma, D. Mondal, N. Singh, N. Trivedi, B. Jit Kurmar and P. Kamalesh, High Concentration DNA Solubility in Bio Ionic Liquids with Long Lasting Chemical and Structural Stability at Room Temperature, *RSC Adv.*, 2015, **5**(9007), 40546–40551.
46. M. Marušič, H. Tateishi-Karimata, N. Sugimoto and J. Plavec, Structural Foundation for DNA Behavior in Hydrated Ionic Liquid: An NMR Study, *Biochimie*, 2015, **108**, 169–177.
47. K. Fujita and H. Ohno, Stable G-Quadruplex Structure in a Hydrated Ion Pair: Cholinium Cation and Dihydrogen Phosphate Anion, *Chem. Commun. (Camb.)*, 2012, **48**(46), 5751–5753.
48. C. Mukesh, D. Mondal, M. Sharma and K. Prasad, Rapid Dissolution of DNA in a Novel Bio-Based Ionic Liquid with Long-Term Structural and Chemical Stability: Successful Recycling of the Ionic Liquid for Reuse in the Process BT - Chemical Communications, *Chem. Commun.*, 2013, **49**(61), 9606.
49. R. Vijayaraghavan, A. Izgorodin, V. Ganesh, M. Surianarayanan and D. R. MacFarlane, Long-Term Structural and Chemical Stability of DNA in Hydrated Ionic Liquids, *Angew. Chem., Int. Ed.*, 2010, **49**(9), 1631–1633.
50. R. R. Mazid, A. Cooper, Y. Zhang, R. Vijayaraghavan, D. R. MacFarlane, C. Cortez-Jugo and W. Cheng, Enhanced Enzymatic Degradation Resistance of Plasmid DNA in Ionic Liquids, *RSC Adv.*, 2015, **5**(54), 43839–43844.

51. E. Stellwagen, Q. Dong and N. C. Stellwagen, Quantitative Analysis of Monovalent Counterion Binding to Random-Sequence, Double-Stranded DNA Using the Replacement Ion Method, *Computer (Long. Beach. Calif).*, 2008, **144**(5), 724–732.
52. H. Tateishi-Karimata and N. Sugimoto, A-T Base Pairs Are More Stable than G-C Base Pairs in a Hydrated Ionic Liquid, *Angew. Chem., Int. Ed.*, 2012, **51**(6), 1416–1419.
53. M. Nakano, H. Tateishi-Karimata, S. Tanaka and N. Sugimoto, Choline Ion Interactions with DNA Atoms Explain Unique Stabilization of A-T Base Pairs in DNA Duplexes: A Microscopic View, *J. Phys. Chem. B*, 2014, **118**(2), 379–389.
54. I. Machado. PhD-Thesis Polymat, University of the Basque Country, 2016.
55. I. Khimji, K. Doan, K. Bruggeman, P.-J. J. Huang, P. Vajha and J. Liu, Extraction of DNA Staining Dyes from DNA Using Hydrophobic Ionic Liquids, *Chem. Commun. (Camb.)*, 2013, **49**(40), 4537–4539.
56. A. Chandran, D. Ghoshdastidar and S. Senapati, Groove Binding Mechanism of Ionic Liquids: A Key Factor in Long-Term Stability of DNA in Hydrated Ionic Liquids?, *J. Am. Chem. Soc.*, 2012, **134**(50), 20330–20339.

Subject Index